Springer Collected Works in Mathematics

Photo by Christopher Campbell

Serge Lang
(ca. 1970s)

Serge Lang

Collected Papers I

1952–1970

Reprint of the 2000 Edition

 Springer

Serge Lang (1927–2005)
Department of Mathematics
Yale University
New Haven, CT
USA

ISSN 2194-9875
ISBN 978-1-4614-6136-4   (Softcover)
       978-0-387-98802-3   (Hardcover)
DOI 10.1007/978-1-4614-7383-1
Springer New York Heidelberg Dordrecht London

Library of Congress Control Number: 2012954381

Mathematics Subject Classification (1991): 11Gxx, 14Kxx, 11Fxx, 11Jxx, 14H

Printed on acid-free paper

Springer is part of Springer Science+Business Media (www.springer.com)

Contents

Contents of Volumes II, III, IV, and V

Volume IV

Volume V (Jorgenson-Lang)

Foreword

I thank Springer-Verlag for publishing my collected papers. Some Springer Lecture Notes are also included, as well as the short book *Introduction to Transcendental Numbers* (Addison Wesley, 1966), which has been out of print for years. The Benjamin book *Rapport sur la cohomologie des groupes* has also been out of print, and an expanded English version was published by Springer a few years ago, but is out of print, so it has been included. Some of the seminar talks (Bourbaki, Théorie des Nombres, . . .) are included, based on *ad hoc* judgment.

The papers roughly group themselves in various periods, with the main focus of interest shifting from one period to the next.

1.	1951–1954	The thesis on quasi algebraic closure and some related matters
2.	1954–1962	Algebraic geometry and abelian (or group) varieties; geometric class field theory; jump to Katz-Lang 1981
3.	1962–1975	Transcendental numbers and diophantine approximations on algebraic groups
4.	1970	First paper on analytic number theory—jump to Jorgenson-Lang
5.	1975	$SL_2(\mathbf{R})$—jump to Jorgenson-Lang
6.	1972–1977	Lang-Trotter Frobenius distributions
7.	1973–1981	Modular curves, Kubert-Lang modular units
8.	1974, 1982–1991	Diophantine geometry, complex hyperbolic spaces and Nevanlinna theory
9.	1985, 1988	Riemann-Roch and Arakelov theory
10.	1992–2000+	Jorgenson-Lang (analytic number theory and connections with spectral analysis, heat kernel, differential geometry, Lie groups, and symmetric spaces)

I take this opportunity to express once more my appreciation for having been Artin's student. I could not have had a better start in my mathematical life.

Looking back at the bibliograpy, I note the three-year hiatus in research papers (1967, 1968, 1969). These are years during which I was active in social and political events, deeply affecting science and the universities. I went back into the woodwork in 1970. I got out of the woodwork again in 1977–1978, and have been challenging ever since, but I did not allow my involvement to be so great that it would interrupt again what is otherwise a continuous production. I thank W.A. Benjamin for publishing my first political book, *The Scheer Campaign*, in 1966. I thank Springer-Verlag again for publishing the other two (*The File* in 1981 and *Challenges* in 1998). I thank and express my greatest appreciation personally for Springer's New York editors Walter Kaufmann-Bühler and Tom von Foerster, who took the decision to publish these books and provided intellectual support when I needed it.

SERGE LANG
Yale, New Haven 1999

Curriculum Vitae

1927	Born
1946	B.S., Caltech
1946–47	U.S. Army
1947–48	Graduate Philosophy, Princeton
1948–1951	Graduate Mathematics, Princeton, Ph.D. 1951
1951–52	Instructor, Princeton
1952–53	Institute for Advanced Study
1953–55	Instructor, University of Chicago
1955–70	Columbia University
1957–58	Fulbright fellow
1970–71	Visiting professor, Princeton
1971, fall	Visiting professor, Harvard
1972–present	Yale

Cole Prize 1959; Prix Carrière 1967; Steele Prize 1999; Humboldt award 1984; elected National Academy of Sciences 1985

Bibliography (through 1999)

(Boldface items are books or Lecture Notes.)

[1952a]   On quasi algebraic closure, *Ann. of Math.* **55** No. 2 (1952) pp. 373–390.

[1952b]   Hilbert's nullstellensatz in infinite dimensional space, *Proc. AMS* **3** No. 3 (1952) pp. 407–410.

[1952c]   (with J. TATE) On Chevalley's proof of Luroth's theorem, *Proc. AMS* **3** No. 4 (1952) pp. 621–624.

[1953]   The theory of real places, *Ann. of Math.* **57** No. 2 (1953) pp. 378–391.

[1954a]   Some applications of the local uniformization theorem, *Am. J. Math.* **76** No. 2 (1954) pp. 362–374.

[1954b]   (with A. WEIL) Number of points of varieties in finite fields, *Am. J. Math.* **76** No. 4 (1954) pp. 819–827.

[1955]   Abelian varieties over finite fields, *Proc. NAS* **41** No. 3 (1955) pp. 174–176.

[1956a]   Unramified class field theory over function fields in several variables, *Ann. of Math.* **64** No. 2 (1956) pp. 285–325.

[1956b]   On the Lefschetz principle, *Ann. of Math.* **64** No. 2 (1956) pp. 326–327.

[1956c]   L-series of a covering, *Proc. NAS* **42** No. 7 (1956) pp. 422–424.

[1956d]   Sur les séries L d'une variété algébrique, *Bull. Soc. Math. France* **84** (1956) pp. 385–407.

[1956e]   Algebraic groups over finite fields, *Am. J. Math.* **78** (1956) pp. 555–563.

[1957a]   (with J.-P. SERRE) Sur les revêtements non ramifiés des variétés algébriques, *Am. J. Math.* **79**, No. 2 (1957) pp. 319–330.

[1957b]   (with W.-L. CHOW) On the birational equivalence of curves under specialization, *Am. J. Math.* **79**, No. 3 (1957) pp. 649–652.

[1957c]   Divisors and endomorphisms on an abelian variety, *Am. J. Math.* **79** No. 4 (1957) pp. 761–777.

[1957d]   Families algébriques de Jacobiennes (d'après IGUSA), *Séminaire Bourbaki* No. 155, 1957/1958.

[1958a]   Reciprocity and correspondences, *Am. J. Math.* **80** No. 2 (1957) pp. 431–440.

[1958b]   (with J. TATE) Principal homogeneous spaces over abelian varieties, *Am. J. Math.* **80** No. 3 (1958) pp. 659–684.

[1958c]   (with E. KOLCHIN), Algebraic groups and the Galois theory of differential fields, *Am. J. Math.* **80** No. 1 (1958) pp. 103–110.

[1958d]   *Introduction to algebraic geometry*, Wiley-Interscience, 1958.

[1959a]   (with A. NÉRON) Rational points of abelian varieties over function fields, *Am. J. Math.* **81** No. 1 (1959) pp. 95–118.

[1959b]   Le théorème d'irreductibilité de Hilbert, *Séminaire Bourbaki* No. 201, 1959/1960.

[1959c]   *Abelian varieties*, Wiley-Interscience, 1959; Springer-Verlag, 1983.

[1960a]   (with E. KOLCHIN) Existence of invariant bases, *Proc. AMS* **11** No. 1 (1960) pp. 140–148.

[1960b]   Integral points on curves, *Pub. IHES* No. **6** (1960) pp. 27–43.

[1960c]   Some theorems and conjectures in diophantine equations, *Bull. AMS* **66** No. 4 (1960) pp. 240–249.

[1960d]   On a theorem of Mahler, *Mathematika* **7** (1960) pp. 139–140.

[1960e]   L'équivalence homotopique tangentielle (d'apres MAZUR), *Séminaire Bourbaki* No. 222, 1960/1961.

[1961]   Review: Elements de géométrie algébrique (A. Grothendieck). *Bull. AMS* **67** No. 3 (1961) pp. 239–246.

[1962a]   A transcendence measure for E-functions, *Mathematika* **9** (1962) pp. 157–161.

[1962b]   Transcendental points on group varieties, *Topology* **1** (1962) pp. 313–318.

[1962c]   Fonctions implicities et plongements Riemanniens, *Séminaire Bourbaki* 1961/1962, No. 237, May 1962.

[1962d]   *Introduction to Differential Manifolds*, Addison Wesley, 1962.

[1962e]   *Diophantine Geometry*, Wiley-Interscience, 1962.

[1963]   *Transzendente Zahlen*, Bonn Math. Schr. No. 21 (1963).

[1964a]   Diophantine approximations on toruses, *Am. J. Math.* **86** No. 3 (1964) pp. 521–533.

[1964b]   Les formes bilinéaires de Néron et Tate, *Séminaire Bourbaki* 1963/64 Fasc. 3 Exposé 274, Paris 1964.

[1964c]   *First Course in Calculus*, Addison Wesley 1964; Fifth edition by Springer-Verlag, 1986.

[1964d]   *Algebraic and Abelian Functions*, W.A. Benjamin Lecture Notes, 1964. See also [1982c].

[1964e]   *Algebraic Numbers*, Addison Wesley, 1964; superceded by [1970d].

[1965a]   Report on diophantine approximations, *Bulletin Soc. Math. France* **93** (1965) pp. 177–192.

[1965b]   Division points on curves, *Annali Mat. pura ed applicata*, Serie IV **70** (1965) pp. 229–234.

[1965c]   Algebraic values of meromorphic functions, *Topology* **3** (1965) pp. 183–191.

[1965d]   Asymptotic approximations to quadratic irrationalities I, *Am. J. Math.* **87** No. 2 (1965) pp. 481–487.

[1965e]   Asymptotic approximations to quadratic irrationalities II, *Am. J. Math.* **87** No. 2 (1965) pp. 488–496.

[1965f]   (with W. ADAMS) Some computations in diophantine approximations, *J. reine angew. Math.* Band **220** Heft 3/4 (1965) pp. 163–173.

[1965g]   Corps de fonctions méromorphes sur une surface de Riemann (d'après ISS'SA), *Séminaire Bourbaki* No. 292, 1964/65.

[1965h]   *Algebra*, Addison Wesley, 1965; second edition 1984; third edition 1993.

[1966a]   Algebraic values of meromorphic functions II, *Topology* **5** (1960) pp. 363–370.

[1966b]   Asymptotic diophantine approximations, *Proc. NAS* **55** No. 1 (1966) pp. 31–34.

[1966c]   *Introduction to transcendental numbers*, Addison Wesley, 1966.

[1966d]   *Introduction to Diophantine Approximations*, Addison Wesley 1966; see [1995d].

[1966e]   *Rapport sur la cohomologie des groupes.* Benjamin, 1966.

[1967]   *Algebraic structures*, Addison Wesley 1967.

[1968]   *Analysis I,* Addison Wesley, 1968; (superceded by [1983c]).

[1969]   *Analysis II*, Addison Wesley, 1969; (superceded by [1993c]).

[1970a]   (with E. BOMBIERI) Analytic subgroups of group varieties, *Inventiones Math.* **11** (1970) pp. 1–14.

[1970b]   Review: L.J. Mordell's *Diophantine Equations, Bull. AMS* **76** (1970) pp. 1230–1234.

[1970c]   *Introduction to Linear Algebra*, Addison Wesley 1970; see also [1986b].

[1970d]   *Algebraic Number Theory*, Addison Wesley 1970; see also [1994c].

[1971a]   Transcendental numbers and diophantine approximations, *Bull. AMS* **77** No. 5 (1971) pp. 635–677.

[1971b]   On the zeta function of number fields, *Invent. Math.* **12** (1971) pp. 337–345.

[1971c]   The group of automorphisms of the modular function field, *Invent. Math.* **14** (1971) pp. 253–254.

[1971d]   *Linear Algebra*, Addison Wesley 1971; see also [1987b].

[1971e]   *Basic Mathematics*, Addison Wesley 1971; Springer-Verlag 1988.

[1972a]   Isogenous generic elliptic curves, *Amer. J. Math.* **94** (1972) pp. 661–674.

[1972b] (with H. TROTTER) Continued fractions for some algebraic numbers, *J. reine angew. Math.* **255** (1972) pp. 112–134.

[1972c] *Differential manifolds*, Addison Wesley, 1972.

[1972d] *Introduction to Algebraic and Abelian Functions*, Benjamin-Addison Wesley, 1972; second edition see [1982c].

[1973a] Frobenius automorphisms of modular function fields, *Amer. J. Math.* **95** (1973) pp. 165–173.

[1973b] *Calculus of Several Variables*, Addison Wesley 1973; Third edition see [1987d].

[1973c] *Elliptic functions*, Addison Wesley 1973; second edition see [1987d].

[1974a] Higher dimensional diophantine problems, *Bull. AMS* **80** No. 5 (1974) pp. 779–787.

[1974b] (with H. TROTTER) Addendum to "Continued fractions of some algebraic numbers," *J. reine angew. Math.* **267** (1974) pp. 219–220.

[1975a] Diophantine approximations on abelian varieties with complex multiplication, *Advances Math.* **17** (1975) pp. 281–336.

[1975b] Division points of elliptic curves and abelian functions over number fields, *Amer. J. Math.* **97** No. 1 (1975) pp. 124–132.

[1975c] (with D. KUBERT) Units in the modular function field I, Diophantine Applications, *Math. Ann.* **218** (1975) pp. 67–96.

[1975d] (with D. KUBERT) Units in the modular function field II, A full set of units, *Math. Ann.* **218** (1975) pp. 175–189.

[1975e] (with D. KUBERT) Units in the modular function field III, Distribution relations, *Math. Ann.* **218** (1975) pp. 273–285.

[1975f] La conjecture de Catalan d'après Tijdeman, *Séminaire Bourbaki* 1975/76 No. 29.

[1975g/85] SL$_2$(**R**), Addison Wesley, 1975; Springer-Verlag corrected second printing, 1985.

[1976a] (with J. COATES) Diophantine approximation on abelian varieties with complex multiplication, *Invent. Math.* **34** (1976) pp. 129–133.

[1976b] (with D. KUBERT) Distribution on toroidal groups, *Math. Z.* **148** (1976) pp. 33–51.

[1976c]   (with D. KUBERT) Units in the modular function field, in *Modular Functions in One Variable* V, Springer Lecture Notes **601** (Bonn Conference) 1976, pp. 247–275.

[1976d]   (with H. TROTTER) *Frobenius distributions in* GL_2-*extensions*, Springer Lecture Notes **504**, Springer-Verlag 1976.

[1976e]   *Introduction to Modular Forms*, Springer-Verlag, 1976.

[1977a]   (with D. KUBERT) Units in the modular function field IV, The Siegel functions are generators, *Math. Ann.* **227** (1997) pp. 223–242.

[1977b]   (with H. TROTTER) Primitive points on elliptic curves, *Bull. AMS* **83** No. 2 (1977) pp. 289–292.

[1977c]   *Complex Analysis*, Addison Wesley; second Edition Springer-Verlag 1985; fourth edition Springer-Verlag 1999.

[1978a]   (with D. KUBERT) The p-primary component of the cuspidal divisor class group of the modular curve $X(p)$, *Math. Ann.* **234** (1978) pp. 25–44.

[1978b]   (with D. KUBERT) Units in the modular function field V, Iwasawa theory in the modular tower, *Math. Ann.* **237** (1978) pp. 97–104.

[1978c]   (with D. KUBERT) Strickelberger ideals, *Math. Ann.* **237** (1978) pp. 203–212.

[1978d]   (with D. KUBERT) The index of Strickelberger ideals of order 2 and cuspidal class numbers, *Math. Ann.* **237** (1978) pp. 213–232.

[1978e]   Relations de distributions et exemples classiques, *Séminaire Delange-Pisot-Poitou (Théorie des Nombres)*, 1978 No. 40 (6 pages).

[1978f]   *Elliptic curves: Diophantine Analysis*, Springer-Verlag 1978.

[1978g]   *Cyclotomic Fields* I, Springer-Verlag, 1978.

[1979a]   (with D. KUBERT) Cartan-Bernoulli numbers as values of L-series, *Math. Ann.* **240** (1979) pp. 21–26.

[1979b]   (with D. KUBERT) Independence of modular units on Tate curves, *Math. Ann.* **240** (1979) pp. 191–201.

[1979c]   (with D. KUBERT) Modular units inside cyclotomic units, *Bull. Soc. Math. France* **107** (1979) pp. 161–178.

[1980]   *Cyclotomic Fields* II, Springer-Verlag 1980.

[1981a]   (with N. KATZ) Finiteness theorems in geometric classfield theory, *Enseignement mathématique* **27** (3–4) (1981) pp. 285–314.

[1981b]   (with DAN KUBERT) *Modular Units*, Springer-Verlag 1981.

[1982a]   Représentations localement algébriques dans les corps cyclotomiques, Séminaire de Théorie des Nombres 1982, Birkhauser, pp. 125–136.

[1982b]   Units and class groups in number theory and algebraic geometry, *Bull. AMS* **6** No. 3 (1982) pp. 253–316.

[1982c]   *Introduction to algebraic and abelian functions, Second Edition*, Springer-Verlag, 1982.

[1983a]   Conjectured diophantine estimates on elliptic curves, in Volume I of *Arithmetic and Geometry*, dedicated to Shafarevich, M. Artin and J. Tate editors, Birkhauser (1983) pp. 155–171.

[1983b]   *Fundamentals of Diophantine Geometry*, Springer-Verlag 1983.

[1983c]   *Undergraduate Analysis*, Springer-Verlag, 1983.

[1983d]   (with GENE MURROW) *Geometry: A High School Course*, Springer Verlag, 1983 Second edition 1988.

[1983e]   *Complex Multiplication*, Springer-Verlag 1983.

[1984a]   Vojta's conjecture, *Arbeitstagung Bonn 1984*, Springer Lecture Notes **1111** 1985, pp. 407–419.

[1984b]   Variétés hyperboliques et analyse diophantienne, *Séminaire de théorie des nombres*, 1984/85, pp. 177–186.

[1985a]   (with W. FULTON) *Riemann-Roch Algebra*, Springer-Verlag, 1985.

[1985b]   *The Beauty of Doing Mathematics*, Springer-Verlag, 1985 originally published as articles in the *Revue du Palais de la Découverte*, Paris, 1982–1984, specifically:
>   Une activité vivante: faire des mathématiques, *Rev. P.D.* Vol. **11** No. **104** (1983) pp. 27–62
>   Que fait un mathématicien pure et pourquoi?, *Rev. P.D.* Vol. **10** No. **94** (1982) pp. 19–44
>   Faire des Maths: grands problèmes de géométrie et de l'espace, *Rev. P.D.* Vol. **12** No. **114** (1984) pp. 21–72.

[1985c]   *Math! Encounters with High School Students*, Springer-Verlag, 1985 (French edition *Serge Lang, des Jeunes et des Maths*, Berlin, 1984).

[1986a]   Hyperbolic and diophantine analysis, *Bulletin AMS* **14** No. 2 (1986) pp. 159–205.

[1986b]   *Introduction to Linear Algebra*, Second Edition, Springer-Verlag 1986.

[1987a]   Diophantine problems in complex hyperbolic analysis, *Contemporary Mathematics AMS* **67** (1987) pp. 229–246.

[1987b]   *Linear Algebra*, Third Edition, Springer-Verlag 1987.

[1987c]   *Undergraduate Algebra*, Springer-Verlag 1987.

[1987d]   *Elliptic functions*, Second Edition, Springer-Verlag 1987.

[1987e]   *Introduction to complex hyperbolic spaces*, Springer-Verlag 1987.

[1988a]   The error term in Nevanlinna theory, *Duke Math. J.* **56** No. 1 (1988) pp. 193–218.

[1988b]   *Introduction to Arakelov Theory*, Springer-Verlag 1988.

[1990a]   The error term in Nevanlinna Theory II, *Bull. AMS* **22** No. 1 (1990) pp. 115–125.

[1990b]   Old and new conjectured diophantine inequalities. *Bull. AMS* **23** No. 1 (1990) pp. 37–75.

[1990c]   *Lectures on Nevanlinna theory*, in *Topics in Nevanlinna Theory*, Springer Lecture Notes **1433** (1990) pp. 1–107.

[1990d]   *Cyclotomic Fields I and II*, combined edition with an appendix by Karl Rubin, Springer-Verlag, 1990.

[1991]   *Number Theory III*, Survey of Diophantine Geometry, Encyclopedia of Mathematical Sciences, Springer-Verlag 1991.

[1993a]   (with J. JORGENSON) On Cramér's theorem for general Euler products with functional equations, *Math. Ann.* **297** (1993) pp. 383–416.

[1993b]   (with J. JORGENSON) *Basic analysis of regularized series and products*, Springer Lecture Notes **1564** (1993).

[1993c]   *Real and Functional Analysis*, Springer-Verlag, 1993.

[1993d]   *Algebra, Third Edition*, Addison Wesley 1993.

[1994a]   (with J. JORGENSON) Artin formalism and heat kernels, *J. reine angew. Math.* **447** (1994) pp. 165–200.

[1994b]   (with J. JORGENSON) *Explicit Formulas for regularized products and series*, in Springer Lecture Notes **1593** pp. 1–134.

[1994c]   *Algebraic Number Theory, Second Edition*, Springer-Verlag 1994.

[1995a]   Mordell's review, Siegel's letter to Mordell, diophantine geometry, and 20th century mathematics, *Notices AMS* March 1995 pp. 339–350.

[1995b]   Some history of the Shimura-Taniyama conjecture, *Notices AMS* November 1995 pp. 1301–1307.

[1995c]   *Differential and Riemannian Manifolds*, Springer-Verlag 1995.

[1995d]   *Introduction to Diophantine Approximations, new expanded edition*, Springer-Verlag 1995.

[1996a]   La conjecture de Bateman-Horn, *Gazette des mathématiciens* January 1996 No. 67 pp. 82–84.

[1996b]   Comments on Chow's works, *Notices AMS* **43** (1996) No. 10 pp. 1117–1124.

[1996c]   (with J. JORGENSON) Extension of analytic number theory and the theory of regularized harmonic series from Dirichlet series to Bessel series, *Math. Ann.* **306** (1996) pp. 75–124.

[1996d]   *Topics in Cohomology of groups*, Springer-Verlag, Springer Lecture Notes **1625**, 1996 (English translation and expansion of *Rapport sur la Cohomologie des Groupes*, Benjamin, 1966).

[1997]   *Survey of Diophantine Geometry*, Springer-Verlag 1997 (same as *Number Theory III*, with corrections and additions).

[1998]   The Kirscher article and HIV: Scientific and journalistic (ir)responsibilities.

[1999a]   (with J. JORGENSON) Hilbert-Asai Eisenstein series, regularized products and heat kernels, *Nagoya Math. J.* **153** (1999) pp. 155–188.

[1999b]   Response to the Steele Prize, *Notices AMS* **46** No. 4, April 1999 p. 458.

[1999c]   *Fundamentals of Differential Geometry*, Springer-Verlag 1999.

[1999d]   *Complex analysis*, fourth edition, Springer-Verlag 1999.

[1999e]   *Math Talks for Undergraduates*, Springer-Verlag 1999.

The Zurich Lectures

I was invited by Wustholz for a decade to give talks to students in Zurich. I express here my appreciation, also to Urs Stammbach for his translations, and for his efforts in producing and publishing the articles in *Elemente der Mathematik*.

Primzahlen, *Elem. Math.* **47** (1992) pp. 49–61

Die abc-vermutung, *Elem. Math.* **48** (1993) pp. 89–99

Approximationssätze der Analysis, *Elem. Math.* **49** (1994) pp. 92–103

Die Wärmeleitung auf dem Kreis und Thetafunktionen, *Elem. Math.* **51** (1996) pp. 17–27

Globaler Integration lokal integrierbarer Vektorfelder, *Elem. Math.* **52** (1997) pp. 1–11

Bruhat-Tits-Raüme, *Elem. Math.* **54** (1999) pp. 45–63

Articles on Scientific Responsibility

Circular A-21. A history of bureaucratic encroachment, J. Society of Research Administrators, 1984

Questions de responsabilité dans le journalisme scientifique, *Revue du Palais de la Découverte* Paris February 1991 pp. 17–46

Questions of scientific responsibility: The Baltimore case, *J. Ethics and Behavior* **3(1)** (1993) pp. 3–72

The Kirschner article and HIV: Scientific and journalistic (ir)responsibilities, Refused publication by the *Notices AMS*, dated 5 January 1998

Books on Scientific Responsibility

The Scheer Campaign, W.A. Benjamin, 1966

The File, Springer-Verlag, 1981

Challenges, Springer-Verlag, 1998

Photo by Henry Verby Photography

Serge Lang
(ca. 1950s)

Annals of Mathematics
Vol. 55, No. 2, March, 1952
Printed in U.S.A.

ON QUASI ALGEBRAIC CLOSURE[1]

By

Serge Lang

(Received May 2, 1951)

Introduction

Artin called a field F quasi algebraically closed if every form in F in n variables and degree d with $n > d$ has a non trivial zero in F. We generalize this concept, and say that a field F is C_i if every form in F with $n > d^i$ has a non trivial zero in F. A C_0 field is therefore algebraically closed, and a C_1 field is quasi algebraically closed.

In Part I we shall discuss this definition, and investigate certain algebraic properties of C_i fields.

In Part II we prove that certain special fields are C_i. For example if F is a function field in k variables over a C_i constant field then F is C_{i+k}. We then investigate the local situation and prove that a field complete under a discrete valuation with algebraically closed residue class field is C_1. From this we deduce a conjecture of Artin's, that the maximal unramified extension of a field complete under a discrete valuation with perfect residue class field is C_1, by means of the following theorem: Let $\bar{F}$ be a field complete under any valuation and let F be a dense subfield of $\bar{F}$, algebraically closed in $\bar{F}$. Let $\eta = (\eta_1, \cdots, \eta_n)$ be a generic point with $\eta_\nu \in \bar{F}$, such that the field $F(\eta)$ is separably generated over F. Then η has arbitrarily close specializations in F.

If the residue class field is a finite field, it has been conjectured that F is C_2. We can prove this only in the case of power series fields, leaving the question open in the case of p-adic fields.

In Part III we follow a remark of Artin's and apply the results of Part II to class field theory, using the language of cohomology [2]. More precisely, if a field F has an extension Ω which is C_1 then the second cohomology group of F is completely determined by Ω: every cocycle splits on a finite subfield of Ω. In the case of complete fields with perfect residue class field, every cocycle has an unramified splitting field. In this way we recover the well known results in the classical case where the residue class field is a finite field. In particular, we get Chevalley's Theorem that any cocycle split by one field E/F of degree n is split by every field of degree n. The group of such cocycles is cyclic of order n and is determined by the unramified extensions.

In the large, function fields turn out to be relatively easy to handle, and Tsen [3] essentially had the following result: If F is a function field in one variable over a constant field $\mathfrak{f}$ then every cocycle has a splitting field which is a finite extension of $\mathfrak{f}$.

---

[1] A thesis presented in partial fulfillment of the requirements for the Ph. D, Princeton University, May 1951.

One hopes eventually to extend the local arithmetic results to number fields, but little is known at present in this direction.

I wish to express here my great appreciation for Artin's constant encouragement and useful suggestions during the preparation of this paper.

PART I. ALGEBRAIC PROPERTIES OF C_i FIELDS

DEFINITION. A field F will be said to be C_0 if every form in F in n variables and degree d, with $n = d$ has a non trivial zero in F.

By a non trivial zero we shall always mean a zero $(\alpha_1, \cdots, \alpha_n)$ such that some α_i is not equal to 0. The trivial zero is $(0, \cdots, 0)$.

One sees immediately that we have just given an alternative definition for algebraic closure. Indeed,

THEOREM 1. *If a field F admits one extension E of degree n then there exists in F a form of degree n, in n variables and having only the trivial zero.*

PROOF. If $\omega_1, \cdots, \omega_n$ is a basis for E/F, the norm of the general element $x_1\omega_1 + \cdots + x_n\omega_n$ is the desired form.

We define algebraic closure as we have done above in order to make more striking the relation which exists between the closure of the field and the existence of non trivial zeros for certain forms. There exist certain interesting fields which are not too far removed from their algebraic closure. This property will be reflected in the existence of non trivial zeros for forms whose number of variables is large compared to the degree.

DEFINITION. A field F will be said to be C_i if every form in F of degree d in n variables, with $n > d^i$ has a non trivial zero in F.

As a matter of notation, we shall from now on use n and d to denote the number of variables and degree of a form respectively.

The C_1 fields, those for which every form with $n > d$ has a non trivial zero, were first defined by Artin who called such fields quasi algebraically closed. He showed that the C_1 property generalizes algebraic closure by the following argument: If a field F admits a non-commutative finite extension E (and F is contained in the center of E), then there exists in F a form with $n > d$ and having only the trivial zero. The proof is the same as that of Theorem 1, replacing the word "norm" by the word "reduced norm". If a field is quasi algebraically closed, it has therefore no finite division algebras over it.

It will become clear later that the C_i property generalizes naturally the C_{i-1} property. It is worth while at this point to make a few comments on the definitions we have chosen above. We may actually strengthen them in two ways: First, we may demand that polynomials without constant term (instead of forms) have the non trivial zero. Secondly, we may demand that several forms (or polynomials without constant term) have a simultaneous non trivial zero. The question arises: Does the weakest of these conditions imply the strongest?

DEFINITION. A field F will be said to be strongly C_i if every polynomial in F without constant term with $n > d^i$ has a non trivial zero in F.

As far as we know there exists no sufficient condition under which we could prove that a C_i field is also strongly C_i. Of course, we can conceive a polynomial f without constant term to be the sum of forms, $f = f_1 + f_2 + \cdots + f_r$, and by this crude analysis a simultaneous solution of $f_1 = 0, \cdots, f_r = 0$ would provide a zero for f. In view of Theorem 3 below, a C_i field would usually be strongly C_{i+1}. This answer to our question is highly unsatisfactory, and we made the remark only to show that the situation is not completely hopeless. In the case of C_0 fields, however, a complete answer can be given, and the following theorem is classical.

THEOREM 2. *Let F be algebraically closed. Let $f_1, \cdots, f_r$ be polynomials in F in n variables, each without constant term. Then they have a non trivial common zero if $n > r$. In fact, the dimension of the variety defined by them is at least $n - r$.*

The degree of the polynomials is irrelevant when F is C_0. This will naturally not be the case in similar theorems concerning C_i fields. Artin found a very useful sufficient condition under which the existence of a zero for one form implies the existence of a simultaneous zero for several forms. We need a

DEFINITION. A form in a field F is called normic of order i if $n = d^i$ and the form has only the trivial zero in F.

If F is not algebraically closed, Theorem 1 asserts the existence of normic forms of order 1: The norm itself.

A corresponding definition can be made for normic polynomials of order i, that is polynomials without constant term, with $n = d^i$ and only the trivial zero. We shall state below certain theorems about C_i fields which will remain true if the terms "C_i" and "form" are replaced throughout by "strongly C_i" and "polynomial without constant term" respectively. In this parallel theory, the proofs will always go over verbatim, and we shall not point out each time that the theorem has an alternative formulation.

We note that a normic form of order i and degree 1 is of type ax, $a \neq 0$ in the field. Such a form exists in any field, and is of a trivial nature. We suppose from now on that any reference we make to normic forms of order i excludes this trivial case. Remark also that a normic form of order 0 is of the same type. Since we shall use normic forms of order i in conjunction with C_i fields, we shall need them only if $i \geq 1$. If $i = 0$ Theorem 2 will always give all the information needed. In particular, in Theorems 3, 4, 5 below either $i = 0$ in which case the theorem reduces to Theorem 2, or $i \geq 1$ and our proofs apply.

The following lemma will be useful:

LEMMA. *If a field F has one normic form of order $i \geq 1$ then it has normic forms of order i of arbitrarily high degree.*

PROOF. Let φ, ψ be two such forms, of degrees d_1, d_2 in n_1, n_2 variables respectively (φ may be ψ itself). Then $\varphi(\psi \mid \psi \mid \cdots \mid \psi)$ has degree $d_1 d_2$ and $n_1 n_2$ variables, with $n_1 n_2 = (d_1 d_2)^i$, and is normic.

The notation $\varphi(\psi \mid \psi \mid \cdots \mid \psi)$ means that new variables are to be used after each occurrence of $\mid$.

We may now state Artin's Criterion:

THEOREM 3. *Let F be a C_i field admitting at least one normic form of order i.*

Let $f_1, \cdots, f_r$ be r forms in F in n common variables each of degree d. If $n > rd^i$ then the forms have a non trivial common zero in F.

PROOF. In view of the lemma, we may assume N is a normic form of order i, of suitably high degree (to be determined below). Put

$$\Phi = N(f_1, \cdots, f_r \mid f_1, \cdots, f_r \mid \cdots \mid f_1, \cdots, f_r \mid 0, \cdots, 0)$$

where zeros occur in $t < r$ places if we insert as many $f_1, \cdots, f_r$ as possible, say s times. Then Φ has degree $d \sqrt[i]{sr + t}$ and has ns variables. We want $ns > d^i(sr + t)$ or equivalently $(n - rd^i) > td^i/s$. By hypothesis, we can achieve this by selecting s large enough. Φ will then have a non trivial zero. Since N is normic, this can happen only if all f_ν vanish simultaneously and at least one batch $f_1, \cdots, f_r$ will vanish non trivially, which proves our theorem.

If F is C_1 we can avoid mentioning the existence hypothesis.

COROLLARY. *Let F be C_1, and let $f_1, \cdots, f_r$ be forms in n variables, each of degree d. If $n > rd$ then $f_1, \cdots, f_r$ have a non trivial zero.*

PROOF. If F is C_0 use Theorem 2. Otherwise, the existence of normic forms is given by Theorem 1.

In Theorem 3, the forms had to be all of the same degree. We can eliminate this restriction by adopting a slightly stronger existence hypothesis, as follows:

THEOREM 4. *Let F be C_i and suppose F admits normic forms of order i of any given degree. If $f_1, \cdots, f_r$ are forms in n variables, degrees $d_1, \cdots, d_r$ such that $n > d_1^i + \cdots + d_r^i$ then they have a non trivial zero in F.*

PROOF. We reduce this theorem to the previous one. Let $N_1, \cdots, N_r$ be normic forms of order i, of degrees $d_2 \cdots d_r$, $d_1 d_3 \cdots d_r$, $\cdots$, $d_1 \cdots d_{r-1}$ respectively and consider the r systems of equations:

1) $N_1(f_1 \mid f_1 \mid \cdots \mid f_1) = 0 \mid N_1(f_1 \mid f_1 \mid \cdots \mid f_1) = 0 \mid \cdots$

2) $N_2(f_2 \mid f_2 \mid \cdots \mid f_2) = 0 \mid N_2(f_2 \mid f_2 \mid \cdots \mid f_2) = 0 \mid \cdots$

$$\cdots$$

r) $N_r(f_r \mid f_r \mid \cdots \mid f_r) = 0 \mid N_r(f_r \mid f_r \mid \cdots \mid f_r) = 0 \mid \cdots$

where each system possesses respectively $d_1^i, \cdots, d_r^i$ equations; furthermore in the first system each new appearance of f_1 is given new variables and thereafter the same variables are used. Then altogether there are $d_1^i + \cdots + d_r^i$ forms, each of degree $(d_1 \cdots d_r)$ and $n(d_1 \cdots d_r)^i$ variables. Since by hypothesis

$$n(d_1 \cdots d_r)^i > (d_1^i + \cdots + d_r^i)(d_1 \cdots d)^i$$

we apply the previous theorem to complete this proof.

Algebraic Extensions of C_i Fields.

If a field F is C_i it is reasonable to suppose that any finite (and consequently any algebraic) extension of F is likewise C_i. For $i \geq 2$ we shall need to assume explicitly a weak existence condition on normic forms to prove this conjecture.

THEOREM 5. *Let F be C_i and suppose F admits at least one normic form of order i. Then any finite extension E of F is also C_i.*

PROOF. Let $f(x_1, \cdots, x_n)$ be a form in E with $n > d^i$. Let the degree of E/F

be r and let $\omega_1, \cdots, \omega_r$ be a basis for E/F. Put

$$x_\nu = x_{\nu 1}\omega_1 + \cdots + x_{\nu r}\omega_r \qquad \nu = 1, \cdots, n.$$

Then $f(x_1, \cdots, x_n) = f_1(x_{ij})\omega_1 + \cdots + f_r(x_{ij})\omega_r$ where each f_ν is a form in F of degree d, and has nr variables. To make these forms all zero simultaneously, we note that $nr > d^i r$ and use Theorem 3.

COROLLARY. *If F is C_1, any finite extension of F is also C_1.*

PROOF. If F is not C_0, Theorem 1 gives the necessary normic forms.

Theorem 5 can be given a formulation for simultaneous forms as well as for strongly C_i. The proof remains the same except that the existence of normic forms of any given degree must be assumed and Theorem 4 must be used instead of Theorem 3.

PART II. FUNCTION FIELDS AND LOCAL FIELDS

In this part we shall give examples of C_i fields, and show that certain fields useful in arithmetic are either C_1 or C_2.

The fields F we shall be dealing with will often be quotient fields of an integral domain D. Given a set of polynomials in F for which we want to find a zero, we may clear all denominators from their coefficients, and assume these lie in D. The problem is then reduced to finding a zero in D, and becomes eventually a problem in diophantine analysis. We shall apply the following procedure systematically:

Investigate the existence of normic forms in F.

Prove that a certain integral domain $D \subset F$ is C_i.

Deduce that its quotient field F is likewise C_i.

Use the existence of normic forms and Theorem 5 to show that finite extensions of F are also C_i.

Existence of Normic Forms

Let F be a field with a discrete valuation, and let $\mathfrak{f}$ be the residue class field. If $\mathfrak{f}$ admits a normic form N^* of order i, then F has a normic form of order $i + 1$. Indeed let N be a representative in F of N^* and let π be an element of order 1 in F. Then $N + \pi N + \cdots + \pi^{d-1} N$ where each occurrence of N receives new variables is normic of order $i + 1$. This form has no zero in F, nor in the completion of F under the valuation. If the residue class field is algebraically closed, we select $N = x^d$, and $x_1^d + \pi x_2^d + \cdots + \pi^{d-1}x_d^d$ is a normic form of order 1. Other normic forms of order 1 are given by Theorem 1. If the residue class field is not algebraically closed, then there always exists a normic form of order 2 in F.

We notice finally that if a field F is the center of a division algebra then the reduced norm of the general element of this division algebra is normic of order 2.

Function Fields

We generalize here a result obtained by Tsen [3] for function fields in one variable over an algebraically closed constant field, and this essentially by Tsen's methods.

5

Let F be a C_i field having normic forms of order i. The rational function field in one variable $F(t)$ has the valuation defined by t, and F is its residue class field. $F(t)$ has consequently normic forms of order $i + 1$. Inductively, the rational field in k variables has normic forms of order $i + k$. We contend that this is the worst that can possibly happen.

LEMMA. *Let F be a C_i field, and suppose that F admits at least one normic form of order i. Then $F[t]$ is C_{i+1}.*

PROOF. Suppose that $f(x_1, \cdots, x_n)$ has $n > d^{i+1}$, and has coefficients in $F[t]$. We shall show that f has a non trivial zero. Let

$$x_1 = \xi_{10} + \xi_{11}t + \cdots + \xi_{1s}t^s$$

$$x_2 = \xi_{20} + \xi_{21}t + \cdots + \xi_{2s}t^s$$

$$\cdots$$

$$x_n = \xi_{n0} + \xi_{n1}t + \cdots + \xi_{ns}t^s$$

where ξ_{ij} are unknowns ranging over F, and s is sufficiently large. If r is the highest degree of the coefficients of f, we see that

$$f(x_1, \cdots, x_n) = f_0 + f_1 t + \cdots + f_{ds+r}t^{ds+r}$$

becomes a polynomial in t where each f_j is a form of degree d in $n(s + 1)$ variables. We must satisfy simultaneously $f_0 = 0$, $f_1 = 0$, $\cdots$, $f_{ds+r} = 0$ and use Theorem 3. To satisfy the inequality

$$n(s + 1) > d^i(ds + r + 1)$$

or equivalently $(n - d^{i+1}) > (d^i(r + 1) - n)/s$ we can select s high enough. This proves our lemma.

This lemma obviously generalizes to strongly C_i or simultaneous polynomials if we assume the existence of normic forms of any given degree.

We may collect the above results in

THEOREM 6. *Let F be a function field in k variables over a constant field $\mathfrak{k}$ which is C_i and admits a normic form of order i. Then F is C_{i+k}. If we assume that $\mathfrak{k}$ admits normic forms of any given degree, then the theorem generalizes to strongly C_i or simultaneous polynomials.*

Taking into account Theorem 2 and Theorem 7 below, we have the most important special cases of Theorem 6 in a

COROLLARY. *Let F be a function field in k variables over an algebraically closed constant field. Let $f_1, \cdots, f_r$ be polynomials without constant term in F, with $n > d_1^k + \cdots + d_r^k$. Then they have a non trivial common zero. If the constant field is a finite field, then they have a non trivial common zero provided $n > d_1^{k+1} + \cdots + d_r^{k+1}$.*

Local Fields

Let F be a field with a valuation. Then F is also a topological space. All operations are continuous so that F is in fact a topological field. The Cartesian product

F^n is also a topological space under the usual product topology. A polynomial $f(x_1, \cdots, x_n)$ in n variables is then a continuous function of F^n into F.

We shall consider a power series field in one variable t as a topological field with the usual topology, in which neighborhoods of 0 are given by powers of t.

Artin has conjectured that if F is a field complete under a discrete valuation with finite residue class field then F is C_2. We are able to prove this only in the case of power series fields, leaving the question open in the case of p-adic fields. We lean heavily on a result of Chevalley [1]:

THEOREM 7. *Let $\mathfrak{k}$ be a finite field. Let $f_1, \cdots, f_r$ be a set of polynomials in $\mathfrak{k}$ without constant term, with $n > d_1 + \cdots + d_r$. Then they have a non trivial common zero in $\mathfrak{k}$.*

We need also an auxiliary statement:

LEMMA. *Let F be a field complete under a non-archimedean valuation and let the integers $\mathfrak{o}$ be compact. Let f be a form in F and suppose that there exists a sequence of forms f_ν converging to f such that each f_ν has a non trivial zero in F. Then f also has a non trivial zero in F.*

PROOF. We may assume all coefficients and zeros are integers. Further, by homogeneity, we may assume that some α_i in a zero $(\alpha_1, \cdots, \alpha_n)$ of f_ν is a unit. We obtain in this way a collection of n-tuples in the Cartesian product $\mathfrak{o}^n$. This product is compact, and our collection must have a point of accumulation, which will obviously be the desired zero. It is non trivial because each element of our collection has at least one unit component.

We may now prove

THEOREM 8. *Let $\mathfrak{k}$ be a finite field. The power series field $k\{t\}$ in one variable is C_2.*

PROOF. Given the form $f(x_1, \cdots, x_n)$ of degree d, $n > d^2$ with integral coefficients, we must show it has a non trivial zero. If we chop off the coefficients of f after a power of t, we obtain a form in $\mathfrak{k}[t]$. By Theorems 6 and 7 this form has a non trivial zero. Doing this for each power of t we obtain an approximating sequence of forms, each having a zero. We can apply the lemma to complete the proof.

It would be interesting to know whether a field complete under a discrete valuation with C_i residue class field is C_{i+1}. We shall treat the case where the residue class field is algebraically closed. Our goal is now to prove the following statement: Let F be complete under a discrete valuation with algebraically closed residue class field. Then F is C_1.

We deal first with a very general situation to be applied later. Let F be complete under any valuation. If a polynomial $f(x_1, \cdots, x_n)$ has a zero in F then in general we can find a whole neighborhood of zeros for f by shifting very slightly $n - 1$ variables and readjusting the last one accordingly. This procedure breaks down when the zero is singular, that is all partial derivatives vanish at the point. If however this calamity does not happen, the procedure will go through by Newton's approximation method. For the real numbers, the proof is well known. In the non-archimedean case, we use the analogous

LEMMA. *Let F be complete under a non-archimedean valuation. Let $f(x)$ have integral coefficients in F and let α_0 be an integer in F such that $\left| \dfrac{f(\alpha_0)}{f'(\alpha_0)^2} \right| < 1$. Then the sequence*

$$\alpha_{i+1} = \alpha_i - \frac{f(\alpha_i)}{f'(\alpha_i)}$$

converges to a root α of $f(x)$. Furthermore, $|\alpha - \alpha_0| \leqq \left| \dfrac{f(\alpha_0)}{f'(\alpha_0)^2} \right| < 1$.

PROOF. Let $\left| \dfrac{f(\alpha_0)}{f'(\alpha_0)^2} \right| = a < 1$. Remark that $|f(\alpha_0)| < 1$ and $|f'(\alpha_0)| \leqq 1$. We show inductively that

1. $|\alpha_i| \leqq 1$,

2. $|\alpha_i - \alpha_0| \leqq a$,

3. $\left| \dfrac{f(\alpha_i)}{f'(\alpha_i)^2} \right| \leqq a^{2^i}$.

These three conditions obviously imply our lemma. For $i = 0$ these are hypotheses. By induction, assume them for i. Then:

1. $\left| \dfrac{f(\alpha_i)}{f'(\alpha_i)^2} \right| \leqq a^{2^i}$ gives $|\alpha_{i+1} - \alpha_i| \leqq a^{2^i} < 1$ whence $|\alpha_{i+1}| \leqq 1$.

2. $|\alpha_{i+1} - \alpha_0| \leqq \max(|\alpha_{i+1} - \alpha_i|, |\alpha_i - \alpha_0|) = a$.

3. By Taylor's expansion, we have

$$|f(\alpha_{i+1})| = \left| f(\alpha_i) - f'(\alpha_i)\frac{f(\alpha_i)}{f'(\alpha_i)} + \left(\frac{f(\alpha_i)}{f'(\alpha_i)}\right)^2 (\text{integer}) \right| \leqq \left| \frac{f(\alpha_i)}{f'(\alpha_i)} \right|^2 .$$

Again by Taylor's expansion:

$$f'(\alpha_{i+1}) = f'(\alpha_i) + \frac{f(\alpha_i)}{f'(a_i)} (\text{integer})$$

whence

$$\frac{f'(\alpha_{i+1})}{f'(\alpha_i)} = 1 + \frac{f(\alpha_i)}{f'(\alpha_i)^2} (\text{integer}).$$

Consequently

$$\left| \frac{f'(\alpha_{i+1})}{f'(\alpha_i)} \right| = 1.$$

Now we get:

$$\left| \frac{f(\alpha_{i+1})}{f'(\alpha_{i+1})^2} \right| \leqq \left| \frac{f(\alpha_i)^2}{f'(\alpha_i)^2 f'(\alpha_{i+1})^2} \right| = \left| \frac{f(\alpha_i)}{f'(\alpha_i)^2} \right|^2 \leqq a^{2^{i+1}},$$

as was to be shown. This concludes the proof.

The meaning of our lemma in n variables is clear. Given $f(x_1, \cdots, x_n)$ suppose we can find an $\alpha = (\alpha_1, \cdots, \alpha_n)$ such that $f(x_1, \alpha_2, \cdots, \alpha_n)$ satisfies the conditions of the lemma, so that α_1 can be refined to a root. If $\beta = (\beta_1, \cdots, \beta_n)$ lies in a sufficiently small neighborhood of α, then $f(x_1, \beta_2, \cdots, \beta_n)$ will also satisfy the conditions of the lemma, and β_1 can be refined to a root of $f(x_1, \beta_2, \cdots, \beta_n)$.

We return to our special case where F is complete under a discrete valuation with algebraically closed residue class field $\mathfrak{f}$. It is known that such fields are finite, completely ramified extensions of an invariantly determined subfield. This subfield may be of two types:

A power series field $\mathfrak{f}\{t\}$ in one variable, which is F itself.

A field of Witt vectors [5] with components in $\mathfrak{f}$. ($\mathfrak{f}$ has then characteristic $p > 0$, but F has characteristic 0.)

We shall refer to either one of these fields as the complete unramified field having $\mathfrak{f}$ as residue class field. We shall deal with the domain of integers, whose elements are given by vectors

$$x_1 = (\xi_{10}, \xi_{11}, \cdots), \qquad x_2 = (\xi_{20}, \xi_{21}, \cdots), \cdots, \qquad x_n = (\xi_{n0}, \xi_{n1}, \cdots)$$

interpreted either as Witt vectors or as power series, $x_i = \sum_0^\infty \xi_{i\nu} t^\nu$ as the case may be.

Given a polynomial $f(x_1, \cdots, x_n)$ we may write

$$f(x_1, \cdots, x_n) = (f_0(\xi_0), f_1(\xi_0, \xi_1), f_2(\xi_0, \xi_1, \xi_2), \cdots)$$

where $\xi_0 = (\xi_{10}, \cdots, \xi_{n0})$, $\xi_1 = (\xi_{11}, \cdots, \xi_{n1})$, $\cdots$ and the polynomials $f_0, f_1, \cdots$ have coefficients in $\mathfrak{f}$.

The problem of finding a zero for $f(x_1, \cdots, x_n)$ amounts to solving the infinite set of equations $f_0 = f_1 = f_2 = \cdots = 0$ in $\mathfrak{f}$. We shall now develop the tools necessary for solving such an infinite system. We first give an outline of our argument. We shall prove:

1. If the ideal $\mathfrak{a} = (f_0, f_1, \cdots)$ in the ring of denumerably many variables is not the unit ideal, then it has a zero in a suitable algebraically closed extension field $\mathfrak{f}_1$ of $\mathfrak{f}$. This will imply that $f(x_1, \cdots, x_n)$ has a zero in the complete field F_1 having $\mathfrak{f}_1$ as residue class field.

2. Given such a zero in F_1, we can specialize it back into F.

3. In the case of a form with $n > d$ we shall prove that we may put one of the variables equal to 1 and that if f is the resulting polynomial then the ideal $\mathfrak{a} = (f_0, f_1, \cdots)$ is not the unit ideal. By steps 1 and 2 we shall conclude that our form has a non trivial zero in F.

Our first step is easily settled. Let $\mathfrak{o} = \mathfrak{f}[\xi_0, \xi_1, \cdots]$. If $\mathfrak{a}$ is a proper ideal of $\mathfrak{o}$ let $\mathfrak{p}$ be a prime ideal containing $\mathfrak{a}$ ($\mathfrak{p}$ maximal will do, for instance). Then $\mathfrak{o}/\mathfrak{p}$ is isomorphic to a ring $\mathfrak{f}[\eta_0, \eta_1, \cdots]$ where the η_ν lie in a suitable extension field of $\mathfrak{f}$. Then $(\eta_0, \eta_1, \cdots)$ is a zero of $\mathfrak{p}$, a fortiori of $\mathfrak{a}$. If we let $\mathfrak{f}_1$ be the algebraic closure of the field $\mathfrak{f}(\eta_0, \eta_1, \cdots)$ and F_1 be the complete unramified field having $\mathfrak{f}_1$ as residue class field, then $F \subset F_1$ and $f(x_1, \cdots, x_n)$ has a zero in F_1.

Our next step is to prove that the zero just obtained can be specialized back into F. We need to settle first two minor technical details, and formulate them in lemmas.

LEMMA 1. *Let* $\mathfrak{f} \subset \mathfrak{f}_1$ *be two algebraically closed fields. Let* $F \subset F_1$ *be the complete, unramified fields having* $\mathfrak{f}$ *and* $\mathfrak{f}_1$ *respectively as residue class fields. Then* F *is algebraically closed in* F_1 *and every intermediate finitely generated subfield is separably generated.*

PROOF. Let E/F be a finite extension contained in F_1. Let e be the ramification index. Since the residue class field of F is algebraically closed, e is the degree of E/F. But since F_1 is also unramified, E cannot be contained in F_1 unless $E = F$.

Concerning the separable generation, the question arises only if $F = \mathfrak{f}\{t\}$ and $F_1 = \mathfrak{f}_1\{t\}$ are power series fields, and $\mathfrak{f}$ has characteristic $p > 0$. In such a case, if an intermediate field were not separably generated, we would have a relation of lowest degree $f(x_1^p, \cdots, x_n^p) = 0$ where f has integral coefficients in F, some coefficient of f is a unit, and $x_1, \cdots, x_n$ are integral power series in F_1. We contend that this cannot happen. The coefficients of f are of the type $a = a_0 + a_1 t + a_2 t^2 + \cdots$ and we may assume some $a_0 \neq 0$. Expand the relation and look at the coefficients of $t^{ps+\nu}$, $0 \leq \nu \leq p - 1$.

If $\nu = 0$, only elements a_j where $p \mid j$ will appear.

If $\nu \neq 0$, only elements a_j where $p \nmid j$ will appear.

In both cases the terms a_j appear linearly and homogeneously. If we put $a_j = 0$ for all j where $p \nmid j$ in the coefficients of f and let g be the resulting polynomial, then g is not identically zero and we have $g(x_1^p, \cdots, x_n^p) = 0$. All coefficients of g are now p^{th} powers, and taking the pth root of this relation would give one of lower degree than that of f, contrary to assumption.

LEMMA 2. *Let* $\mathfrak{f}$ *be an algebraically closed field. Let* $\eta = (\eta_1, \cdots, \eta_n)$ *be a generic point over* $\mathfrak{f}$ *and let* $h_\nu(x_1, \cdots, x_n)$ *be a finite set of polynomials in* $\mathfrak{f}$ *none of which vanishes on* η. *Then there exists a specialization* η' *of* η *into* $\mathfrak{f}$ *such that none of the* h_ν *vanish on* η'.

PROOF. Let $h(x) = \prod_\nu h_\nu(x)$. Then $h(\eta) \neq 0$. Let $\mathfrak{p}$ be the ideal in $\mathfrak{f}[x_1, \cdots, x_n]$ vanishing on η. Then $h(x) \notin \mathfrak{p}$. By the Hilbert Nullstellensatz $h(x)$ does not vanish on some specialization of η.

THEOREM 9. *Let* $\mathfrak{f} \subset \mathfrak{f}_1$ *be two algebraically closed fields. Let* $F \subset F_1$ *be the unramified complete fields having* $\mathfrak{f}$ *and* $\mathfrak{f}_1$ *as residue class fields. Let* $\eta = (\eta_1, \cdots, \eta_n)$ *be a generic point over* F *with* $\eta_\nu \in F_1$. *Then* η *has a specialization in* F.

PROOF. By Lemma 1, we have $F(\eta_1, \cdots, \eta_n) = F(t_1, \cdots, t_r, \alpha)$ where $t = (t_1, \cdots, t_r)$ is a separating transcendence base and α is separable over $F(t)$. We may assume that $t_1, \cdots, t_r$ and α are integers (multiply them if necessary by a high power of a prime in F). We can then write

$$\eta_i = \sum_\nu \frac{\varphi_{i\nu}(t)}{\psi_{i\nu}(t)} \alpha^\nu$$

where $\varphi_{i\nu}$ and $\psi_{i\nu}$ are polynomials with coefficients in F

α satisfies an irreducible equation $g(y, t)$ over $F(t)$ and by clearing denominators we may assume that the coefficients of g are all in $F[t]$, that is are polynomials in $t_1, \cdots, t_r$. Then $\partial g/\partial \alpha \neq 0$ because α is separable.

Let $\{f_j(x_1, \cdots, x_n)\}$ be a finite basis for the prime ideal of $F[x_1, \cdots, x_n]$ vanishing on η. Then $f_j(\eta) = 0$ means that we can write

$$(1) \qquad f_j\left(\cdots, \sum_\nu \frac{\varphi_{i\nu}(t)}{\psi_{i\nu}(t)} y^\nu, \cdots\right) = g(y, t)h_j(y, t)$$

where $h_j(y, t)$ is a polynomial in y with coefficients in $F(t)$.

Let $t_1, \cdots, t_r$ and α have the vector expansions

$$t_1 = (t_{10}, t_{11}, t_{12}, \cdots)$$
$$t_2 = (t_{20}, t_{21}, t_{22}, \cdots)$$
$$\cdots$$
$$t_r = (t_{r0}, t_{r1}, t_{r2}, \cdots) \qquad t_{ij} \in \mathfrak{f}_1$$
$$\alpha = (\alpha_0, \alpha_1, \alpha_2, \cdots) \qquad \alpha_i \in \mathfrak{f}_1 .$$

We consider $(\alpha_0, \alpha_1, \cdots, \alpha_s, t_{10}, \cdots, t_{r0}, t_{11}, \cdots, t_{r1}, \cdots, t_{1s}, \cdots, t_{rs})$ denoted briefly by $(\alpha_i, t_{ij})_s$ as a generic point over $\mathfrak{f}$. Expand $g(y, t)$ and $\partial g/\partial y$ in vectors so we get

$$g(\alpha, t) = (g_0(\alpha_i, t_{ij}), g_1(\alpha_i, t_{ij}), \cdots)$$

$$\partial g/\partial \alpha = (g_0'(\alpha_i, t_{ij}), g_1'(\alpha_i, t_{ij}), \cdots).$$

Then $g_\nu(\alpha_i, t_{ij}) = 0$ for all ν, and say $g_m'(\alpha_i, t_{ij}) \neq 0$.

We shall denote by $(\beta_0, \beta_1, \cdots, \beta_s, u_{10}, \cdots, u_{r0}, \cdots, u_{is}, \cdots, u_{rs}) = (\beta_i, u_{ij})_s$ a suitable specialization of $(\alpha_i, t_{ij})_s$ into $\mathfrak{f}$, and we let

$$u_1 = (u_{10}, u_{11}, \cdots, u_{1s}, 0, 0, 0, \cdots)$$
$$\cdots \qquad\qquad u_{ij} \in \mathfrak{f}$$
$$u_r = (u_{r0}, u_{r1}, \cdots, u_{rs}, 0, 0, 0, \cdots)$$
$$\beta = (\beta_0, \beta_1, \cdots, \beta_s, 0, 0, 0, \cdots) \qquad \beta_i \in \mathfrak{f}$$

so $u_1, \cdots, u_r, \beta$ lie in F.

Our aim is to find a specialization $(\beta_i, u_{ij})_s$ of $(\alpha_i, t_{ij})_s$ into $\mathfrak{f}$ such that:

1. $g(\beta, u)$ satisfies the hypotheses of the Newton approximation lemma.
2. The denominators involving t's in (1) do not vanish under this specialization.

There is only a finite number of such denominators, appearing in $\psi_{i\nu}(t)$ and $h_j(y, t)$. If we expand these denominators into vectors, their components will be typically represented by vectors like $(\psi_0(t_{ij}), \psi_1(t_{ij}), \cdots)$ and in each expansion some $\psi_\nu(t_{ij}) \neq 0$. We let the finite set of polynomials in Lemma 2 consist of these ψ_ν and of g_m'. Then by selecting first s sufficiently large and secondly by using Lemma 2, we can find the specialization $(\beta_i, u_{ij})_s$ such that the denominators do not vanish, $|g(\beta, u)|$ is very small, and since $g_m'(\beta_i, u_{ij})$ does not vanish, $|\partial g/\partial \beta|$ is large compared to $|g(\beta, u)|$.

β can then be refined to a root γ of $g(y, u)$ and putting

11

$$\bar{\eta}_i = \sum_\nu \frac{\varphi_{i\nu}(u)}{\psi_{i\nu}(u)} \gamma^\nu$$

we see that $\bar{\eta} = (\bar{\eta}_1, \cdots, \bar{\eta}_n)$ is a specialization of η.

The u's are in F and γ is consequently algebraic over F. Since γ is also in F_1, we conclude by Lemma 1 that γ is in F. Hence all $\bar{\eta}_i$ lie in F, and our theorem is proved.

COROLLARY. *Let $\mathfrak{f}$ be an algebraically closed field and let F be the unramified complete field with $\mathfrak{f}$ as residue class field. Let $\mathfrak{p}$ be the valuation prime. Let $f(x_1, \cdots, x_n)$ be a polynomial in F such that $f \equiv 0$ (mod $\mathfrak{p}^\nu$) is solvable in F for all ν. Then f has a zero in F.*

PROOF. Since $f \equiv 0$ (mod $\mathfrak{p}^\nu$) is solvable for all ν, this means that each partial system $f_0 = 0, \cdots, f_\nu = 0$ is solvable in $\mathfrak{f}$. The ideal $\mathfrak{a} = (f_0, f_1, \cdots)$ cannot be the unit ideal for otherwise 1 could be expressed by means of a finite expression in $f_0, \cdots, f_\nu$. In a previous discussion we saw that $\mathfrak{a}$ has a zero in an extension field $\mathfrak{f}_1$ of $\mathfrak{f}$ and f has therefore a zero in the unramified field F_1 having $\mathfrak{f}_1$ as residue class field. Theorem 9 now shows that f has a zero in F.

All that remains to be done is to carry out the third step of the procedure outlined above. We shall then have

THEOREM 10. *Let F be a field complete under a discrete valuation with algebraically closed residue class field. Then F is C_1.*

PROOF. It is sufficient to prove the theorem for the unramified complete field because a finite extension of a C_1 field is also C_1.

Given a form $f(x_1, \cdots, x_n)$ in F with $n > d$, we must show that we can satisfy simultaneously (and non trivially) the infinite system $f_0 = 0, f_1 = 0, \cdots$. We shall consider partial systems $f_0 = 0, f_1 = 0, \cdots, f_r = 0$. These equations define a variety in the $n(r + 1)$ space of the variables $(\xi_{10}, \cdots, \xi_{n0}, \xi_{11}, \cdots, \xi_{n1}, \cdots, \xi_{1r}, \cdots, \xi_{nr})$ or briefly $(\xi_0, \xi_1, \cdots, \xi_r)$. We shall first prove that the point set theoretic projection of this variety on $(\xi_{10}, \cdots, \xi_{n0})$ is not zero for any r.

Consider for the moment only the Witt vectors. If the form f has degree d and all x_i have first component zero, that is $\xi_{10} = \xi_{20} = \cdots = \xi_{n0} = 0$ we may put $x_i = py_i$ where $y_i = (\xi_{i1}^{1/p}, \xi_{i2}^{1/p}, \cdots)$. Then $f(x_1, \cdots, x_n) = f(py_1, \cdots, py_n) = p^d f(y_1, \cdots, y_n) = V^d f(y_1, \cdots, y_n)^{p^d}$. Consequently we have the relations

$$
\begin{aligned}
& f_r(0, \xi_1, \cdots, \xi_r) = 0 && r < d \\
& f_r(0, \xi_1, \cdots, \xi_r) = f_{r-d}^{p^d}(\xi_1^{p^{d-1}}, \cdots, \xi_{r-d+1}^{p^{d-1}}) && r \geqq d
\end{aligned}
$$

(2)

where $f_{r-d}^{p^d}$ means that the coefficients of f_{r-d} are to be raised to the $p^{d\,\text{th}}$ power. For power series the same relations clearly hold without the extra exponents p^d and p^{d-1}.

We shall argue by contradiction. Suppose that for some system $f_0 = 0, \cdots,$ $f_r = 0$ the only solutions are trivial on ξ_0, so that $\xi_{10} = \xi_{20} = \cdots = \xi_{n0} = 0$. Then any larger system also has this property, in particular the system $f_0 = 0, \cdots, f_{r+d} = 0$. Furthermore, solving this larger system amounts to solving

$$f_0(0) = 0, f_1(0, \xi_1) = 0, f_2(0, \xi_1, \xi_2) = 0, \cdots, f_{r+d}(0, \xi_1, \cdots, \xi_{r+d}) = 0.$$

Because of relations (2) this means that we must solve

$$f_0(\xi_1) = 0, f_1(\xi_1, \xi_2) = 0, \cdots, f_r(\xi_1, \cdots, \xi_{r+1}) = 0.$$

By assumption we know that this system must have solutions trivial on ξ_1. Inductively it follows that $f_0 = 0, \cdots, f_{r+kd} = 0$ has solutions trivial on $\xi_0, \cdots, \xi_k$. We now consider the dimension of the solution of the system $f_0 = 0, \cdots, f_{r+kd} = 0$ in $n(r + kd + 1)$ space. By Theorem 2, the dimension is at least $n(r + kd + 1) - (r + kd + 1)$. By the above remarks, it is at most $n(r + kd + 1) - n(k + 1)$. Comparing these two numbers we must have

$$(n - d)k \leqq r + 1 - n$$

for all k. This is a contradiction and therefore each partial system has zeros which are non trivial on ξ_0.

This means that if $\mathfrak{p}$ is the prime valuation ideal, $f(x_1, \cdots, x_n) \equiv 0 \pmod{\mathfrak{p}^\nu}$ is solvable for each ν with some x_i being a unit. By homogeneity we may assume this x_i to be 1. One of the x_i, say x_1, must occur for infinitely many ν. If we put $x_1 = 1$ in f we can solve $f(1, x_2, \cdots, x_n) \equiv 0 \pmod{\mathfrak{p}^\nu}$ for all ν. By the corollary to Theorem 9 we can solve $f(1, x_2, \cdots, x_n) = 0$ in F and this proves our theorem.

Let now F be the field complete under a discrete valuation with perfect residue class field $\mathfrak{f}$. Let T be the maximal unramified extension of F. T is in general not complete. Let $\bar{\mathfrak{f}}$ be the algebraic closure of $\mathfrak{f}$, and let E range over all intermediate fields $\mathfrak{f} \subset E \subset \bar{\mathfrak{f}}$ such that $E/\mathfrak{f}$ is finite. Then T is quotient field of the ring of vectors $x = (\xi_0, \xi_1, \xi_2, \cdots)$ where $\xi_i \epsilon E$ for some E. If $\mathfrak{f}$ is not algebraically closed, or equivalently if F admits a proper unramified extension, then we know that F has normic forms of order 2. Likewise, so does every proper subfield of T containing F, so that no such proper subfield can be C_1. We shall prove however that T itself is C_1. It will be a consequence of

THEOREM 11. *Let $\bar{F}$ be a field complete under any valuation. Let F be a dense subfield of $\bar{F}$, algebraically closed in $\bar{F}$. Let $\eta = (\eta_1, \cdots, \eta_n)$ be a generic point with $\eta_\nu \epsilon \bar{F}$, such that the field $F(\eta_1, \cdots, \eta_n)$ is separably generated over F. Then η has arbitrarily close specializations in F.*

(By "arbitrarily close specialization" we mean: Given a neighborhood of η, there exists a specialization of η in that neighborhood.)

PROOF. Let $f_j(x_1, \cdots, x_n)$ be a (finite) set of polynomials in $F[x_1, \cdots, x_n]$ generating the prime ideal vanishing on η. By assumption, there exists a separating transcendence base $t = (t_1, \cdots, t_r)$ in $\bar{F}$ such that $F(\eta) = F(t, \alpha)$ and α is separable over $F(t)$. We can write

$$\eta_i = \sum_\nu \frac{\varphi_{i\nu}(t)}{\psi_{i\nu}(t)} \alpha^\nu.$$

Let $g(y, t_1, \cdots, t_r)$ be the irreducible polynomial of α over $F(t)$. Then $\partial g / \partial \alpha \neq 0$, and by the Newton approximation lemma, we may select u_i, β in F close to t_i, α such that $g(\beta, u) = 0$.

We have

$$f_j\left(\cdots, \sum \frac{\varphi_{i\nu}(t)}{\psi_{i\nu}(t)}\, y^\nu, \cdots\right) = g(y; t)h_j(y, t).$$

Let

$$\bar{\eta}_i = \sum_\nu \frac{\varphi_{i\nu}(u)}{\psi_{i\nu}(u)}\, \beta^\nu.$$

We can choose u_i close enough to t_i that $\bar{\eta}_i$ is defined and $h(\beta, u)$ is defined (namely, denominators do not vanish). Then $f_j(\bar{\eta}) = 0$ and consequently $\bar{\eta}$ is the desired specialization.

COROLLARY. *Let F and $\bar{F}$ be as in Theorem 11, and suppose in addition that any finitely generated intermediate field is separably generated. Then any set of polynomials $f_1, \cdots, f_r$ with coefficients in F having a zero in $\bar{F}$ also have a zero in F (non trivial if the zero in $\bar{F}$ is non trivial). Hence if $\bar{F}$ is C_i so is F.*

The assumption that $F(\eta_1, \cdots, \eta_n)$ is separably generated in Theorem 11 is necessary, as the following counterexample shows. Suppose that we could find in $\bar{F}$ a set of elements which generate over F an extension which is not separably generated. Then by a well known criterion [4] we could find elements $a_1, \cdots, a_s \in F$ linearly independent over F^p but dependent over $\bar{F}^p$. In other words the form $a_1 x_1^p + \cdots + a_s x_s^p = 0$ would have no zero in F even though it had a zero η in $\bar{F}$. It follows that η has no specialization in F. The example can actually be given explicitly: Let $\mathfrak{f}$ be a finite field, and put $\bar{F} = \mathfrak{f}\{t\}$. Select $a_1, \cdots, a_s, \alpha_1, \cdots, \alpha_s$ algebraically independent in $\bar{F}$. Let

$$a_{s+1} = a_1 \alpha_1^p + \cdots + a_s \alpha_s^p$$

and let F be the algebraic closure in $\bar{F}$ of $\mathfrak{f}(a_1, \cdots, a_{s+1})$. We may suppose $a_1 = t$, so F is dense in $\bar{F}$. The equation $a_1 x_1^p + \cdots + a_s x_s^p - a_{s+1} = 0$ has no zero in F, for if it did we would get by subtraction

$$a_1(x_1 - \alpha_1)^p + \cdots + a_s(x_s - \alpha_s)^p = 0$$

which contradicts the algebraic independence of $a_1, \cdots, a_s, \alpha_1, \cdots, \alpha_s$.

We now have

THEOREM 12. *The following fields and their algebraic extensions are C_1 :*

The maximal unramified extension of a field complete under a discrete valuation, with perfect residue class field.

The absolutely algebraic subfield of the maximal unramified extension of an ordinary p-adic field.

The convergent power series over an algebraically closed valuated constant field.

PROOF. In each case the verification that the field in question satisfies the conditions of the Corollary to Theorem 11 is either trivial or well known. The completions of these fields were proved to be C_1 in Theorem 10, and our theorem is therefore proved.

Actually in the proof of the above theorem we could have made a short circuit and by-passed Theorem 11. Namely, in the proof of Theorem 9 the u_i could actually have been replaced by elements of any dense subfield. Hence the specialization $\bar{\eta}$ could have been made to lie in a subfield depending only on the coefficients of the polynomial $f(x_1, \cdots, x_n)$. This short circuit is possible only because we are dealing throughout with the special case of a discrete valuation with algebraically closed residue class field. Since the Specialization Theorem 11 holds for any complete field (including the real numbers) it seemed preferable to follow the slightly longer path we have selected here.

PART III.

APPLICATIONS TO CLASS FIELD THEORY

We show how the results of Part II can be applied to class field theory. We use the language of cohomology and refer to [2] for a complete treatment of the subject. We recall here merely the definitions and notation, and the fundamental algebraic properties.

Let F be a field, C/F a normal (possibly infinite) extension, and G its Galois group considered as a topological group with the Krull topology: Neighborhoods of 1 are the subgroups of finite index. A 1-cochain is a continuous map of G into C^*, denoted by a_σ where $\sigma \, \epsilon \, G$. A 2-cochain is a continuous map of $G \times G$ into C^* denoted by $a_{\sigma,\tau}$. (The * here denotes the multiplicative group of non zero elements.) A 2-cocycle is a 2-cochain whose boundary

$$\frac{a_{\tau,\rho}^\sigma \, a_{\sigma,\tau\rho}}{a_{\sigma,\tau} \, a_{\sigma\tau,\rho}} = 1,$$

and the 2 cocycles form a multiplicative group Z. There is a subgroup B consisting of all 2-cochains of type

$$\frac{a_\sigma \, a_\tau^\sigma}{a_{\sigma\tau}},$$

the boundaries of 1-cochains. The second cohomology group $H^2(C, F)$ is defined to be the factor group Z/B. If $a_{\sigma,\tau} \equiv b_{\sigma,\tau} \pmod{B}$ we write $a_{\sigma,\tau} \sim b_{\sigma,\tau}$. We say that a cocycle $a_{\sigma,\tau}$ splits if $a_{\sigma,\tau} \sim 1$. As we shall deal only with 2-cocycles, we omit the 2- and speak only of cocycles.

We shall hereafter let C be the separable part of the algebraic closure of F and all fields considered will be subfields of C. G is the group of C/F. The group $H^2(C, F)$ will be written $H^2(F)$. Let $a_{\sigma,\tau}$ be a cocycle. Let E/F be any extension and let H be the group of C/E. By the restriction of $a_{\sigma,\tau}$ to H, written $\mathrm{res}_H(a_{\sigma,\tau})$, we mean the cocycle $a_{\sigma,\tau}$ itself where σ, τ range over H only. We shall also write $\mathrm{res}_E(a_{\sigma,\tau})$ instead of $\mathrm{res}_H(a_{\sigma,\tau})$. We say that a cocycle $a_{\sigma,\tau}$ splits on H, or splits on E if $\mathrm{res}_E(a_{\sigma,\tau}) \sim 1$. It follows from continuity that every cocycle has a finite splitting field and in fact a normal splitting field.

Let E/F be finite, and let c be a cocycle split by E. Then c has a representative $a_{\sigma,\tau}$ such that $a_{\sigma,\tau\gamma} = a_{\sigma,\tau}$ for all $\gamma \, \epsilon \, H$.

Let E/F be normal, and let c by a cocycle split by E. Then c has a representative $a_{\sigma,\tau}$ depending only on G/H, i.e. $a_{\sigma,\tau} = a_{\sigma,\gamma_1\tau\gamma_2}$ for all $\gamma_1, \gamma_2 \, \epsilon \, H$, and further-

388 SERGE LANG

more $a_{\sigma,\tau} \in E$. Consequently we can think of c as a cocycle in $H^2(E, F)$. If Ω/F is arbitrary then $\mathrm{res}_\Omega(a_{\sigma,\tau})$ is a cocycle of $H^2(\Omega)$ and is split by $E\Omega$. We can consider it as a translation of c over Ω.

If E/F is cyclic of order n with group generated by σ and if c is a cocycle split by E then c has a normalized representative of the following type:

$$a_{\sigma^\nu,\sigma^\mu} = \alpha^{[(\nu+\mu)/n]-[\nu/n]-[\mu/n]}$$

for some $\alpha \in F$, and $c \sim 1$ if and only if $\alpha \in N_{E/F}$. (By $N_{E/F}$ we denote the norm group of non-zero norms from E to F.) We shall write more briefly $c \sim (\alpha)$, and it is understood that a generator σ has already been fixed.

If Ω/F is arbitrary and $(E \cap \Omega/F) = m$, then $\mathrm{res}_\Omega(c) = (\alpha)$ where the group generator is now σ^m. Moreover, (α) is split by $E\Omega/\Omega$.

We shall need one special property of cocycles, given as a

LEMMA. *If c is a cocycle in $H^2(F)$ and is split by a field E with $(E/F) = n$, then $c^n \sim 1$. Conversely if c is a cocycle of period m and p is a prime $p \mid m$ and E is a splitting field of c, then $p \mid (E/F)$.*

PROOF. Let H be the group of C/E. Let $G = \bigcup_1^n \tau_\nu H$. We have

$$a_{\sigma,\tau} a_{\sigma\tau,\rho} = a_{\tau,\rho}^\sigma a_{\sigma,\tau\rho}.$$

Put $\rho = \tau_\nu$ and take the product over ν. Let $f_\sigma = \prod a_{\sigma,\tau\nu}$. Then $a_{\sigma,\tau}^n f_{\sigma\tau} = f_\tau^\sigma f_\sigma$ shows that $a_{\sigma,\tau}^n \sim 1$.

Conversely, let $(E/F) = n$ and suppose that $p \nmid n$. Then by the above, $a_{\sigma,\tau}^r \sim 1$ where $r = (m, n)$ and $p \nmid r$, a contradiction.

The problem is to determine the structure of the second cohomology group $H^2(F)$. Class field theory solves this problem, among others, for number fields, function fields in one variable over a finite constant field, and the completions of these fields at some prime.

In general, let F be any field and suppose F admits an extension Ω which is C_1. Then $H^2(F)$ is completely determined by Ω. The source of this result lies in the following

THEOREM 13. *Let F be C_i and let N be a normic form of order i in F. Then N represents every element of F. This means: The equation $N(x_1, \cdots, x_n) = \alpha$ is solvable in F for all $\alpha \in F$.*

PROOF. The form $N(x_1, \cdots, x_n) - \alpha x^d$ has more than d^i variables and has therefore a non trivial zero. Furthermore, $x \neq 0$ because N is normic. Let $y_1 = x_1/x, \cdots, y_n = x_n/x$. Then $N(y_1, \cdots, y_n) = \alpha$ as was to be shown.

In particular if F is C_1 and E/F is any finite extension then every element of F is norm of an element of E.

Let F be any field again, and suppose F admits an extension Ω which is C_1. We shall show that every cocycle splits on a finite subfield of Ω. First in the cyclic case:

LEMMA. *Let c be a cocycle split by a cyclic extension E/F. Let Ω/F be any extension and let Ω be C_1. Then c is also split by some subfield K/F of Ω, and K/F is finite.*

PROOF. We know that $c \sim (\alpha)$ for some $\alpha \in F$. We may suppose $E \cap \Omega = F$. Let N be the norm form of E/F. For any extension $K \subset \Omega$ N will also be the norm form of EK/K. By Theorem 13 there exists a finite extension $K \subset \Omega$ such that $\alpha \in N_{EK/K}$. Consequently $\mathrm{res}_K(\alpha) \sim 1$ and the factor set splits on K, which proves the lemma.

THEOREM 14. *Let F be any field and let Ω be an extension of F which is C_1. Then every cocycle is split by a finite subfield of Ω.*

PROOF. Let c be a cocycle of $H^2(F)$. It is split by a finite normal extension E/F. We may suppose $E \cap \Omega = F$. Let c have exponent n and let $p \mid n$. Let E_0 be the fixed field of a p-Sylow group of the group of E/F. We can find a tower $E_0 \subset E_1 \subset E_2 \subset E$ such that E_2/E_1 is cyclic, and such that E_2 splits c but E_1 does not. By the lemma, there exists a finite extension $K \subset \Omega$ such that c splits on E_1K. Since E_1K/E_0K is a p-tower we can repeat this procedure until we have found a K such that c splits on E_0K.

It follows that $\mathrm{res}_K(c)$ has an exponent $m \mid n$, and $p \nmid m$; for if $p \mid m$ then (E_0K/K) would have to be divisible by p, and this is not the case. To complete the proof, we repeat the performance for the rest of the primes dividing n.

COROLLARY. *If F is C_1 then $H^2(F)$ is trivial.*

If we consider the fields of Theorem 12 as special cases of Theorem 14, we get

THEOREM 15. *Let F be complete under a discrete valuation with perfect residue class field. Then every cocycle has a finite unramified splitting field.*

We see that the structure of $H^2(F)$ is completely determined by the unramified extensions. In case the residue class field is a finite field, this structure is well known and very simple: Let T_n be the unramified extension of degree n over F. Let σ be the Frobenius substitution. The cocycles split by T_n are all cyclic, and can be represented by (α) for some $\alpha \in F$. Since every unit is a norm, $(\alpha) \sim (\pi^\nu)$ where $\nu = \mathrm{ord}\ \alpha$. The group of cocycles split by T_n is therefore cyclic of order n, generated by the canonically determined cocycle (π) which we shall denote by $c_{1/n}(F)$. In fact, the only invariant of a cocycle (α) is the rational number $r = \mathrm{ord}\ \alpha/n$ (mod 1). Any cocycle can therefore be represented by some $c_r(F)$ and we have $c_r(F) \cdot c_s(F) \sim c_{r+s}(F)$. The group $H^2(F)$ is therefore naturally isomorphic with the rationals (mod 1).

We may say more, however. If E/F is any extension of degree n, then it is well known and not difficult to show that $\mathrm{res}_E(c_r(F)) = c_{nr}(E)$. Consequently $c_{1/n}(F)$ is split by any extension of degree n. Conversely, let c be any cocycle split by E. By Theorem 15, c has an unramified splitting field. We contend that c is split by the unramified field of degree n: We have $c \sim c_r(F)$ for some r. Let $r = s/m$ where s, m are relatively prime integers. We know $c^n \sim 1$ and consequently $c_{rn}(F) = c_{sn/m}(F) \sim 1$. Hence $m \mid n$ and c is split by T_m and a fortiori by T_n because $T_m \subset T_n$. Summarizing these results, we have

THEOREM 16. *Let F be complete under a discrete valuation with finite residue class field. Then $H^2(F)$ is naturally isomorphic with the rationals (mod 1). If c is any cocycle of exponent n, then c is split by any field of degree n over F. The group of cocycles split by an extension E/F of degree n is cyclic of order n, generated by $c_{1/n}(F)$.*

We can state an analogous theorem in the large in the case of function fields by applying Theorem 14 to Theorem 6, namely

THEOREM 17. *Let F be a function field of one variable over a constant field $\mathfrak{k}$. Then every cocycle has a splitting field which is a finite extension of $\mathfrak{k}$.*

PRINCETON UNIVERSITY

REFERENCES

[1] C. CHEVALLEY, *Démonstration d'une hypothèse de M. Artin*, Abh. Math. Sem. Hansischen Univ. **11** (1935) p. 73.

[2] G. HOCHSCHILD, *Local Class field theory*, Ann. of Math. Vol. 51 (1950) p. 331.

[3] C. TSEN, *Divisionsalgebren über Funktionenkörper*, Nachr. Ges. Wiss. Gottingen (1933) p. 335.

[4] A. WEIL, *Foundations of Algebraic Geometry*, Amer. Math. Soc. Colloquium Publications, Vol. XXIX.

[5] E. WITT, *Zyklische Körper und Algebren der Characteristik p vom Grad p^n*, J. Reine Angew. Math. Band **176** (1936).

Additional Comment, 1999. I learned much later that some of the results of Part I of the above paper are due to Tsen:

[Tse 36] C. C. TSEN, Zur Stufentheorie der Quasi-Algebraische-Abgeschlossenheit Commutativer Körper, *J. Chinese Math. Soc.* (1936), pp. 81–92.

Proceedings AMS
Vol. 3 No. 3
June 1952

HILBERT'S NULLSTELLENSATZ IN
INFINITE-DIMENSIONAL SPACE

SERGE LANG

1. **The theorem.** Let k be an algebraically closed field. Let $A = \{\alpha\}$ be an indexing set. Let $\mathfrak{o} = k[x_\alpha]$ be the polynomial ring generated over k by an indexed set of variables x_α, $\alpha \in A$. Let $\mathfrak{a}$ be an ideal of $\mathfrak{o}$. A zero of $\mathfrak{a}$ is a set (ξ_α) of elements ξ_α in some extension field of k such that $f(\xi_\alpha) = 0$ for all $f \in \mathfrak{a}$. (Of course, a polynomial f involves only a finite number of the variables x_α.) A zero of $\mathfrak{a}$ will be called algebraic if all ξ_α lie in k. The set of all algebraic zeros of an ideal $\mathfrak{a}$ will be called the variety defined by $\mathfrak{a}$.

The Hilbert Nullstellensatz is in general not valid if A is an infinite set. We shall prove however the following theorem:

THEOREM. *The following three statements are equivalent:*

S1. *If $\mathfrak{a}$ is an ideal of $\mathfrak{o}$ and $f \in \mathfrak{o}$ vanishes on the variety defined by $\mathfrak{a}$, then $f^p \in \mathfrak{a}$ for some integer p.*

S2. *If $\mathfrak{a}$ is an ideal of $\mathfrak{o}$ and $\mathfrak{a} \neq \mathfrak{o}$, then $\mathfrak{a}$ has an algebraic zero.*

S3. *A ring extension $k[\xi_\alpha]$ by elements ξ_α in some extension field of k is a field if and only if all ξ_α lie in k.*

Furthermore the three statements hold if and only if one of the following two conditions is satisfied:

(i) *A is a finite set.*

(ii) *Let A have cardinality a. Let the transcendence degree of k over the prime field have cardinality b. Then $a < b$.*

The above two conditions can be replaced by the single condition that the cardinality of the field k itself should be greater than the cardinality of A.[1] However, our proof will apply to the case in which a is finite only under condition (ii) if we interpret $a < b$ to mean that b is infinite.

It is possible to interpret our theorem geometrically. Let k and A satisfy one of our two conditions. Let V_ν be a collection of varieties in the space of $k[x_\alpha]$. If the V_ν have the finite intersection property (that is, finite intersections $V_{\nu_1} \cap \cdots \cap V_{\nu_r}$ are never empty), then $\bigcap_\nu V_\nu$ is not empty. This follows immediately from the fact that the union of all ideals $\mathfrak{a}_\nu$ defining V_ν is not the ring $\mathfrak{o}$ and has therefore an algebraic zero which lies in all varieties V_ν. If neither (i) nor (ii) is satisfied, then the intersection of all V_ν may be empty.

Received by the editors October 2, 1951.

[1] Professor I. Kaplansky has informed me that he has obtained the theorem independently. His proof is different from the one presented here, however.

When A is a finite set, the above is the usual Hilbert theorem and is well known. A proof was given recently by Zariski [1] to whom Statement 3 is due.

2. **The proof.** We first prove the equivalence of the three statements.

S1→S2. Let $\mathfrak{a}$ be an ideal without algebraic zero. Then the polynomial 1 vanishes (vacuously) on the variety of $\mathfrak{a}$. Hence

$$1 \in \mathfrak{a} \text{ and } \mathfrak{a} = \mathfrak{o}.$$

S2→S1. We reproduce here a well known argument due to Rabinowitsch. The statement is well known if A is a finite set, so we shall assume that A is an infinite set. Let $\mathfrak{a}$ be an ideal of $\mathfrak{o}$ and let f vanish on the variety of $\mathfrak{a}$. Let t be a new variable. In the ring $\mathfrak{o}[t]$ we consider the ideal $(\mathfrak{a}, 1-tf)$. The ring $\mathfrak{o}[t]$ is isomorphic to $\mathfrak{o}$, and $(\mathfrak{a}, 1-tf)$ has no algebraic zero. Hence by S2 there is an expression of the type

$$A_1 f_1 + \cdots + A_r f_r + A(1 - tf) = 1$$

where A_ν, $A \in \mathfrak{o}[t]$ and $f_1, \cdots, f_\nu \in \mathfrak{a}$. We may now let $t = 1/f$ and clear denominators to give $f^\rho \in \mathfrak{a}$ for some integer ρ.

S2→S3. Let $k[\xi_\alpha]$ be a field. Let $\mathfrak{p}$ be the kernel of the map $k[x_\alpha] \to k[\xi_\alpha]$ so $\mathfrak{p}$ is a maximal ideal. $\mathfrak{p}$ has an algebraic zero (η_α) and is the kernel of the map $k[x_\alpha] \to k[\eta_\alpha]$. Hence

$$(x_\alpha - \eta_\alpha) \in \mathfrak{p} \text{ (all } \alpha),$$

and therefore $\xi_\alpha = \eta_\alpha$ lies in k.

S3→S2. Let $\mathfrak{a}$ be an ideal $\neq \mathfrak{o}$. Let $\mathfrak{p}$ be a maximal ideal containing $\mathfrak{a}$. Then $\mathfrak{o}/\mathfrak{p}$ is a field $k[\xi_\alpha]$ and all ξ_α lie in k. It follows that (ξ_α) is an algebraic zero of $\mathfrak{a}$.

We shall now prove that the three statements hold precisely under the above-mentioned conditions. We assume familiarity with cardinal arithmetic, and the following fact: If a field F has transcendence degree b over the prime field, and b is not finite, then the set consisting of all elements in the algebraic closure of F has cardinality b also.

If A is a finite set, the theorem is well known. We suppose that A is not finite and that $a < b$. We shall prove that S2 holds, that is: A proper ideal $\mathfrak{a}$ has an algebraic zero.

LEMMA. *If $\mathfrak{a}$ is an ideal of $\mathfrak{o}$, then $\mathfrak{a}$ has a basis consisting of at most a elements.*

PROOF. We consider the set of all subrings $k[x_{\alpha_1}, \cdots, x_{\alpha_r}]$ generated by a finite number of indeterminates. This set has cardinality a, and we may index it by A again. Let these rings be denoted by $\mathfrak{o}_\alpha$ and let $\mathfrak{a}_\alpha = \mathfrak{a} \cap \mathfrak{o}_\alpha$. Then $\mathfrak{a} = \bigcup_\alpha \mathfrak{a}_\alpha$. Each $\mathfrak{a}_\alpha$ has a finite basis and therefore the union of these basis elements is a basis for $\mathfrak{a}$ having cardinality at most a, as was to be shown.

Let $\mathfrak{a}$ be a proper ideal. Let $\mathfrak{p}$ be a prime ideal containing $\mathfrak{a}$. (A maximal such ideal will always exist by Zorn's Lemma.) It has a basis of at most a elements. Let k_0 be the algebraic closure of the field obtained by adjoining to the prime field all coefficients of elements in such a basis of $\mathfrak{p}$. Then k_0 has cardinality at most a, and by hypothesis k/k_0 has transcendence degree of cardinality at least a.

Let $\mathfrak{p}_0 = k_0[x_\alpha] \cap \mathfrak{p}$. It has the same basis as $\mathfrak{p}$. Any algebraic zero of $\mathfrak{p}_0$ will therefore be an algebraic zero of $\mathfrak{p}$. We shall now find an algebraic zero of $\mathfrak{p}_0$. The residue class ring $k_0[x_\alpha]/\mathfrak{p}_0$ is isomorphic to a ring extension $k_0[\xi_\alpha]$ under the natural map, and (ξ_α) is a zero of $\mathfrak{p}_0$. It is easy to see that there exists an isomorphism of $k_0[\xi_\alpha]$ into k which is the identity on k_0. If (ξ'_α) is the image of (ξ_α) under this isomorphism, then (ξ'_α) is the desired algebraic zero. (The above-mentioned isomorphism may be constructed as follows: Consider the field $k_0(\xi_\alpha)$. Select a transcendence base. By our cardinality assumption we may map the rational field generated over k_0 by this base isomorphically into k. It is now well known that this map may be extended to the algebraic closure, and its restriction to the ring $k_0[\xi_\alpha]$ is the desired isomorphism.)

In order to complete the proof we need only show that our conditions (i) and (ii) are necessary. In other words, if neither condition is satisfied, then there exists a proper ideal without algebraic zero. We show how to construct such an ideal.

Suppose that A is an infinite set and that $b \leq a$. Then the set of elements of k has cardinality $c \leq a$. Select an element α_0 in A. Let $C = \{\gamma\}$ be a subset of $A - \{\alpha_0\}$ having cardinality c, and index the elements of k by C, so $k = \{\xi_\gamma\}$. Let y be transcendental over k. Map $k[x_\alpha]$ as follows:

$$x_\alpha \;\rightarrow\; 0, \qquad\qquad \alpha \notin C, \alpha \neq \alpha_0,$$

$$x_{\alpha_0} \;\rightarrow\; y,$$

$$x_\gamma \;\rightarrow\; \frac{1}{y - \xi_\gamma}, \qquad\qquad \gamma \in C,$$

and let the map be identity on k. Then $k[x_\alpha]$ is mapped onto $k[y, 1/(y - \xi_\gamma), \cdots]$ which is easily seen to be a field. The kernel of our map is a maximal ideal and we see that Statements S2 and S3 are violated. This concludes the proof of our theorem.

REFERENCE

1. O. Zariski, *A new proof of Hilbert's Nullstellensatz*, Bull. Amer. Math. Soc. vol. 53 (1947) p. 362.

PRINCETON UNIVERSITY

Proceedings AMS, Vol. 3 No. 4, August 1952

ON CHEVALLEY'S PROOF OF LUROTH'S THEOREM

SERGE LANG AND JOHN TATE

Let k be a function field in one variable over a constant field k_0, and let g be its genus. By a subfield of k we shall always mean a subfield k' properly containing k_0 so that k' is likewise a function field with k_0 as constants. We let g' be the genus of k'.

If k/k' is separable, then the classical formula

$$2g - 2 = n(2g' - 2) + \mu$$

where μ is a non-negative integer and $n = (k:k')$ shows that $g' \leqq g$. If k/k' is inseparable, then g' may be greater than g. Nevertheless, we have:

THEOREM 1. *If k is separably generated over k_0 then $g' \leqq g$.*

PROOF. In view of the above remarks we may assume k/k' is purely inseparable. Let p be the characteristic. Then k/k' is a p-tower in which each step is of degree p and is inseparable. We may further assume that $(k:k') = p$ because a subfield of a separably generated field is also separably generated. (This is an immediate consequence of MacLane's criterion that k is separably generated over k_0 if and only if k is linearly disjoint from $k_0^{1/p}$ over k_0.)

Let x be a separating variable for k over k_0 so that we may write $k = k_0(x, y)$ where y is separable over $k_0(x)$. Then we also have $k = k_0(x, y^p)$. We see that $k_0(x^p, y^p) \subset k'$ and in fact we must have $k' = k_0(x^p, y^p)$ because

$$(k : k_0(x^p, y^p)) = (k_0(x, y^p) : k_0(x^p, y^p)) \leqq p = (k : k').$$

Thus $k' = k_0 k^p$. But k^p/k_0^p is an isomorphic image of k/k_0, and therefore the genus of k^p (considered as function field over the constant field k_0^p) is g. Since k' may be regarded as a constant field extension of k^p its genus g' is at most g, as was to be shown.

That the genus cannot increase in a constant field extension is proved in [1] and [2].

Our theorem generalizes the argument used by Chevalley [2, p. 106] to prove Luroth's theorem. Namely, a rational field R is a separably generated field of genus zero. By Theorem 1 any subfield R' is of genus zero. A prime of degree 1 in R induces a prime of degree 1 in R' and hence, by a well known criterion, R' is a rational field.

If the field k is not separably generated, however, the behavior of

---

Received by the editors December 3, 1951.

621

23

its subfields may be much more pathological and for fields of genus zero we can prove the converse of Theorem 1. In fact we prove more:

THEOREM 2. *A field of genus zero which is not separably generated over its constant field contains subfields of arbitrarily high genus.*

PROOF. Let k be a field of genus zero. It is well known and easy to show [1, chap. XVI, 4] that k is either a rational field, or $k = k_0(x, y)$ where x, y satisfy a quadratic equation

$$F(x, y) = ay^2 + (bx + c)y + dx^2 + ex + f = 0.$$

If k is not separably generated, then the characteristic of the field must be 2 and the partial derivatives $\partial F/\partial x$ and $\partial F/\partial y$ must both vanish. Consequently $k = k_0(x, y)$ where x, y satisfy an equation of the type

$$(1) \qquad\qquad y^2 = ax^2 + b, \qquad\qquad a, b \in k_0.$$

Furthermore, $k_0(a^{1/2}, b^{1/2})$ has degree 4 over k_0. Suppose otherwise, that is, $(k_0(a^{1/2}, b^{1/2}):k_0) \leq 2$, and say $a^{1/2}$ is a generator of $k_0(a^{1/2}, b^{1/2})$. Then we can write $b^{1/2} = c + da^{1/2}$ with c, d in k_0. In a suitable extension we have $y = a^{1/2}x + b^{1/2}$, and hence $y = a^{1/2}(x+d) + c$. This shows that y and $a^{1/2}$ generate the same field over $k_0(x)$, and that k is rational, contrary to assumption.

We shall now construct hyperelliptic subfields k' of k of arbitrarily high genus.

Let $k' = k_0(z, w)$ where $z = x^2$ and $w = x^{2n+1} + y$, $n \geq 1$. Then $w^2 = z^{2n+1} + az + b$. We shall prove that k' has genus n by developing the theory of inseparable quadratic extensions of a rational field in analogy with the classical separable theory. We need a lemma.

LEMMA. *Let k_0 be any field of characteristic 2. Let $k_0(x)$ be the rational field in the variable x, and let $k/k_0(x)$ be an inseparable extension of degree 2. Let $f(x)$ be a polynomial in $k_0[x]$ of least degree such that $k = k_0(x, y)$ where $y^2 = f(x)$. (Such a polynomial will be called minimal.) Then $\{1, y\}$ is a minimal basis for the integers of k over $k_0[x]$.*

PROOF. Suppose $(r(x) + s(x)y)/t(x)$ is integral over $k_0[x]$ with $r(x)$, $s(x)$, $t(x)$ in $k_0[x]$. We may assume deg r and deg $s < $ deg t. We must then show that $r = s = 0$. For some polynomial g we have

$$r^2 + s^2 f = t^2 g.$$

If $s \neq 0$, then g competes with f as a field generator, so deg $g \geq$ deg f. This yields deg $r^2 =$ deg $t^2 g$, which is impossible. Hence $s = 0$ and therefore $r=0$ also, by comparing degrees again. This proves that $\{1, y\}$ is a minimal basis.

THEOREM 3. *Let $k = k_0(x, (f(x))^{1/2})$ be the field defined in the preceding lemma, with $f(x)$ minimal. Then if $f(x)$ is of degree $n > 0$, the genus of k is $-[-n/2]-1$ in exact analogy with the classical case.*

PROOF. We first note that $n > 0$ implies that k_0 is the constant field of k. Otherwise $k/k_0(x)$ would be generated by $c^{1/2}$ where c lies in k_0, and this would mean $n = 0$.

Let $\mathfrak{a}$ be the divisor of the poles of x in k. Then $\mathfrak{a}$ has degree 2 in k. We now determine the dimension $l(\mathfrak{a}^{-\nu})$ of the vector space of multiples of $\mathfrak{a}^{-\nu}$ in two ways.

First by the Riemann-Roch Theorem we have for large ν

$$\text{(2)} \qquad l(\mathfrak{a}^{-\nu}) = 2\nu + 1 - g.$$

Secondly, using the fact that $\{1, y\}$ is a minimal basis,

an integer $r(x) + s(x)y$ is a multiple of $\mathfrak{a}^{-\nu}$

$$\leftrightarrow \mathfrak{a}^{-2\nu} \mid r^2 + s^2 f$$

$$\leftrightarrow \deg (r^2 + s^2 f) \leqq 2\nu$$

$$\leftrightarrow \deg r \leqq \nu \quad \text{and} \quad \deg s \leqq \nu + [-n/2].$$

Each of the preceding equivalences is trivial except possibly the last. But we assumed that $f = a_n x^n + \cdots + a_0$ is minimal. It follows that $a_n x^n$ is not a square, and therefore

$$\deg (r^2 + s^2 f) = \max (\deg r^2, \deg s^2 f).$$

This immediately implies the last equivalence.

For ν large $(> n/2)$ we obtain

$$\text{(3)} \qquad l(\mathfrak{a}^{-\nu}) = \nu + 1 + \nu + 1 + [-n/2].$$

From (2) and (3) we solve for the genus, and get

$$g = -[-n/2] - 1$$

which proves Theorem 3.

In order to complete the proof of Theorem 2 it suffices to show that the polynomial $f(z) = z^{2n+1} + az + b$ is minimal for the extension $k'/k_0(z)$. If this is not the case, let $g(z)$ be minimal. By the lemma we can write

$$(f(z))^{1/2} = r(z) + s(z)(g(z))^{1/2}$$

and squaring we get

$$f(z) = r(z)^2 + s(z)^2 g(z).$$

Differentiating formally with respect to z we get

$$(4) \qquad f'(z) = z^{2n} + a = (z^n + a^{1/2})^2 = s(z)^2 g'(z).$$

This shows that in the polynomial domain $k_0(a^{1/2})[z]$, $g'(z)$ is a square: $g'(z) = (l(z) + a^{1/2} m(z))^2$, where the polynomials l and m have coefficients in k_0. Substituting back in (4) we obtain

$$z^n + a^{1/2} = s(z)(l(z) + a^{1/2} m(z)).$$

Comparing coefficients of $a^{1/2}$ we see that $s(z)m(z) = 1$, and that $s(z)$ must be a constant. But in this case deg $g'(z) = 2n$ and therefore deg $g(z) \geq 2n + 1 = \deg f(z)$; $f(z)$ is minimal, and Theorem 2 is proved.

Actually we have not yet shown the existence of inseparably generated fields of genus zero, but this gap is easily filled. Let k_0 be a field of characteristic 2 which contains elements a and b such that $(k_0(a^{1/2}, b^{1/2}) : k_0) = 4$. Then the field $k = k_0(x, y)$ defined by equation (1)

$$y^2 = ax^2 + b$$

is of genus zero, is not separably generated, and has k_0 as its field of constants. Indeed, $k/k_0(x)$ is of degree 2. If k_0 were not the constant field, then k would be $k_0(x, c^{1/2})$ where $c \in k_0$, and would therefore be a rational field over $k_0(c^{1/2})$. Then y could be expressed as a rational function in x with coefficients in $k_0(c^{1/2})$; this rational function must in fact be a polynomial because its square is a polynomial. We have $y = a^{1/2}x + b^{1/2}$. This means that $k_0(a^{1/2}, b^{1/2}) \subset k_0(c^{1/2})$ has degree not greater than 2 over k_0, contrary to assumption.

By Theorem 3 we now know that k has genus zero. In the proof of Theorem 2 we have seen that such a field contains hyperelliptic subfields of arbitrarily high genus. By Theorem 1 the field cannot be separably generated, a fact which could of course be established directly.

REFERENCES

1. E. Artin, *Algebraic numbers and algebraic functions.* I, Notes, New York University, 1950–1951.

2. C. Chevalley, *Introduction to the theory of algebraic functions of one variable,* Mathematical Surveys no. 6, New York, American Mathematical Society, 1951.

3. J. Tate, *Genus change in inseparable extensions of function fields,* Proceedings of the American Mathematical Society vol. 3 (1952) pp. 400–406.

PRINCETON UNIVERSITY

Annals of Mathematics
Vol. 57, No. 2, March, 1953
Printed in U.S.A.

THE THEORY OF REAL PLACES

By Serge Lang

(Received March 7, 1952)

1. Introduction

A field F is *real* if -1 cannot be expressed as a sum of squares in F, or equivalently if the form $x_1^2 + \cdots + x_n^2$ has only the trivial zero in F for all n. F is *real closed* if it is maximal with respect to this property in its algebraic closure. For instance, the ordinary real numbers are real closed, and so are the real algebraic numbers. Real closed fields will be denoted by P. They were characterized by Artin and Schreier [1], [2], [4], [5] as those fields which are of finite degree under their algebraic closure. This degree must then be 2, and the algebraic closure of P can be obtained by adjunction of $\sqrt{-1}$. If P is real closed it can be ordered, the positive elements being the squares in P. Any real field can therefore be ordered since it can be imbedded in a real closure.

Artin showed [2] how such a characterization can be applied to treat certain problems of algebraic geometry, and in particular the problems concerning the existence and nature of real specializations.

We shall consider anew this question, using the language of modern algebraic geometry. We recall some relevent definitions.

Let K be a field. We shall assume that the reader is familiar with the natural one–one correspondence which exists between (a) the valuation rings of K, (b) the classes of equivalent valuations on K, and (c) the classes of equivalent places of K. Let L, L' be two fields. Let ϕ, ϕ' be two places of K mapping K onto L, ∞ and L', ∞. Then ϕ is *equivalent* to ϕ' ($\phi \sim \phi'$) if there exists an isomorphism λ of L onto L' such that $\phi' = \lambda\phi$ on K if we put $\lambda(\infty) = \infty$. Let $\mathfrak{o}$, $\mathfrak{p}$ and $\mathfrak{o}'$, $\mathfrak{p}'$ be the valuation rings and ideals of ϕ, ϕ'. We see that $\phi \sim \phi'$ if and only if $\mathfrak{o} = \mathfrak{o}'$ and $\mathfrak{p} = \mathfrak{p}'$. We have $\mathfrak{o}/\mathfrak{p} \simeq L$ and $\mathfrak{o}'/\mathfrak{p}' \simeq L'$. We shall also assume familiarity with the Theorem on the Extension of Specializations. For a discussion of places and valuations and a proof of the Extension Theorem, we refer the reader to Artin [3] or Weil [7].

Let L be a field. Let ϕ be a place mapping K into L, ∞. We shall often say that ϕ is L *valued*, or takes on its values in L (instead of L, ∞). Let F be a subfield of K. We shall call ϕ a *place of K over F* (briefly K/F) if ϕ is an isomorphism on F. We call ϕ *zero dimensional* (or *algebraic*) *over F* if ϕ is an isomorphism on F and if ϕ maps K into an algebraic closure of $\phi(F)$. We often identify $\phi(F)$ with F if ϕ is an isomorphism on F. If ϕ takes on its values in $\phi(F)$, we say ϕ is *rational* over F.

If the field L is real, ϕ will be called a *real place*.

We shall prove in real closed fields an analogue to the Extension Theorem in algebraically closed fields. Specifically: Let K be a field. If K has a real place ϕ, taking its values in a real closed field P, then K is real. Furthermore, there exists

378

27

an extension of ϕ to a P valued place of a real closure of K. We shall consider the existence of real places in the case of function fields. Let K be a function field (in several variables) over a real closed constant field P and suppose that K is real. K will be called a real function field over P. Then we shall prove that there exists a rational place of K over P. In fact, we shall prove that there exist enough such places to distinguish the elements of K. As an immediate consequence of this, we obtain a generalization of the Hilbert Problem solved by Artin [2].

Next, we shall characterize the places of a real field by showing how they are intimately related to the possible orderings of the field.

Let K be an ordered real field. Let F be a subfield with the induced ordering. As usual, we put $|\alpha| = \alpha$ if $\alpha \geqq 0$, and $|\alpha| = -\alpha$ if $\alpha < 0$. An element α in K is *infinitely large* over F if $|\alpha| > a$ for all $a \in F$, and *infinitely small* over F if $0 \leqq |\alpha| < |a|$ for all $a \neq 0$ in F. α is infinitely large over F if and only if α^{-1} is infinitely small. K is *archimedean over* F if K has no elements which are infinitely large over F. An intermediate field $F_1 \supset F$ is *maximal archimedean over* F if it is archimedean over F, and no other intermediate field $E \supset F_1$ is archimedean over F. If a field F_1 is archimedean over F, and F_2 is archimedean over F_1, then F_2 is archimedean over F. Hence by Zorn's Lemma there always exists a maximal archimedean field F_1 over F in K. F will be said to be *maximal archimedean in K* if it is maximal archimedean over itself in K. We remark also that if K is archimedean over F then any real algebraic extension of K is also archimedean over F. Indeed, an algebraic element α lies in an interval bounded essentially by the sum of the coefficients of its equation over K. From this it follows that if F is maximal archimedean in K then F is algebraically closed in K.

Concerning the places of a field K, we can prove: Let P be a real closed field, and K/P any extension field. Let ϕ be a rational place of K/P. Then K is real and there exists an ordering of K such that P is maximal archimedean in K. Furthermore the valuation ring of ϕ consists of all elements in K which are not infinitely large over P, and the valuation ideal consists of all elements which are infinitely small over P.

These results give us insight into the nature of real function fields and their places. In particular we see that to describe the places of real function fields over the real numbers we need the theory of abstract real fields and their non archimedean orderings.

Finally, we consider function fields over real fields from the point of view of quasi algebraic closure. One can prove theorems analogous to those of [6] by restricting oneself to forms of odd degree. For instance: Let K be a function field in s variables over a real closed field P. Let $f_1, \cdots, f_r$ be r forms in K, of odd degrees $d_1, \cdots, d_r$, in n variables. If $n > d_1^s + \cdots + d_r^s$ then the forms have a non trivial common zero in K. It seems therefore that the condition of reality which prevents a field from being C_i for any i, affects only the forms of even degree. It is a reasonable conjecture that if K is a function field in s variables over a real closed constant field, and K is not real then K is actually C_s. A similar conjecture can be made for power series.

2. Theory of Real Places

Most of the definitions which we shall use have been made in the introduction. We may therefore proceed immediately with a theorem.

THEOREM 1. *Let K be a field. If there exists a real place of K then K is real.*

PROOF. Suppose that we have a non trivial relation

$$0 = \sum a_\nu^2 \qquad\qquad a_\nu \neq 0,\ a_\nu \,\epsilon\, K.$$

Let ϕ be a place of K. Then ϕ gives rise to a valuation on K. Say a_i has maximal value among the a_ν. We divide the relation by a_i^2. This gives

$$-1 = \sum_{\nu \neq i} b_\nu^2 \qquad\qquad b_\nu = a_\nu/a_i$$

where no b_ν is mapped into ∞ by ϕ. The image under ϕ of this relation shows that ϕ cannot be a real place.

We shall now give an important example of a real place. Let K be an ordered real field and let F be a subfield. Let $\mathfrak{o}$ be the ring of elements of K which are not infinitely large over F. It is clear that for any $\alpha \,\epsilon\, K$, either α or α^{-1} lies in $\mathfrak{o}$, which is consequently a valuation ring. Let $\mathfrak{p}$ be the ideal of all $\alpha \,\epsilon\, K$ which are infinitely small over F. Then $\mathfrak{p}$ is the unique maximal ideal, because any element in $\mathfrak{o}$ which is not in $\mathfrak{p}$ has an inverse in $\mathfrak{o}$. $\mathfrak{o}$ and $\mathfrak{p}$ induce a place of K which will be called the *canonical place of K with respect to F*. This place may be trivial, i.e. may be an isomorphism on K. It is in any case an isomorphism on F, i.e. is trivial on F.

THEOREM 2. *Let K be an ordered field and let F be a subfield. Then the canonical place of K with respect to F is real.*

PROOF. Otherwise we could write

$$-1 = \sum \alpha_\nu^2 + a$$

where $a \,\epsilon\, \mathfrak{p}$ and $\alpha_\nu \,\epsilon\, \mathfrak{o}$. Since $\sum \alpha_\nu^2$ is positive and a is infinitely small, such a relation is clearly impossible.

The next theorem shows that under certain circumstances the canonical place is not trivial.

THEOREM 3. *Let K be an ordered real field and let F be a subfield which is maximal archimedean in K. Then the canonical place of K with respect to F is algebraic over F.*

PROOF. We know that F is algebraically closed in K. Let $t \,\epsilon\, K$, $t \,\epsilon\, F$. Then $F(t)$ contains an element u which is infinitely small over F and which must map into 0 under the canonical place ϕ. t is algebraic over $F(u)$ and $F(u, t)$ is a function field in one variable over F. Since $\phi(u) = 0$, it follows that $\phi(t) = \infty$ or $\phi(t)$ is algebraic over F. Hence ϕ is zero dimensional, as was to be shown.

The following lemma will be useful in Theorem 4.

LEMMA 1. *Let Γ be a real closed field and let P be a subfield which is algebraically closed in Γ. Then P is real closed.*

PROOF. Let $f(x)$ be an irreducible polynomial over P. It splits in Γ into linear and quadratic factors. Its coefficients in Γ are algebraic over P and must therefore lie in P. Hence $f(x)$ is either linear itself, or quadratic irreducible already over P. It is clearly positive definite and by a linear transformation may be brought into the form $x^2 + a^2 = 0$. Any root of this polynomial will bring $\sqrt{-1}$ with it and therefore the only extension of P is $P(\sqrt{-1})$. This proves that P is real closed.

The next theorem shows that the canonical place in Theorem 3 can be extended to the real closure of K.

THEOREM 4. *Let K be an ordered real field and let F be a subfield which is maximal archimedean in K. Let Γ be the real closure of K (preserving the ordering) and let P be the real closure of F contained in Γ. Let Φ be the canonical place of Γ with respect to P and put $\bar{P} = \Phi(P)$. Then Φ is algebraic over $\bar{P}$ and consequently $\bar{P}$ valued. The place induced by Φ on K is equivalent to the canonical place ϕ of K with respect to F.*

PROOF. Let $\mathfrak{o}$, $\mathfrak{p}$ be the valuation ring and ideal of ϕ, and let $\mathfrak{O}$, $\mathfrak{P}$ be the valuation ring and ideal of Φ. For $\alpha \in \mathfrak{O}$, denote by $\bar{\alpha}$ the image of α under the natural map

$$\Phi : \mathfrak{O} \to \mathfrak{O}/\mathfrak{P}.$$

We see easily that $\mathfrak{O} \cap K = \mathfrak{o}$ and $\mathfrak{P} \cap K = \mathfrak{p}$. Consequently $\mathfrak{o}/\mathfrak{p}$ is contained naturally in $\mathfrak{O}/\mathfrak{P}$ and the place induced by Φ on K is therefore equivalent to ϕ.

All that remains to be shown is that Φ is $\bar{P}$ valued. We apply Φ to $\mathfrak{O}$, and put $\bar{\mathfrak{O}} = \Phi(\mathfrak{O})$. We must show that $\bar{\mathfrak{O}} = \bar{P}$, and it suffices to show that $\bar{\mathfrak{O}}$ is algebraic over $\bar{P}$, because $\bar{\mathfrak{O}}$ is real (Theorem 2).

An element $\alpha \in \mathfrak{O}$ is algebraic over K and satisfies an equation

$$a_n \alpha^n + \cdots + a_0 = 0 \qquad\qquad a_\nu \in K.$$

We may assume in fact that $\Phi(a_\nu) \neq \infty$ and some $\Phi(a_i) \neq 0$. (If necessary, divide the equation by the coefficient having maximal value under Φ.)

Applying Φ to the equation gives a non trivial relation

$$\bar{a}_n \bar{\alpha}^n + \cdots + \bar{a}_0 = 0$$

from which we see that $\bar{\mathfrak{O}}$ is algebraic over $\bar{P}$. This concludes the proof of Theorem 4.

We shall now prove a uniqueness statement. Under certain circumstances the real places are characterized by the ordering of the field.

We shall say that a real field K is *quadratically closed* if for any $\alpha \in K$ either $\sqrt{\alpha}$ or $\sqrt{-\alpha}$ lies in K. The ordering of a quadratically closed real field K is then uniquely determined, and so is the real closure of such a field (up to an isomorphism over K).

THEOREM 5. *Let K be a real field which is quadratically closed (and consequently ordered). Let F be a subfield, and suppose that F is maximal archimedean in K. Let ϕ be a zero dimensional real place of K over F. Then ϕ is equivalent to the canonical place of K/F.*

PROOF. Let $\mathfrak{o}$ be the ring of elements of K which are not infinitely large over F and $\mathfrak{p}$ the maximal ideal of elements which are infinitely small over F. Let $\mathfrak{o}_1$ and $\mathfrak{p}_1$ be the valuation ring and valuation ideal of ϕ. We contend that $\mathfrak{o} = \mathfrak{o}_1$ and $\mathfrak{p} = \mathfrak{p}_1$, i.e. that ϕ is the canonical place.

ϕ is real, and the real closure P of F is uniquely determined because F is quadratically closed. We may suppose that ϕ is P valued.

Let $\alpha \in K$ and $a \in F$. If $\alpha > a$ then either $\phi(\alpha) > a$ or $\phi(\alpha) = \infty$. Indeed, $\alpha - a = \beta^2$, $\beta \in K$, and therefore $\phi(\alpha) = a + \phi(\beta)^2$.

(A) $\mathfrak{p}_1 \subset \mathfrak{p}$: Suppose $\phi(\alpha) = 0$. Then α is infinitely small. Otherwise $|\alpha| > a > 0$ for some $a \in P$. We may suppose α is positive. Then either $\phi(\alpha) > a$ or $\phi(\alpha) = \infty$ contrary to assumption.

(B) $\mathfrak{o} \subset \mathfrak{o}_1$: Suppose that α is not infinitely large. If $\phi(\alpha) = \infty$ then $\phi(\alpha^{-1}) = 0$. Hence α^{-1} is infinitely small by (A) and α is infinitely large, contradiction.

(C) $\mathfrak{p} \subset \mathfrak{p}_1$: Let α be infinitely small. Then $\alpha \in \mathfrak{o}$ and $\phi(\alpha) \neq \infty$ by (B). We may assume $\phi(\alpha) \geq 0$. For any positive $a \in F$, $a - \alpha$ is positive and therefore $a - \alpha = \beta^2$, $\beta \in K$. Hence $a - \phi(\alpha) = \phi(\beta)^2$ and $\phi(\beta) \neq \infty$. We see that in the ordering of P, $\phi(\alpha) < a$ for all positive $a \in F$. P is archimedean over F, and therefore $\phi(\alpha) < a$ for all positive $a \in P$. Hence $\phi(\alpha) = 0$.

(D) $\mathfrak{o}_1 \subset \mathfrak{o}$: Let $\phi(\alpha) \neq \infty$. If α were infinitely large then α^{-1} would be infinitely small. Hence $\phi(\alpha^{-1}) = 0$ by (C) and therefore $\phi(\alpha) = \infty$, contradiction.

COROLLARY. *Let K be a quadratically closed real field. Let ϕ be a real place of K, taking its values in a real closed field P. Let F be a maximal subfield such that ϕ is an isomorphism on F and identify F with $\phi(F)$. F exists and is maximal archimedean in K. ϕ is zero dimensional over F, and is induced by the canonical place of K/F.*

PROOF. Since ϕ is real, it must be an isomorphism on the rational numbers (which are contained in K). By Zorn's Lemma, F exists, and can be identified with its image under ϕ. F is clearly algebraically closed in K, and is therefore quadratically closed.

F is maximal archimedean in K. Otherwise, there would exist a subfield E of K which is archimedean over F. ϕ is not an isomorphism on E and there exists an element α of E such that $\phi(\alpha) = 0$. We may suppose $\alpha > 0$. Since E is archimedean over F, we have $0 < a < \alpha$ for some $a \in F$. Then $\alpha - a = \beta^2$ for some $\beta \in K$. Applying ϕ yields $-a = \phi(\beta)^2$, a contradiction.

ϕ is zero dimensional over F. Otherwise it would map an element $t \in K$, $t \notin F$ on a transcendental element over F and would be an isomorphism on the field $F(t)$, contrary to assumption.

By Theorem 5, ϕ is then the canonical place of K/F.

In contrast to the preceding theorem, let K be quadratically closed, and let F be a subfield which is algebraically closed in K, $F \neq K$. Suppose that K is archimedean over F. Then there does not exist a zero dimensional place of K/F. Otherwise such a place ϕ would map some $\alpha \in K$ into 0. This is easily seen to be impossible by an argument similar to (A).

The preceding theorems give us an insight into the real places of a field K, and show how one may obtain a place if one starts with a suitable ordering of K. Conversely, starting with a real place ϕ of a field K, we shall attempt to recover naturally an ordering for K such that the place ϕ is induced by the canonical place arising from the ordering. In order to achieve this goal we shall obtain in real closed fields an analogue to the Extension Theorem, namely we shall prove that a real place ϕ of a field K can always be extended to a real place of a real closure of K.

Our first step in this direction is to show that ϕ can always be extended to a real place of a quadratic closure of K. This will yield automatically an ordering and will clear the path for an extension of ϕ to a real closure of K by means of the canonical place.

LEMMA. *Let K be a real field. Let ϕ be a real place of K, taking its values in a real closed field* P. *Then there exists an extension of ϕ to a* P *valued place of a quadratic closure of K.*

PROOF. By the Extension Theorem, we know that ϕ can always be extended, if we allow the extension to take its values in an algebraic closure of P.

Let $\alpha \in K$. We shall prove that ϕ can be extended to a P valued place of $K(\sqrt{\alpha})$ or $K(\sqrt{-\alpha})$. The lemma follows from this by Zorn's Lemma.

We prove first that if $\phi(\alpha) \neq \infty$ and $\phi(\alpha) > 0$ then any extension of ϕ to $K(\sqrt{\alpha})$ is P valued. Indeed, we have $\sqrt{\phi(\alpha)} = \phi(\sqrt{\alpha}) \in$ P. Let $a, b \in K$. We must show that if $\phi(a + b\sqrt{\alpha})$ is finite then it lies in P. If $\phi(a) = 0$ this is clear. If $\phi(a) \neq 0$ write

$$\phi(a + b\sqrt{\alpha}) = \phi\left(a\left(1 + \frac{b}{a}\sqrt{\alpha}\right)\right) = \gamma.$$

Then $\phi(1 + (b/a)\sqrt{\alpha})$ is finite and it is equal to $\phi(1) + \phi(b/a)\phi(\sqrt{\alpha})$ which lies in P. If it is not equal to 0 then γ clearly lies in P. If it is equal to 0, then $\phi(1 - (b/a)\sqrt{\alpha}) = 2$ and we get $\phi(a(1 - (b^2/a^2)\alpha)) = 2\gamma$ which must lie in P. Hence γ lies in P.

We now distinguish two cases. First, there exists an element a in K such that $\phi(a^2\alpha) \neq 0, \neq \infty$. Then $\pm a^2\alpha$ may be used as generator instead of $\pm\alpha$. Either $\phi(a^2\alpha) > 0$ or $\phi(-a^2\alpha) > 0$ and we can apply the preceding argument to $K(\sqrt{a^2\alpha})$ or $K(\sqrt{-a^2\alpha})$.

Secondly, suppose that for all $a \in K, \phi(a^2\alpha) = 0$ or ∞. Then any extension of ϕ to $K(\sqrt{\alpha})$ is P valued. In fact let $a, b \in K$, and consider $\phi(a + b\sqrt{\alpha})$ which we may write in one of the following forms:

$$\phi(a + b\sqrt{\alpha}) = \phi\left(a\left(1 + \frac{b}{a}\sqrt{\alpha}\right)\right) \text{ or } = \phi\left(b\sqrt{\alpha}\left(1 + \frac{a}{b\sqrt{\alpha}}\right)\right).$$

We see from this that it must equal either $\phi(a)$ which lies in P, or $\phi(b\sqrt{\alpha})$ which is 0 or ∞, whence the theorem.

We note that the quadratic closure of K thus obtained is real because it has a real place (Theorem 1).

THEOREM 6. *Let K be a real field. Let ϕ be a real place of K taking its values in* *a real closed field* P. *Then* ϕ *can be extended to a* P *valued place of a real closure of* K.

PROOF. By the lemma we may assume that K is quadratically closed. We are then in the situation of the Corollary to Theorem 5, and by Theorem 4 the canonical place extends to the real closure in a natural way.

COROLLARY. *Let* P *be a real closed field. Let* K *be any field containing* P, *and let* ϕ *be a rational place of* $K/$P. *Then there exists an ordering of* K *such that* P *is maximal archimedean in* K *and such that* ϕ *is induced by the canonical place of* K *with respect to* P.

PROOF. ϕ extends to a P valued place of a real closure Γ of K. P is obviously maximal archimedean in Γ and ϕ is induced by the canonical place according to Theorem 5.

It was essential in Theorem 6 to start with a place of a field, and not a ring. This is shown by the following example. Let P be real closed, and let P$[x, y]$ be the ring of polynomials in two variables. The polynomial $x^2 + y^2$ is irreducible over P and generates a prime ideal. The residue class ring P$[\xi, \eta]$ is real in the sense that -1 is not a sum of squares in it. (Otherwise, write

$$-1 = \sum f_\nu(x, y)^2 + h(x, y)(x^2 + y^2).$$

This is absurd for $x = y = 0$.)

However, -1 is a square in the quotient field. P$[\xi, \eta]$ has a real zero dimensional specialization, namely $\phi(\xi) = \phi(\eta) = 0$, but an extension of this specialization to the quotient field is not real.

The essential feature of the preceding example is that $(0, 0)$ is an isolated real point of the curve $x^2 + y^2 = 0$. Another example is given by the curve $y^2 = -x^2(1 - x) = x^3 - x^2$, which we assume defined over the real numbers R. This curve also has an isolated real point $(0, 0)$. No extension of the specialization $(\xi, \eta) \to (0, 0)$ to a place of $R(\xi, \eta)$ is real. On the other hand, the function field $R(\xi, \eta)$ is a real field, and has many real places as will be shown in Theorem 7. These places extend specializations of (ξ, η) to points on the real branch passing through $(1, 0)$.

If we start with any real place ϕ of a field K, Theorem 6 asserts that ϕ extends to a real closure of K, but nothing is said concerning the uniqueness of this real closure. If an ordering is originally given on K, the real closure Γ to which ϕ extends may induce another ordering on K. In the situation of Theorem 4, we see on the other hand that the canonical place extends to a real closure which does indeed preserve the ordering.

We have now seen how to get a place from an ordering of a real field, and conversely how to get an ordering from a place. There is still one question concerning the existence of real places which we want to consider. Let P be real closed and let $K \supset$ P be a real field. When does there exist a place of K rational, over P? If K is itself real closed and archimedean over P then no such place exists. On the other hand, in the important case of function fields such a place always exists, and we now proceed to construct it.

It is convenient to introduce the language of point set topology. The ordering of P gives rise to a metric $|\alpha|$ as mentioned in the introduction. It is clear that P is a topological field under this metric. In particular, rational functions are continuous.

As pointed out by Artin and Schreier [4] the ordinary theorems of real algebra hold in P. The most important of these is Sturm's Theorem, and for the convenience of the reader we recall its statement.

Let $f(y) = y^s + a_{s-1}y^{s-1} + \cdots + a_0$ be a polynomial with coefficients in P. (We assume that it has no double roots.) Let α, β be two elements of P, $\alpha < \beta$. There exists a sequence of polynomials $f_1(y) = f(y)$, $f_2(y)$, $\cdots$, $f_r(y)$ such that the distribution of signs in the sequence $f_1(\alpha)$, $\cdots$, $f_r(\alpha)$, $f_1(\beta)$, $\cdots$, $f_r(\beta)$ gives the exact number of roots of $f(y)$ in P between α and β. Furthermore the coefficients of f_i are rational functions of the coefficients of $f(y)$, with rational coefficients.

We know that the roots of f in P lie between $\pm\alpha$ where $\alpha = 1 + |a_0| + \cdots + |a_{s-1}|$. We can therefore write $\alpha = 1 \pm a_0 \pm \cdots \pm a_{s-1}$, choosing $+$ if $a_i \geqq 0$ and $-$ if $a_i < 0$.

Now let $f^*(y) = y^s + A_{s-1}y^{s-1} + \cdots + A_0$ have indeterminate coefficients. We shall be interested in the roots of the polynomial $f(y)$ obtained from f^* by specializing A_i into P. If we know before hand the signs of the specializations a_i of A_i we may put $\alpha^* = 1 \pm A_0 \pm \cdots \pm A_{s-1}$. Then the polynomials $f_i^*(\alpha^*)$ and $f_i^*(-\alpha^*)$ will constitute a sequence of rational functions $G_\nu(A)$ in the A_i with rational coefficients having the following property: For special values a_i of A_i in P with the predetermined sign, the distribution of signs in this sequence will give the exact number of roots of $f(y)$ in P.

Such a sequence $G_\nu(A)$ will be called a Sturm Sequence.

We shall prove the existence of a place for K/P by showing that K may be mapped isomorphically into any given real closed field $\Omega \supset P$ of sufficiently large transcendence degree over P by a mapping which is identity on P. It will then suffice to use a field Ω in which P is maximal archimedean. The canonical place of Ω/P will induce a place of K/P. Actually this procedure will be carried out only when K has transcendence degree 1 over P, because one can then use induction to complete the proof.

We construct first the auxiliary field Ω as follows. We adjoin a variable t to P and order $P(t)$ in such a way that t is infinitely large over P. Let Ω be a real closure of $P(t)$ which preserves the ordering of $P(t)$. Then P is maximal archimedean in Ω. Indeed, P is algebraically closed in Ω. Suppose that there exists an element $u \,\epsilon\, \Omega$ such that $P(u)$ is archimedean over P. Then Ω is algebraic over $P(u)$. In particular t satisfies an equation with coefficients in $P(u)$. t would therefore have to lie in an interval determined by elements of $P(u)$, and consequently by elements of P since $P(u)$ is archimedean over P, contradicting the hypothesis that t is infinitely large with respect to P.

LEMMA. *Let* P *be a real closed field. Let* $g_\nu(t)$ *be a finite set of rational functions in one variable with coefficients in* P. *Suppose the rational field* $P(t)$ *ordered in some* *way, so that each $g_\nu(t)$ has a sign attached to it. Then there exist infinitely many special values τ of t in P such that $g_\nu(\tau)$ has the same sign as $g_\nu(t)$.*

PROOF. We may write $g_\nu(t) = c \cdot \prod(t - b) \cdot \prod p(t)$ where the first product is extended over all linear factors, and the second over positive definite quadratic factors, and $b, c \in$ P. For any $a \in$ P, $p(a)$ is positive. It suffices therefore to show that the signs of $(t - b)$ can be preserved under a specialization. We order all values of b and of t and obtain

$$\cdots < b_1 < t < b_2 < \cdots$$

where possibly b_1 or b_2 is omitted if t is larger or smaller than any b. Any value τ of t in P selected between b_1 and b_2 will then satisfy the requirements of our lemma.

PROPOSITION. *Let K be a real function field in one variable over a real closed constant field* P. *Let* $P_1 \supset$ P *be another real closed field whose transcendence degree over* P *is at least 1. Then there exists an isomorphism of K into* P_1 *which is identity on* P.

PROOF. Let $K =$ P(t, α) where t is transcendental over P. Give K any ordering. Let

$$f(y, t) = y^n + a_{n-1}(t)y^{n-1} + \cdots + a_0(t)$$

be the irreducible equation of α over P(t). Let $G_\nu(a_i(t))$ be a Sturm sequence for $f(y, t)$. Let the set $g_\nu(t)$ of the lemma consist of all coefficients $a_i(t)$ and of $G_\nu(a_i(t))$. There exists a special value τ of t in P such that $g_\nu(\tau)$ has the same sign in P as $g_\nu(t)$ in K. Let $\lambda \in P_1$ be transcendental over P and so small that the polynomials $g_\nu(\tau + \lambda)$ have the same sign in P_1 as they had in P. (They are continuous functions.) Then the polynomial $f(y, \tau + \lambda)$ must have a root α_1 in P_1, and P(t, α) is isomorphic to P$(\tau + \lambda, \alpha_1)$. (The existence of the element λ is easy to prove. By taking a reciprocal if necessary, we may assume that a given transcendental element of P_1 is not infinitely large over P. One can make it as small as one pleases by multiplying it by small elements of P.)

We shall apply the proposition in case $P_1 = \Omega$ and P is maximal archimedean in Ω. We obtain a result somewhat stronger than the mere existence of a place by making the following observation. Given a finite set of elements $\xi_1, \cdots, \xi_n \in K$. We want the place ϕ to be such that $\phi(\xi_\nu) \neq \infty$. We need a lemma.

LEMMA 2. *Let K be a field,* ϕ *a real place, and* $\xi_1, \cdots, \xi_n$ *a finite set of non zero elements of K such that* $\phi(\xi_1^2 + \cdots + \xi_n^2) \neq \infty$. *Then* $\phi(\xi_\nu) \neq \infty$.

PROOF. Say ξ_1 has maximum value under ϕ. Put $\eta_\nu = \xi_\nu/\xi_1$. Then $\phi(\eta_\nu) \neq \infty$. Write

$$\phi(\xi_1^2 + \cdots + \xi_n^2) = \phi(\xi_1^2(1 + \eta_2^2 + \cdots + \eta_n^2)).$$

Then $\phi(1 + \sum \eta_\nu^2) = 1 + \sum \phi(\eta_\nu)^2 \neq \infty, \neq 0$ because it is positive. Hence $\phi(\xi_1) \neq \infty$. Since we assumed ξ_1 had maximal value, we see that $\phi(\xi_\nu) \neq \infty$, as contended.

Returning to the field K/P of the proposition, we let $t = \xi_1^2 + \cdots + \xi_n^2$. If $t \in P$ then any place will be finite on ξ_ν. If $t \notin P$, t is transcendental over P and we use this t as a transcendence basis in the proof of the proposition. The place we have constructed is clearly finite on t because the image of t in Ω is not infinitely large over P. By the lemma, the place is finite on ξ_ν, as desired.

We now prove the general theorem.

THEOREM 7. *Let K be a real function field in a finite number of variables over a real closed constant field P. Let ξ_ν be a finite set of elements of K. Then there exists a place ϕ of K rational over P such that all $\phi(\xi_\nu)$ are finite.*

PROOF. Let K/P have transcendence degree r. We shall prove that there exists a place of K taking its values in a real function field of transcendence degree $r - 1$ over P which is finite on ξ_ν. Since K is real, we may suppose K ordered, and imbedded in a real closure Γ.

Let $t_1, \cdots, t_r$ be a transcendence basis of K/P. Let P' be the real closure of $P(t_1, \cdots, t_{r-1})$ in Γ. Then $K/K \cap P'$ has transcendence degree 1, and we consider K as a function field in 1 variable over the constant field $K \cap P'$. We note that KP'/P' has also transcendence degree 1. By the proposition, there exists a place ϕ of KP'/P' which induces a zero dimensional real place of K over $K \cap P'$ such that all $\phi(\xi_\nu)$ are finite. The image of K is a finite algebraic extension of $K \cap P'$ which may be considered as a function field of transcendence degree $r - 1$ over P. The procedure may now be repeated to yield our theorem.

By feeding Theorem 7 back into itself we can derive a seemingly stronger theorem, in which we impose certain algebraic conditions on the place.

THEOREM 8. *Let K be a real function field (in several variables) over a real closed constant field P. We suppose that K is ordered. Let $\xi = (\xi_1, \cdots, \xi_n)$ be a finite set of elements of K with $\xi_1 < \xi_2 < \cdots < \xi_n$. Let $G_\nu(X)$ be a finite set of polynomials in $P[X]$ such that $G_\nu(\xi) \neq 0$. Then there exists a place ϕ of K/P and rational over P such that all $\phi(\xi_i)$ are finite, and*

1. $$\phi(\xi_1) < \phi(\xi_2) < \cdots < \phi(\xi_n),$$

2. $$G_\nu(\phi(\xi)) \neq 0.$$

PROOF. Imbed K in a real closure Γ which preserves its ordering. We can write $\xi_\nu - \xi_{\nu-1} = \eta_\nu^2$ in Γ. $K(\eta_\nu)$ is again a real function field over P. We apply Theorem 7. Let $G(\xi) = \prod G_\nu(\xi)$. There exists a place ϕ of $K(\eta_\nu)$ over P such that ϕ does not map any of the elements ξ_ν, η_ν, η_ν^{-1}, $G(\xi)^{-1}$ on ∞. Theorem 8 is now obvious.

We see that the place may be chosen so as to preserve the signs of ξ_ν.

We may apply Theorem 8 to a generalization of the Hilbert Problem solved by Artin in [2].

THEOREM 9. *Let K be a real function field over a real constant field F. Let ξ be an element of K such that for all zero dimensional real places of K/F we have either $\phi(\xi) = \infty$ or $\phi(\xi) \geq 0$ for any ordering of $\phi(K)$. Then ξ is a sum of squares in K.*

PROOF. Suppose that ξ were negative in some ordering of K. Let Γ be a real closure of K preserving this ordering. Γ contains a real closure P of F (Lemma 1) and a compositum KP which may be viewed as a function field over P. There exists a place ϕ of KP/P such that $\phi(\xi)$ is negative. ϕ induces a zero dimensional place of K/F and $\phi(\xi) < 0$. Hence ξ is positive in all orderings of K. By a theorem of Artin [2], ξ is a sum of squares in K.

Using Theorem 8 it is now easy to generalize to several variables the proposition preceding Theorem 7.

THEOREM 10. *Let K be a real function field of transcendence degree r over a real closed constant field* P. *Let* P_1/P *be a real closed field of transcendence degree at least r over* P. *Then there exists an isomorphism of K into* P_1 *which is identity on* P.

PROOF. Let $K = P(t_1, \cdots, t_r, \alpha)$ where $t = (t_1, \cdots, t_r)$ is a transcendence basis. Let

$$f(y, t) = y^s + a_{s-1}(t)y^{s-1} + \cdots + a_0(t)$$

be the irreducible equation of α over $P(t)$. Let $G_\nu(a_i(t))$ be a Sturm sequence for f. Let $g_\nu(t)$ consist of t_i, $a_i(t)$, $G_\nu(a_i(t))$. These rational functions in t are elements of K and have a definite sign attached to them. By Theorem 8 there exists a specialization $\tau = (\tau_1, \cdots, \tau_r)$ of t in P such that $g_\nu(\tau)$ has the same sign as $g_\nu(t)$. Furthermore, the $g_\nu(\tau)$ will keep their sign in a neighborhood of τ. Let $\lambda = (\lambda_1, \cdots, \lambda_r)$ be a set of algebraically independent elements of P_1 which are so small that $g_\nu(\tau + \lambda)$ has the same sign as $g_\nu(\tau)$. Then $f(y, \tau + \lambda)$ has a root α_1 in P_1 and $P(\tau + \lambda, \alpha_1)$ is isomorphic to $P(t, \alpha)$.

It follows in particular that any real function field over a real algebraic number field has an isomorphic image in any real closed field having the same transcendence degree over the rational numbers. This shows that such a function field can be ordered in a large variety of ways.

Given a real closed field P, we construct a real closed field $\Omega \supset P$ having arbitrarily high degree of transcendence over P and such that P is maximal archimedean in Ω. Such a field will find many applications: It gives a convenient tool to construct places over P. The following Theorem will be needed.

THEOREM 11. *Let* $P \subset P_1 \subset P_2$ *be three real closed fields. Let* P *be maximal archimedean in* P_1, *and* P_1 *maximal archimedean in* P_2. *Then* P *is maximal archimedean in* P_2.

PROOF. Let $\mathfrak{o}_1$, $\mathfrak{o}_2$, $\mathfrak{o}$ and $\mathfrak{p}_1$, $\mathfrak{p}_2$, $\mathfrak{p}$ be the canonical valuation rings and their respective maximal ideals for P_1/P, P_2/P_1, and P_2/P. We must prove that $\mathfrak{o}/\mathfrak{p}$ is isomorphic to P, i.e. every residue class in $\mathfrak{o}$ (mod $\mathfrak{p}$) has a representative in P. We have trivially

$$\mathfrak{p}_2 \subset \mathfrak{p} \subset \mathfrak{o} \subset \mathfrak{o}_2$$

$$\mathfrak{p}_1 \subset \mathfrak{p}.$$

Let $\alpha \in \mathfrak{o}$. Then by hypothesis

$$\alpha \equiv \beta \pmod{\mathfrak{p}_2}$$

where $\beta \in P_1$. In fact, we contend that $\beta \in \mathfrak{o}_1$. Indeed, $\beta = \alpha + a$ where a is infinitely small over P. Since α is not infinitely large over P, β cannot be infinitely large.

By hypothesis we now have also

$$\beta \equiv \rho \pmod{\mathfrak{p}_1}$$

where $\rho \in P$. We weaken these congruences to congruences mod $\mathfrak{p}$ and we see that

$$\alpha \equiv \rho \pmod{\mathfrak{p}}$$

as was to be shown.

It is now clear how to construct Ω. We had already constructed such a field having transcendence degree 1 over P in the proof of Theorem 7. Given a well ordered set of variables t_i, we suppose that the construction has been carried out for all $i < j$. Let P_1 be the real closed field so obtained, P being maximal archimedean in P_1. Perform the construction for the 1 variable case, i.e. for $P_1(t_j)$, thereby obtaining a real closed field P_2 in which P_1 is maximal archimedean. By Theorem 10 we can conclude that P is maximal archimedean in P_2. We let Ω be the union of all these fields, and it is clear that P is maximal archimedean in Ω.

3. Oddly C_i Fields

A field F is said to be *oddly C_0* if every form of odd degree d, in n variables, with $n = d$, has a non trivial zero in F.

In analogy with algebraic closure, we can prove that a field is oddly C_0 if and only if its algebraic extensions are of even degree. Namely, if the field is oddly C_0 then the norm of the general element of an extension of odd degree would provide a form of odd degree having only the trivial zero. Conversely, if the field has only extensions of even degree, then every polynomial of odd degree has a root in the field.

A field F will be said to be *oddly C_i* if every form of odd degree d in n variables, with $n > d^i$ has a non trivial zero in F.

As with the C_i fields, we try to prove that the existence of a zero for one form implies the existence of a simultaneous zero for several forms. We note that the theory as it is developed in [6] is multiplicative, and that consequently all the algebraic theorems pertaining to C_i fields will hold for oddly C_i fields. We shall repeat the statements of the theorems, but omit their proofs which are analogous to those of [6].

An *odd normic form* is a normic form of odd degree.

LEMMA. *If a field F has one odd normic form of order $i \geqq 1$ then it has odd normic forms of order i of arbitrarily high degree.*

THEOREM 12. *Let F be an oddly C_i field ($i \geqq 1$) admitting at least one odd normic form of order i. Let $f_1, \cdots, f_r$ be r forms in n common variables, each of odd degree d. If $n > rd^i$ then the forms have a non trivial common zero in F. If we suppose that F admits normic forms of order i of any given odd degree and f_ν are forms in n variables of odd degrees d_ν such that $n > d_1^i + \cdots + d_r^i$, then they have a non trivial zero in F.*

THEOREM 13. *Let F be oddly C_i and suppose that F admits at least one odd normic form of order i. Then any algebraic extension of F is also oddly C_i.*

If F has a discrete valuation, whose residue class field has odd normic forms of order i, then F has odd normic forms of order $i + 1$.

By exactly the same procedure that is used in [6] we can prove

THEOREM 14. *Let F be a function field in n variables over a constant field $\mathfrak{k}$ which is oddly C_i and admits an odd normic form of order i. Then F is oddly C_{i+n}.*

What makes the above theorems interesting is the fact that they can be applied to real fields. Namely, we can prove

THEOREM 15. *Let P be a real closed field. Let $f_1, \cdots, f_r$ be r forms of odd degrees, in n common variables. If $n > r$ then the forms have a common zero in P.*

PROOF. Given a form f of degree d in n variables with coefficients in P. We consider all possible monomials

$$\mu = x_1^{\nu_1} \cdots x_n^{\nu_n}$$

of degree d in n variables. Let Ω be the field constructed at the end of Part II, and let λ_μ be algebraically independent elements of Ω over P which are infinitely small over P. A form f in P can be written

$$f = \sum a_\mu \mu$$

where $a_\mu \in$ P. For such a form f we construct the form f^* by putting

$$f^* = \sum (a_\mu + \lambda_\mu)\mu.$$

Then f^* has algebraically independent coefficients in Ω and can be considered as a generic form. We do this construction for each one of the forms $f_1, \cdots, f_r$ in $r + 1$ variables using each time new indeterminates λ_μ from Ω. Then $f_1^*, \cdots, f_r^*$ can be considered as a set of generic forms in Ω.

By Bezout's theorem these forms have exactly $d_1 \cdots d_r$ zeros in $\Omega(\sqrt{-1})$ and $d_1 \cdots d_r$ is odd. If ξ is a zero then so is the conjugate of ξ. Hence the number of zeros can be odd only if some zero lies in Ω. Let $\xi = (\xi_1, \cdots, \xi_n)$ be such a zero. In the ordering of Ω, say $|\xi_1| \geq |\xi_\nu|$ all ν. Then by homogeneity, we can divide by ξ_1 all the components of ξ and obtain in this way a zero $(1, \eta_2, \cdots, \eta_n)$ such that no η_ν is infinitely large over P. We have

$$f_\nu^*(1, \eta_2, \cdots, \eta_n) = 0.$$

Taking the canonical place of Ω over P and putting $\bar{\eta}_\nu = \phi(\eta_\nu)$ we see that

$$f_\nu(1, \bar{\eta}_2, \cdots, \bar{\eta}_n) = 0$$

as was to be shown.

COROLLARY. *Let F be a function field in n variables over a real closed field. Then F is oddly C_i.*

In the special case where P is the field of real numbers R, we can avoid the use of Ω, and use the compactness of projective space instead. Namely, the elements λ_μ can actually be selected in the reals, and we can obviously find forms

$f_1^*, \cdots, f_r^*$ arbitrary close to the given special forms $f_1, \cdots, f_r$, and generic, say over the rational numbers. By Bezout's Theorem, the generic forms have a common real zero, which we may view as lying in real projective n-space. Letting the generic forms converge to the special forms, their zeros must have a point of accumulation in projective space. This point of accumulation will be a zero of the special forms by continuity, thereby proving the theorem in the special case of real numbers.

PRINCETON UNIVERSITY

BIBLIOGRAPHY

Note: All the fundamental theorems concerning real fields which we have taken for granted in this paper can be found in the chapter on Real Fields, of Van der Waerden's *Moderne Algebra*.

[1] E. ARTIN, *Kennzeichnung des Korpers der reellen algebraischen Zahlen*, Abh. Math. Sem. Hansischen Univ., Bd. 3 (1924).

[2] E. ARTIN, *Über die Zerlegung definiter Funktionen in Quadrate*, Abh. Math. Sem. Hansischen Univ., Bd. 5 (1926).

[3] E. ARTIN, Lectures on Algebraic Numbers and Algebraic Functions, Princeton, 1951, Ch. 16.

[4] ARTIN AND SCHREIER, *Algebraische Konstruktion reeller Körper*, Abh. Math. Sem. Hansischen Univ., Bd. 5 (1926).

[5] ARTIN AND SCHREIER, *Eine Kennzeichnung der Reell abgeschlossenen Körper*, Abh. Math. Sem. Hansischen Univ., Bd. 6 (1927).

[6] S. LANG, *On Quasi Algebraic Closure*, Ann. of Math., vol. 55 (1952), pp. 373–390.

[7] A. WEIL, *Arithmetic on Algebraic Varieties*, Ann. of Math., vol. 53 (1951), pp. 412–444.

Added in Proof. I became acquainted with Krull's fundamental and beautiful paper "Allgemeine Bewertungstheorie", J. Reine u. Angew. Math. 1932, some time after the present paper was completed. In particular, Krull was the first to point out explicitly that the ring of elements which are not infinitely large is a valuation ring, and his Satz 22 follows immediately from the Corollary to our Theorem 6.

Am. J. Math., LXXVI No. 2, 1954

SOME APPLICATIONS OF THE LOCAL UNIFORMIZATION THEOREM.*

By Serge Lang.

---

1. Introduction. Let K be a function field over a constant field k. We shall always assume that K/k is regular: k is algebraically closed in K and K/k is separably generated. If $K = k(x_1, \cdots, x_n) = k(x)$, we may view (x) as the generic point (over k) of a variety V, which is then called a model of K. If $\mathfrak{p}$ is a place of K/k, i. e. a place which is an isomorphism, or identity, on k, and if we put $\bar{x}_i = x_i(\mathfrak{p}) \neq \infty$, then $(\bar{x})$ is a point of V and we say that $(\bar{x})$ is at the center of $\mathfrak{p}$.

Zariski's Theorem of Local Uniformization [10] states that given a place $\mathfrak{p}$ of K/k (characteristic zero) it is always possible to find a model of the field for which the point at the center of $\mathfrak{p}$ is non singular.

In the classical case where k is the complex numbers, this implies that the place has a certain neighborhood.

It is our purpose to show that it is not the algebraic closure of the complex numbers which is essential for the existence of this neighborhood, but rather its topological completeness.

More precisely: Suppose that the constant field k of K is complete under a valuation, and of characteristic zero. We say that a place of K/k is rational over k if it is k-valued. Let $\mathfrak{M}$ consist of all rational places of K/k, and assume that $\mathfrak{M}$ is not empty. $\mathfrak{M}$ may be called the rational Riemann manifold of K/k, but as we shall deal only with rational places, we shall call it the Riemann manifold. We topologize $\mathfrak{M}$ (as it is done classically) by giving it the weakest topology such that the map $x \to x(\mathfrak{p}) \,\varepsilon\, k_\infty$ is continuous for each x in K.

One can then prove that $\mathfrak{M}$ contains infinitely many places, and in fact enough places to distinguish between elements of K. If k is locally compact, then $\mathfrak{M}$ is a compact space, and in the case where K has dimension 1 over k (the case of curves) then $\mathfrak{M}$ is locally homeomorphic to k itself. Our proofs are closely related to Chevalley's [2].

Kawada [4] has also considered the abstract manifold in the case of curves, but restricted himself to the case where k is algebraically closed. In

---

* Received March 16, 1953.

362

that case, the complex numbers are the only locally compact field. If however we do not assume k algebraically closed, then we can include the real numbers, and p-adic fields. In fact, in the case of curves, we can also treat the power series over finite fields because the uniformization theorem is true irrespective of the characteristic.

If k is the real numbers, then $\mathfrak{M}$ consists of a finite number of components, each homeomorphic to a circle. It is a classical theorem of Harnack [3] that the number of these components is at most equal to $g + 1$, where g is the genus of K. We shall give a simple new proof of this theorem, based on the Riemann-Roch Theorem.

2. The Riemann manifold. We let k be a field complete under a rank one valuation, and do not make any restrictions on the characteristic. We view the value group as a subgroup of the positive reals. If $a \,\varepsilon\, k$, we let $|\,a\,|$ denote its value.

We suppose that K/k is a function field with the properties stated in the introduction. We assume that there exists at least one rational place of K/k, and let $\mathfrak{M}$ be the set of all such places. For each $x \,\varepsilon\, K$, we let Γ_x be the "k-sphere" k_∞ obtained from k by the formal adjunction of ∞, and topologized in the obvious way: Neighborhoods of ∞ are the outsides of circles around 0. We let $\Gamma = \prod_x \Gamma_x$ be the Cartesian product taken over all $x \,\varepsilon\, K$, with the product topology.

Let $\mathfrak{p} \,\varepsilon\, \mathfrak{M}$. Then $\mathfrak{p}$ maps K into k_∞. Let Φ be the map of $\mathfrak{M}$ into Γ defined by $\Phi : \mathfrak{p} \to (\cdots, x(\mathfrak{p}), \cdots)$ and denote $\Phi(\mathfrak{M})$ by $\mathfrak{M}'$. Then Φ is obviously 1-1.

We topologize $\mathfrak{M}'$ as a subspace of Γ, and give $\mathfrak{M}$ the weakest topology so that Φ is continuous. In other words, the open sets of $\mathfrak{M}$ consist of the sets $\Phi^{-1}(U)$ where U is open in $\mathfrak{M}'$

The following theorem and its proof are identical with Chevalley's [2], except for the fact that only the completeness of k is used, and not its algebraic closure.

THEOREM 1. *$\mathfrak{M}'$ is closed in* Γ.

Proof. Let $u = (\cdots, u_x, \cdots)$ be in the closure of $\mathfrak{M}'$. Let $\mathfrak{o} = \{x \,\varepsilon\, K \mid u_x \neq \infty\}$. We shall prove that $\mathfrak{o}$ is a valuation ring (or v-ring) whose induced place $\mathfrak{p}$ is such that $x(\mathfrak{p}) = u_x$.

We first note that $k \subset \mathfrak{o}$, because $a(\mathfrak{p}) = a$ for all $a \,\varepsilon\, k$, and all $\mathfrak{p} \,\varepsilon\, \mathfrak{M}$. Let x and y lie in $\mathfrak{o}$. Then $u_x, u_y \neq \infty$. Since u is a limit point of elements in $\mathfrak{M}'$, there exists a prime $\mathfrak{p} \in \mathfrak{M}$ such that $x(\mathfrak{p})$, $y(\mathfrak{p})$, and $(x + y)(\mathfrak{p})$ are arbitrarily close to u_x, u_y, and u_{x+y} respectively. For such $\mathfrak{p}$, we see that $x(\mathfrak{p})$ and $y(\mathfrak{p}) \neq \infty$, whence $(x + y)(\mathfrak{p}) = x(\mathfrak{p}) + y(\mathfrak{p}) \neq \infty$. Hence $x + y \in \mathfrak{o}$. Similarly, $x - y$ and xy lie in $\mathfrak{o}$, which is therefore a ring.

Furthermore, the map $x \to u_x$ is a homomorphism of $\mathfrak{o}$ into k and is identity on k. This follows from a continuity argument as above.

Finally, $\mathfrak{o}$ is a v-ring. Suppose $x \notin \mathfrak{o}$, Then $u_x = \infty$. Let $y = x^{-1}$. There exists a prime $\mathfrak{p} \in \mathfrak{M}$ such that $x(\mathfrak{p})$ is close to u_x and $y(\mathfrak{p})$ is close to u_y. But $x(\mathfrak{p})$ close to u_x means $|x(\mathfrak{p})|$ is very large. Hence $|y(\mathfrak{p})|$ is very small. Hence $u_y = 0$, so $y \in \mathfrak{o}$. Thus $\mathfrak{o}$ is a v-ring.

Since $x \to u_x$ is a homomorphism of $\mathfrak{o}$ into k, and since $\mathfrak{o}$ is a v-ring, it follows that $x \to u_x$ is a place of K, which is rational because $u_x \in k_\infty$. Our theorem is thereby proved.

In particular, if k is locally compact, then k_∞ is the ordinary one-point compactification, and k_∞ is therefore compact. By Tychonoff's theorem, Γ is compact. Since $\mathfrak{M}'$ is closed in Γ, it is also compact. We obtain therefore the following.

COROLLARY. *If k is locally compact, then $\mathfrak{M}$ is compact.*

We shall now investigate more closely the special one-dimensional case. For the rest of this section, we suppose that K/k has transcendence degree 1, and is regular. For the convenience of the reader, we reproduce a proof of the local uniformization theorem.

Let $\mathfrak{p}$ be a prime of K, whose residue class field is separable over k. Let $x \in K$ be a separating variable, such that $\mathrm{ord}_\mathfrak{p} x = 1$. (It is well known that such an x can always be found. See for instance [1], Ch. 17, 4.) If p is the rational prime induced by $\mathfrak{p}$ on $k(x)$, then p is unramified in K. Hence there exists a generator y of K over $k(x)$ which is integral at $\mathfrak{p}$ and whose local $\mathfrak{p}$-different is not $\equiv 0 \pmod{\mathfrak{p}}$. This means: If $f(x, y) = 0$ is the equation for x and y over k, and if we put $\bar{x} = x(\mathfrak{p})$, $\bar{y} = y(\mathfrak{p})$ then $f_y(\bar{x}, \bar{y}) \neq 0$. Hence $(\bar{x}, \bar{y})$ is non-singular at the center of $\mathfrak{p}$ as was to be shown.

We recall one more useful theorem from the theory of curves.

Let $K = k(x, y)$ be the function field of a curve, and let $f(x, y) = 0$ be its equation over k. Let $(\bar{x}, \bar{y})$ be a non-singular point on the curve, so that $f(\bar{x}, \bar{y}) = 0$ but $f_y(\bar{x}, \bar{y}) \neq 0$. Then $(\bar{x}, \bar{y})$ is at the center of exactly one place of K/k. In other words, there exists one and only place $\mathfrak{p}$ of K/k such that $\bar{x} = x(\mathfrak{p})$, $\bar{y} = y(\mathfrak{p})$. Furthermore, the residue class field $\bar{K}$ of K under $\mathfrak{p}$ is naturally isomorphic to $k(\bar{x}, \bar{y})$.

For a proof, see Artin [1], Ch. 16, 2.

In particular, if $(\bar{x}, \bar{y})$ is a non-singular rational point (that is if its coordinates lie in k), then the place above it is also rational.

Let $\mathfrak{p}$ be a given prime of $\mathfrak{M}$. We shall find a neighborhood for $\mathfrak{p}$. Let x be a local uniformizing parameter at $\mathfrak{p}$, separating for K/k. In other words, assume $\operatorname{ord}_{\mathfrak{p}} x = 1$, and $K/k(x)$ is separable. According to the local uniformizing theorem, we can find a model $K = k(x, y)$ such that the point $(\bar{x}, \bar{y})$ at the center of $\mathfrak{p}$ is non-singular: $f_y(\bar{x}, \bar{y}) \neq 0$. By multiplying x and y by a constant $a \,\varepsilon\, k$ having sufficiently small value, we may assume that $\bar{x}$ and $\bar{y}$ are integral in k. This will not affect the non-singularity of the point. We may also assume that f has integral coefficients by clearing denominators if necessary. These preparations are necessary in order to be able to apply the Newton approximation method later. (Cf. [5].) They apply of course only if the valuation of k is non-archimedean.

Split the polynomial $f(\bar{x}, Y)$ in the algebraic closure $\bar{k}$ of k. Then $\bar{y}$ is a root of multiplicity 1, so we have $f(\bar{x}, Y) = (Y - \bar{y})(Y - y_1)^{e_1} \cdots (Y - y_r)^{e_r}$. We recall that the roots of an algebraic equation are continuous functions of the coefficients ([1], Ch. 2, 6). Given a neighborhood $\bar{M}$ of $\bar{y}$ in $\bar{k}$ not containing $y_1, \cdots, y_r$ there exists a neighborhood N_1 of x in k such that for all values $\xi \,\varepsilon\, N_1$, the equation $f(\xi, Y) = 0$ has exactly one root in M.

Let $M = \bar{M} \cap k$. There exists a neighborhood $N \subset N_1$ of x such that for any value $\xi \,\varepsilon\, N$, we can refine $\bar{y}$ to a root η of $f(\xi, Y)$ with $\eta \,\varepsilon\, M$. This can be done with the Newton approximation method, as given in [5]. Since the partial derivative is continuous, we also have $f_y(\xi, \eta) \neq 0$. Hence there exists a unique rational place $\mathfrak{q}$ of K such that $x(\mathfrak{q}) = \xi$ and $y(\mathfrak{q}) = \eta$. This proves that the map $\mathfrak{q} \to x(\mathfrak{q})$ is 1-1 for all $\mathfrak{q}$ such that $(x(\mathfrak{q}), y(\mathfrak{q}))$ lies in the neighborhood (N, M) of $(\bar{x}, \bar{y})$. By what has been said above, this latter condition is equivalent to the single condition that $x(\mathfrak{q}) \,\varepsilon\, N$.

The map is continuous. We may select N and M to be closed "discs" around $\bar{x}$ and $\bar{y}$ respectively, in which case the set of primes $\mathfrak{q}$ mapping x and y into (N, M) is closed.

If k is locally compact, this set of primes is compact. Hence the map is a homeomorphism. Summarizing, we have

THEOREM 2. *If k is locally compact, and if K/k is a regular extension of dimension one, then $\mathfrak{M}$ is locally homeomorphic with k. Given a prime $\mathfrak{p} \,\varepsilon\, \mathfrak{M}$, and a separating local uniformizing parameter $x \,\varepsilon\, K$ at $\mathfrak{p}$, the map $\mathfrak{q} \to x(\mathfrak{q})$ gives a homeomorphism of a certain neighborhood of $\mathfrak{p}$ with a disc of center 0 in k.*

If $k = R$ is the real numbers, then $\mathfrak{M}$ is the union of a finite number of components, which have classically been called the real branches of the curve. It is obvious (and in fact can easily be proved) that each component of $\mathfrak{M}$ is homeomorphic to a circle. (One uses the facts that it is compact, connected, and locally homeomorphic to real segments.) Harnack [3] has proved that the number of these branches cannot exceed $g + 1$, where g is the genus of the curve. We give below a proof depending on the Riemann-Roch theorem.

We begin by a remark. Let B be one of the branches, and let $x \, \varepsilon \, K$ be a function having no pole on B. We assert that x cannot have only a single zero on B. In other words, x has either no zero on B, or has at least two zeros, counting multiplicities. Indeed, if x has only a single zero $\mathfrak{p}$, then $x(\mathfrak{q}) > 0$ and $x(\mathfrak{q}') < 0$ for $\mathfrak{q}$ and $\mathfrak{q}'$ in a small neighborhood of $\mathfrak{p}$. Naively speaking, x changes sign as it runs through $\mathfrak{p}$. But $B - \{\mathfrak{p}\}$ is connected. Since x gives a continuous function of $B - \{\mathfrak{p}\}$, it must therefore adopt the value 0 somewhere in $B - \{\mathfrak{p}\}$, contradiction.

Now suppose there are $g + 2$ branches. Select one place on each branch, and let $\mathfrak{p}_1, \cdots, \mathfrak{p}_{g+2}$ be these places. Let $\mathfrak{q}$ be a place of K of degree 2 (a complex place). Then $\mathfrak{q}$ does not lie on any one of the real branches. The divisor $\mathfrak{p}_1 \cdots \mathfrak{p}_{g+2}/\mathfrak{q}^{g+1}$ has degree $g + 2 - 2(g + 1) = -g$. By the Riemann-Roch theorem, there exists a function $x \, \varepsilon \, K$ which is multiple of this divisor. Thus x has g more zeros in addition to $\mathfrak{p}_1, \cdots, \mathfrak{p}_{g+2}$, and has no pole on any of the real branches. This implies that on at least two branches, x will have only a single zero, a contradiction.

Harnack proved his theorem by taking a model of the curve, and assuming that the singularities were not too horrible. The classical proof is obtained by viewing the real branches as lying on the complex Riemann surface, by cutting the latter along these real branches, and using a symmetry argument: the real branches consist precisely of those places on the surface which are invariant under the complex conjugate automorphism.

Our proof is of course much more algebraic. In fact, it is very likely that one can define branches for a curve over an arbitrary real closed field, and that one can prove in that case also that there are at most $g + 1$ branches. As we have not been able to carry out all the details of this idea, we shall leave it as it stands for the moment.

If one takes a model for the curve, $K = k(x, y)$, then it is extremely difficult, if not impossible, to give a rigorous definition of a branch in the (x, y) plane without leaving the plane. For instance, we want to say that a lemniscate is one branch, but that two tangent circles are two branches.

The correct way of defining a branch in the (x, y) plane is to let it consist of the projection on this plane of one of the components B of $\mathfrak{M}$. This projection will be the set of all points $(x(\mathfrak{p}), y(\mathfrak{p}))$ with $\mathfrak{p} \, \varepsilon \, B$. We note that this projection on the plane may not be connected (as in the case of a hyperbola, where the connection is at infinity), and may also intersect the projection of another branch. On the real manifold $\mathfrak{M}$, however, two branches can never intersect.

3. Function fields over complete fields.

In this section, we return to the general case, but assume that k has characteristic zero. It is again complete under a rank one valuation. We assume that k is contained in a universal domain of infinite degree of transcendence over k, and in general follow the conventions of Weil [8]. However, we do not distinguish between fields and abstract fields, as the context will always make our meaning clear.

We say that an extension E/F of a field F is regular if every finitely generated intermediate field E' with $F \subset E' \subset E$ is regular over F.

A subfield k_1 of k will be called admissible if it is dense in k, if k/k_1 is regular, and if k/k_1 has infinite degree of transcendence.

Our main result is concerned with function fields over admissible fields.

THEOREM 3. *Let k_1 be admissible. Let K/k_1 be a finitely generated extension, regular over k_1. There exists a rational place of K/k_1 if and only if there exists an isomorphism of K into k which is identity on k_1. If this is the case, then there exists infinitely many rational places. In fact, given a finite set of quantities $(z) = (z_1, \cdots, z_n)$ with $z_i \, \varepsilon \, K$, and a finite set of polynomials $g_j(Z) \, \varepsilon \, k_1[Z]$ such that $g_j(z) \neq 0$, there exists a rational place $\mathfrak{p}$ of K/k_1 such that all $\bar{z}_i = z_i(\mathfrak{p})$ are finite and such that $g_j(\bar{z}) \neq 0$.*

Proof. Suppose that there exists a rational place $\mathfrak{p}$ of K/k_1. By the local uniformization theorem, we can find a model of K whose center at $\mathfrak{p}$ is non-singular. By a sufficiently general projection, we may assume that this model V is a non-singular hypersurface: $K = k_1(x_1, \cdots, x_r, y)$ where (x, y) satisfy the irreducible equation $f(x, y) = 0$ over $k_1, x_1, \cdots, x_r$ are algebraically independent over k_1, and since V is non-singular, $f_y(\bar{x}, \bar{y}) \neq 0$.

By multiplying if necessary the elements x_i, y by an element $a \, \varepsilon \, k_1$ having sufficiently small value, we may assume that $\bar{x}_i, \bar{y}$ are integral in k_1. Namely we have $(ax)(\mathfrak{p}) = a \cdot x(\mathfrak{p})$ because k_1 is fixed under $\mathfrak{p}$. Such a transformation does not affect the non-singularity of the point. In addition we may assume that f has integral coefficients, by clearing its denominators if necessary.

46

We can now apply the Newton approximation method as in [5]. Select $\xi_1, \cdots, \xi_r$ very close to $x_1, \cdots, x_r$ in k, but algebraically independent over k_1. Our assumptions on k allow us to do this. Then y can be refined to a root η of $f(\xi, Y)$ in k, and it is clear that $k_1(\xi, \eta)$ is isomorphic to $k_1(x, y)$. This proves the first part of our theorem. We note that we have a tremendous amount of freedom in the isomorphism of K into k.

Assume now that $k_1 \subset K \subset k$. We must show that there exists rational places of K/k_1. This will be done by an induction on the transcendence degree of K over k_1. If this transcendence degree is 1, let $K = k_1(x, y)$, and let $f(x, y) = 0$ be the irreducible equation for x and y over k_1. Then $f_y(x, y) \neq 0$ because (x, y) is generic. As before we may assume that x and y are integral in k and that f has integral coefficients. We then use the Newton method in the reverse direction. We can select ξ in k_1 arbitrarily close to x so that y can be refined to a root of $f(\xi, Y)$. This root will lie in k, and will be algebraic over k_1, hence in k_1. This procedure yields infinitely many points on the curve $f(x, y) = 0$, and almost all of them will be non-singular. We had pointed out that a non-singular point of a curve lies at the center of exactly one place and that the residue class field of this place is then generated by the coordinates of the point over the constant field. Since the coordinates obtained in our proof are rational, we have proved what we wanted.

Our construction shows in addition that any finite set of quantities (z) will not be mapped into 0 or ∞ by almost all the places. In particular, we can find places such that none of the quantities $g_j(z)$ are mapped into 0 or ∞. Furthermore, if we view (z) as a generic point, infinitely many of the places just constructed will induce specializations of (z) into k_1 which are arbitrarily close to (z) itself. For a proof and an application of this, see [5].

It is now easy to complete the induction. Let $k_1 \subset E \subset K$, where K/E has dimension 1. Let E_1 be the algebraic closure of E in k. Then E_1 is admissible. Let $L = KE_1$. Then L may be viewed as a function field in one variable over the constant field E_1. There exists a place $\mathfrak{p}$ of L rational over E_1, and $\mathfrak{p}$ induces a place on K/E which is E_1-valued. Since K/E has dimension 1, the image $\bar{K}$ of K under $\mathfrak{p}$ is finitely generated over E_1, and consequently over k_1. Since $\bar{K}$ has smaller dimension over k_1 than K we conclude by induction that there exists a place $\mathfrak{q}$ of $\bar{K}$ which is rational over k_1, and satisfying all the extra algebraic requirements that we desire. The composite of the places $\mathfrak{p}$ and $\mathfrak{q}$ is a rational place of K over k_1 satisfying all our conditions. This concludes our proof.

Theorem 2 can be used to study function fields over admissible fields k_1

which are interesting in themselves, such as the maximal unramified extension of a p-adic field, or the convergent power series over a valuated field. It can also be extended to function fields over k itself. The fields k_1 then play the role of fields of definition. This is done as follows.

Let K/k be a function field, and assume that there exists one rational place $\mathfrak{p}$ of K/k. Let V be a hypersurface which is a model of K and is nonsingular at the center of $\mathfrak{p}$. Let $K = k(x)$ where (x) is a generic point of V. Let k_1 be an admissible subfield of k which contains a field of definition for V (i. e. all the coefficients of the irreducible polynomial $f(X)$ vanishing on (x)), and all the values $x_i(\mathfrak{p})$. Such a field may be obtained by adjoining to an admissible field the finite set of quantities just mentioned, and taking the relative algebraic closure in k.

Let $K_1 = k_1(x)$. By a procedure similar to the one carried out in the first part of the proof of Theorem 3, there exists an isomorphism of K_1 into k which is identity on k_1. Hence we can find infinitely many places $\mathfrak{q}$ of K_1 which are rational over k_1. All that remains to be shown is that such places extend in a natural way to rational places of K over k.

The field K_1 is linearly disjoint from k over k_1 by Weil [8], Ch. I, 6, Th. 3. Our statement now follows from the following general lemma.

LEMMA. *Let K, L be two extensions of a field F and suppose that they are linearly disjoint over F. Let $\mathfrak{q}$ be a rational place of K/F. Then $\mathfrak{q}$ extends uniquely to a rational place of KL/L.*

Proof. Let $K_\mathfrak{q}$ be the ring of elements of K which are finite under $\mathfrak{q}$. The specialization on $K_\mathfrak{q}$ induced by $\mathfrak{q}$ extends uniquely to a specialization of the ring $L[K_\mathfrak{q}]$ generated over L by the elements of $K_\mathfrak{q}$. This is well known for a finite set of quantities, and the generalization to an infinite set is immediate.

Extend this specialization to a place $\mathfrak{q}$ of KL over L. We contend that it is uniquely determined, and that it is L-valued.

An element of KL can be written

$$\gamma = (a_1\alpha_1 + \cdots + a_r\alpha_r)/(b_1\beta_1 + \cdots + b_s\beta_s)$$

with $a_i, b_j \in K$, and $\alpha_i, \beta_j \in L$, and with the β_j linearly independent over F.

Say b_1 has maximal value under $\mathfrak{q}$, or equivalently under $\mathfrak{Q}$. Divide numerator and denominator by b_1. We get

$$\gamma = ((a_1/b_1)\alpha_1 + \cdots + (a_r/b_1)\alpha_r)/(\beta_1 + \cdots + (b_s/b_1)\beta_s)$$

where each a_i/b_1 and b_j/b_1 is finite under $\mathfrak{Q}$, and lies in F. Since $\mathfrak{Q}$ is identity

7

on L, and since the β_j are linearly independent over F, the denominator is now finite under $\mathfrak{D}$, and is not $\equiv 0 \pmod{\mathfrak{D}}$. The numerator is finite, and clearly both are L-valued. This concludes our proof.

We may summarize the preceding results:

THEOREM 4. *Let K/k be a function field over a field k complete under a valuation, and of characteristic zero. Assume that there exists one rational place of K/k. Then there exists infinitely many such places. In fact, given a finite set of algebraic conditions as in Theorem 3, we can find a rational place satisfying these conditions.*

In order to take care of the algebraic conditions on a finite set of elements (z) in K, we must be sure that all the elements of this set will lie in the field K_1. As they can be expressed as rational functions of (x) with a finite number of coefficients from k, all we need to do is to take a field k_1 containing these coefficients.

We shall now make a few additional remarks on function fields over complete fields, which admit rational places.

If we work over the complete field k itself, there is in general no natural extension of k in which we can imbed the function field K having a rational place, as we could do if we worked over an admissible subfield. However, there is an interesting special case where such an extension exists.

Namely, let k be complete under a discrete valuation, whose residue class field is algebraically closed. Then k can be represented as a power series field or a field of Witt vectors [9] with components in a field $\mathfrak{f}$ isomorphic to the residue class field. Denote such a complete field by $F(\mathfrak{f})$. Let $\mathfrak{K}$ be an algebraically closed field containing $\mathfrak{f}$ and assume $\mathfrak{K}/\mathfrak{f}$ has infinite degree of transcendence. Then $F(\mathfrak{f})$ is contained naturally in $F(\mathfrak{K})$. It turns out that $F(\mathfrak{K})$ plays towards $F(\mathfrak{f})$ the same role that k played towards k_1 in Theorem 4. In fact, one can replace k and k_1 by $F(\mathfrak{K})$ and $F(\mathfrak{f})$ everywhere in Theorem 4 without losing the validity of the theorem. The proof is entirely similar, and we shall not reproduce the details of the argument (cf. [5]).

The results which have just been obtained may be interpreted as follows: A complete field may be viewed as a sort of universal domain for those function fields over it which admit a rational place.

The admissible subfields behave as the natural fields of definition for varieties defined over k. However, the restriction to admissible subfields was partly done for convenience. For instance, if a function field K over the rational numbers Q has a rational place, it is clear from our theorems that there exists an isomorphism of K/Q into each p-adic field, and into the real numbers.

4. Additional remarks. It is perhaps worth while to note the analogy of Theorem 3 with the isomorphism theorem for real function fields K over real closed fields P, as proved in [6]. In the latter case, the internal criterion of reality suffices to yield the existence of a real, zero-dimensional place, and to prove that K can be mapped isomorphically into every real closed field of sufficiently high transcendence degree over P. In the case of complete fields, no such internal criterion seems to be available: One has to assume a priori the existence of at least one rational place in order to get enough of them. It would of course be of great interest to find such an internal criterion, if it exists.

Addendum.* We take this opportunity to give another proof of the theorem for real function fields, following as closely as possible the proof of Theorem 3.

THEOREM 5. *Let K be a real function field over a real closed constant field* P. *Let $z_j \varepsilon K$, $z_j \neq 0$ $(j = 1, \cdots, m)$ be a finite set of functions of K. Then there exists a rational place $\mathfrak{p}$ of $K/$P such that each $z_j(\mathfrak{p})$ is finite and $\neq 0$.*

Proof. By essentially the same inductive step used in Theorem 3, and in the corresponding theorem of [6], we can assume that $K/$P has transcendence degree 1. It suffices then to prove that for a suitable model of K, we can find infinitely many points on the curve with coordinates in P. We proceed to do this. We suppose that K is ordered, and distinguish two cases.

Suppose first the ordering of $K/$P is non archimedean. Then the canonical place $\mathfrak{p}$ of $K/$P is rational over P. Let $\mathrm{P}(x, y)$ be a non-singular model of K at the center of $\mathfrak{p}$. Then $f(\bar{x}, \bar{y}) = 0$ but $f_y(\bar{x}, \bar{y}) \neq 0$ and hence for small δ in P $f(\bar{x}, \bar{y} + \delta) = \delta f_y(\bar{x}, \bar{y}) + \alpha$ where α is much smaller than δ. This means that $f(\bar{x}, Y)$ changes sign at $\bar{y} + \delta$ and $\bar{y} - \delta$. For all $\xi \varepsilon$ P sufficiently close to $\bar{x}$, $f(\xi, Y)$ will likewise change sign at $\bar{y} + \delta$ and $\bar{y} - \delta$. It will therefore have a zero in P, and we get in this way infinitely many zeros of $f(X, Y)$, as desired.

If the ordering of $K/$P is archimedean, we take any model $K = \mathrm{P}(x, y)$.

---

* Received April 9, 1953.

Then $f_y(x, y) \neq 0$ because (x, y) is generic. The same procedure as above can now be applied to get the rational points in this case. This proves our theorem.

The isomorphism theorem can also be obtained in a similar way:

THEOREM 6. *Let K/P be a real function field of transcendence degree r. Let Γ be a real closed field containing P, and having transcendence degree $\geq r$ over P. Then there is an isomorphism of K into Γ which is identity on P.*

Proof. Let $K = P(x_1, \cdots, x_r, y)$ where (x, y) satisfies over P an irreducible equation $f(x, y) = 0$, $f_y(x, y) \neq 0$. Using Theorem 5 we can find a rational place $\mathfrak{p}$ of K/P such that the point $(\bar{x}, \bar{y})$ at the center of $\mathfrak{p}$ is non-singular: $f(\bar{x}, \bar{y}) = 0$ but $f_y(\bar{x}, \bar{y}) \neq 0$. This means that $f(\bar{x}, Y)$ changes sign in P around $\bar{y}$. For a point (ξ) with coordinates in Γ sufficiently close to those of $(\bar{x})$, the polynomial $f(\xi, Y)$ will change sign in Γ. We may clearly select (ξ) algebraically independent over P. If $f(\xi, \eta) = 0$ for $\eta \, \varepsilon \, \Gamma$, we see that $P(\xi, \eta)$ is isomorphic to $P(x, y)$.

The proofs of Theorems 3 and 5 are essentially alike, differing only in that the existence of rational points on a curve in the neighborhood of a given non-singular point obtained in one case by the Newton method, is obtained in the case of real closed fields by using the intermediate value theorem. This procedure avoids Sturm's Theorem, and uses instead a more naive property of real closed fields. It also has the advantage of unifying the local techniques used in both cases, so that the ordinary real numbers appear as a common field for both cases. It must be remarked finally that although the full force of the local uniformization theorem was used in Theorem 3, only the special case of curves was needed to handle the real fields. This means that the proofs of Theorems 5 and 6 remain elementary.

An interesting possibility for generalization lies in the following direction: How far do the results concerning specializations remain valid if one replaces the local fields by the quotient fields of complete local domains? The trouble arising here is partly due to the fact that the topology on the local domains has no natural extension to its quotient field, and consequently makes the handling of specializations over such fields more difficult. In particular, we know of no analogue to the Newton approximation method.

One might for instance make the following conjecture: Let $\mathfrak{o}_1$ be the ring of convergent power series in several variables over the complex numbers, and let $\mathfrak{o}$ be the ring of formal power series. Let k_1 and k be the respective quotient fields. Let $(x) = (x_1, \cdots, x_n)$ be a generic point over k_1, with components in $\mathfrak{o}$. Does there exist a specialization of (x) into $\mathfrak{o}_1$?

It is actually possible to prove the following result. Let $\mathfrak{o}_1 = \mathfrak{k}\{t\}$ be the ring of power series in several variables, over an algebraically closed field $\mathfrak{k}$ whose cardinality is greater than denumerable. Let $\mathfrak{R}$ be an algebraically closed field containing $\mathfrak{k}$, and let $\mathfrak{o} = \mathfrak{R}\{t\}$ be the power series over $\mathfrak{R}$. Then $\mathfrak{o}_1 \subset \mathfrak{o}$ in a natural way. Let F_1 and F be the respective quotient fields.

Let $x_i(t) \, \varepsilon \, \mathfrak{R}\{t\}$ be a finite set of power series, and put $(x) = (x_1, \cdots, x_n)$. View (x) as a generic point over F_1. Then (x) has a specialization $(\bar{x})$ over F_1, with $\bar{x}_i \, \varepsilon \, \mathfrak{o}_1$.

Indeed, let $f_j(X)$ be a basis for the ideal in $F_1[X]$ vanishing on (x). By clearing denominators, we may assume that all coefficients of f_j lie in $\mathfrak{o}_1$.

Write each variable X_i as a power series $X_i(t)$ with indeterminate coefficients: $X_i = \sum_\mu \xi_{i\mu}\mu$ where μ ranges over all monomials $t_1{}^{\mu_1} \cdots t_s{}^{\mu_s}$ in t's. Then we can write $f_j(X) = \sum_\mu f_{j\mu}(\xi)\mu$ where $f_{j\mu}(\xi) \, \varepsilon \, \mathfrak{k}[\xi]$. That all f_j vanish on (x) implies that the denumerable system of equations $f_{j\mu}(\xi) = 0$ has a solution in $\mathfrak{R}$, and hence that every finite subsystem is consistent. By the Hilbert Nullstellensatz in infinite dimensional space [7], it follows that the system also has a solution in $\mathfrak{k}$, and such a solution yields a specialization of (x) into $\mathfrak{o}_1$.

By an analogous argument, we can also prove the following statement: Let $\mathfrak{m}$ be the maximal ideal of $\mathfrak{o}_1$. Let $f(X)$ be a polynomial with coefficients in $\mathfrak{o}_1$. The necessary and sufficient condition that f has a zero in $\mathfrak{o}_1$ is that $f \equiv 0 \pmod{\mathfrak{m}^\nu}$ is solvable in $\mathfrak{o}_1$ for each positive integer ν.

The problem is of course to relax the cardinality assumption, but we see no way of doing this at present.

THE INSTITUTE FOR ADVANCED STUDIES.

REFERENCES.

[1] E. Artin, "*Algebraic Numbers and Algebraic Functions,*" Princeton and New York Universities, 1951.

[2] C. Chevalley, "Introduction to the Theory of Algebraic Functions of One Variable," *Mathematical Surveys* No. VI, American Mathematical Society, New York, 1951.

[3] A. Harnack, "Uber die Vieltheiligkeit der ebene algebraische Curve," *Mathematische Annalen*, vol. 10 (1876), pp. 189-198.

[4] Y. Kawada, "Uber die Riemannsche Fläche algebraischer Funktionen," *Proceedings of the Imperial Academy*, Tokyo, vol. 14 (1938), pp. 160-165.

[5] S. Lang, "On quasi algebraic closure," *Annals of Mathematics*, ser. 2, vol. 55 (1952), pp. 373-390.

[6] ———, "The theory of real places," *Annals of Mathematics*, ser. 2, vol. 57 (1953), pp. 378-391.

[7] ———, "Hilbert's Nullstellensatz in infinite dimensional space," *Proceedings of the American Mathematical Society*, vol. 3 (1952), pp. 407-410.

[8] A. Weil, "Foundations of algebraic geometry," *American Mathematical Society Colloquium Publications*, New York, 1946.

[9] E. Witt, "Zyklische Körper und Algebren der Characteristik p vom Grad p^n," *Journal fur Reine und Angewandte Mathematik*, vol. 176 (1937), pp. 126-140.

[10] O. Zariski, "Local uniformization on algebraic varieties," *Annals of Mathematics*, ser. 2, vol. 41 (1940), pp. 852-896.

Am. J. Math., LXXVI No. 4, 1954

NUMBER OF POINTS OF VARIETIES IN FINITE FIELDS.*

By Serge Lang and André Weil.

---

1. Statement and proof of the main theorem. Let V be a variety in a projective space P^n. If V has dimension r and degree d, we shall indicate this by writing $V = V_{n,d,r}$. A point of V can be represented by homogeneous coordinates $(x) = (x_0, \cdots, x_n)$, and we shall also say that (x) is a point of V. We use k to denote a finite field and q to denote the number of elements in k. Let V be defined over k. A point (x) of V is said to be rational over k, or more briefly to be in k, if its coordinate ratios x_i/x_j $(x_j \neq 0)$ are in k. We intend to give an estimate for the number of points of V which are in k. Denoting this number by N, we prove

THEOREM 1. *There exists a constant $A(n, d, r)$ depending only on n, d, r such that for any variety $V = V_{n,d,r}$ defined over a finite field k we have*

$$| N - q^r | \leq \delta q^{r-\frac{1}{2}} + A(n, d, r) q^{r-1}$$

where $\delta = (d-1)(d-2)$.

If $r = 1$, i.e. if V is a curve, the above diophantine statement is a reformulation of the Riemann hypothesis in function fields [2]. Indeed, let V_1 be a non singular projective curve, defined over k, birationally equivalent to V over k. Let g be the genus of V and V_1. Then we know that the number N_1 of rational points of V_1 satisfies $| 1 + q - N_1 | \leq 2gq^{\frac{1}{2}}$. Each non singular point of V corresponds to exactly one point of V_1. The number of singular points on V is bounded by a constant depending only on d, and each singular point of V corresponds at most to d points on V_1. Hence $| N_1 - N | \leq A$ where A is a constant depending only on d. This shows that $| N - q | \leq 2gq^{\frac{1}{2}} + (A + 1)$. Using the fact that $g \leq \frac{1}{2}(d-1)(d-2)$, we see that Theorem 1 is true for $r = 1$.

The proof of Theorem 1 will now be carried out by induction on the dimension r, and the arguments will be of an elementary nature.

We begin by two lemmas and use throughout the terminology of [1],

---

* Received November 30, 1953.

819

except that a bunch of varieties will be called an algebraic set. We shall say that an algebraic set is defined over k if it is defined by algebraic equations $f_\alpha = 0$ with coefficients in k. If the algebraic set is in projective space, then the f_α can be chosen to be forms, and the algebraic set is identified with the rays of zeros of these forms. For example, a hyperplane H in P^n is defined by an equation $\sum w_i X_i = 0$. If $w_i \, \epsilon \, k$, then H is defined over k, or as we shall say more briefly, H is in k.

Let Z be a positive cycle in P^n, of degree $d > 0$, and dimension r. We can then write $Z = \sum a_i V_i$ as a formal sum of distinct varieties V_i, with integer coefficients $a_i > 0$. We then have $d = \deg Z = \sum a_i \cdot \deg V_i$, and $\dim V_i = r$ for each i. By a point of Z we shall mean a point of any of the varieties V_i. The set of points of Z is an algebraic set denoted by $|Z|$. If Z is rational over k, then $|Z|$ is defined over k. We let N_Z be the number of points of Z in k.

LEMMA 1. *There exists a constant $A_1(n, d, r)$ depending only on n, d, r such that for any positive cycle Z in P^n, of degree d, dimension r, and rational over k, we have $N_Z \leqq A_1 q^r$.*

Proof. By induction on the dimension r. Suppose first $r = 0$. Then $|Z|$ consists of at most d points, and the lemma is trivial. Assume now $r \geqq 1$. If we express Z as a sum of prime rational cycles over k, then there will be at most d such cycles in the sum. Hence it suffices to prove the result for a prime rational cycle. Let $(X_0, \cdots, X_n)$ be the variables of P^n. For some pair of indices, say $(0, 1)$, Z (which we assume to be prime rational over k) intersects properly the hyperplanes H_ξ defined by the equations $X_0 - \xi X_1 = 0$ where ξ ranges over k, and the hyperplane H defined by $X_1 = 0$. Every point of Z in k is contained in one of the cycles $Z \cdot H_\xi$ or $Z \cdot H$, each of which has degree d by Bezout's Theorem ([1], App. I) and is of dimension $r - 1$. According to the induction hypothesis, there exists a constant $B(n, d, r - 1)$ such that $N_{Z \cdot H_\xi}$ and $N_{Z \cdot H} \leqq B(n, d, r - 1) q^{r-1}$. As we have exactly $q + 1$ cycles, we see that there are at most $B(q + 1) q^{r-1}$ points in Z, and this is certainly $\leqq 2Bq^r$. The constant $2B$ is what we are looking for.

Our second lemma will be concerned with the following situation. Let V be a variety in P^n, of dimension $r \geqq 2$, and not contained in a hyperplane. Let P'^n be the projective space dual to P^n, and let $(w) = (w_0, \cdots, w_n)$ be the homogeneous coordinates of a point in P'^n. Let H_w be the hyperplane defined by $\sum w_i X_i = 0$. Then the set R of points (w) of P' such that the cycle $V \cdot H_w$ is not a variety is an algebraic set, defined over every field of definition of V, and $R \neq P'$. (Cf. [4], § 1, 6). The hyperplanes H such that $V \cdot H$ is not a variety may therefore be viewed as forming an algebraic set R in P'. Lemma 2 gives an estimate for the number of such hyperplanes defined over a finite field k.

LEMMA 2. *There exists a constant $A_2(n, d, r)$ depending only on n, d, r having the following property. If $V = V_{n,d,r}$ is any variety defined over k, not contained in a hyperplane, and if N_R is the number of hyperplanes H in k such that $V \cdot H$ is not a variety, then $N_R \leqq A_2 q^{n-1}$.*

Proof. Let Z be a positive cycle of dimension r and degree d in P^n. Let $F(U^{(0)}, \cdots, U^{(r)})$ be the associated form of Z. It is of degree d in each set of variables $U^{(0)}, \cdots, U^{(r)}$, and its coefficients are called the Chow coordinates of Z. These are viewed as the homogeneous coordinates of a point in a projective space P^M. The coordinates $(c) = (c_0, \cdots, c_M)$ of a cycle which is not a variety form an algebraic set C in P^M (cf. [4], § 1, 6), and C is defined over the prime field. We let $\phi_\alpha = 0$ be a finite system of equations with coefficients in the prime field, defining C. The forms ϕ_α depend only on n, r, and d.

Let now $V = V_{n,d,r}$ be a variety defined over k, not contained in a hyperplane. Let (w) be a point of P', and let $c(w) = (c_0(w), \cdots, c_M(w))$ be the Chow coordinates of the cycle $V \cdot H_w$. If $F(U^{(0)}, \cdots, U^{(r)})$ is the associated form of V, then it can be verified that the associated form of $V \cdot H_w$ is $F(w, U^{(0)}, \cdots, U^{(r-1)})$. Hence each $c_i(w)$ is a form of degree d in the quantities (w), with coefficients in k.

Let R be the algebraic set of points (w) in P' such that $V \cdot H_w$ is not a variety. This algebraic set R is defined over k, and we have $(w) \, \epsilon \, R$ if and only if $c(w) \, \epsilon \, C$, i. e. if and only if $\phi_\alpha(c(w)) = \phi_\alpha(c_0(w), \cdots, c_M(w)) = 0$ for all α. If $(W) = (W_0, \cdots, W_n)$ are the variables of P', at least one of the polynomials $\phi_\alpha(c(W))$ does not vanish identically. (If they all did, R would be the entire space P', which is not true.) The degree of this $\phi_\alpha(c(W))$ depends only on the degree of ϕ_α and on the degree d of each $c_i(W)$. Since the ϕ_α depend only on n, d, and r we have proved that R is contained in a hypersurface $\phi_\alpha(c(W)) = 0$ with coefficients in k, whose degree is a constant e depending only on n, d, r. This hypersurface defines a positive cycle Z in P^n, of degree e and dimension $n - 1$, rational over k. By Lemma 1, we have $N_R \leqq N_Z \leqq A_1(n, e, n-1) q^{n-1}$ where A_1 is the constant of Lemma 1. Putting $A_2(n, d, r) = A_1(n, e, n-1)$, we see that Lemma 2 is proved.

Remark. All the constants depending on n, d, r which we are finding in this paper can easily be estimated, except for e in the proof of the preceding lemma.

We are now in a position to prove the main theorem.

Let $V = V_{n,d,r}$ be defined over k. We may assume that V is not contained in a hyperplane. Indeed, if V is contained in a hyperplane, or in a linear variety, then it is easily verified that the smallest linear variety L_0 containing V is also defined over k. (The linear forms defining L_0 are precisely the linear forms contained in the prime homogeneous ideal in $k[X]$ defining V.) L_0 can then be viewed as a projective space of dimension $n_0 \leqq n$, and it would suffice to prove our theorem for V as a subvariety of L_0.

If V is not contained in a hyperplane, then the cycle $V \cdot H$ is defined for every hyperplane H, and Lemma 2 will be applicable.

We consider the pairs $((x), H)$ consisting of a point (x) of V in k and a hyperplane H in P'^n defined by an equation $\sum w_i X_i = 0$ with $w_i \epsilon k$ and passing through (x), i. e. such that $H(x) = \sum w_i x_i = 0$. We shall count these pairs in two ways.

We shall denote by κ_{n+1} the number of points in k in the projective space P^n. This is given by $\kappa_{n+1} = q^{n+1} - 1/q - 1 = 1 + q + \cdots + q^n$. This is also the number of hyperplanes H of P' in k. Similarly, the number of hyperplanes in k passing through a given point (x) in k is κ_n. From these remarks, it follows that the number of pairs $((x), H)$ is given by the following two equal numbers:

$$N \kappa_n = \sum_H N_{V \cdot H}$$

where N is the number of points of V in k, $N_{V \cdot H}$ is the number of points of the cycle $V \cdot H$ in k, and the sum is taken over all hyperplanes H in k. Solving for N, and using the same notation as in Lemma 2, we see that N is equal to

$$(1) \qquad \frac{1}{\kappa_n} \sum_{H \notin R} N_{V \cdot H} + \frac{1}{\kappa_n} \sum_{H \epsilon R} N_{V \cdot H}$$

where R is the algebraic set of hyperplanes H such that $V \cdot H$ is not a variety.

If N_R is the number of terms in the second sum, i. e. the number of hyperplanes H in R, the number of terms in the first sum is $\kappa_{n+1} - N_R$. We shall now estimate the two sums, and begin with the first one.

For $H \notin R$, $V \cdot H$ is a variety of degree d, dimension $r - 1$, defined over k. It follows from the induction hypothesis that for $H \notin R$, we have $| N_{V \cdot H} - q^{r-1} | \leqq \delta q^{r-3/2} + A_3 q^{r-2}$ where $A_3 = A(n, d, r - 1)$ is the constant determined inductively, depending only on n, d, and $r - 1$. Summing over $H \notin R$, and dividing by κ_n we get

$$\left| \frac{1}{\kappa_n} \sum_{H \notin R} N_{V \cdot H} - Q \right| \leqq \frac{\kappa_{n+1} - N_R}{\kappa_n} (\delta q^{r-3/2} + A_3 q^{r-2})$$

where $Q = \kappa_n^{-1}(\kappa_n - N_R)q^{r-1}$. We note that $\kappa_{n+1} = q\kappa_n + 1$, and $q^{n-1} \leqq \kappa_n$. Also we know by Lemma 2 that $N_R \leqq A_2 q^{n-1} \leqq A_2 \kappa_n$. This shows that $|Q - q^r| \leqq (1 + A_2)q^{r-1}$ and that $\kappa_n^{-1}(\kappa_{n+1} - N_R) \leqq \kappa_n^{-1}\kappa_{n+1} \leqq q + 1$. From this we conclude immediately that

$$(2) \qquad \left| \frac{1}{\kappa_n} \sum_{H \notin R} N_{V \cdot H} - q^r \right| \leqq \delta q^{r-\frac{1}{2}} + A_4 q^{r-1}$$

with a suitable A_4.

We now turn to our second sum, $\sum_{H \in R} N_{V \cdot H}$, and we shall prove that it can be absorbed by the error term involving q^{r-1} only.

If $H \in R$, then $V \cdot H$ is a cycle of degree d, dimension $r - 1$, and rational over k. By Lemma 1, it follows that $N_{V \cdot H} \leqq A_1 q^{r-1}$ for a suitable constant A_1. By Lemma 2, we know that $N_R \leqq A_2 q^{n-1}$. Hence

$$(3) \qquad \frac{1}{\kappa_n} \sum_{H \in R} N_{V \cdot H} \leqq \frac{q^{n-1}}{\kappa_n} A_1 A_2 q^{r-1} \leqq A_5 q^{r-1}$$

because $q^{n-1} \leqq \kappa_n$, as we have already remarked.

Combining (1), (2), and (3) we see that there exists a constant A_6 (the desired constant $A(n, d, r)$) such that $|N - q^r| \leqq \delta q^{r-\frac{1}{2}} + A_6 q^{r-1}$. This concludes the proof of the theorem.

2. Corollaries and applications.

We list some of the immediate corollaries to Theorem 1.

In the first place, our theorem is of the nature of an asymptotic result. If k_ν is the extension of degree ν over k, and $N^{(\nu)}$ is the number of points of V in k_ν then

$$N^{(\nu)} = q^{\nu r} + O(q^{\nu(r-\frac{1}{2})}) \quad \text{for} \quad \nu \to \infty.$$

In particular, if $\nu \to \infty$ then $N^{(\nu)} \to \infty$ also. We shall now see that this asymptotic behavior can be stated more generally for abstract varieties.

Let $V = V_{n,d,r}$ be a variety defined over k, and let F be a frontier on V (i. e. an algebraic set properly contained in V) also defined over k. Let $N^{(\nu)}_{V-F} = N^{(\nu)}_V - N^{(\nu)}_F$ be the number of points of $V - F$ which are in k_ν. Then from Lemma 1 we see immediately that $|N^{(\nu)}_{V-F} - N^{(\nu)}_V| \leqq Bq^{\nu(r-1)}$ where B is a constant depending on V and on F.

If V and V' are two varieties of dimension r in P^n, defined over k, and if T is a birational correspondence between them, also defined over k, then there exist frontiers F and F' defined over k such that T is everywhere biregular in $V - F$ and $V' - F'$. T gives a $1 - 1$ correspondence between the points of $V - F$ and those of $V' - F'$, so $N^{(\nu)}_{V-F} = N^{(\nu)}_{V'-F'}$.

Let now $V = (V_\alpha, F_\alpha, T_{\beta\alpha})$ be an abstract variety, represented by varieties V_α in projective space, with frontiers F_α, and birational correspondences $T_{\beta\alpha}$, all of which are defined over k. If P is a point of V, then P has representatives P_α in some of the V_α. If one of these P_α is rational over k, then so are all the other representatives, and hence in this case we may say that P is rational over k.

If K is a function field of dimension r over k, then we shall say that an abstract variety V is a model of K if V is defined over k, and if $K = k(P)$ where P is a generic point of V over k. The following two corollaries of Theorem 1 are immediate consequences of the preceding discussion.

COROLLARY 1. *Let K be a function field of dimension r over a finite constant field k. If V is any abstract model of K, and $N^{(\nu)}$ the number of points of V in k_ν, then $N^{(\nu)} - q^{\nu r}$, mod $O(q^{\nu(r-1)})$, is a birational invariant (i. e. does not depend on the choice of the model V of K).*

COROLLARY 2. *Let K/k be as in Corollary 1. There exists a constant γ such that for any abstract model V of K, we have*

$$| N^{(\nu)} - q^{\nu r} | \leqq \gamma q^{\nu(r-\frac{1}{2})} + B q^{\nu(r-1)}$$

where B is a constant depending on V. If K has a projective model of degree d, then we can take $\gamma \leqq (d-1)(d-2)$.

The smallest constant γ that can be selected in Corollary 2 is obviously a birational invariant, and we shall return to discuss it below. For the moment, we note that an abstract model of the function field K can be projective or affine, and can also be chosen without singularities, because the singular points on a variety defined over k are a frontier F, properly contained in V and defined over k. Hence our corollaries show that as ν increases indefinitely, the number of simple points on V in k_ν also increases indefinitely.

COROLLARY 3. *Let V be an abstract variety defined over k. Let m be a given integer. There exists a zero dimensional positive cycle on V, rational over k, and of degree prime to m. Hence there exist cycles on V which are zero dimensional, rational over k, and of degree 1.*

Proof. From the above remark and the corollaries, there exist non singular points of V in the field k_ν when ν is sufficiently large and prime to m.

The preceding result generalizes the well known theorem that a function field in one variable over a finite constant field always has a divisor of degree 1. In this vein, we can state the following invariant result:

COROLLARY 4. *Let K be a function field over a finite constant field k. Let $x_1, \cdots, x_n$ be a finite set of non zero elements of K. For all ν sufficiently large, there exists a place ϕ of K which is k_ν-valued, and such that $\phi(x_i) \neq 0$, $\neq \infty$ for any i.*

Proof. After enlarging our set (x) by a finite number of elements if necessary, we may assume that $k(x) = K$, and that for each i, x_i^{-1} appears in the set (x). Under these assumptions it suffices to prove that there exists a place ϕ of K which is k_ν-valued and such that $\phi(x_i)$ is finite for all x_i. We view (x) as the generic point of an affine variety V defined over k. By Corollary 1, for all sufficiently large ν, there exists a point (x') of V in k_ν which is simple on V. If $\mathfrak{o}$ is the specialization ring of (x') in K and $\mathfrak{p}$ the maximal prime ideal, then $\mathfrak{o}/\mathfrak{p}$ is isomorphic to the subfield $k' = k(x')$ of k_ν. (The isomorphism is induced by the map $f(x) \to f(x')$, $f \epsilon k[X]$). Since k is perfect, the completion of $\mathfrak{o}$ is isomorphic to a power series ring $k'\{t\} = k'\{t_1, \cdots, t_r\}$ and K can be identified with a subfield of the quotient field Ω of $k'\{t\}$. It is clear that there exists a k'-valued place ϕ of Ω mapping each t_j on 0. (For instance, view Ω as a subfield of the repeated power series field $\Gamma = k'\{t_1\}\{t_2\} \cdots \{t_r\}$. By mapping successively each t_j $(j = r, \cdots, 1)$ on 0, we get a k'-valued place ϕ of Γ.) Since the quantities x_i are in $\mathfrak{o}$ for all i, the place ϕ is finite on the x_i. The restriction of ϕ to K gives the desired place.

We now turn to the applications of Corollaries 1 and 2 to the zeta function. Let V be an abstract variety of dimension r, defined over k. We associate with V the analytic function $Z(U)$ defined by

$$d \log Z(U)/dU = \sum_{\nu=1}^{\infty} N^{(\nu)} U^{\nu-1}.$$

If V is complete and without multiple points, then $Z(U)$ is the zeta function of V [5] and Weil's conjectures state among other things that $Z(U)$ is a rational function, satisfying a functional equation of the usual type. If V has singular points, or is not complete, then the function $Z(U)$ is probably still a rational function, but it need not satisfy a functional equation.

We shall not at this point go into a detailed analysis of the relation between the functions $Z(U)$ associated with arbitrary varieties, and those associated with complete varieties without multiple points, because the results which we shall now prove as a consequence of Corollary 2 will be applicable to any variety.

COROLLARY 5. *Let V be an abstract variety of dimension r, defined over k. Then the associated function $Z(U)$ has no pole or zero in the circle*

$|U| < q^{-r}$, and has exactly one pole of order 1 in the circle $|U| < q^{-(r-\frac{1}{2})}$, namely at $U = q^{-r}$.

Proof. According to Corollary 2, we can write $N^{(\nu)} = q^{\nu r} + A_\nu q^{\nu(r-\frac{1}{2})}$, where A_ν is a constant, bounded in absolute value by a fixed constant A. This gives

$$d \log Z(U)/dU = q^r \sum_{\nu=0}^{\infty} (q^r U)^\nu + \sum_{\nu=1}^{\infty} A_\nu (q^{r-\frac{1}{2}} U)^\nu.$$

The first of these series is the geometric series and defines the function $1/(1 - q^r U)$. Since $-q^r/(1 - q^r U) = d \log (1 - q^r U)/dU$ we see that

$$d \log [Z(U)(1 - q^r U)]/dU = \sum_{\nu=1}^{\infty} A_\nu (q^{r-\frac{1}{2}} U)^\nu.$$

Hence $Z(U)(1 - q^r U)$ has no pole or zero in the circle $|U| < q^{-(r-\frac{1}{2})}$ because the latter series converges in that circle. Furthermore q^{-r} is a pole, and is the only pole inside the circle, as was to be shown.

COROLLARY 6. *Let V and V' be two varieties which are birationally equivalent, and let $Z(U)$ and $Z'(U)$ be the associated functions. Then the function $Z(U)/Z'(U)$ has no zero or pole in the circle $|U| < q^{-(r-1)}$. Hence the zeros and poles of $Z(U)$ in this circle are birational invariants.*

Proof. It is clear from Corollaries 1 and 2 that the series for

$$d \log Z(U)/dU - d \log Z'(U)/dU = d \log (Z(U)/Z'(U))/dU$$

converges in the circle $|U| < q^{-(r-1)}$, and hence that $Z(U)/Z'(U)$ has no zero or pole in the circle.

Concerning the behavior of $Z(U)$ for $|U| \geqq q^{-(r-\frac{1}{2})}$ we can only make the following conjectural statements, which complement the conjectures of Weil [5].

Let P be the Picard variety of V. P is an abelian variety, whose dimension g is the irregularity of V. Let ι be the endomorphism induced on P by the automorphism $(x \rightarrow x^q)$ of the universal domain, and let

$$F(U) = \prod_{i=1}^{2g} (1 - \alpha_i U)$$

be the characteristic polynomial of ι as defined in [3], § 67. Then the series for

$$d \log Z(U)/dU + d \log (1 - q^r U)/dU - d \log F(U)/dU$$

converges in the circle $|U| < q^{-(r-1)}$, and the function $Z(U)(1 - q^r U)/F(U)$ has no zero and no pole inside this circle, and has at least one pole on the circle $|U| = q^{-(r-1)}$, provided of course that V is complete and without multiple points.

Furthermore, if K/k is a function field as in Corollary 2, and if γ is the smallest constant for which Corollary 2 holds, then $\gamma = 2g$. (This gives the explicit determination of γ as a birational invariant.)

If V is a curve, then $P = J$ is the Jacobian variety, and g is the genus of the curve. For this case, all the preceding statements are well known and are contained in [2], except for the last, i. e. that $\gamma = 2g$ is actually the best possible constant for the inequality $|N^{(\nu)} - q^\nu| \leqq \gamma q^{\nu/2} + B$ of Corollary 2. As to this, it obviously suffices to prove it in the case that V is complete and without multiple points, and then it amounts to proving that $\gamma = 2g$ is the best possible constant for the inequality $|1 + q^\nu - N^{(\nu)}| \leqq \gamma q^{\nu/2}$. Actually, this last statement is implicit in [2], as the following argument shows.

We have ([2], § 22, p. 71) $|1 + q^\nu - N^{(\nu)}| = \sigma(\iota^\nu) = \sum_{i=1}^{2g} \alpha_i^\nu$ and $|\alpha_i| = q^{\frac{1}{2}}$, so that we can write $\alpha_i = q^{\frac{1}{2}} \exp(2\pi i \theta_i)$ where θ_i is a real number between 0 and 1. Hence we can write

$$\sigma(\iota^\nu) = q^{\frac{1}{2}} \sum_{i=1}^{2g} \exp(2\pi i \nu \theta_i).$$

For infinitely many ν, $\sum_{i=1}^{2g} \exp(2\pi i \nu \theta_i)$ will be as close to $2g$ as we please, and hence $2g$ is indeed the best possible constant.

UNIVERSITY OF CHICAGO.

REFERENCES.

[1] A. Weil, "Foundations of algebraic geometry," American Mathematical Society Colloquium Publications, Vol. XXIX, New York, 1946.

[2] ———, "Sur les courbes algébriques et les variétés qui s'en déduisent," Hermann et Cie., Paris, 1948.

[3] ———, "Variétés abéliennes et courbes algébriques," Hermann et Cie., Paris, 1948.

[4] ———, "Critères d'Equivalence," to appear in *Mathematische Annalen*.

[5] ———, "Number of solutions of equations in finite fields," *Bulletin of the American Mathematical Society*, vol. 55 (1949), no. 5, pp. 497-508.

Reprinted from the Proceedings of the National Academy of Sciences,
Vol. 41, No. 3, pp. 174–176.   March, 1955

ABELIAN VARIETIES OVER FINITE FIELDS

By Serge Lang

DEPARTMENT OF MATHEMATICS, UNIVERSITY OF CHICAGO

Communicated by S. Mac Lane, December 17, 1954

We shall generalize to Abelian varieties the well-known fact that an elliptic curve over a finite field always has a rational point (see Theorem 3).   Our first theorem is purely algebraic and generalizes parts of Chow's Theorem 3.[1]   Its proof uses techniques of Matsusaka,[2] to whom I am indebted for valuable discussions concerning this matter.   Given a separable extension k_1/k and a function field K_1/k_1, we give a sufficient condition for the existence of a function field K/k which lifts to K_1 over k_1.

THEOREM 1.   *Let k be a field, k_1/k a finite separable extension, and k^*/k any extension linearly disjoint from k_1 over k.   Let K_1/k_1 and K^*/k^* be two function fields such that $K_1 \subset K^*k_1$.   (We assume, of course, that K_1 is free from k^* and K^* is free from k_1 over k.)   If $K_1K_1{}^\sigma = K_1k_1{}^\sigma$ for every isomorphism σ of K^*k_1 over K^*, then there exists a unique function field K/k such that $K \subset K_1$, $K \subset K^*$, and $K_1 = Kk_1$.*

A geometric formulation of Theorem 1 will be given below.   For applications to finite constant fields, we shall not need Theorem 1.   However, the construction used in its proof will be needed to prove the following essential result.

THEOREM 2.   *Let K/k be a function field, k_1/k a finite separable extension.   If Kk_1 is an Abelian function field over k_1, then there exists a model V of K over k which becomes biregularly equivalent to an Abelian variety over k_1.*

We now prove Theorem 1, beginning with the uniqueness.   If K, K' are two function fields over k satisfying the required conditions, then $KK' \subset K_1$, $KK' \subset K^*$; hence KK' is regular over k.   If $k_1 = k(\alpha)$, then α has the same degree over KK' as over k, and hence $KK' = K$.

To prove the existence of K, we shall use the geometric language.   Our hypotheses can then be stated as follows.   We can write $K_1 = k_1(x_1)$ and $K^* = k^*(x^*)$, where x_1 is the generic point of a variety V_1/k_1 and x^* is the generic point of a variety V^*/k^*.   There is a rational map $T_1\colon\ V^* \to V_1$ defined over k_1k^* such that $T_1(x^*) = x_1$.   Without loss of generality, we can assume that V_1 and V^* are in the same projective space, that k_1 is the smallest field of definition for V_1 containing k, and k_1k^* the smallest for T_1 containing k^*.   Then the condition on the isomorphism can be expressed as follows: If k_i, V_i, T_i $(i = 1, \ldots, m)$ are the conjugates of k_1, V_1, T_1, respectively, over k, k, k^*, there exist birational correspondences $R_i\colon$ $V_1 \to V_i$ defined over k_1k_i such that commutativity holds in the following diagram:

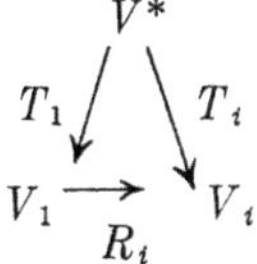

If (x_i) are the conjugates of (x_1) over $k^*(x^*)$, then the 0-cycle $\sum (x_i)$ is rational over $k^*(x^*)$.   Let ξ be the Chow point of this cycle.   The following statements will prove that ξ has a locus W over k and that $K = k(\xi)$ satisfies our requirements.

174

1. $k(\xi) \subset k^*(x^*)$.   This follows immediately from what has been said above.

2. $k_1(x_1) \subset k_1(\xi)$.   The cycle $\sum(x_i)$ is rational over $k_1(\xi)$, and (x_1) is separable over $k_1(\xi)$.   Let y be a conjugate of (x_1) over $k_1(\xi)$, and let σ be an isomorphism of the universal domain mapping $x_1 \to y$ over $k_1(\xi)$.   Then σ leaves $\sum(x_i)$ invariant because the cycle is rational over $k_1(\xi)$.   Hence $y = x_j$ for some j.   But the fact that σ leaves k_1 fixed implies $V_1^\sigma = V_1$; hence x_j lies in V_1, and that is the case only if $j = 1$ (because the conjugates of V_1 are distinct and so have distinct generic points).   Having proved that x_1 is separable over $k_1(\xi)$ and that it is fixed under all isomorphisms leaving $k_1(\xi)$ fixed, we conclude that it is rational over $k_1(\xi)$.

3. $k_1(\xi) \subset k_1(x_1)$.   We must prove that $\sum(x_i)$ is rational over $k_1(x_1)$, and for this it suffices to prove that $\sum R_i$ is rational over k_1 because $\sum(x_i) = \sum(R_i(x_1))$.   Hence it suffices to prove that if σ is an automorphism of the smallest normal extension k' of k containing k_1 which leaves k_1 invariant, then σ leaves $\sum R_i$ invariant.   But σ can be extended to an automorphism σ^* of $k'k^*$ leaving k_1k^* fixed, and σ^* leaves $\sum T_i$ invariant because $\sum T_i$ is rational over k^*.   Recalling the commutativity of the above diagram, we see that σ leaves $\sum R_i$ invariant, as desired.

4. $k(\xi)/k$ is a regular extension.   This follows immediately from statements 1 and 3.   This concludes the proof of Theorem 1.

We can now prove Theorem 2.   Starting with $k^* = k$ in Theorem 1 and V^* a model of $K^* = K$, we have constructed another model W of K and a birational map $F: V_1 \to W$ such that $F(x_1) = \xi$.   It is clear that F^{-1} has no fundamental points on W.   Taking the model V_1 of $K_1 = Kk_1$ to be an Abelian variety, we shall use an argument due to Matsusaka[2] to prove that F is everywhere defined on V_1.   Let k' be the smallest normal extension of k containing k_1.   We note that the product $x_1 \times \ldots \times x_m$ has a locus over k'.   By Weil's Theorem 6[3] and the fact that $k'(x_1) = k'(x_1, \ldots, x_m)$, we conclude that our locus is itself an Abelian variety, biregularly equivalent to V_1 over k'.   Hence the map F is everywhere defined on V_1.   If we now take V to be the normalization of W over k, then the induced correspondence between V_1 and V (over k_1) is biregular.   This proves Theorem 2.

COROLLARY.   *Let K/k be a function field.   Let k_s be the separable closure of k. If Kk_s is an Abelian function field over k_s, then there exists a model V of K over k having the following property: If k_1/k is finite separable, and Kk_1 is an Abelian function field over k_1, then V is an Abelian variety over k_1.*

This follows immediately from Theorem 2 and the following lemma.

LEMMA.   *Let V/k be a variety which becomes biregularly equivalent to an Abelian variety over the separable closure k_s of k.   Then V is an Abelian variety over k if and only if V has a rational point in k.*

Proof: If V is an Abelian variety over k, then the unit element is a rational point.   Conversely, let x be a rational point.   Let $f: V \times V \to V$ be a law of composition defined over k_s for which x is the zero element.   For any automorphism σ of k_s/k, f^σ is a law of composition on V and x is again a zero element for f^σ.   Since two laws of composition differ by a translation, we must have $f^\sigma = f$, i.e., f is defined over k.

From now on, we assume that k is a finite field with q elements.

THEOREM 3.   *Let K/k be a function field over the finite field k.   If $K\bar{k}$ is an Abelian function field over $\bar{k}$, then K is an Abelian function field over k.   In particular, K has a rational place.*

Geometrically, the theorem states that if a variety U/k becomes birationally equivalent to an Abelian variety over the algebraic closure of k, then it is birationally equivalent to an Abelian variety over k. (If k is a field with a discrete valuation, and π is a prime element, then the curve defined by the equation $X^n + \pi Y^n + \pi^2 Z^n = 0$, with $n = 3$, shows that the conclusion of the theorem does not remain valid. In fact, if the curve has a rational point in a separable extension k_1 of k, then n divides $[k_1:k]$ because two terms of the equation must have the same absolute value in k_1.)

Proof of Theorem 3:[4]   Let V be the model of K given by Theorem 2. According to the lemma, it suffices to prove that V has a rational point. Let x be a generic point of V/k, and let x^q be obtained from x by raising all co-ordinates of x to the qth power. Then x^q is also a generic point of V/k, and $(x \times x^q)$ has a locus over k which is the graph of a rational map $\varphi\colon V \to V$ (onto). But over $\bar{k}$, V is an Abelian variety, and hence $\varphi = \varphi_0 + c$, where φ_0 is an endomorphism of V and c is a constant. Hence the map $x \to x^q - c$ is an endomorphism of V. The map $x \to x^q - x - c$ is an endomorphism of V, which is easily seen to have finite kernel. It is therefore onto V, and hence $\xi^q - \xi - c = -c$ for some $\xi \in V$. Hence we have $\xi^q = \xi$, and ξ is therefore a rational point.

COROLLARY. *If a variety U/k becomes birationally equivalent to an Abelian variety over any extension of k (algebraic or not), then U is birationally equivalent to an Abelian variety over k.*

Proof: Standard specialization techniques.

[1] W. L. Chow, "Abelian Varieties over Function Fields," *Trans. Am. Math. Soc.* (to appear).

[2] T. Matsusaka, "Some Theorems on Abelian Varieties," *Natural Sci. Rept. Ochanomizu University,* **4,** No. 1, 1953.

[3] A. Weil, *Variétée Abéliennes et courbes algébriques,* (Paris: Hermann et cie, 1948).

[4] I am indebted to A. Weil for a simplification of the proof.

Annals of Mathematics
Vol. 64, No. 2, September, 1956
Printed in U.S.A.

UNRAMIFIED CLASS FIELD THEORY OVER FUNCTION FIELDS IN SEVERAL VARIABLES

By Serge Lang

(Received November 30, 1955)   (To Artin)

Introduction

Class field theory in its classical form is the theory of abelian extensions of fields of absolute dimension 1: number fields and function fields in one variable over a finite constant field. It involves the following central themes:

An existence and characterization theorem for these extensions, which are described by an arithmetic group canonically associated with the ground field. (Takagi)

The reciprocity law of Artin, which gives a canonical isomorphism between this group and the Galois group of the maximal abelian extension by means of the Frobenius automorphism.

An analytic translation of the reciprocity law in terms of the splitting of the zeta function into L-series. This translation is entirely formal once the reciprocity law has been proved.

In addition to these, we have also a cohomological theme, independent of the abelian extensions, and having its nucleus in a fundamental 2-cocycle. This 2-cocycle can be seen only when one deals with the ramified as well as the unramified extensions of the given ground field.

In this paper, we extend the first three to the unramified extensions of function fields in several variables over a finite constant field. Since we deal only with unramified extensions, it is not surprising that we meet no cohomology of any importance. Extensions which may be due to torsion in the group of divisors are left untreated, but one knows by Néron's Theorem that there can only be a finite number of those. For a complete discussion of problems raised by these extensions, see below, §4 and §5.

We begin with the general algebraic theory of coverings (especially the unramified ones) and we have to distinguish two kinds of coverings: constant field extensions, and geometric coverings belonging to a rational map. By using the language of local rings, these two types can be discussed in a unified manner. After describing certain basic concepts in terms of local rings and residue class fields, we then analyse these concepts in the two cases separately. In the geometric case, we translate local ring statements into the geometric language. This expository pattern will be followed throughout. However many proofs are given later by discussing the two cases separately.

This algebraic theory of coverings is essentially contained in the literature, but we have taken the opportunity of assembling here all the results we need. For the proofs we refer to the literature.

We have considered coverings of non-complete varieties and given the local

285

statements whenever possible. This is not really needed in the present paper, but will be useful in the generalizations to the ramified coverings.

In §2 we discuss the Galois theory of local rings, which is due to Krull. We consider Galois extensions of a function field over a finite field k, and the splitting of a prime $\mathfrak{p}$. If L/K is an abelian extension, unramified above $\mathfrak{p}$, then we catch the Frobenius automorphism (Artin symbol) $(\mathfrak{p}, L/K)$ in a manner entirely similar to the classical one.

We then introduce in §3 the group of cycle classes of a variety V/k, as follows. ("Cycle" will always mean cycle of dimension 0 in this paper.)

Let $\alpha: V \to A$ be the rational map of V into its Albanese variety. If $\mathfrak{a}$ is a cycle on V of degree 0, rational over k, and $\mathfrak{a} = \sum n_i(Q_i)$ then we associate with $\mathfrak{a}$ the point $\pi(\mathfrak{a}) = \sum n_i \alpha(Q_i)$ on A. It is a rational point of A in k. The kernel of the mapping π is called the kernel of Albanese, and plays a role similar to the divisors linearly equivalent to 0 on curves. The factor group of cycles modulo the kernel of Albanese is called the group of cycle classes, and is a birational invariant of the function field K of V. We denote it by C_K.

Given an abelian, divisor unramified extension of K which is not of the exceptional type mentioned above, we can define a homomorphism of the cycles of V rational over k onto the Galois group, by means of the Frobenius automorphism, and the kernel of Albanese is contained in the kernel of our homomorphism. Hence we get a homomorphism of C_K onto the Galois group, whose kernel is a trace group defined in an obvious manner.

The precise formulation of the class field theory is given in §5.

We first give proofs in the special case of constant field extensions (§6) and then over Abelian varieties (§7) (in fact over arbitrary commutative group varieties). This case is interesting for its own sake, but it is also used as an auxiliary tool in the general proofs. Let A be an abelian variety, and x a generic point of A/k. If z is a generic point of A/k such that $z^q - z = x$, then $k(z)$ is the generalization of the Hilbert Class Field of $k(x)$.

Consider now for simplicity the case of a curve which we may assume contained in its Jacobian. Then the unramified extensions of J can be specialized over the curve, and by this method we obtain class fields of the curve for which the reciprocity law is verifiable directly. (For ramified extensions, one can probably use generalized Jacobians in a similar capacity.)

For arbitrary varieties, the same idea works, provided the specialization of a covering of the Albanese variety A over the image of V in A is then followed up by a lifting process, described in §4. In this manner, we construct Hilbert Class Fields over arbitrary function fields. We use the theory of the Picard and Albanese variety developed by Matsusaka and Chow, and the precise results of this theory needed for our applications are stated in §4.

Having proved the reciprocity law in the field constructed above, we prove in §9 that every divisor-unramified abelian extension (not due to exceptional types) is contained in the composite of a constant field extension with an extension obtained by a specialization-lifting from the Albanese variety. After this, it is a routine matter to conclude the proofs in §10.

In the course of our work, we need from time to time a lemma on the existence of primes in generalized arithmetic progressions. These lemmas are collected together in an Appendix. We have here proved the bare minimum to carry us through the class field theory. We shall return elsewhere to this question, in order to obtain the existence of primes having given Frobenius substitutions (in any Galois extension of the given function field K).

Finally we have added a short paragraph on the nth power residue symbol. If an abelian unramified extension of suitable type happens to be also Kummer, then there is a duality between a subgroup of the rational points on the Picard variety and a factor group of the rational points on the Albanese variety. This duality is expressed by means of the nth power residue symbol, is well known in dimension 1, and gives nothing surprisingly new in the higher dimensional case. The classical theory is a little clarified because we have separated the phenomena due to divisor classes and cycle classes.

The possibility of defining L-series (abelian and non-abelian) is then obvious, since they are always defined formally in the classical theory. Over an Abelian variety we prove that the L-series are trivial (end of §6). For arbitrary (non-singular, complete) varieties, the conjecture of Artin can also be formulated in our general context. Given the Weil zeta function $Z_V(t)$ associated with V/k, we know (conjecturally) that it is of the form

$$\frac{F_1(t) \cdots F_{2r-1}(t)}{(1 - t)F_2 \cdots F_{2r-2}(1 - q^r t)}$$

where $r = \dim V$. The L-series will then look like a zeta function, except that the terms $(1 - t)$ and $(1 - q^r t)$ will be absent.

If one follows the formal arguments given by Artin in his second paper on L-series [6], then one sees that our reciprocity law again gives the equality between two L-series in the unramified abelian case under consideration. (If there is divisor torsion on the variety, we are led to some conjectures in §5.) Namely one of them is defined by a character of the Galois group, and the other by a character of the cycle classes. The classical arguments show that the zeta function of the covering is the product of the standard L-series of the base variety. These formal arguments go over word for word from Artin's papers, and hence there is no need to repeat them here.

It should be remarked that the formal statements concerning zeta functions and L-series are complemented (in the one dimensional case) by non-formal statements concerning the existence and density of primes in generalized arithmetic progressions. These and related questions will be considered in a subsequent paper.

We note finally that our theory contains a mixture of birational properties and biregular properties: It gives the decomposition laws of regular local rings in extension fields. However, we never need a non-singular model of the ground field. At the very end, if we want to have a non-singular unramified covering of a non-singular model of the ground field then of course we have to make the extra hypothesis that we can resolve singularities. Still, the fact that we can carry out the work with an arbitrary normal (projective) model is significant for the following reason: If we allow ramifications in the extension, then it is not usually true that the normalization of a non-singular variety in a finite separable extension of its function field is also non-singular. From this point of view (i.e. the class field theory in ramified extensions) we are therefore led to problems of algebraic geometry, in any characteristic, 0 or p, and over any ground field, namely the search of finer equivalence relations in the group of cycles of dimension 0 and in the group of divisors. For curves, Rosenlicht's generalized Jacobians will probably prove adequate.

I wish to conclude this introduction by expressing my great and sincere appreciation to Chow, Matsusaka, and Weil for the numerous and profitable discussions which we have had on the Picard and Albanese varieties. In particular, without the work of Chow and Matsusaka on the algebraic aspects of these varieties, this paper could not have existed. We have used especially the following important results of Chow:

1. The general theory of algebraic systems of abelian varieties (the k-trace and k-image) [10].

2. The fact that when a variety (projective, normal) V is defined over a perfect field k, then its Albanese variety is also defined over k, and V and A have the same Picard variety, under a natural isomorphism [11].

CONTENTS

§1. Coverings

The general algebraic theory of the splitting of local rings in extension fields has been systematically studied by Abhyankar [1, 2, 3, 4], Chevalley [9], Krull [13], [14] Nagata [18] and Zariski [27]. Our purpose here is to summarize briefly the facts needed in the arithmetic theory to be developed below. We shall be mostly concerned with unramified splitting.

Let V be a variety, abstract and normal, defined over a perfect field k. We follow throughout the terminology of Weil's Foundations, except when otherwise specified.

A cycle of codimension 1 will be called a divisor. The word cycle will be used only to denote cycles of dimension 0. We recall that by definition, the components of a cycle are assumed to be simple on their ambient variety.

If x is a generic point of V/k, then $k(x)$ will be called a function field of V/k. It is naturally isomorphic with the field of functions (in the sense of F-VIII) on V, rational over k.

Let K/k be a function field for V. By the K/k-skeleton (or briefly the k-skeleton) $\mathfrak{B}_k$ of V, we shall mean the set of all local rings in the function field K belonging to points of V which are algebraic over k. Each local ring $\mathfrak{o}$ has a maximal ideal $\mathfrak{p}$. Following the classical terminology of arithmetic, we shall refer to $\mathfrak{p}$ as a *prime*. Each local ring being associated with exactly one prime we can identify formally the elements of $\mathfrak{B}_k$ with primes. The regular local rings of $\mathfrak{B}_k$ are in $1 - 1$ correspondence with the prime rational cycles of V over k. Again such a cycle will be called a prime rational cycle, or simply a prime cycle, and denoted by the same letter $\mathfrak{p}$. The context will always make our meaning clear.

We emphasize that if k' is a finite extension of k and not equal to k, then $\mathfrak{B}_{k'}$ and $\mathfrak{B}_k$ are distinct, even though arising from the same variety V.

Let L/K be a finite separable algebraic extension of degree $n = [L:K]$. Then each local ring $\mathfrak{o}$ in $\mathfrak{B}_k$ splits in L into local rings $\mathfrak{O}_i$ with maximal ideals $\mathfrak{P}_i$. Namely, if R is the integral closure of $\mathfrak{o}$ in L, then R has only a finite number of maximal ideals, and if M_i denote these maximal ideals then $\mathfrak{O}_i$ are the local rings of R with respect to M_i. The residue class field $\mathfrak{o}/\mathfrak{p}$ is canonically imbedded in $\mathfrak{O}_i/\mathfrak{P}_i$ and the residue class field degree $[\mathfrak{O}_i/\mathfrak{P}_i : \mathfrak{o}/\mathfrak{p}]$ is denoted by m_i. The set of all local rings of L derived from the splitting of a local ring in $\mathfrak{B}_k$ will be denoted by $\mathfrak{U}$.

The finite extension L/K can be obtained by a constant field extension, followed by an extension preserving the constant field. The latter type will be called a *geometric extension*. If k' is the algebraic closure of k in L, then L/Kk' is geometric, and will be called the *geometric part of* L/K.

Suppose k'/k is finite algebraic, and $L = Kk'$ is a constant field extension. Then the set $\mathfrak{U}$ above is obviously the k'-skeleton of the same variety V.

Suppose on the other hand that L/K is geometric, and let U be the model of L derived from V by normalization in L. (See F-App. II, pages 274–275 where the normalization process extends without change to finite extensions of the function field. This was first noted by Matsusaka.) The set $\mathfrak{U}$ is exactly the k-skeleton $\mathfrak{U}_k$ of the variety U. We shall say that $\mathfrak{U}/\mathfrak{B}_k$ is a *covering*.

In general, if k' is the algebraic closure of k in L, and $\mathfrak{U} = \mathfrak{U}_{k'}$ is the k'-skeleton of the variety U/k' which is the normalization of V in L/k', then we say again that $\mathfrak{U}/\mathfrak{B}_k$ is a *covering*.

If L/K is geometric, then we have a rational map $f: U \to V$ of degree $n = [L:K] = [U:V]$ associated with it and defined over k. We can use the geometric language in this case, but the need for dealing eventually with constant field extensions prevents us from using this language throughout. We shall say (in the geometric case) that U/V is a *geometric covering*, defined over k, and of degree n. It is more convenient to assume our varieties normal, and the extension L/K separable at the outset, even though more general definitions might have been given.

To summarize, we see that a rational map $f: U \to V$ of a normal variety U onto (generically) a normal variety V is a covering of degree n if and only if:

a. The map is defined at each point of U.

b. There is only a finite number of points P on U such that $f(P)$ is a given point Q on V. In other words, if Γ_f is the graph of f, and Q is on V, then $\Gamma_f \cap (U \times Q)$ is a finite set of points.

c. If Q is a generic point of V then the cycle $f^{-1}(Q)$ consists precisely of n distinct points of U. This condition corresponds to our assumption that L/K is separable.

d. U lies properly above every point of V.

Let $\mathfrak{p}$ be a prime rational cycle of V/k, and for each point Q in $\mathfrak{p}$ assume that all points in $\Gamma_f \cap (U \times Q)$ are simple on $U \times V$. This means that each point in the set theoretic inverse image of $\mathfrak{p}$ by f is simple on U. Using intersection theory, we can define the inverse image of $\mathfrak{p}$ by

$$f^{-1}(\mathfrak{p}) = \mathrm{pr}_U \left(\Gamma_f \cdot (U \times \mathfrak{p})\right).$$

Since $\mathfrak{p}$ is rational over k, so is $f^{-1}(\mathfrak{p})$, and we can write

$$f^{-1}(\mathfrak{p}) = \sum e_i \mathfrak{P}_i$$

where each $\mathfrak{P}_i$ is a prime rational cycle of U/k. The prime rational cycles $\mathfrak{P}_i$ are in natural $1 - 1$ correspondence with the local rings $\mathfrak{O}_i$ into which $\mathfrak{o}$ splits, and the intersection theory has allowed us to introduce the multiplicities e_i. The degree m_i discussed before is equal to $[k(\mathfrak{P}_i):k(\mathfrak{p})]$ where $k(\mathfrak{p})$ is the field obtained from k by adjoining to it any set of coordinates for a point of $\mathfrak{p}$ (and similarly for $k(\mathfrak{P}_i)$). By the principle of conservation of number, applied locally, we see immediately that $\sum e_i m_i = n$ is the degree $[U:V]$(F-VII$_5$ Th. 8).

We return to an arbitrary covering $\mathfrak{U}$ over $\mathfrak{V}_k$ (not necessarily geometric) obtained from L/K, and say that this covering is *unramified over* $\mathfrak{o}$ if $\sum m_i = n$. It is *unramified* if it is unramified over every local ring of $\mathfrak{V}_k$.

Since we assumed k perfect, it follows that a constant field extension is unramified.

Suppose that we are in the geometric case, and that $\mathfrak{U}/\mathfrak{V}_k$ comes from the geometric covering U/V. Let $\mathfrak{o}$ be a local ring in $\mathfrak{V}_k$ and let Q be any one of the points on V having $\mathfrak{o}$ as local ring. (The others are conjugates of Q over k.) Then one sees immediately that $\mathfrak{U}/\mathfrak{V}_k$ is unramified over $\mathfrak{o}$ if and only if there are exactly n distinct points in the set theoretic inverse image of Q on U, i.e. if there are exactly n points in $\Gamma_f \cap (U \times Q)$. In that case, we shall say that U/V is *unramified above Q*. We have just considered points which are algebraic over k. It is easily seen that if U/V is unramified over every point which is algebraic over k, then U/V is unramified over every point of V. Hence the property that a geometric covering be unramified is a geometric property: It does not depend on fields of definition. Note finally that if U/V is unramified above $\mathfrak{p}$ then all the multiplicities e_i above equal 1.

We shall now list the basic facts concerning unramified coverings, and begin by a local statement. See Krull [14] and Nagata [18]. We assume that $\mathfrak{U}/\mathfrak{B}_k$ is an arbitrary covering.

PROPOSITION 1. *Let $\mathfrak{o}$ be a local ring in $\mathfrak{B}_k$, $\mathfrak{p}$ its maximal ideal, and $\mathfrak{O}$, $\mathfrak{P}$ any one of the local rings and maximal ideals lying above $\mathfrak{o}$, $\mathfrak{p}$ in L. If $\mathfrak{U}/\mathfrak{B}_k$ is unramified above $\mathfrak{o}$ then $\mathfrak{P} = \mathfrak{p}\mathfrak{O}$.*

Geometrically, this means that local parameters downstairs can also be used as parameters in the covering. As an application of this fact, we get:

COROLLARY. *If $\mathfrak{o}$ is a regular local ring and if $\mathfrak{U}/\mathfrak{B}_k$ is unramified above $\mathfrak{o}$, then $\mathfrak{O}$ is regular.*

This is due to the fact that the maximal ideal of $\mathfrak{O}$ is generated by exactly the correct number of parameters. Geometrically speaking, we can say that if Q is a simple point on the variety V, and if U/V is a geometric covering unramified above Q, then each one of the n distinct points on U lying above Q is simple on U.

The converse of the above corollary is also true, and is due to Chow [12]. Namely, we have in the geometric case:

PROPOSITION 2. *Let $f: U \to V$ be a geometric covering, which we assume unramified over a point Q of V. If P is a point of U such that $f(P) = Q$ and if P is simple on U, then Q is simple on V.*

The proof of this proposition consists in identifying the situation we are considering with the one considered by Chow. This is the purpose of the following discussion.

Since we are working locally, we could assume U and V are affine. However, we assume them projective, in order to be able to speak about Chow points. The following proposition then shows how V can be recovered from U.

PROPOSITION 3. *Let $f: U \to V$ be a geometric covering, defined over k and of degree n. Assume U and V are projective. Let x be a generic point of U/k, let $y = f(x)$, and let ξ be the Chow point of the cycle $\sum (x_i)$, where (x_i) are the n distinct conjugates of (x) over $k(y)$. Let V_0 be the normalization (in the field $k(y)$) of the locus of ξ over k. Then V and V_0 are isomorphic. Furthermore, if W is the locus of ξ over k, and $g: U \to W$ is the rational map given by $g(x) = \xi$, then g is everywhere defined on U.*

PROOF. Given a point Q_0 on V_0 , it determines a unique point on W, and hence a unique specialization of the cycle $\sum (x_i)$. Hence it determines a finite number of points on V. By Zariski's Main Theorem it follows that the birational correspondence $T: V_0 \to V$ is everywhere defined on V_0 . The converse follows by a similar argument. That g is everywhere defined on U is clear.

At a simple point ξ' of W, the varieties W and V_0 are locally isomorphic to each other and to V. We now see that our proposition is a special case of Chow's Theorem.

We shall now be exclusively concerned with geometric coverings. Let $f: U \to V$ be such a covering. Let W be a prime divisor on V (i.e. a subvariety of codimension 1, necessarily non-singular on V since V is assumed normal). If k is an algebraically closed field of definition for everything, and K/k is a function field for V/k, then we can associate with W a discrete valuation of K in the usual manner, and we shall say that U/V is *unramified over* W, or that W is unramified, if all extensions of this valuation in the function field L of U are unramified. If U/V is unramified over every prime divisor of V, then we shall say that U/V is *divisor-unramified*. Note that the property of U/V to be unramified over W can be expressed geometrically by the condition that every component in the divisor $f^{-1}(W)$ on U appear with multiplicity 1.

It is trivially verified that if U/V is unramified (i.e. unramified over every point of V) then U/V is divisor unramified.

The converse is deeper, and is due to Zariski (Theorem 1 of [1]). More precisely, we have:

PROPOSITION 4. *Let $f: U \to V$ be a covering which is divisor unramified. If Q is a simple point of V, then U/V is unramified above Q.*

We consider finally the birational invariance of unramified coverings.

PROPOSITION 5. *Let V be complete. Let $f: U \to V$ be an unramified covering, and let L/K be the corresponding function field over the constant field k. Then every valuation of K/k is unramified in L.*

PROOF. I am indebted to Mr. Abhyankar for the proof of this proposition, which is a direct consequence of Krull's general discriminant theorem [13], whose statement is as follows:

THEOREM. *Let E be a finite separable extension of a field F, of degree n. Let R be an integrally closed domain whose quotient field is F, and assume R has a unique maximal ideal M. Let R_i be the local rings lying above R in E, and let M_i be their respective maximal ideals. Let m_i be the separable factor of the residue class degree $[R_i/M_i : R/M]_s$. Then $\sum m_i = n$ if and only if there exists a linear base $w_1, \cdots, w_n$ of E/F such that each w_j is integral over R and the discriminant $D(w_1, \cdots, w_n)$ is a unit in R. If that is the case, then each residue class field extension R_i/M_i over R/M is separable.*

Our Proposition 5 follows from the theorem as follows. Let R be any valuation ring in K. Then R has a center on the given normal complete model V. This center will be a point which may be transcendental over k. We let S be its local ring in K. By Krull's theorem, there exists a linear base $w_1, \cdots, w_n$ whose discriminant D lies in S. Since S is contained in R, D lies in R and hence by Krull's theorem again (applied in the opposite direction) R is unramified.

COROLLARY. *The notation being that of Proposition 5, if V_1 is another complete, normal model of K/k, and U_1 its normalization in L, then U_1/V_1 is divisor-unramified, and hence unramified over every simple point of V_1.*

The above results now allow us to speak of unramified extensions of fields:

Given once more a function field K/k (k perfect) and L/K a finite separable extension, involving possibly a constant field extension, we shall say that L/K is *unramified* if there exists one complete normal model V/k of K with skeleton $\mathfrak{B}_k$ such that its covering $\mathfrak{U}$ in L is unramified over $\mathfrak{B}_k$. The corollary shows that any other covering with the same function field will be divisor unramified, and unramified over any regular local ring in $\mathfrak{B}_k$, by Zariski's theorem.

In certain cases of abelian divisor unramified extensions, we can prove (using the birational invariance of the Picard variety) that divisor-unramified is a birationally invariant notion (see below, Kummer theory). For our purposes in this paper, this is enough, and we never need Proposition 5 or its Corollary, but since the proof above is very simple, it was worth including here.

§2. The decomposition group and the Frobenius automorphism

In case the extension L/K is a Galois extension with group G, the covering $\mathfrak{U}/\mathfrak{B}_k$ will be said to be Galois, and abelian if G is abelian. The classical statements concerning the decomposition group and inertia group have been generalized to local rings by Krull [14] and will now be summarized. They are entirely local and semilocal. Hence we start with a regular local ring $\mathfrak{o}$ in $\mathfrak{B}_k$, and let $\mathfrak{O}_i$ be the local rings in L lying above $\mathfrak{o}$. We *assume* that all $\mathfrak{O}_i$ are regular. (This is certainly the case if the extension is unramified above $\mathfrak{o}$.)

The group G permutes the rings $\mathfrak{O}_i$ transitively: For each i there exists an automorphism σ of G such that $\mathfrak{O}_1^\sigma = \mathfrak{O}_i$. To each $\mathfrak{O}_i$ we have a decomposition group D_i consisting of all elements σ in G leaving $\mathfrak{O}_i$ invariant (as a set, not necessarily element-wise). Each σ in D_i can be viewed as inducing an automorphism on the residue class field extension $\mathfrak{O}_i/\mathfrak{P}_i$ over $\mathfrak{o}/\mathfrak{p}$ (which is necessarily Galois) and if $\mathfrak{U}/\mathfrak{B}_k$ is unramified over $\mathfrak{o}$, then D_i can be identified with the Galois group of this local extension.

If $\mathfrak{O}_i = \mathfrak{O}_1^\sigma$ then $D_i = D_1^\sigma = \sigma D_1 \mathfrak{o}^{-1}$. Hence if L/K is abelian, all D_i coincide and to each prime $\mathfrak{p}$ there is associated a unique decomposition group $D(\mathfrak{p})$, independent of the various rings into which $\mathfrak{o}$ splits. The fixed field of $D(\mathfrak{p})$ is the maximal subfield of L containing K into which $\mathfrak{p}$ splits completely.

(As usual, we say $\mathfrak{p}$ *splits completely* if the number of distinct local rings into which $\mathfrak{o}$ splits is n, the degree of the covering. We say $\mathfrak{p}$ *remains prime*, if there is only one local ring above $\mathfrak{o}$ in L. In the unramified case, this means that the residue class degree m is equal to n.)

From now on, we assume that $\mathfrak{U}/\mathfrak{B}_k$ is unramified above $\mathfrak{o}$. We also assume that k is a finite field with q elements. For each prime $\mathfrak{p}$ we let $N\mathfrak{p} = q^d$ where d is the absolute degree of $\mathfrak{p}$, i.e. the degree of the residue class field extension $\mathfrak{o}/\mathfrak{p}$ over the constant field k. We can therefore lift the Frobenius automorphism into the Galois group in the usual manner:

PROPOSITION 6. *Let $\mathfrak{U}/\mathfrak{B}_k$ be a Galois covering, unramified over the regular local ring $\mathfrak{o}$. For each $\mathfrak{O}_i$, $\mathfrak{P}_i$ lying above $\mathfrak{o}$, $\mathfrak{p}$ there exists a unique automorphism $\sigma_i \in D_i$ whose effect on the residue class field $\mathfrak{O}_i/\mathfrak{P}_i$ is to raise each element to the q^dth power ($q^d = N\mathfrak{p}$). If the covering is abelian, all σ_i are equal, and hence depend only on $\mathfrak{o}$, $\mathfrak{p}$.*

The last statement follows from the fact that if two primes are conjugate in L, then the Frobenius automorphisms are conjugate.

Assume that the covering is abelian. The Frobenius automorphism associated with the local ring $\mathfrak{o}$ in $\mathfrak{V}_k$ will be denoted by $(\mathfrak{o}, L/K)$. In terms of primes, we write $(\mathfrak{p}, L/K)$.

The first property we note concerning the Frobenius automorphism is its consistency: If $L \supset E \supset K$, then $(\mathfrak{p}, L/K)$ restricted to E is $(\mathfrak{p}, E/K)$.

Let us now consider the Frobenius automorphism in the two cases: constant field extensions and geometric extensions.

If $L = Kk'$ is a constant field extension, then for any prime $\mathfrak{p}$ of degree d, it is clear that $(\mathfrak{p}, L/K)$ is the unique automorphism of L/K which raises every element of k' to the q^d-power.

Let L/K be geometric, and let us translate our preceding statements into the geometric language. Let $f: U \to V$ be our Galois geometric covering defined over k, unramified over the simple point Q. Let $K = k(x)$ where x is a generic point of V/k, and $L = k(z)$ where z is a generic point of U/k such that $f(z) = x$. To each σ in the Galois group G there corresponds a birational transformation T_σ of U such that $T_\sigma(z) = z^\sigma$. Each T_σ is everywhere defined on U, and the set of these birational transformations is a group isomorphic to G in a natural fashion. It operates on the points of U, and for any point P lying above Q on U we have

$$f^{-1}(Q) = \sum_{\sigma \in G} T_\sigma(P).$$

The n points $T_\sigma(P)$ are all distinct, because U/V is assumed unramified over Q. The properties of G can therefore be summarized as follows:

PROPOSITION 7. *Let $f: U \to V$ be a Galois covering of degree n, unramified above a simple point Q. Let G be the Galois group of birational transformations of U/V. Then each T in G is everywhere defined on U, is consistent with the projection $f: U \to V$, and G permutes transitively the n points of U lying above Q.*

The Galois covering is defined over a field k if k is a field of definition for f, U, V and each one of the transformations of G.

Let $\mathfrak{p}$ be a prime rational cycle of V/k. Then $f^{-1}(\mathfrak{p}) = \sum \mathfrak{P}_i$ (always assuming U/V unramified above the points of $\mathfrak{p}$) and the group of transformations permutes the $\mathfrak{P}_i$ transitively.

Concentrating further on the special case of a finite field k, we note first that we have trivially for T in G

$$T(P^q) = (T(P))^q \qquad \text{and} \qquad f(P^q) = (f(P))^q$$

where the (symbolic) power P^q is the point obtained by raising all coordinates of P to the qth power. This comes from the fact that T and f are defined over k, hence invariant under the automorphism $x \to x^q$ of the universal domain. If $\mathfrak{P}$ is any prime on U appearing in $f^{-1}(\mathfrak{p})$ then the decomposition group $D(\mathfrak{P})$ of $\mathfrak{P}$ consists of those birational transformations T such that $T(\mathfrak{P}) = \mathfrak{P}$. In the geometric case, we can therefore characterize the Frobenius automorphism as follows:

PROPOSITION 8. *Let $f: U \to V$ be a geometric covering, abelian with group G, and defined over a finite field k. Assume it is unramified over a prime rational cycle $\mathfrak{p}$ of V/k. Let $d = \deg \mathfrak{p}$. Then there exists a unique transformation T in G such that for any point Q in $\mathfrak{p}$ and any P in $f^{-1}(Q)$ we have $T(P) = P^{q^d}$.*

The transformation of Proposition 8 will be denoted by $(\mathfrak{p}, U/V)$. If L/K is the function field belonging to U/V over k, then $(\mathfrak{p}, U/V)$ can be identified with the automorphism $(\mathfrak{p}, L/K)$ of L/K in a natural manner.

In the geometric case, one can give an independent and simple proof for the existence of the Frobenius transformation. Suppose that L/K is abelian and geometric, corresponding to the covering $f: U \to V$. Assume it is unramified above a prime cycle $\mathfrak{p}$ of V/k. Let Q be any point in $\mathfrak{p}$. If P is any point in $f^{-1}(Q)$, then $f(P) = Q$, and applying the automorphism $x \to x^{q^d}$ of the universal domain, we see that $f(P^{q^d}) = Q^{q^d} = Q$. Hence P^{q^d} is also in $f^{-1}(Q)$. Using the assumption that U/V is unramified above Q, and hence that the Galois group of covering transformations permutes the points above Q transitively, we conclude that there exists some T in G such that $T(P) = P^{q^d}$. If P_1 is another point in $f^{-1}(Q)$, there is some T_1 such that $T_1(P) = P_1$. We see that $T_1 T T_1^{-1}$ maps P_1 on $P_1^{q^d}$, and since G is assumed abelian, T itself maps P_1 on $P_1^{q^d}$. Hence T does not depend on the P selected. It is unique, because if a covering transformation leaves one point P above Q fixed, then it must be the identity (U/V is unramified). It is also trivially seen that the T above does not depend on the choice of Q in the prime cycle $\mathfrak{p}$ (apply the automorphism $x \to x^q$ to the universal domain).

§3. The trace and cotrace, and the cycle classes

Let K be a function field over the finite constant field k, and let L/K be a finite separable extension as before. Let V be a projective, normal model of K/k, and let $\mathfrak{V}_k$ be its k-skeleton. Let $\mathfrak{U}$ be the covering of $\mathfrak{V}_k$ in L. We assume that the geometric covering U/V defined over the algebraic closure $\bar{k}$ of k in L is divisor unramified. If $\mathfrak{p}$ is a prime cycle of V/k, belonging to a regular local ring $\mathfrak{o}$ in $\mathfrak{V}_k$, then all the local rings $\mathfrak{O}_i$ in $\mathfrak{U}$ lying above $\mathfrak{o}$ are regular by Prop. 1 of §1. Geometrically, U/V is unramified over every simple point of V.

We denote by $Z(\mathfrak{V}_k)$ the free abelian group generated by the prime cycles of $\mathfrak{V}_k$. We thus exclude the primes which are not simple on V. We let $Z(\mathfrak{U})$ be the free abelian group generated by the primes of $\mathfrak{U}$ which lie above some prime cycle of V. This may be *smaller* than the group of cycles of $\mathfrak{U}$. It is a subgroup of the group of cycles on U, rational over k'. There is an obvious isomorphism of $Z(\mathfrak{V})$ into $Z(\mathfrak{U})$ defined in the usual manner:

If $L = Kk'$ is a constant field extension, then each rational cycle of V/k is a fortiori rational over k', so $Z(\mathfrak{V}_k)$ is in fact contained in $Z(\mathfrak{V}_{k'})$.

If L/K is geometric, then each prime rational cycle $\mathfrak{p}$ of V/k maps on $f^{-1}(\mathfrak{p}) = \sum \mathfrak{P}_i$ in $Z(\mathfrak{U}_k)$, and the $\mathfrak{P}_i$ appear with coefficient 1 because our covering is unramified over the prime cycles. Each prime cycle $\mathfrak{P}$ of U/k occurs in exactly one such inverse image of a prime cycle $\mathfrak{p}$ of V, and in the old arithmetical terminology, one writes $\mathfrak{P} \mid \mathfrak{p}$. So our map is an isomorphism.

REMARK. In case one considers ramified extensions, one can restrict $Z(\mathfrak{V}_k)$ in the following manner. According to Zariski's theorem, the ramification locus above simple points is a divisor (all its components have codimension 1). Say D is this locus. Then one can take the free abelian group $Z_D(\mathfrak{V}_k)$ generated by the prime cycles of V over k whose support does not lie in the support of D. We let $Z(\mathfrak{U})$ be again the group generated by the prime cycles of $\mathfrak{U}$ lying above some prime cycle of $Z_D(\mathfrak{V}_k)$, and in the map f^{-1}, all the $\mathfrak{P}_i$ appear again with multiplicity 1.

Let $\alpha: V \to A$ be a canonical map of V into its Albanese variety. Such maps are defined up to translations, but over a finite field k, it is always possible to find one over the given k. (Proof: Weil [23] has defined a canonical map of V into a principal homogeneous space over A, everything being defined over k. According to [15], a principal homogeneous space over A has a rational point and hence is isomorphic to A over k.)

Let $Z_0(\mathfrak{V}_k)$ be the subgroup of $Z(\mathfrak{V}_k)$ consisting of the cycles of degree 0. Then α induces a homomorphism of $Z_0(\mathfrak{V}_k)$ into the group A_k of rational points of A in k in the following manner: Given a cycle $\mathfrak{a}$ on V of degree 0, then

$$\alpha(\mathfrak{a}) = \mathrm{pr}_A \left(\Gamma_\alpha \cdot (\mathfrak{a} \times A) \right)$$

is a cycle of A. (Recall that α is defined at every simple point of V.) The sum of the points of $\alpha(\mathfrak{a})$ is a rational point of A in k, and will be denoted by $\pi(\mathfrak{a})$. (In his "Variétés Abéliennes," Weil writes $S(\alpha(\mathfrak{a}))$, but we shorten the notation here.) The kernel of the map

$$\pi: Z_0(\mathfrak{V}_k) \to A_k$$

will be denoted by $Z_\alpha(\mathfrak{V}_k)$ and will be called the *kernel of Albanese*. We have the following inclusions:

$$Z(\mathfrak{V}_k) \supset Z_0(\mathfrak{V}_k) \supset Z_\alpha(\mathfrak{V}_k).$$

The factor group Z/Z_0 (omitting the $\mathfrak{V}_k$ for simplicity) is infinite cyclic, and has a representative cycle of degree 1 (Corollary 3 of [16]). The factor group Z_0/Z_α is isomorphic to a subgroup of A_k, and in fact to A_k itself, because according to Proposition 2 of the Appendix, given a point $a \in A_k$, there exists a cycle $\mathfrak{a}$ of degree 0 on V, rational over k, such that $\pi(\mathfrak{a}) = a$. The factor group Z/Z_α will be denoted by C_K, because it is a birational invariant, depending only on the function field K. It will be called the group of *cycle classes*. The factor group Z_0/Z_α will be denoted by C_K^0, and will be called the *cycle classes of degree* 0.

If K is the function field of a curve, then the cycle classes coincide with the divisor classes because the Albanese variety coincides with the Picard variety.

Let us expand on the statement that C_K is a birational invariant. Let W be (say) an affine normal model of K/k, and let $K = k(v) = k(w)$ where w is a generic point of W/k and v a generic point of V/k. Let $\alpha(v) = \xi$ be the image of

v in A. Then the rational map $\beta: W \to A$ given by $\beta(w) = \xi$ may be called a map of W into its Albanese variety. Given a cycle $\mathfrak{b}$ on W rational over k, the map $\pi_W(\mathfrak{b})$ is defined just as $\pi = \pi_V$ was defined. According to Proposition 3 of the Appendix, it is clear how to find a cycle $\mathfrak{b}_1$ of degree 1 on W, rational over k, such that $\pi_W(\mathfrak{b}) = 0$. Let $\mathfrak{a}_1$ be a cycle on V rational over k and of degree 1 such that $\pi_V(\mathfrak{a}) = 0$. Using the fact that there are cycles on V and W which map on any given point of A_k under π, we see that we can establish a natural isomorphism between the cycle classes of V and those of W, mapping the class of $\mathfrak{b}_1$ on that of $\mathfrak{a}_1$.

We could also introduce here the notion of a rational map $\alpha: K \to A$ of the function field itself into its Albanese variety, and consider the free abelian group generated by *all* regular local rings of K (arising from any algebraic point on any affine model of K). The cycle classes could then be defined for this group, and the terminology could be made completely invariant. However, for this paper, we shall remain within the usual geometric terminology.

We make one additional remark. The group $Z(\mathfrak{U})$ defined above is not necessarily the entire group of cycles of the variety U/k', rational over k'. However, the prime cycles missing are obviously contained in a proper algebraic subset of U. Hence as far as the cycle classes are concerned, it is just as good as the entire group of cycles, by the same arguments that have been mentioned above in our discussion of the birational invariance of the cycle classes. (See Proposition 2 of the Appendix.)

Let us now consider more closely the imbedding of $Z(\mathfrak{V}_k)$ into $Z(\mathfrak{U})$ in the two types of extensions.

If $L = Kk'$ is a constant field extension, then

$$Z_\alpha(\mathfrak{V}_k) = Z_\alpha(\mathfrak{V}_{k'}) \cap Z(\mathfrak{V}_k).$$

In other words, the kernel of Albanese is geometric, and we have a natural isomorphism of C_K into C_L induced by the inclusion of the cycles.

If L/K is geometric, then the inverse mapping f^{-1} gives a (rational) isomorphism of the cycles of V into those of U, and of $Z(\mathfrak{V}_k)$ into $Z(\mathfrak{U})$. Hence by Chow's definition of the Albanese variety (universal mapping property) we have an induced rational homomorphism (contravariant)

$$f^*: A(V) \to A(U)$$

of the Albanese variety of V into that of U, defined over k. We also obviously have a homomorphism of C_K into C_L.

If L/K is a combination of both types of extensions, then by composing the above maps, we have an induced isomorphism of $Z(\mathfrak{V}_k)$ into $Z(\mathfrak{U})$ and a homomorphism of C_K into C_L called the *cotrace*.

We shall make little use of the cotrace mapping on the cycle classes. In class field theory, it enters in the formulation of theorems such as the transfer theorem and Principal Ideal theorem, which can be proved formally from the reciprocity law. All such theorems will be omitted from the present paper, and left to the reader to work out.

Let us now study the trace mapping.

Let L/K be Galois, with group G. Let V be a normal, projective model of K/k, and assume $\mathfrak{U}/\mathfrak{B}_k$ divisor-unramified for simplicity. This means the covering U/V (defined over $\bar{k}$) is divisor-unramified. Let $\mathfrak{o}$ be a regular local ring in $Z(\mathfrak{B}_k)$. We have seen that G permutes the local rings $\mathfrak{O}_i$ of $\mathfrak{U}$ lying above $\mathfrak{o}$ transitively. Thus G may be viewed as a group of operators on the module $Z(\mathfrak{U})$. This means that the notion of trace is defined, namely the trace of an element is the sum of that element and its conjugates under G. It follows immediately from the assumption that $\mathfrak{U}/\mathfrak{B}_k$ is divisor unramified, hence unramified over any prime cycle of $\mathfrak{B}_k$, that the fixed module under G is precisely the group $f^{-1}(Z(\mathfrak{B}_k))$. From now on we shall always identify $Z(\mathfrak{B}_k)$ with its inverse image in $Z(\mathfrak{U})$. We may therefore say that $Z(\mathfrak{B}_k)$ is the submodule of $Z(\mathfrak{U})$ fixed under G. This is in complete analogy with the same fact for divisor classes on curves.

The trace maps $Z(\mathfrak{U})$ into $Z(\mathfrak{B}_k)$, and furthermore if k' is the constant field of L maps $Z_\alpha(\mathfrak{U}_{k'})$ into $Z_\alpha(\mathfrak{B}_k)$ (i.e. maps kernal of Albanese into kernel of Albanese). This will be discussed in greater detail below, by decomposing the extension L/K into a geometric part and a constant field extension. Assuming it for the moment, we see that the trace induces a homomorphism

$$S_K^L : Z(\mathfrak{U})/Z_\alpha(\mathfrak{U}) \to Z(\mathfrak{B}_k)/Z_\alpha(\mathfrak{B}_k).$$

Even though $Z(\mathfrak{U})$ may not be the full group of cycles on U, rational over k', it differs from it by cycles whose supports are contained in a proper algebraic subset of U. Hence we see that the trace actually gives a well defined homomorphism of the full group C_L into C_K, i.e. we have a homomorphism

$$S_K^L : C_L \to C_K$$

of C_L into C_K which will again be called the trace.

Finally, we defined the *trace group* (or trace subgroup of $Z(\mathfrak{B}_k)$) to be the subgroup of $Z(\mathfrak{B}_k)$ generated by those cycles which are traces, and by the kernel of Albanese. The image $S_K^L C_L$ in C_K will be called the *trace group of cycle classes*. We omit the words cycle classes when the context makes our meaning clear.

The transitivity of the trace holds here as usual.

Let us consider the trace mapping in the two kinds of extensions. Suppose first that L/K is geometric, and let $f : U \to V$ be the rational map derived from L/K. Let $\mathfrak{p}$ be a prime rational cycle of V/k, and $\mathfrak{P}$ any one of the primes of U lying above it. Let $m = [k(\mathfrak{P}) : k(\mathfrak{p})]$ be the relative degree. Then the image

$$f(\mathfrak{P}) = \mathrm{pr}_V \left(\Gamma_f \cdot (\mathfrak{P} \times V) \right)$$

defined by intersection theory is immediately seen to be the cycle $m\mathfrak{p}$, and we also see that

$$f^{-1}(m\mathfrak{p}) = \sum_{T \in G} T(\mathfrak{P}) = S_K^L(\mathfrak{P}).$$

We had agreed to identify $\mathfrak{p}$ with $f^{-1}(\mathfrak{p})$ in $Z(\mathfrak{U}) = Z(\mathfrak{U}_k)$, and hence this identification shows that the trace mapping in the case of geometric extensions can be identified with the rational homomorphism

$$f: Z(\mathfrak{U}_k) \to Z(\mathfrak{V}_k)$$

on the cycles, induced by the rational map f.

If we look at the cycles of degree 0, namely $Z_0(\mathfrak{V}_k)$, then we note that f maps the kernel of Albanese on U into the kernal of Albanese on V (universal mapping property). Hence f induces a homomorphism (covariant) $f_*: A(U) \to A(V)$ of the Albanese variety of U into the Albanese variety of V, and all the above objects are defined over k. (This is precisely Chow's definition.)

To simplify the notation, put $A(V) = A$ and $A(U) = B$.

The group C_K^0 of cycle classes of degree 0 (i.e. $Z_0(\mathfrak{V}_k)/Z_a(\mathfrak{V}_k)$) can be identified with the group A_k of rational points of A in k, and we see from the preceding discussion that the trace subgroup $S_K^L C_L^0$ corresponds to the subgroup $f_* B_k$ under this identification. In the case of geometric coverings, we see therefore that the trace factor group $C_K^0/S_K^L C_L^0$ is canonically isomorphic to $A_k/f_* B_k$.

The map f_* of course maps B onto A, but given a rational point $a \in A_k$, it may not be possible to find a rational point $b \in B_k$ such that $f_*(b) = a$. Hence in general, $f_* B_k$ is not equal to A_k . Class field theory will establish an isomorphism between this trace factor group and the Galois group of certain abelian unramified extensions of K.

Let us finally consider the case of a constant field extension: $L = Kk'$. If $\mathfrak{a}$ is any cycle on V rational over k', then the trace of $\mathfrak{a}$ is simply the cycle

$$\mathfrak{a} + \mathfrak{a}^q + \cdots + \mathfrak{a}^{q^{n-1}}$$

where $\mathfrak{a}^q$ means that all points in $\mathfrak{a}$ are to be raised to the (symbolic) qth power. The plus sign here refers to the formal addition of cycles. The group C_L^0 is canonically isomorphic to $A_{k'}$, and the Galois group of k'/k operates on these points. If $a \in A_{k'}$ then the trace of a is $a + a^q + \cdots + a^{q^{n-1}}$, the plus sign referring to the addition on the abelian variety A. We note that $\pi(\mathfrak{a})^q = \pi(\mathfrak{a}^q)$, because the map α of V into A and the law of composition of A are defined over k, and so commute with the automorphism $x \to x^q$ of the universal domain. This implies in particular that π commutes with the trace, and that the trace sends the kernel of Albanese $Z_a(\mathfrak{V}_{k'})$, into the kernel of Albanese $Z_a(\mathfrak{V}_k)$.

§4. Unramified extensions

We start with a variety V, projective, normal, over the field k. Let $\alpha: V \to A$ be a mapping of V into an abelian variety (we do not yet assume A is the Albanese variety of V) with both α and A defined over k. Let $\lambda: B \to A$ be a finite covering of A with an abelian variety B, defined over k. Then λ may be assumed to be a homomorphism (by Weil's Th. 9 of [26]). If B/A is abelian over k, i.e. if the kernel of λ consists of rational points of B in k, then we shall say that k is *complete for* λ. We assume that this is the case. We shall show how to pull back the covering B/A to a covering of V.

Let x be a generic point of A/k, and y a generic point of B/k such that $\lambda(y) = x$. The extension $k(y)$ over $k(x)$ is abelian, and its Galois group is the group of translations on B given by the kernal $\mathfrak{g}$ of λ, which is assumed contained in the rational points B_k. As before, we let its order be n.

Note that every intermediate field between $k(x)$ and $k(y)$ is the function field of an abelian variety. Indeed, these are obtained by taking B modulo subgroups of $\mathfrak{g}$.

Let v be a generic point of V/k and put $K = k(v)$. Let $\xi = \alpha(v)$ be the image of v in A. Then $x \to \xi$ is a specialization. Let η be any specialization of y over $x \to \xi$.

The automorphisms of $k(y)$ over $k(x)$ are represented by translations: If b_i are the points in $\mathfrak{g}$ then $y \to y + b_i$ gives all automorphisms of $k(y)$ over $k(x)$. The cycle $\sum (\eta + b_i)$ is the unique specialization of the cycle $\sum (y + b_i)$ over $x \to \xi$. In fact we have $\lambda^{-1}(x) = \sum (y + b_i)$ and $\lambda^{-1}(\xi) = \sum (\eta + b_i)$. Hence $\sum(\eta + b_i)$ is rational over $k(\xi)$. Furthermore, $\eta + b_i$ is a point of B and the fields $k(\eta + b_i)$ are all equal to $k(\eta)$, which is therefore a Galois abelian extension of $k(\xi)$. The group of the specialized extension may be a proper subgroup of $\mathfrak{g}$.

The extension $k(v, \eta)$ over $k(v)$ obtained by lifting the extension $k(\xi, \eta)$ of $k(\xi)$ will be called the extension *derived*, or *pulled back* from the extension $k(y)/k(x)$.

If A is the Albanese variety of V, then the following statements hold:

1. The extension $k(\eta)$ over $k(\xi)$ is an abelian extension of degree n, and k is algebraically closed in $k(\eta)$.

2. If b_i ranges over $\mathfrak{g}$, then the n distinct conjugates of η over $k(\xi)$ are $\eta + b_i$. Hence the Galois group of $k(\eta)/k(\xi)$ can be identified with the group of translations $\mathfrak{g}$.

3. The field $k(\eta)$ is linearly disjoint from $k(v)$ over $k(\xi)$.

4. The extension $L = k(v, \eta)$ over $k(v) = K$ is geometric and abelian. Let U be the normalization of V in L, and $f:U \to V$ the associated covering, $L = k(u)$ where u is a generic point of U/k and $f(u) = v$. Let $g:U \to B$ be the rational map defined by $g(u) = \eta$. Let G be the Galois group of birational transformations of U/V corresponding to the Galois group of L/K. Then a transformation T in G and a translation τ in $\mathfrak{g}$ determine each other uniquely by the formula

$$gT = \tau g,$$

and the correspondence $T \leftrightarrow \tau$ is an isomorphism between G and $\mathfrak{g}$.

The proof of the above statements will now be given. In the original proof of Statement 1, the Kummer theory was used, and hence we needed to assume that the degree $[B:A]$ is prime to the characteristic. I am much indebted to J. P. Serre for the following proof, valid in all cases. Since the rest of the arguments worked without characteristic trouble, this means that the class field theory can be carried out without distinguishing Albanese p-extensions from the others.

Let us now go to the proof of Statement 1.

Going to the language of local rings, if R is the local ring of ξ in $k(x)$ and M its maximal ideal then R/M is canonically isomorphic to $k(\xi)$. The ring R splits in $k(y)$ into local rings (R_j^*, M_j^*) and the decomposition groups D_j all coincide because $\mathfrak{g}$ is abelian. If we denote them by D, then D can be interpreted canonically as the Galois group of $k(\eta)/k(\xi)$ (see Krull [14]). The fixed field of D is the function field of an abelian variety C intermediate to A and B, and is the largest subfield of $k(y)$ into which (R, M) splits completely. This subfield can be written $k(w)$ where w is a generic point of C/k. Since (R, M) splits completely in $k(w)$, it follows that any specialization $w \to \omega$ over $x \to \xi$ must be rational over $k(\xi)$, and in fact we have $k(\xi) = k(\omega)$ because $\xi = \lambda(\omega)$. Since ω is the generic point of a subvariety of C, we get a map of V into C which contradicts the universal mapping property of A unless D is all of $\mathfrak{g}$.

In particular, $k(\eta)$ has degree $n = [k(y):k(x)]$ over $k(\xi)$. Furthermore, k must be algebraically closed in $k(\eta)$. To see this, note that the above arguments could have been carried over a larger constant field than k, and that we would have found the same degree n for our specialized extension $k(\eta)/k(\xi)$. Hence a constand field extension could not have contributed to the preservation of the degree in $k(\eta)/k(\xi)$. This concludes the proof of Statement 1, and also of Statement 2 because we see also that the n conjugates of η over $k(\xi)$ must be its translations of $\eta + b_i$. The Galois group of $k(\eta)$ over $k(\xi)$ will be identified with $\mathfrak{g}$ from now on.

We come to the problem of raising $k(\eta)/k(\xi)$ over to $k(v)$. We contend that $k(\eta)$ is linearly disjoint from $k(v)$ over $k(\xi)$.

PROOF. Every intermediate field between $k(x)$ and $k(y)$ is the function field of an abelian variety lying between A and B. If we take a generic point w for one of these, and a simultaneous specialization $x \to \xi$, $y \to \eta$, and $w \to \omega$ then $k(\omega)$ is an abelian extension of $k(\xi)$ of degree $[k(w):k(x)]$ and it is clear that every intermediate extension between $k(\xi)$ and $k(\eta)$ can be obtained by specializing such an intermediate w. Hence each intermediate extension $k(\omega)$ is the function field of a subvariety of an abelian variety. If $k(v)$ is not disjoint from $k(\eta)$ then a subfield of $k(\eta)$ properly containing $k(\xi)$ would be contained in $k(v)$ (because $k(\eta)$ is abelian over $k(\xi)$). This subfield would give rise to a rational map of V into an abelian variety, and this would contradict the universal mapping property of the Albanese variety. Hence $k(v)$ is disjoint from $k(\eta)$ over $k(\xi)$. This proves Statement 3.

The composite field $k(v, \eta)$ is therefore a Galois, abelian extension of $k(v)$, whose Galois group G is isomorphic in a natural way with that of $k(\eta)$ over $k(\xi)$ by the well known lifting process (sometimes called the theorem of natural irrationalities). Let U be the normalization of V in $k(v, \eta)$. Then Statement 4 is simply an obvious geometric expression of the isomorphism between G and $\mathfrak{g}$.

A covering $f: U \to V$ for which there exists a field of definition over which it can be obtained by the preceding pull back process from the Albanese variety of V will be said to be of *Albanese type*.

PROPOSITION 9. *If $f: U \to V$ is a covering of Albanese type, then it is divisor unramified, and hence unramified over every simple point of V.*

PROOF. We preserve the notations used above. We have a commutative diagram:

$$U \xrightarrow{\;g\;} B$$
$$f \Big\downarrow \qquad\quad \Big\downarrow \lambda$$
$$V \underset{\alpha}{\;\longrightarrow\;} A$$

Since both U and V are assumed normal, the generic point of a divisor on U or V is simple on U or V respectively. We recall g and α are defined at simple points. If u' is the generic point of a prime divisor W on U then $g(u') = \eta'$ is defined. Let $f(u') = v'$ and $\alpha(v') = \xi'$. Then v' is the generic point of a prime divisor on V below W, and the specialization $v \to v'$ determines uniquely the specialization $\xi \to \xi'$. Furthermore, if $\sum (u_i) = f^{-1}(v)$, then $v \to v'$ also determines uniquely the specialization of the cycle $\sum (u_i)$ to a cycle $f^{-1}(v') = \sum(u_i')$. Hence u' must appear among the u_i'. The cycle $\lambda^{-1}(\xi')$ being uniquely determined and consisting of n distinct points, it follows that the cycle $f^{-1}(v')$ must also consist of n distinct points, and hence that U/V is unramified over the given divisor. That it is unramified over simple points follows from Zariski's theorem.

We shall now discuss the converse of Proposition 1.

We start therefore with a given geometric covering $f : U \to V$, abelian with group G and defined over k. We assume that it is of exponent n (i.e. every element of the Galois group has period dividing n) and that the nth roots of unity lie in the constant field k (n being now assumed prime to the characteristic p).

Suppose U/V is divisor-unramified. If it is cyclic, we can find a function φ in the function field K of V/k such that $L = K(\varphi^{1/n})$ and such that $(\varphi) = nX$ for some divisor X on V, rational over k. If X is algebraically equivalent to 0, the covering will be said to be of *Picard type*. If in general the covering is abelian, corresponding to a function field extension L/K, then we shall say that U/V is of *Picard type* if L is a compositum of cyclic fields of Picard type.

It is not always true that if the covering is cyclic and divisor unramified then it has to be of Picard type, for there may be torsion in the Néron Severi group, i.e. there may be divisors X such that mX is algebraically equivalent to $0 (m > 0)$ but X is not. The following example due to Godeaux illustrates this situation.

Consider the surface U defined by $t_0^5 + t_1^5 + t_2^5 + t_3^5 = 0$ in projective 3-space. It is non-singular and its Picard variety is trivial (first Betti number 0 if we are in characteristic 0). We assume of course that $p \neq 5$. We can define a transformation T operating without fixed points on U in the following manner: If ω is a primitive fifth root of unity then a point $(t_0, \cdots, t_3)$ on U is mapped on $(t_0\omega, t_1\omega^2, t_2\omega^3, t_3\omega^4)$. This transformation generates a cyclic group of order 5. According to Chow's Theorem (Prop. 2 of §1) the quotient variety V is non-singular and U/V is unramified. Since V is a rational image of U its Picard variety is also trivial. The function t_3/t_2 generates the extension of the function field. The torsion group on V is cyclic of order 5 and produces the unramified covering.

In general let D be the group of divisors on V. Then we have the usual inclusions

$$D \supset D_n \supset D_a \supset D_l .$$

If $X \in D_n$ and $mX \in D_a$ then by Corollary 3 of Prop. 4 of [24] there exists X_1 algebraically equivalent to X such that $mX_1 \in D_l$. This means that the torsion group D_n/D_a can be put (essentially) as a direct factor of D_n/D_l (but not canonically).

If V is non-singular, then the torsion group D_n/D_a is a birational invariant in the classical case. This has not yet been verified in the abstract case.

In addition to the divisor unramified Kummer extensions of Picard type, we get a finite number of divisor unramified extensions due to Néron-Severi torsion.

If our covering is of Albanese type, we know that it is divisor unramified. Conversely, if there is no torsion, we still are unable to show that every abelian divisor unramified covering of V of degree p^μ ($p =$ characteristic) is of Albanese type.

Excluding coverings due to torsion, or of degree p^μ, we can however prove the following result, which is simply another interpretation of one of Chow's fundamental results in the theory of Picard and Albanese varieties.

PROPOSITION 10. *A covering $f: U \to V$ is of Picard type if and only if it is of Albanese type, and of degree prime to p.*

PROOF. We begin by two remarks of a general nature.

REMARK 1. Given a variety V/k as usual, we say that a divisor class on V is rational over k if it is left fixed by every automorphism of $\bar{k}$ over k. (We have assumed that the class contains divisors which are rational over $\bar{k}$.) It is conceivable that a divisor class rational over k does not contain any divisors rational over k. It contains such divisors if k is a finite field. This is more or less well known and we reproduce a proof in the appendix. This shows that given a point a on the Picard variety of V rational over k, it is always possible to find a divisor X rational over k whose class belongs to a. The same is also true if k is algebraically closed. In the following discussion, we shall assume that k behaves like a finite field or an algebraically closed field in this respect.

REMARK 2. This is really a lemma. Let $f: W \to W'$ be a rational map of a variety W into a variety W' (both complete normal) such that $f(W)$ is simple on W'. Assume all objects defined over k. If Y' is a divisor on W', then there exists a divisor Y'_1 linearly equivalent to Y' and rational over k if Y' is rational over k, such that $f^{-1}(Y'_1)$ is defined.

The proof is very simple. Let q be a prime rational cycle of W', of dimension 0 over k, whose support $|\,q\,|$ is contained in $f(W)$. Using the unique factorization in the local ring of q in the function field of W' over k, one sees immediately the existence of a function φ' on W', rational over k, such that no point in q is contained in the support of the divisor $Y' - (\varphi')$. This divisor is the required Y'_1. (Our lemma simply expressed the well known fact that at a simple point a divisor is locally linearly equivalent to zero, with the added remark that we can take a simple prime instead of a point.)

Our proposition will now become almost obvious, after we state the following statement due to Chow [11].

Let $\alpha: V \to A$ be a canonical mapping of V into its Albanese variety with α, V, A defined over k. Then V and A have the same Picard variety. Actually, all we need is the fact that the mapping α^{-1} gives a rational isomorphism of the Picard variety of A onto that of V.

If one has a rational point on the Picard variety of A, then one can select in its class a divisor also rational over k. Using the lemma, we can select the divisor Y' such that $\alpha^{-1}(Y')$ is defined.

Given the covering $f: U \to V$, which we assume cyclic for simplicity, and the function field extension $L = K(\varphi^{1/n})$ where $(\varphi) = nX$ with X algebraically equivalent to zero and rational over k, there is a divisor Y on A, rational over k such that $\alpha^{-1}(Y)$ is defined and linearly equivalent to X. If n is the exact period of the class of X on V, then n is also the exact period of the class of Y on A. If $nY = (\psi)$ for some function ψ on A rational over k, then (by F-VIII$_2$ Th. 4 Cor. 2) the function φ_1 on V given by $\varphi_1(\xi) = \psi(\xi)$ has the divisor nX_1 where $X_1 = \alpha^{-1}(Y)$. Since X_1 is linearly equivalent to X we have $X_1 = (\varphi_2) + X$ with φ_2 rational over k. Then $c\varphi_1 = \varphi_2^n \varphi$ for some constant c in k. From this we see that we could have taken the function ψ before such that $c = 1$. In that case the function φ_1 can be used instead of φ to generate the Kummer extension, i.e. we have $K(\varphi^{1/n}) = K(\varphi_1^{1/n})$. Furthermore, we have

$$k(x, \psi^{1/n}):k(x)] = [k(v, \varphi^{1/n}):k(v)] = n.$$

Since the extension $k(\xi, \varphi_1^{1/n})$ over $k(\xi)$ has degree $\leq n$ and lifts to an extension of degree n over $k(v)$, it follows that its degree is also n and in addition to the above equalities we have also $n = [k(\xi, \varphi_1^{1/n}):k(\xi)]$.

Put $\Re = k(x)$ and $\Im = \Re(\psi^{1/n})$. Then $\Im$ is the function field of an abelian variety B/k (see for instance Prop. 32 of [26]) and the extension $\Im/\Re$ corresponds to a homomorphism $\lambda: B \to A$ whose kernel is a subgroup $\mathfrak{g}$ of B_k. Let y be a generic point of B/k such that $\Im = k(y)$ and let $\mu = \mu(y)$ be a function on B such that $\mu^n = \psi$. Let η be any specialization of y over $x \to \xi$. Since $\psi(\xi)$ is defined, $\mu(\eta)$ is also defined and is contained in $k(\eta)$. However, the degree $[k(\eta):k(\xi)]$ is obviously $\leq n$. Hence it must be equal to n, and we get $k(\eta) = k(\xi, \varphi_1^{1/n})$ and $L = k(v, \varphi_1^{1/n}) = k(v, \eta)$.

This proves that every covering of Picard type is of Albanese type. That every covering of Albanese type and degree prime to p is also of Picard type can be seen in a similar fashion.

The main purpose of Proposition 10 was to show that coverings of Albanese type give, from the algebraic and geometric point of view, almost all divisor unramified coverings of a variety V. From now on, however, we shall forget about this aspect of the unramified extensions and work exclusively with coverings of Albanese type. We need never again use the Picard variety.

Given a function field K over a constant field k, let L/K be a finite extension, separable, but not necessarily Galois, and not necessarily preserving the constant field. We shall say that L/K is of *Albanese type* if some covering U/V defined over $\bar{k}$ and belonging to the geometric extension $L\bar{k}$ over $K\bar{k}$ is of Albanese type. In particular, over a suitable constant field extension k', which can always be taken finite over k, L/K becomes abelian, and can be obtained by a pull back, as described at the beginning of this section, from an abelian extension $k'(y)/k'(x)$ of the function field of the Albanese variety. The constructions of coverings of Albanese type shows clearly that the notion is birationally invarient and that our property of an extension to be of Albanese type does not depend on the choice of (projective, normal) models U/V selected. As we have mentioned once previously, we could use throughout a language where we would deal only with function fields, and never mention models.

We shall prove two routine auxiliary results. We recall the fact already mentioned that if L/K is of Albanese type, then any intermediate subfield E with $L \supset E \supset K$ is also of Albanese type.

PROPOSITION 11. *Let K/k be a function field and L_1, L_2 two finite separable extensions of Albanese type. Then the compositum $L_1 L_2$ is of Albanese type.*

PROOF. We may assume k is algebraically closed. Let x be, as usual, a generic point of the Albanese variety A of K/k, and y_1, y_2 generic points of B_1, B_2 such that L_1, L_2 are derived by pulling back the extensions $k(y_1)/k(x)$ and $k(y_2)/k(x)$ respectively. It suffices to prove that the composite field $k(y_1, y_2)$ is the function field of an abelian variety, because it is then clear that $L_1 L_2$ is contained in the extension of K obtained by pulling back $k(y_1, y_2)$.

For some integer m, and any t on A such that $mt = x$, the points y_1 and y_2 are both rational over the field $k(t)$. This shows that $k(y_1, y_2)$ is between two abelian function fields, and hence that the dimension of its Albanese variety must be equal to the dimension of A. On the other hand, there is a birational map of a model of $k(y_1, y_2)$ into $B_1 \times B_2$ (given by (y_1, y_2)). This implies that the mapping into the Albanese variety is birational, by the universal mapping property.

PROPOSITION 12. *Let K/k be a function field, and E any finitely generated extension of K. Let L/K be a finite separable extension of Albanese type. Then LE is of Albanese type over E.*

PROOF. We may again assume k algebraically closed. As usual, we have $k(x)/k$, the function field of the Albanese variety of K/k, and now we have an induced homomorphism $h: A^* \to A$ of the Albanese variety of E/k onto that of K. We let $x = h(x^*)$ where x^* is a generic point of A^*/k. The extension L/K can be pulled back from a covering B/A, with a corresponding extension $k(y)/k(x)$ where y is a generic point of B/k. Again, if we can show that the field $k(x^*, y)$ is an abelian function field, belonging to an abelian variety B^*, then the extension LE/E will be contained in the extension of Albanese type over E obtained by pulling back B^* over E.

For some integer m and any point t on A such that $mt = x$, we must have y rational over $k(t)$. Let t^* on A^* be such that $mt^* = x^*$. Then $h(mt^*) = mh(t^*) = x$. Putting $h(t^*) = t$, we see that y must be rational over $k(t^*)$. Hence $k(x^*, y)$ is again squeezed between two abelian function fields namely $k(t^*)$ and $k(x^*)$. As before, there is a birational map into an abelian variety $A^* \times B$ given by the point (x^*, y). This proves that $k(x^*, y)$ is an abelian function field, as desired.

§5. Statement of the class field theory

We formulate the class field theory.

Let K/k be a function field over a finite field k. Let L/K be an abelian extension of Albanese type. Let V be any projective normal model of K/k. Let $\mathfrak{p}$ be a prime cycle in $Z(\mathfrak{B}_k)$ and let $(\mathfrak{p}, L/K)$ be its Frobenius automorphism in the Galois group G of L/K. Then by linearity we can extend the map $\mathfrak{p} \to (\mathfrak{p}, L/K)$ to a homomorphism $\omega: Z(\mathfrak{B}_k) \to G$ of the cycles on V rational over k into the Galois group. This will be called the *reciprocity mapping* and its kernel will be called the reciprocity kernel. The following results will be proved.

The reciprocity mapping is surjective, and the kernel of Albanese is contained in the reciprocity kernel. In fact the reciprocity kernel is equal to the trace group defined in §3. Hence the reciprocity mapping induces an isomorphism of the trace factor group $C_K/S_K^L C_L$ onto G. This is the reciprocity law. Note that the decomposition laws are expressed locally. Any normal model of K could have been selected.

If the extension L/K happens to be geometric, belonging to a covering $f: U \to V$ defined over k, let $f_*: A(U) \to A(V)$ be the induced homomorphism of the Albanese varieties. Then the map $\mathfrak{a} \to (\mathfrak{a}, U/V)$ from the cycles of degree 0 into G induces an isomorphism of $C_K^0/S_K^L C_L^0 = A_k(V)/f_* A_k(U)$ onto G.

On the other hand if L/K is a constant field extension, then $(\mathfrak{a}, L/K) = 1$ for every cycle $\mathfrak{a}$ of degree 0, and hence C_K^0 is contained in the reciprocity kernel.

Given a subgroup M of C_K of finite index this subgroup is the trace group of one and only one abelian extension of K of Albanese type (*existence* and *uniqueness theorem*) which is called the class field to this subgroup. The correspondence between M and its class field L_M reverses the inclusion lattice, i.e. if L_1 is class field to M_1 and L_2 class field to M_2 then L_1 contains L_2 if and only if M_1 is contained in M_2, $M_1 \cap M_2$ belongs to the composite $L_1 L_2$ while $M_1 M_2$ belongs to the intersection $L_1 \cap L_2$.

Finally, if E/K is (for simplicity) a Galois extension, then the class field L of K belonging to the subgroup $S_K^E C_E$ of C_K is contained in E. In particular, if E/K is Galois and is such that $(C_K : S_K^E C_E) = [E:K]$ then E is an abelian extension of K, of Albanese type.

The above statement is known classically as the *limitation theorem*. It gives us another characterization for the abelian extensions of Albanese type. Namely, we see that these extensions can be characterized arithmetically by the fact that their trace index is equal to their degree. If K is the function field of a curve and L/K is an *arbitrary* unramified abelian extension then J. Tate has given a very short and elegant proof that $(C_K : S_K^L C_L) \geqq [L:K]$. (See [8].) From this and the limitation theorem one sees immediately that L is actually equal to the class field of Albanese type belonging to $S_K^L C_L$.

We end this section with further remarks on the torsion. Assume that V is non-singular. If we are faced with an abelian unramified extension which is of torsion type, then the reciprocity law mapping is of course defined for such an extension. We are therefore led to conjecture the existence of a subgroup $Z_n(V)$ contained in $Z_a(V)$ such that the factor group is dually isomorphic to the torsion group in the divisors of V. This subgroup should have a geometric definition over any field (characteristic 0 or p) and it should turn out that over a finite field the corresponding group $Z_n(\mathfrak{V}_k)$ should have torsion class fields.

The subgroup $Z_n(\mathfrak{V}_k)$ would then allow us to define non-formal L-series with each character of $Z(\mathfrak{V}_k)/Z_n(\mathfrak{V}_k)$, and we would get once again a decomposition of a zeta function in a covering in terms of these L-series of V. Although one cannot detect the torsion on V by means of the zeta function alone, we see now that it could be detected in the L-series.

Finally, note that this matter has some connection with a theorem of Severi (see [28], Appendix A, 4). Indeed, if the subgroup Z_n is assumed to be between Z_α and Z_l then we can make the following statement: Whenever linear equivalence coincides with Albanese equivalence on a non-singular projective variety, then there can be no torsion in the divisors. Severi's theorem would be a special case, in which in addition Albanese equivalence is the same as algebraic equivalence for cycles of dimension 0 (i.e. $Z_0 = Z_\alpha$).

Remarks added in proof.

1. The class field theory as it is stated here and Theorem 5 of §9 hold under the following more general conditions: A is a commutative algebraic group subject to the sole condition that the pull back over V of its separable isogenies is non degenerate. By this we mean that if $\lambda: B \to A$ is a separable isogeny and U/V is the covering of V obtained by pull back then the degrees $[U:V]$ and $[B:A]$ are equal. The proofs go over essentially without change, replacing "of Albanese type" by "of type (α, A)". Further details, and the necessary lemmas on the existence of primes in arithmetic progressions are given in a forthcoming paper, "Sur les séries L d'une variété algébrique", to appear in the Bulletin de la Société Mathématique de France.

2. We have noted above that all unramified coverings of a complete non singular curve defined over a finite field come from a pull back of separable isogenies of its Jacobian. The class field theory was used to prove this. In the paper immediately following this one, we show how the general algebraic theorem for a curve defined over an arbitrary field can be reduced easily to the arithmetical case.

3. A propos of the questions concerned with torsion (mainly p-torsion), Serre has shown recently that there is no p-torsion on an abelian variety, and that any unramified covering of an abelian variety is an isogeny. This settles the question raised below concerning the generalization of Theorem 4.

Furthermore, Serre has also given an example of a projective non singular variety without torsion, having a trivial Albanese variety, but having an unramified cyclic covering of degree p. The question whether the existence of torsion (especially p-torsion) gives rise to a group $Z_n(V)$ and has any relation to the existence of unramified coverings of degree p remains open.

§6. Class field theory in constant field extensions

Let L/K be a constant field extensions, $L = Kk'$. We let V be a normal, complete model of K. For any cycle $\mathfrak{a}$ in $Z(\mathfrak{V}_k)$ of degree d, one sees directly from the definitions that $(\mathfrak{a}, L/K)$ induces on k' the automorphism $x \to x^{q^d}$. Hence the cycles of degree 0 are contained in the kernel, and $(\mathfrak{a}, L/K) = 1$ if and only if n divides the degree of $\mathfrak{a}$. Since there exists cycles on V rational over k and of degree 1 (Cor. 3 of [16]) the reciprocity law mapping is surjective. We shall prove that its kernel is the trace group. For this we need a theorem which holds for arbitrary commutative algebraic groups. We first make some remarks. Let A/k be an algebraic commutative group (see Nakano [19] and Weil [23]). If k' is a finite extension of k, then the Galois group of k'/k operates on the group of points of A rational over k', which is denoted by $A_{k'}$. Given a point a' in $A_{k'}$, then its conjugates are simply its (symbolic) q^ith powers.

THEOREM 1. *Let A be a commutative algebraic group defined over a finite field k. Let k' be the extension of k of degree n. Then the trace $S = S_k^{k'}$ maps $A_{k'}$ onto A_k. In other words, given a in A_k, there exists a' in $A_{k'}$ such that*

$$a = a' + a'^q + \cdots + a'^{q^{n-1}}.$$

PROOF. Let x be a generic point of A/k, and consider the rational map $x \to x + x^q + \cdots + x^{q^{n-1}}$ of A into itself, defined over k. It is an endomorphism of A. We contend that its kernel is finite. If this were not the case, there would be a group subvariety B of A of dimension ≥ 1 and mapping onto 0. Let y be a generic point of this subvariety over $\bar{k}$. Then $y + y^q + \cdots + y^{q^{n-1}} = 0$ implies that y can be expressed rationally in terms of y^q. This is obviously impossible if the dimension of y is ≥ 1, and our contention is proved. The kernel being finite it follows that our rational map is generically surjective. Hence by Nakano's Theorem 1 [19] it is surjective. Let a be a given point of A_k. There exists b in A such that $a = b + b^q + \cdots + b^{q^{n-1}}$, and our theorem will be proved if we can show that b is in $A_{k'}$. Raising both sides to the qth power and subtracting, we get

$$0 = a^q - a = b^{q^n} - b.$$

This implies that $b^{q^n} = b$, and hence that b is rational over k'.

We have used the fact that given a homomorphism of an algebraic group $f : G \to H$ which is generically surjective, then it is surjective. I am indebted to Kolchin for the following short and elegant proof. The set of points in H which have inverses in G certainly contains an open set (Zariski topology). The cosets are therefore contained in a closed set, and this is obviously absurd, unless they are empty.

COROLLARY. *All cohomology groups of $G(k'/k)$ with coefficients in $A_{k'}$ are trivial. In particular, if $b + b^q + \cdots + b^{q^{n-1}} = 0$ for $b \in A_{k'}$, there exists $c \in A_{k'}$ such that $b = c^q - c$.*

PROOF. This is well known abstract cohomology for cyclic groups. The special case corresponding to the additive form of Hilbert's Theorem 90 can easily be proved independently of cohomology, and is essentially Herbrand's Lemma. We reproduce a proof for the convenience of the reader. Let G be a cyclic group acting on a finite module M. Let σ be a generator of G. Consider the following submodules of M:

$$M \supset M_s \supset M^{1-\sigma} \supset 0$$

$$M \supset M^G \supset SM \supset 0$$

where M_s is the kernel of the trace, M^G the fixed elements under G, SM the elements of M which are traces, and $M^{1-\sigma}$ those of type $m - \sigma m$ for $m \, \epsilon \, M$. Then by the elementary homomorphism theorems, one sees that M/M_s and SM are isomorphic, and that M/M^G and $M^{1-\sigma}$ are isomorphic. This implies in particular that $M_s/M^{1-\sigma}$ and M^G/SM have the same number of elements. In our corollary, we know that $M^G = A_k$ and that M^G/SM is trivial. Hence $M_s/M^{1-\sigma}$ is also trivial, and the special case of our corollary is therefore proved. (See Artin-Tate, [8].)

We return to the class field theory and the kernel of the reciprocity law map. We let A be the Albanese variety of V. Let $\mathfrak{p}_i$ be prime rational cycles of V/k, whose degrees are relatively prime to n, and such that $\pi(\mathfrak{p}_i) = 0$. Such primes can always be found by Proposition 3 of the Appendix. Then a suitable linear combination $\mathfrak{b} = \sum n_i \mathfrak{p}_i$ is of degree 1, and the trace of $\mathfrak{b}$ is clearly $n\mathfrak{b}$. If deg $\mathfrak{a} = d$ then $\mathfrak{a} - d\mathfrak{b}$ has degree 0, and $\pi(\mathfrak{a} - d\mathfrak{b}) = a$ is some point in A_k. By Theorem 1, there exists a' in $A_{k'}$ such that the trace of a' is a. Let $\mathfrak{a}'$ be a cycle rational over k' and of degree 0 such that $\pi(\mathfrak{a}') = a'$ (Proposition 2 of Appendix) and let $\mathfrak{c}$ be the trace of $\mathfrak{a}'$. Then $\pi(\mathfrak{c}) = a$, and we see that $\mathfrak{a} - d\mathfrak{b} - \mathfrak{c}$ is in the kernel of Albanese. This proves that the reciprocity kernel is equal to the trace group, and concludes the proofs of the class field theory in constant field extensions.

We end this section by defining L-series for an arbitrary commutative algebraic group A/k, and showing that they are trivial.

THEOREM 2. *Let χ be a character of A_k. For any cycle $\mathfrak{a}$ of A, rational over k, define $\chi(\mathfrak{a})$ to be $\chi(\pi(\mathfrak{a}))$. Define the L-series in the usual manner as a product or as a sum:*

$$\sum \chi(\mathfrak{a})t^{\deg \mathfrak{a}} = \prod \frac{1}{1 - \chi(\mathfrak{p})t^{\deg \mathfrak{p}}}$$

the sum being taken over the positive cycles and the product over all the primes. If χ is a non-trivial character, then the L-series is equal to the constant 1.

PROOF. Take the logarithmic derivative of the product and collect the coefficients of t^n in the power series thus obtained. Then the coefficient of t^n is

$$\sum_{P \, \epsilon \, A_{k_n}} \chi(S_n(P))$$

where the sum is taken over all points P of A rational over the field k_n (unique extension of k of degree n), and $S_n(P)$ is the trace: $S_n(P) = P + P^q + \cdots + P^{q^{n-1}}$. By Theorem 1, the trace is surjective. Hence if χ is a non-trivial character of A_k, the composite map $\chi \circ S_n$ is a non-trivial character of A_{k_n} and the sum over all points in A_{k_n} is therefore 0. Hence the L-series is equal to 1, as desired.

COROLLARY. *Let $s_n(a)$ be the number of positive cycles $\mathfrak{a}$ on A of degree n, rational over k, such that $\pi(\mathfrak{a}) = a$, for a given point a in A_k. Then given two points a and b in A_k, we have $s_n(a) = s_n(b)$. Thus we have an exact equidistribution of cycles.*

PROOF. If one collects the coefficients of t^n in the additive expression for the L-series, one sees that for a non-trivial character, they must be 0. This is true for every non-trivial character of A_k, and hence by the orthogonality relations, the statement of our corollary follows.

§7. Proofs over abelian varieties

Our next step is to carry out the proofs for the class field theory over abelian varieties. They are interesting for their own sake, but they will also be used in an essential way in the proofs over arbitrary varieties.

Actually the proofs go over to arbitrary commutative group varieties, and hence will be given in that case.

As usual, k is assumed throughout to be finite, with q elements.

Let A/k be a commutative group variety. Let z be a generic point of A/k, and let $\rho = \rho_A$ be the rational map defined by $\rho(z) = z^q - z$. (Weil calls this $\iota - \delta$.) Then ρ is separable, i.e. $k(z)$ is a separable algebraic extension of $k(\rho z)$. In fact the extension $k(z)$ over $k(x)$ is Galois and abelian, the kernel of ρ consists precisely of the group of rational points A_k of A in k, and the Galois group can be identified with the group of translations A_k. The degree $\nu(\rho)$ is the number of rational points. Conversely, if a generic point x of A/k is given, then there exists a point z such that $\rho z = x$, and the cycle $\rho^{-1}(x)$ is simply $\sum(z + a_i)$ where a_i ranges over the points of A_k. We see that $k(z)$ over $k(x)$ is abelian with group A_k.

If A is an abelian variety, then the extension $k(z)$ over $k(x)$ will be called a *Hilbert Class Field* (abbreviated by HCF). We cannot extend this definition reasonably to the function field of an arbitrary commutative group variety, because a function field (say the rational field in one variable) may have several distinct group models (additive or multiplicative groups in one variable) and the extensions derived from the mapping ρ in each case are usually different.

Let $\lambda: B \to A$ be a homomorphism of a commutative group variety B onto a commutative group variety A, all these objects being defined over k. Assume that λ has finite kernel (but not necessarily that the kernel is a subgroup of the rational points of A). Then trivially

$$\rho_A \lambda = \lambda \rho_B .$$

Taking the degree ν of both sides we get

$$\nu(\rho_A)\nu(\lambda) = \nu(\rho_A\lambda) = \nu(\lambda\rho_B) = \nu(\lambda)\nu(\rho_B).$$

This shows that $\nu(\rho_A) = \nu(\rho_B)$ and hence A_k and B_k have the same number of elements.

If A and B are abelian varieties, then the above is a simple proof that if they are defined over k and isogenous over k, then they have the same number of rational points. In particular, the Albanese variety and the Picard variety of a variety have the same number of rational points, which is called the *class number*.

Our next task will be to prove the reciprocity law for commutative group varieties. To simplify the notation, we shall use *group* to mean commutative group variety for the rest of this section. *Homomorphism* will mean a rational map which is a homomorphism.

Let $\lambda : B \to A$ be a homomorphism of a group B onto a group A, defined over k, with finite kernel. Assume that the kernel is a subgroup $\mathfrak{g}$ of the rational points B_k, and assume that λ is separable. Then the above means: If y is a generic point of B/k and $x = \lambda y$ then $k(y)/k(x)$ is an abelian extension and the Galois group is the group of rational translations $\mathfrak{g}$.

Observe that $k(y)$ is contained in the field $k(z)$ such that $z^q - z = x$. Indeed, let $y_1 = y^q - y$ and $x_1 = x^q - x$. Then

$$k(y) \supset k(x) \supset k(y_1) \supset k(x_1),$$

because $\mathfrak{g}$ is a subgroup of B_k and y_1 is obtained by factoring out B_k, while x is obtained by factoring out $\mathfrak{g}$ from B; similarly for $k(x_1)$ contained in $k(y_1)$. The extension $k(y_1)/k(x_1)$ is just like $k(y)/k(x)$, and we see that it is intermediate to $k(x)$ over $k(x_1)$. Looking at the situation backwards, we conclude that $k(y)$ is contained in $k(z)$.

Let us now look at the reciprocity law mapping.

Let $\mathfrak{p}$ be a prime rational cycle of A over k, of degree d. Then for any point Q in $\mathfrak{p}$, we can write $\mathfrak{p} = \sum_{i=0}^{d-1} (Q^{q^i})$. Either by general principles, or by a simple direct argument, one shows that there is a unique translation a in $\mathfrak{g}$ such that for any point P in $\lambda^{-1}(Q)$ we have

$$P + a = P^{q^d}.$$

This translation does not depend on the choice of Q or P, and is the Frobenius automorphism $(\mathfrak{p},\ B/A)$. By linearity we get the reciprocity law mapping $\mathfrak{a} \to (\mathfrak{a},\ B/A)$ of cycles on A into $\mathfrak{g}$.

If λ happens to be the homomorphism ρ, then we note that the reciprocity law mapping is the identity on the rational points of A. In other words, a rational point a in A_k can be viewed as a prime of degree 1, and its associated translation is simply a itself. In particular, the reciprocity law mapping is surjective in this case. Taking into account an obvious consistency, it follows that it is surjective for any homomorphism λ discussed above (because $k(y)$ is contained in the field $k(z)$).

If $\mathfrak{a}$ is a cycle on A, rational over k, and $\mathfrak{a} = \sum m_j(Q_j)$ (the sum being formal) we define as usual $\pi(\mathfrak{a}) = \sum m_j Q_j$ (the sum being the group-sum).

THEOREM 3. *Let* $\lambda : B \to A$ *be a homomorphism of a group* B *onto a group* A, *defined over* k, *whose kernel* $\mathfrak{g}$ *is a subgroup of the rational points* B_k *of* B/k. *Assume also as usual that* B/A *is separable, hence abelian with group* $\mathfrak{g}$. *Then the kernel of* π *is contained in the reciprocity kernel. The reciprocity law mapping induces a homomorphism of* A_k *onto* $\mathfrak{g}$, *and its kernel is* λB_k.

PROOF. We begin by showing that the kernel of π is contained in the reciprocity kernel. Let $\mathfrak{a} = \sum n_j \mathfrak{p}_j$ be any cycle on A, rational over k. With each prime $\mathfrak{p}_j$ we have associated a point Q_j in $\mathfrak{p}_j$, a point P_j in the inverse image of Q_j, and a unique translation a_j such that

$$P_j^{q^{d_j}} - P_j = a_j .$$

Assume $\pi(\mathfrak{a}) = 0$. This means that

$$\sum n_j(Q_j + Q_j^q + \cdots + Q_j^{q^{d_j-1}}) = 0$$

the sum referring to the sum of points on the group variety. The points which are in the inverse image of 0 under the homomorphism λ are simply the points in $\mathfrak{g}$. Hence if we replace Q_j by P_j in the above equation, we must get one of these rational points, i.e. we have

$$\sum n_j(P_j + P_j^q + \cdots + P_j^{q^{d_j-1}}) = b$$

for some b in $\mathfrak{g}$. Raising both sides to the qth power and subtracting, we get

$$\sum n_j(P_j^{q^{d_j-1}} - P_j) = 0$$

and hence $\sum n_j a_j = 0$. But by linearity, $\sum n_j a_j$ is precisely $(\mathfrak{a}, B/A)$. This proves that $\pi(\mathfrak{a}) = 0$ implies $(\mathfrak{a}, B/A) = 0$, as desired.

The reciprocity mapping being surjective, we see from the result just proved that we get an induced homomorphism of A_k onto $\mathfrak{g}$. We must prove that the kernel is λB_k. Let Q be any rational point in A_k, so it is a prime of degree 1. Let P be any point in the inverse image $\lambda^{-1}(Q)$. Then $(Q, B/A) = 0$ if and only if $P^q - P = 0$ by definition of the reciprocity mapping. Hence $(Q, B/A) = 0$ if and only if Q is of type λP for P rational over k. This proves what we wanted.

We shall now be concerned with abelian varieties again.

THEOREM 4. *Let A/k be an abelian variety, and let $f: U \to A$ be an abelian unramified geometric covering of degree prime to p, defined over k. Then U is an abelian variety. Up to a rational constant, f is a homomorphism, and the Galois group is a group of translations on U, rational over k.*

PROOF. The degree being prime to p, it follows that U/A is of Picard type. Hence over the algebraic closure of k, U becomes an abelian variety (by Kummer theory). According to a recent theorem [15] whose proof will be reproduced below for the convenience of the reader, this implies that U is already an abelian variety over k. The map f is therefore a homomorphism by Weil's Theorem 9 of [26] up to a translation, rational over k. After a suitable translation we may assume f is a homomorphism. Since the covering is assumed Galois (even abelian) the Galois group is a group of translations of U, which must be rational over k. This proves our theorem.

We reproduce the proof of the fact that when a variety U/k becomes an abelian variety over $\bar{k}$, then it has a rational point over k, whence a law of composition is defined over k having this rational point as origin.

Let z be a generic point of U/k and consider the rational map $\varphi: U \to U$ defined by $\varphi(z) = z^q$. We know that over $\bar{k}$, U is an abelian variety. According to Weil's Theorem 9, we must have $\varphi = \varphi_0 + c$ where φ_0 is an endomorphism of U and c is a constant. Hence the map $z \to z^q - c$ is an endomorphism of U. The map $z \rightarrow z^q - z - c$ is an endomorphism of U, and we contend that it has finite kernel. Otherwise, there would be an abelian subvariety B of U mapping onto 0, and of dimension ≥ 1. Let y be a generic point of B (say over $\bar{k}$). Then $y^q - y - c = 0$, and hence $y = y^q - c$. This implies that y can be expressed rationally in terms of y^q, and this is absurd. Our endomorphism $z \rightarrow z^q - z - c$ is therefore surjective, and there exists a point P on U such that $P^q - P - c = -c$. Hence $P^q = P$, and P is a rational point.

It would be interesting to know whether the conclusion of Theorem 4 remains true when the degree is not prime to p, or when the covering is not assumed abelian but only Galois. This question is closely related to the problem concerning the existence of p-torsion on the Abelian variety A.

Let A/k be again an abelian variety, and let V denote the underlying variety of A without its added structure of abelian variety. Then every rational point Q in V_k gives rise to a law of composition on V, defined over k, which makes V into an abelian variety having this point as origin. If h is the class number (equal to the number of rational points), then we get h rational maps of type ρ, depending on the origin selected, and thus we get h Hilbert Class Fields of a given function field K/k of A. Each one of them is canonically determined by a given origin.

If $L_1, \cdots, L_h$ denote the h HCF, then they become all equal over a suitable constant field extension. More precisely, if K_h denotes the constant field extension of degree h, then

$$K_h L_i = K_h L_j = K_h L_1 \cdots L_h$$

for any i, j. This is easily seen in the following manner. Start with the abelian variety A endowed with a definite origin. Let $\rho(z) = z^q - z = x$ be as above. If $\rho(w) = x + a$ for some generic point w and some a in A_k, then $k(w)$ over $k(x)$ is the HCF determined by the origin a on A. Let $b = z - w$. Then $k(z, b) = k(z, w)$, and $b^q - b = a$, so b is algebraic over k. However, $[k(w):k(x)] = h$ implies that $[k(z, w):k(z)]$ divides h. Since $k(z, w) = k(z, b)$, we see that $[k(b):k]$ divides h. This proves what we wanted.

§8. Construction of Hilbert class fields and the reciprocity law in those fields

In a preceding section, we have constructed algebraically extensions of Albanese type. We shall now carry out the class field theory in these extensions, assuming that the constant field is finite.

Our notations are the same as those at the beginning of §4. We have $\alpha: V \rightarrow A$, where A is the Albanese variety of V, and $\lambda: B \rightarrow A$. All the objects are defined over k. The origins on A and B are selected so that λ is a homomorphism. Let x be a generic point of A/k, and $\mathfrak{K} = k(x)$. We can write $\mathfrak{L} = k(y)$, where y is a generic point of B/k and $\lambda y = x$. The field $k(y)$ is then contained in the HCF of $k(x)$ determined by our given origin.

Let v be a generic point of V/k, and put $K = k(v)$. As before, let $\xi = \alpha(v)$. Let η be any specialization of y over $x \rightarrow \xi$. The four algebraic statements proved before will now be supplemented by arithmetical statements.

5. Let $\mathfrak{p}$ be a prime rational cycle of V/k and $\pi(\mathfrak{p}) = a$ its associated point in A_k. Then referring to the isomorphism of Statement 4, we have

$$(\mathfrak{p},\ U/V) \leftrightarrow (a,\ B/A).$$

Hence $\pi(\mathfrak{a}) = 0$ implies $(\mathfrak{a},\ U/V)$ is the identity. The kernel of Albanese is contained in the reciprocity kernel.

6. The reciprocity law mapping $\mathfrak{a} \to (\mathfrak{a},\ U/V)$ from the rational cycles of degree 0 into the Galois group is actually surjective, and its exact kernel is the trace group.

In other words, given T in G, there exists a cycle $\mathfrak{a}$ of degree 0 on V, rational over k, such that $(\mathfrak{a},\ U/V) = T$. If for any cycle $\mathfrak{a}$ rational over k we have $(\mathfrak{a},\ U/V) =$ identity, then there is a cycle $\mathfrak{b}$ rational over k which is a trace and such that $\mathfrak{a} - \mathfrak{b}$ is in the kernel of Albanese. In particular, if $\mathfrak{a}$ is of degree 0, so is $\mathfrak{b}$.

Since we are in a geometric situation, the last statement yields the following result: Let $f_*: A(U) \to A(V)$ be the induced homomorphism. Then the Galois group G is isomorphic to $(A_k(V):f_*A_k(U))$ under the reciprocity law mapping.

We shall now prove these arithmetical statements.

Since U/V is divisor unramified, it is unramified over any simple point, and hence for each prime (rational cycle) of V/k, we have the Frobenius automorphism $(\mathfrak{p},\ U/V)$. Recall also that $\alpha: V \to A$ is defined at every simple point.

Let $\mathfrak{p}$ be a prime cycle of V/k, $\mathfrak{p} = \sum_{j=1}^{d} Q^{q^j}$. Then $\alpha(\mathfrak{p}) = \sum \alpha(Q)^{q^j}$ and from this one sees that $\alpha(\mathfrak{p})$ is of type $m\mathfrak{q}$ where $\mathfrak{q}$ is a prime of A/k. Let $T = (\mathfrak{p},\ U/V)$. Then T is characterized by the fact that for any P in $f^{-1}(Q)$ we have $T(P) = P^{q^d}$. Let $g: U \to B$ be the rational map defined by $g(u) = \eta$. Since Q is simple, P is also simple, and $g(P)$ is defined. There is a unique translation $\tau \in \mathfrak{g}$ such that $gT = \tau g$. Put $g(P) = \eta'$. Then $\tau(\eta') = \eta'^{q^d}$. One sees immediately from the definitions that $\tau = (m\mathfrak{q},\ B/A)$. In other words, we get

$$(\mathfrak{p},\ U/V) \leftrightarrow (m\mathfrak{q},\ B/A) = (\pi(\mathfrak{p}),\ B/A),$$

referring to the isomorphism of Statement 4.

In particular, we conclude that the kernel of Albanese is contained in the reciprocity kernel: If $\mathfrak{a}$ is a cycle on V rational over k, and if $\pi(\mathfrak{a}) = 0$, then $(\pi(\mathfrak{a}),\ B/A) = 0$ and hence $(\mathfrak{a},\ U/V)$ is the identity.

We didn't need to assume that $\deg \mathfrak{a} = 0$ because our extension and the origin of the Albanese variety are normalized. We have proved Statement 5.

From the existence of cycles on V mapping on any point of A_k (Prop. 2 of Appendix) we conclude that our reciprocity law mapping is onto the Galois group $G(U/V)$. (We know that it is surjective for B/A.)

There remains to prove that the exact kernel is the trace group.

Suppose $\mathfrak{a}$ is a cycle on V, rational over k, and such that $(\mathfrak{a},\ U/V) = 1$. Let $\pi(\mathfrak{a}) = a$. If s is the period of a, then there exists a prime $\mathfrak{p}$ on V such that $\pi(\mathfrak{p}) = ma$ where m is prime to s. Take μ such that $\mu m \equiv 1 \bmod s$. Then $\pi(\mu\mathfrak{p}) = a$. Furthermore, we see that $(\mathfrak{p},\ U/V)$ is equal to 1 by Statement 5.

Hence $\mathfrak{p}$ splits completely in U/V, so that it is a trace. Hence $\mu\mathfrak{p}$ is a trace, and $\pi(\mathfrak{a} - \mu\mathfrak{p}) = 0$. We can find primes $\mathfrak{p}_i$ and integers n_i such that $\pi(\mathfrak{p}_i) = 0$ (whence $\mathfrak{p}_i$ splits completely) and such that $\sum n_i \deg (\mathfrak{p}_i) = \deg (\mathfrak{a} - \mu\mathfrak{p})$. Then $-\mu\mathfrak{p} + \sum n_i\mathfrak{p}_i$ is a trace, and $\mathfrak{a} - \mu\mathfrak{p} + \sum n_i\mathfrak{p}_i$ is of degree 0 and in the kernel of Albanese. This concludes the proof of Statement 6, and hence of the reciprocity law in the special class fields derived from a covering B/A of the Albanese variety.

§9. Characterization of unramified abelian extensions

In the preceding section we have constructed class fields over a given function field K by using a specialization lifting process on the class fields of the Albanese variety. Here we shall prove that these fields, and constant field extensions, give all abelian divisor-unramified extensions of K, with the exception of torsion class fields and possibly other extensions whose degree is a power of p.

Let x be a generic point of A/k, and $\rho(y) = y^q - y = x$, so $k(y)/k(x)$ is the HCF determined by our given origin on A. The field $k(v, \eta)$ obtained by the pulling back of $k(y)$ will be called the *HCF of $k(v)$ derived from the HCF $k(y)/k(x)$*, or determined by our given origin on A.

THEOREM 5. *If L' is an abelian extension of K, of Albanese type, then L' is contained in the composite field of a HCF of K and a constant field extension.*

PROOF. The proof will take place in two steps. In the first we reduce our theorem to the case where L'/K is geometric, and L' contains the HCF H of K. In the second, we deduce from the fact that L' and H have the same trace group of cycle classes that they must be equal, by passing to a suitable constant field extension, and comparing the various indices of the trace factor groups.

Let H/K be as above the HCF of K determined by the given origin on A, and the rational map $\alpha: V \to A$. We note especially that any prime $\mathfrak{p}$ on V rational over k such that $\pi(\mathfrak{p}) = 0$ splits completely in H by Statement 5 of the preceding section.

Let $L_1 = HL'$. Then L_1/K is abelian and of Albanese type. Let k_1 be its constant field, and $K_1 = Kk_1$. Let k_2 be a constant field extension of k_1 such that $L_2 = L_1k_2$ becomes a subfield of a HCF over K_2. Such an extension can always be found because of our assumption that our extension is of Albanese type. Let k_3 be a constant field extension of k_2 such that all the HCF of K_2 become equal over k_3 and let L_3/K_3 be the lifting of L_2/k_2 over k_3.

Let $\mathfrak{p}$ be a prime cycle of V rational over k such that $\pi(\mathfrak{p}) = 0$, and such that $\deg \mathfrak{p}$ is relatively prime to the degree $[k_3:k]$. Such a prime can always be found by Prop. 3 of the Appendix. Then $\mathfrak{p}$ remains prime in the constant field extension K_2/K and then splits completely in the HCF of K_2 determined by our given origin. According to the choice of k_3, we see that $\mathfrak{p}$ remains prime in the constant field extension K_3/K and then splits completely in L_3. The decomposition group D of any prime $\mathfrak{P}$ in L_3 extending $\mathfrak{p}$ is therefore precisely of order $[k_3:k]$ and its fixed field is clearly an extension L of K having the same constant field as K (namely k) and such that $Lk_3 = L_3$. In addition, its fixed field is the largest subfield of L_3 into which $\mathfrak{p}$ splits completely, and hence L contains H. This concludes our first step, because L' is contained in $L_3 = Lk_3$.

We are now faced with an abelian extension L/K which is geometric, of Albanese type and such that $L \supset H \supset K$, where H is the HCF determined by the given origin on A. We shall prove that $L = H$ by some index computations.

The inclusion $L \supset H \supset K$ corresponds to homomorphisms $B \to C \to A$ on the Albanese varieties (defined over k). Let $f_*:B \to A$ and $g_*:C \to A$ be the homomorphisms. Then by class field theory in H (Statement 6 of the preceding section) the trace group of H is reduced to 0 and so is the trace group of L, i.e. we have

$$g_* C_k = f_* B_k = 0, \quad \text{and} \quad [H:K] = (A_k:0).$$

Let k' be a constant field extension such that $L' = Kk'$ is exhibited as an extension of Albanese type of $K' = Kk'$, i.e. is contained in a HCF of K'. Then by class field theory in L' over K' we get

$$[L':K'] = (A_{k'}:f_* B_{k'}).$$

We have trivially

$$[L:K] = [L':K'].$$

In order to conclude the proof we combine the preceding equalities with the following lemma.

LEMMA. *Let K/k be a function field over the finite field k and L/K an abelian geometric extension of Albanese type, belonging to the rational map $f:U \to V$. Let A/k and B/k be the Albanese varieties of V and U respectively, and $f_*:B \to A$ the homomorphism induced by f, so that f_*, B, and A are defined over k. Let k'/k be a finite extension such that $L' = Lk'$ is contained in a HCF of $K' = Kk'$. Let $S:A_{k'} \to A_k$ be the trace. Then the inverse (under S) of $f_* B_k$ is exactly $f_* B_{k'}$, and in particular,*

$$(A_k:f_* B_k) = (A_{k'}:f_* B_{k'}).$$

PROOF. The group $f_* B_{k'}$ is obviously contained in the inverse image of the trace, because f_* commutes with the trace. We recall that the trace is surjective (Th. 1 of §6). Hence the only thing to show is that if $S(a') = 0$ then there exists b' in $B_{k'}$ such that $a' = f_* b'$. By the Corollary of Th. 1 in §6 we know that there exists c' in $A_{k'}$ such that $a' = c'^q - c'$. Let $\mathfrak{p}'$ be a prime cycle of V rational over k'. We consider the behaviour of $\mathfrak{p}'$ in the extension L'/K'. Let T be the birational transformation of U/V corresponding to the automorphism $(\mathfrak{p}', L'/K')$ and let T_1 be the transformation corresponding to the automorphism $(\mathfrak{p}'^q, L'/K')$. If we can show that $T = T_1$ then it follows that for any cycle $\mathfrak{a}'$ of V, of degree 0 and rational over k' we have $(\mathfrak{a}', L'/K') = (\mathfrak{a}'^q, L'/K')$ and hence by the class field theory which we know holds in L'/K' (according to §8) we can conclude that $c'^q - c' = f_* b'$ for some b' in $B_{k'}$. This would prove what we want.

Having assumed that $f:U \to V$ is defined and abelian over k, it follows that the birational transformations T and T_1 are also defined over k (even though $\mathfrak{p}'$ and $\mathfrak{p}'^q$ are not rational over k). Let Q be any point in $\mathfrak{p}'$, and P a point in $f^{-1}(Q)$. Let $d = \deg \mathfrak{p}'$. Using the hypothesis that U/V is abelian and unramified over $\mathfrak{p}'$, we know that T is the unique transformation such that

$$T(P) = P^{q'^d}$$

where q' is the number of elements in k'. Since $\deg (\mathfrak{p}') = \deg (\mathfrak{p}'^q)$ we know likewise that T_1 is the unique transformation such that

$$T_1(P^q) = P^{qq'^d}.$$

From this, one sees that T and T_1 have the same effect on P^q. Hence $T = T_1$ because U/V is unramified over $\mathfrak{p}'$. This proves our lemma, and concludes the proof of the theorem.

§10. End of the proofs

The arguments which now follow and which will conclude the proofs of the class field theory as stated in §5 are essentially routine.

Let L/K be an arbitrary abelian extension of Albanese type. Let V be a normal model of K, projective. Let $\mathfrak{a}$ be a cycle on V, rational over k. The reciprocity law mapping $\mathfrak{a} \rightarrow (\mathfrak{a}, L/K)$ gives a homomorphism of $Z(\mathfrak{B}_k)$ onto the Galois group G. This is easily seen as follows: We know that L is contained in Hk' where H is a HCF, and k' a constant field extension. In each type of extension, the reciprocity law mapping is surjective. Hence it is surjective on the composite, and using its consistency, it is surjective in any smaller extension.

The trace group is obviously contained in the kernel. To prove that the trace group is exactly the kernel, we have to prove the inequality

$$(C_K : SC_L) \leqq [L:K]$$

where S denotes the trace from L to K.

Now the inclusion $C_K \supset S_K^E C_E \supset S_K^L C_L$ implies that

$$(C_K : S_K^L C_L) = (C_K : S_K^E C_E)(S_K^E C_E : S_K^L C_L)$$

and since $S_K^L C_L = S_K^E S_E^L C_L$, we have $(S_K^E C_E : S_K^L C_L) \leqq (C_E : S_E^L C_L)$. This shows that to prove our inequality for L/K, it suffices to prove it in each step of a tower, L/E and E/K.

If we take E to be that part of L which is obtained by constant field extension, and use the results of §6, we see that it suffices to prove our inequality for geometric extensions.

Given $c \in C_K$ we can obviously define the degree of c to be the degree of any representative cycle rational over k. Consider the homomorphism of a cycle class on its degree. It is onto the integers $\mathbf{Z}$, because there are cycle classes of degree 1 (Appendix, Prop. 3). By elementary homomorphism theorems, we have

$$(C_K : S_K^L C_L) = (C_K^0 : S_K^L C_L^0)\,(\mathbf{Z} : d\mathbf{Z})$$

where d is the greatest common divisor of the degrees of all cycle classes $c \in C_K$ which are traces from C_L. By the lemma at the end of §9, we know that the first factor $(C_K^0 : S_K^L C_L^0) = [L:K]$. Hence it will suffice to prove that $d = 1$. By Proposition 4 of the Appendix, there are prime cycles $\mathfrak{p}$ of degree relatively prime to any given integer, which split completely in L, i.e. which are traces. Hence there is a cycle of degree 1 which is a trace, and this concludes the proof of our inequality.

The uniqueness theorem is now obvious.

As for the existence theorem, given a subgroup M of C_K, it is clear that the compositum L of a HCF with a suitable constant field extension has the property that $S_K^L C_L$ is contained in M. Let G be the Galois group of L/K and let $\omega : C_K \to G$ be the reciprocity law mapping. Let G_0 be the subgroup of B obtained by taking the images under ω of all elements in M. Let $m = (C_K : M)$. Then $(G : G_0) = m$. Let L_0 be the subfield of L fixed under G_0. Then $[L_0 : K] = m$. From this and the inequality proved earlier, it follows that M is the trace group belonging to L_0, and that L_0 is the class field belonging to M.

We have now proved everything except the limitation theorem. For this we need a lemma (*Translation Theorem*).

LEMMA 1. *Let L/K be abelian, of Albanese type. Let E/K be a Galois extension. Let $c \in C_E$. Then $(c, LE/E) = (S_K^E(c), L/K)$.*

PROOF. We suppose that we have selected a model of K, thereby giving rise to models of L, E, and LE. It suffices to prove our lemma for cycles representing the cycle class c, and in fact for cycles whose points lie outside a given proper algebraic subset of the chosen model for E. Therefore, it suffices to prove: Given a prime cycle $\mathfrak{q}$ in E such that the prime $\mathfrak{p}$ in K lying below $\mathfrak{q}$ is non-singular and unramified in LE, we have

$$(\mathfrak{q}, LE/E) = (S_K^E(\mathfrak{q}), L/K).$$

As usual, the Galois group of LE/E has been identified as a subgroup of the Galois group of L/K, and strictly speaking, the automorphism on the right hand side is the restriction to L of the automorphism on the left hand side.

Let $\mathfrak{P}$ be a prime in LE lying above $\mathfrak{q}$. Let m be the residue class degree of $\mathfrak{q}$ relative to $\mathfrak{p}$. Then $S_K^E(\mathfrak{q}) = m\mathfrak{p}$. Let $\sigma = (\mathfrak{q}, LE/E)$. If $\mathfrak{o}$ is the local ring of $\mathfrak{P}$ in LE, then σ is uniquely determined as the automorphism of LE/E such that

$$x^\sigma \equiv x^{N\mathfrak{q}} \ (\mathrm{mod}\ \mathfrak{P})$$

for all x in $\mathfrak{O}$. But $N\mathfrak{q} = N\mathfrak{p}^m$. This makes our lemma obvious.

The proof of the limitation theorem is now easily concluded: Suppose E/K is Galois. By the existence theorem we know that its trace group is also the trace group of an abelian class field L/K of Albanese type: For any $c \in C_E$ we must have $(S_K^E(c), L/K) = 1$, and hence by Lemma 1, $(c, LE/E) = 1$. Since LE/E is abelian of Albanese type by Prop. 12 of §4 it follows by class field theory over E that $LE = E$, hence that L is contained in E. This concludes the proofs of the class field theory.

§11. The nth power residue symbol

If our abelian extension of Albanese type happens to be Kummer, then on the one hand, we have a duality between its Galois group G and a subgroup of the rational points on the Picard variety given by Kummer theory, and on the other hand, we have a direct isomorphism between G and a factor group of the rational points on the Albanese variety given by the reciprocity law mapping. The pairing between the points of the Picard variety and those of the Albanese variety is easily given by the nth power residue symbol in the classical manner, which we now describe.

Assume k contains the nth roots of unity, n prime to p.

Let X be a divisor on the variety V which is algebraically equivalent to 0, rational over k, and such that $nX = (\varphi)$ is linearly equivalent to 0. Let ψ be any function such that $\psi^n = \varphi$. Let $L = K(\varphi^{1/n})$. Let $\mathfrak{p}$ be a prime rational cycle of V/k, and $\sigma = \sigma(\mathfrak{p}) = (\mathfrak{p}, L/K)$ the Frobenius automorphism. Then $\psi^{\sigma-1}$ is an nth root of unity in k which depends only on the class of $\varphi \bmod K^{*n}$, and so depends only on the linear equivalence of X. By linearity, we get a pairing

$$(X, \mathfrak{a}) \rightarrow \psi^{\sigma(\mathfrak{a})-1}$$

of the divisor classes of period n, rational over k, and the cycles of degree 0 on V, rational over k. (If $\mathfrak{a} = \sum n_i \mathfrak{p}_i$ we put $\sigma(\mathfrak{a}) = \prod \sigma(\mathfrak{p}_i)^{n_i}$.) By the consistency of the Frobenius automorphism, and by class field theory, we can conclude that the kernel of Albanese is contained in the right hand kernel and hence we have a pairing between the points of period n on the Picard variety, rational over k, and the rational points on the Albanese variety.

The root of unity can be described entirely in the ground field in the following manner. Let us first state a lemma.

LEMMA. *Let X on V be rational over k, and $nX = (\varphi)$ be the divisor of a function φ rational over k. Let $\mathfrak{p}$ be a prime rational cycle of V/k. Then there exists X_1 linearly equivalent to X, and rational over k, such that the function φ_1 given by $nX_1 = (\varphi_1)$ is defined at every point of $\mathfrak{p}$.*

PROOF. If $\mathfrak{o}$ is the local ring of $\mathfrak{p}$ in K, then the unique factorization in $\mathfrak{o}$ implies the existence of a function φ_0, rational over k, such that $X = (\varphi_0) + X_0$, and such that no point of $\mathfrak{p}$ is contained in the support $|X_0|$ of X_0. The lemma is now obvious.

Returning to our pairing, given a prime $\mathfrak{p}$, we may assume that the function φ with which we are dealing is defined at every point of $\mathfrak{p}$. Indeed, if $\psi_1^n = \varphi_1$, then $\psi^{\sigma(\mathfrak{p})-1} = \psi_1^{\sigma(\mathfrak{p})-1}$.

Let $\mathfrak{p} = \sum (Q^{q^i})$ and define $\varphi(\mathfrak{p}) = \prod \varphi(Q^{q^i})$. Let $\mathfrak{o}$ be the local ring of $\mathfrak{p}$ in K, and let $\mathfrak{O}$, $\mathfrak{P}$ be any local ring in L lying above $\mathfrak{o}$, $\mathfrak{p}$. By definition, we have

$$\psi^{\sigma-1} \equiv \psi^{(N\mathfrak{p}-1)/n} \pmod{\mathfrak{P}}$$

where $N\mathfrak{p} = q^d$ and $d = \deg \mathfrak{p}$. It is then trivially verified that

$$\psi^{\sigma-1} = \varphi(\mathfrak{p})^{(q-1)/n},$$

which is therefore described entirely in the ground field K.

To summarize: Given a point of order n on the Picard variety, rational over k, we select any divisor X rational over k, algebraically equivalent to 0, and belonging to this point. Let $\mathfrak{a}$ be a cycle of degree 0 on V belonging to a rational point in A_k. Let $nX = (\varphi)$. Then we get a pairing

$$(X, \mathfrak{a}) \to \varphi(\mathfrak{a})^{(q-1)/n}.$$

Denoting by A_k^* the rational points on the Picard variety, the above pairing induces a well-defined pairing between the elements of period n in A_k^* and the elements of A_k.

APPENDIX: EXISTENCE OF CYCLES IN GENERALIZED ARITHMETIC PROGRESSIONS

In this section we prove the auxiliary statements concerning the existence of rational cycles and primes in what corresponds in our theory to arithmetic progressions. We begin by a statement concerning divisors.

PROPOSITION 1. *Let V be a complete normal variety over a finite field k. Let ξ be a rational point (over k) on the Picard variety of V, representing a divisor class on V. Then there exists a divisor X in that class which is rational over $k(\xi)$. In other words, if a divisor class is left invariant by all automorphisms of the universal domain leaving k fixed, then there is a divisor in it which is rational over k.*

PROOF. The proof is of a standard technical nature in function fields in one variable. There certainly is finite extension k_1 of k and a divisor X rational over k_1 in the class. Let G be the Galois group of k_1/k, let K be the function field of V/k, and $K_1 = Kk_1$. By a * we denote the multiplicative group of a field. Then G acts on K_1^*, k_1^*, and on the group D_1 of divisors on V rational ove k_1. Any divisor in D_1 fixed under G is rational over k. We have an exact sequence

$$0 \to K_1^*/k_1^* \to D_1 \to C_1 \to 0$$

where C_1 are the divisor classes, and we have another exact sequence

$$0 \to k_1^* \to K_1^* \to K_1^*/k_1^* \to 0.$$

From the first we get

$$H^0(D_1) \to H^0(C_1) \to H^1(K_1^*/k_1^*)$$

and our goal is to prove that $H^0(D_1)$ maps onto $H^0(C_1)$. This is certainly the case if we can prove that $H^1(K_1^*/k_1^*)$ is trivial. This is proved by using the second exact cohomology sequence

$$H^1(K_1^*) \to H^1(K_1^*/k_1^*) \to H^2(k_1^*),$$

the fact that the first Galois cohomology group of fields is trivial, together with the triviality of the second Galois cohomology group of a finite field (Wedderburn's theorem).

There are of course many other proofs for the above proposition, and the reader is invited to use another one if he is allergic to exact sequences.

Actually, given any point ξ on the Picard variety, transcendental or not, there is always a divisor rational over $k(\xi)$ (k finite as before), but this stronger statement is not needed.

Let us now prove the existence of cycles in given arithmetic progressions.

PROPOSITION 2. *Let V be a projective, normal variety, defined over the finite field k. Let $\alpha: V \to A$ be a given canonical map of V into its Albanese variety, defined over k. Let F be a proper algebraic subset of V. Let a be a given rational point in A_k. Then there exists a cycle $\mathfrak{a}$ on V, rational over k, and of degree 0, such that none of its points lies in F, and such that $\pi(\mathfrak{a}) = a$.*

PROPOSITION 3. *The notation being as in Proposition 2, let s be a given integer. Then there exists a prime rational cycle $\mathfrak{p}$ of V/k such that $\pi(\mathfrak{p}) = ma$ where m is relatively prime to s, such that the degree of $\mathfrak{p}$ is relatively prime to s also, and such that no point in $\mathfrak{p}$ lies in the algebraic set F. If $a = 0$ is the origin on A, then we can take $m = 1$ in the preceding statement.*

In the case of dimension 1, our propositions are known to the analytic number theorist (in characteristic p) and are immediate consequences of the prime number theorem. They follow from the behaviour of the zeta function and L-series at the point $s = 1$, or $t = 1/q$, putting as usual $t = 1/q^s$. For the convenience of the reader we recall these results, and I am indebted to John Tate for the following exposition (using the Riemann hypothesis).

We assume that V/k is a curve (complete, non-singular) over the finite field k, and $\alpha: V \to J$ a canonical mapping into the Jacobian, defined over k. If $\mathfrak{a}$ is a cycle on V rational over k, and χ a character of J_k, then $\chi(\mathfrak{a})$ is defined to be $\chi(\pi(\mathfrak{a}))$. The L-series $L(t, \chi)$ belonging to this character is the analytic function taking the value 1 at $t = 0$, and having the following logarithm:

$$\log L(t, \chi) = \sum_{\mathfrak{p}} \sum_{\mu=1}^{\infty} \frac{\chi(\mu\mathfrak{p})}{\mu} t^{\deg(\mu\mathfrak{p})}.$$

According to the Riemann hypothesis (Weil [26], page 82) the L-series is a polynomial $\prod_{i=1}^{B(\chi)} (1 - \alpha_i t)$ with $|\alpha_i| = q^{\frac{1}{2}}$, whenever χ is not the principal character χ_1. Otherwise, it is the zeta function itself. In the first case, we have therefore

$$\sum_{\deg(\mu\mathfrak{p})=n} \frac{\chi(\mu\mathfrak{p})}{\mu} = - \sum_{i=1}^{B(\chi)} \frac{\alpha_i^n}{n} = O(q^{\frac{1}{2}n})$$

for $n \to \infty$. If χ is the principal character, then

$$\sum_{\deg(\mu\mathfrak{p})=n} \frac{\chi_1(\mu\mathfrak{p})}{\mu} = \frac{q^n + 1}{n} + O(q^{\frac{1}{2}n}).$$

Let a be a given rational point on J, and $\mathfrak{a}$ a cycle rational over k such that $\pi(\mathfrak{a}) = a$. Multiplying by $\chi(-a)$, summing over all characters, and using the orthogonality relations, we get

$$\sum_{\substack{\deg(\mu\mathfrak{p})=n \\ \pi(\mu\mathfrak{p})=a}} \frac{h}{\mu} = \frac{q^n + 1}{n} + O(q^{\frac{1}{2}n}).$$

Since deg $(\mu\mathfrak{p}) \leqq n$ implies deg $(\mathfrak{p}) \leqq \frac{1}{2}n$, for $\mu \geqq 2$, one sees immediately from the above that

$$\sum_{\substack{\deg(\mu\mathfrak{p})=n \\ \mu>1}} (1/\mu) = O(q^{\frac{1}{2}n}).$$

This yields

$$\sum_{\substack{\deg(\mathfrak{p})=n \\ \pi(\mathfrak{p})=a}} 1 = \frac{1}{h}\left(\frac{q^{n}+1}{n}\right) + O(q^{\frac{1}{2}n}).$$

Hence as n increases indefinitely, so does the number of primes of degree n, and lying in a given arithmetic progression (i.e. such that $\pi(\mathfrak{p}) = a$).

Propositions 2 and 3 will be proved by considering an algebraic system of curves on the variety, and using the known results for curves over finite fields.

The generic curve C_u on V has the following properties (cf. Chow's treatment of the Picard and Albanese varieties [11]): It is non-singular, defined over a regular extension $k(u) = k_u$ of k, and there is a canonical mapping $\varphi_u : C_u \to J_u$ of this curve into its Jacobian, both φ_u and J_u being defined over k_u. (From now on, an index u, or t will always mean that the object in question is defined or rational over the field $k(u)$ or $k(t)$.) The Albanese variety A of V is then the k_u/k-image of J_u, and the homomorphism $\lambda_u : J_u \to A$ defined over k_u is primary, i.e. the inverse image of a point of A is set theoretically a variety, by Corollary 2 of Theorem 6, [10]. This means that for any point b in A, $\lambda_u^{-1}(b) = p^\mu W_u(b)$ where $W_u(b)$ is a subvariety of J_u, depending of course on b, and defined over a purely inseparable extension of $k_u(b)$.

We would like commutativity: $\alpha i_u = \lambda_u \varphi_u$ where $i_u : C_u \to V$ is the inclusion. However, we started with a definite map α of V into A, and the canonical map φ_u defined over k_u is determined by a base point Q_u of C_u. Hence as things stand, we cannot guarantee commutativity if we insist that λ_u be a homomorphism, because Q_u is rational over k_u but not necessarily over k. Let $\xi_u = \alpha(Q_u) = \alpha i_u(Q_u)$. Then the following diagram

$$
\begin{array}{ccc}
C_u & \xrightarrow{\varphi_u} & J_u \\
{\scriptstyle i_u}\downarrow & & \downarrow{\scriptstyle \lambda_u - \xi_u} \\
V & \xrightarrow{\alpha} & A
\end{array}
$$

has the following properties:

1. i_u is the inclusion.

2. φ_u is a canonical map of C_u into its Jacobian, sending the point Q_u on the origin of J_u.

3. λ_u is the induced homomorphism of J_u onto A, and ξ_u is the constant which makes the diagram commutative.

4. λ_u is primary, i.e. the inverse image $\lambda_u^{-1}(b)$ for any point b of A is of type $p^\mu W_u(b)$ where $W_u(b)$ is a variety defined over a purely inseparable extension of $k(u, b)$.

Now for almost all specializations t of u in the algebraic closure $\bar{k}$ of k, the diagram specializes to a diagram

$$
\begin{array}{ccc}
C_t & \xrightarrow{\ \varphi_t\ } & J_t \\
{\scriptstyle i_t}\downarrow & & \downarrow{\scriptstyle \lambda_t - \xi_t} \\
V & \xrightarrow[\ \alpha\ ]{} & A
\end{array}
$$

where C_t is non-singular, simple on V, the four properties enumerated above continue to hold provided we replace u by t, and C_t is not contained in a given proper algebraic subset of V.

All of our varieties are assumed imbedded in a projective space, and their projective degrees do not change under specialization. In particular, the degrees of the varieties $W_t(b)$ are all equal to a fixed constant. The field $k(t)$ being a finite field, it follows that each $W_t(b)$ is defined over the finite field $k(t, b)$ (taking b algebraic over k, of course). By the results of [16] and [21] it follows that if $[k(t, b):k]$ is sufficiently large, the variety $W_t(b)$ will have a rational point over $k(t, b)$.

In particular, assume first that $b = \xi_t$, let $[k(t):k]$ be large, and let w_t be a point in $W_t(\xi_t)$ rational over $k(t)$. If we let $\psi_t = \varphi_t - w_t$, then ψ_t is a canonical map of C_t into its Jacobian J_t, and we get a commutative diagram:

$$
\begin{array}{ccc}
C_t & \xrightarrow{\ \psi_t\ } & J_t \\
{\scriptstyle i_t}\downarrow & & \downarrow{\scriptstyle \lambda_t} \\
V & \xrightarrow[\ \alpha\ ]{} & A
\end{array}
$$

Since C_t is simple on V, almost all of its points will be simple on V, and the above diagram is commutative not only at generic points, but at every point of C_t which is simple on V. By a cycle on C_t, we shall from now on mean a *cycle all of whose points are also simple on V*.

Let π_t be the map on the cycles of C_t obtained by taking first the image under ψ_t in J_t, and then taking the sum on J_t. Then for any cycle $\mathfrak{a}_t$ on C_t rational over k_t, (which is also a cycle on V) we have the following commutativity:

$$
\pi(\mathfrak{a}_t) = \pi(i_t(\mathfrak{a}_t)) = \lambda_t(\pi_t(\mathfrak{a}_t)).
$$

Let s be a given integer, divisible by the order of A_k. We can select t such that $[k(t):k]$ is large and prime to s. Let a be a given point in A_k. Let z_t be a rational point of $W_t(a)$ over k_t. By the theory of curves over a finite field, one knows that there exists a cycle $\mathfrak{a}_t$ of C_t of degree 0 such that $\pi_t(\mathfrak{a}_t) = z_t$. Since we have commutativity in diagrams, we have

$$
\pi(\mathfrak{a}_t) = \pi(i_t(\mathfrak{a}_t)) = \lambda_t(\pi_t(\mathfrak{a}_t)) = a.
$$

Identifying $\mathfrak{a}_t$ with its image under inclusion, we see that $\pi(\mathfrak{a}_t^q) = \pi(\mathfrak{a}_t)^q = a^q = a$ implies that

$$
\pi(\mathfrak{a}_t + \mathfrak{a}_t^q + \cdots + \mathfrak{a}_t^{q^{n-1}}) = na
$$

where $n = [k(t):k]$. The cycle $\mathfrak{a}_t + \mathfrak{a}_t^q + \cdots + \mathfrak{a}_t^{q^{n-1}}$ is rational over k. Since n is relatively prime to the order of A_k, hence to the period of a, we see that a suitable multiple of our rational cycle over k will have the image a under the π mapping. This proves Proposition 2.

For Proposition 3, we argue essentially in the same way, and use the result from the theory of curves that there exists actually a prime cycle $\mathfrak{p}_t$ of C_t rational over k_t and of degree prime to s such that $\pi_t(\mathfrak{p}_t) = z_t$, hence such that $\pi(\mathfrak{p}_t) = a$. Now the prime may be rational over a smaller field than $k(t)$ containing k, but the degree of that field over k will again be relatively prime to the period of a. Taking the trace of $\mathfrak{p}_t$ relative to this smaller subfield, we get a prime rational cycle of V over k, of degree prime to s, which maps on ma, with m prime to s. This proves Proposition 3.

(Actually, most of the primes $\mathfrak{p}_t$ will not be rational over a smaller field. We shall return to this and related questions of density in a subsequent paper.)

PROPOSITION 4. *Let* $f: U \to V$ *be a Galois covering, defined over a finite field* k. *Then there exists infinitely many primes* $\mathfrak{p}$ *of* V, *of degree relatively prime to any given integer, and which split completely in* U.

PROOF. We have by [16]

$$N_\mu(V) = q^{\mu r} + O(q^{\mu(r-\frac{1}{2})})$$

$$N_\mu(U) = q^{\mu r} + O(q^{\mu(r-\frac{1}{2})})$$

where $N_\mu(U)$ and $N_\mu(V)$ are the number of simple points of U and V respectively in the field k_μ. The correspondence between U and V is $(n, 1)$ except on a proper algebraic subset of U and V, where n is the degree of the covering. This shows that the points of V in k_μ which split completely (i.e. such that the n points of U above them are rational over k_μ) are at least $(1/n)q^{\mu r}$ in number, with an error term of lower order. These must therefore increase indefinitely as μ increases indefinitely, and such points of degree μ give rise to primes of the desired kind.

COLUMBIA UNIVERSITY

REFERENCES

[1] ABHYANKAR, S., *On the ramification of algebraic functions*, Amer. J. Math., Vol. LXXVII (1955), pp. 575–592.

[2] ———, *Local Uniformization on algebraic surfaces over ground fields of characteristic* $p \neq 0$, Ann. of Math., Vol. 63 (1956), pp. 491–526.

[3] ———, *Splitting of valuations in extensions of local domains*, Proc. Nat. Acad. Sci. U.S.A., Vol. 41 (1955), pp. 220–223.

[4] ———, and ZARISKI, O., *Splitting of valuations in extensions of local domains*, Proc. Nat. Acad. Sci. U.S.A., Vol. 41 (1954), pp. 84–90.

[5] ARTIN, E., *Über eine neue Art von L-Reihen*, Abh. Math. Sem. Hamburg, Bd. 3 (1924), pp. 89–108.

[6] ———, *Zur Theorie der L-Reihen mit allgemeinen Gruppencharakteren*, Abh. Math. Sem. Hamburg, Bd. 8 (1930), pp. 292–306.

[7] ———, *Beweis des Allgemeinen Reziprozitätsgeseztes*, Abh. Math. Sem. Hamburg, Bd. 5 (1927), pp. 353–363.

[8] —— AND TATE, J., *Class Field Theory*, Mimeographed Notes, New York University and Princeton University, to appear.

[9] CHEVALLEY, C., *La notion d'anneau de décomposition*, Nagoya Math. J., Vol. 7 (June 1954), pp. 21–33.

[10] CHOW, W. L., *Abelian varieties over function fields*, Trans. Amer. Math. Soc., Vol. 78 (1955), pp. 253–275.

[11] ——, *Abstract Theory of the Picard and Albanese varieties*, to appear.

[12] ——, *Algebraic systems of positive cycles*, Amer. J. Math., Vol. LXXII (1950), pp. 247–283.

[13] KRULL, W., *Der allgemeine Diskriminantensatz. Unverzweigte Ringerweiterungen*, Math. Zeit., Bd. 45 (1939), pp. 1–19.

[14] ——, *Galoissche Theorie der ganz abgeschlossene stellenringe*, S.-B. Phys. Med. Soz. Erlangen, Bd. 67/68 (1935–36), pp. 324–328.

[15] LANG, S., *Abelian varieties over finite fields*, Proc. Nat. Acad. Sci. U.S.A., Vol. 41 (1955), pp. 174–176.

[16] —— and WEIL, A., *Number of points of varieties in finite fields*, Amer. J. Math., Vol. LXXVI (1954), p. 819–827.

[17] MATSUSAKA, T., *On the algebraic construction of the Picard variety*, I and II, Jap. J. Math., Vol. XXI (1951), pp. 217–235 and Vol. XXII (1952), pp. 51–62.

[18] NAGATA, M., *On the theory of Henselian rings*, Nagoya Math. J., Vol. 5 (Feb. 1953), pp. 45–57.

[19] NAKANO, S., *Note on group varieties*, Mem. Coll. Sci., Univ. Kyoto, Vol. XXVII (1952), pp. 55–66.

[20] NÉRON, A., *Problèmes arithmétiques et géométriques rattachés à la notion de rang d'une courbe algébrique dans un corps*, Bull. Soc. Math. France, 80 (1952), pp. 101–166.

[21] NISNEVIC, L. B., *Über die Anzahl der Punkte einer algebraische Mannigfaltigkeit in einem endlichen primkorper*, (in Russian) Dokl. Akad. Nauk, SSSR, Vol. XCIX (1954), pp. 17–20.

[22] ROSENLICHT, M., *Generalized Jacobian varieties*, Ann. of Math., Vol. 59 (1954), pp. 505–530.

[23] WEIL, A., *Algebraic groups and homogeneous spaces*, Amer. J. Math., Vol. LXXVII (1955), pp. 493–512.

[24] ——, *Critères d'équivalence en géométrie algébrique*, Math. Ann., Bd 128 (1954), pp. 95–127.

[25] ——, *Numbers of solutions of equations in finite fields*, Bull. Amer. Math. Soc., Vol. 55 (1949), pp. 497–508.

[26] ——, *Variétés abeliennes et courbes algébriques*, Hermann et Cie, Paris, 1948.

[27] ZARISKI, O., See Abhyankar [1] Theorem 1, and Abhyankar [4].

[28] ——, *Algebraic surfaces*, Ergeb. der Math., 3. Bd., 5, Berlin, 1935.

ANNALS OF MATHEMATICS
Vol. 64, No. 2, September, 1956
Printed in U.S.A.

ON THE LEFSCHETZ PRINCIPLE

By Serge Lang

(Received June 5, 1956)

The principle is well known, and states that a certain type of algebraic theorem concerning a variety V in the algebraic geometry of characteristic zero can be proved by embedding V into complex space and then using the methods of topology and analysis. There is no reason why we should not turn in the other direction, and why it should not prove equally fruitful to prove a theorem in characteristic zero or $p > 0$ by first proving it for varieties defined over finite fields, using the powerful tools of arithmetic, and then lifting back to an arbitrary variety. We shall illustrate this by an example.

THEOREM. *Let C be a non singular curve (complete). Then all unramified abelian coverings of C can be obtained by pull back from the separable isogenies $\lambda \colon B \to J$ of its Jacobian variety.*

The pull back process has been described in [1], and simply means in the present case the restriction of the covering to C.

For curves defined over the algebraic closure of a finite field (hence over a finite field), one knows this result by class field theory [1]. We shall now show how to deduce it in general from this special case.

Let p be a prime number (not necessarily the characteristic). We consider coverings of exponent p^n (i.e. each element of the Galois group has period dividing p^n). If p^{nr} is the number of points of period p^n on the Jacobian, then we get an abelian covering of degree p^{nr} over J, corresponding to the maximal separable extension in the field obtained by the division of the periods (i.e. the isogeny $p^n\delta$). The restriction of such a covering to C gives an abelian unramified covering of C of the same degree p^{nr}.

Suppose that there were other abelian unramified coverings of exponent p^n. Then their compositum with those derived from the Jacobian would yield an abelian unramified covering $f \colon W \to C$ of exponent p^n, but of degree p^μ with $\mu > nr$.

Let k be a field of finite type over which all the above objects are defined. We may view k as a field of parameters, subject to specialization into an absolutely algebraic field of characteristic > 0 (a finite field in fact).

Let Γ be the graph of $p^n\delta$ on J. Then the number of points on J of exponent p^n is the number of distinct points in the cycle $\Gamma \cdot (J \times e)$. Each point appears with a multiplicity equal to the degree of inseparability. For almost all specializations of the parameters, C, J, Γ, and $\Gamma \cdot (J \times e)$ specialize in a well behaved manner. (This is an immediate consequence of Shimura's specialization theory [2].) Hence the number of points of exponent p^n on almost all specializations of J remains the same, namely p^{nr}. On the other hand, the covering W of C would specialize to an abelian unramified covering W' of C', of the same degree $> p^{nr}$.

326

(The property of being unramified is expressed algebraically by the intersection $(T \cdot \Delta) = 0$ for every covering transformation $T \neq$ identity. Here Δ is the diagonal on $W \times W$.) This would contradict the known theorem for curves over finite fields.

It should be noted that the theorem has previously been proved "algebraically" by J.-P. Serre, using the cohomology theory of sheaves, and more recently by J. Tate, using the cohomology theory of groups.

Columbia University

References

[1] S. Lang, *Unramified class field theory over function fields in several variables*, Ann. of Math., vol. 64 (1956), pp. 285–325.

[2] G. Shimura, *Reduction of algebraic varieties with respect to a discrete valuation of the basic field*, Amer. J. Math., vol. 77, No. 1 (1955), pp. 134–176. (See also forthcoming papers of Shimura and Taniyama concerning complex multiplication, where the specialization of the ring of endomorphisms of an abelian variety is discussed in greater detail.)

Reprinted from the Proceedings of the National Academy of Sciences,
Vol. 42, No. 7, pp. 422–424. July, 1956.

L-SERIES OF A COVERING

By Serge Lang

DEPARTMENT OF MATHEMATICS, COLUMBIA UNIVERSITY

Communicated by Deane Montgomery, March 29, 1956

Let U be a topological space of a reasonable type, and let $G = \{T\}$ be a finite group of homeomorphisms of U. Let V be the quotient space and $\pi\colon U \to V$ the projection. Many theorems are known concerning the relations between the Betti numbers of U and V. In fact, in a recent issue of these PROCEEDINGS Conner[1] has proved the following result: If U_T is the set of fixed points of the transformation T in G, then the Euler characteristic of V is equal to the average of the Euler characteristics of the U_T, taken over all T in G (including the identity).

Here we shall show that the L-series defined by Artin in algebraic number theory[2] can be transplanted to the topological situation and that the new L-series satisfy the same formalism as Artin's. We reduce the proof of this formalism to topological statements (Statements A and B) which are proved elsewhere by Conner in connection with his above-mentioned theorem. In order to recover Conner's theorem, one needs another topological result (Statement C) due to him. Finally, we translate Statements A and B into statements concerning the representation of the Frobenius substitution in the arithmetic case over finite fields.

Let $F\colon U \to U$ be a transformation of U into itself, and assume that F commutes with all T in G. Then F determines a transformation $f\colon V \to V$ by projection.

We denote by $H_r(U)$ the rth homology group of U. Since we do not consider torsion problems, we could assume that we have rational coefficients, but for convenience we assume that we have an algebraically closed field of characteristic zero.

F induces a linear transformation $F_r\colon H_r(U) \to H_r(U)$. Let $P_{F}^{(r)}(t) = \Pi\,(t - \alpha_i^{(r)})$ be the characteristic polynomial of F_r. We define the *zeta function* (in dimension r) associated with F to be the polynomial

$$Z_F^{(r)}(t) = \Pi\,(1 - \alpha_i^{(r)}t),$$

for reasons motivated by arithmetic, and the full zeta function to be the alternating product: $Z(t) = \Pi\,(Z^{(r)}(t))^{(-1)^{r+1}}$. Note that if F is the identity transformation, then its zeta function is $(1 - t)^{B_r(U)}$, where $B_r(U)$ is the rth Betti number of U, i.e., the dimension of $H_r(U)$.

Taking our cue from the form given by Weil to the Artin L-series in the case of curves over a finite field,[3] we shall now give the definition of the L-series. First, we define a linear transformation of $H_r(U)$, as follows: Let χ be a character of G. Let T_r be the transformation on $H_r(U)$ induced by T. Consider the following linear transformation of $H_r(U)$:

$$e_\chi = \frac{m_\chi}{N} \sum_{T\,\epsilon\,G} \chi(T^{-1})T_r,$$

where m_χ is the degree of χ and N is the order of G. It is an immediate consequence of the orthogonality relations that e_χ is idempotent, i.e., $e_\chi^2 = e_\chi$. Since F commutes with each T, it follows that F_r commutes with e_χ, and for any positive integer

μ we have $(e_\chi \circ F_r)^\mu = e_\chi \circ F_r^\mu$. Consider the analytic function that takes the value 1 at $t = 0$ and the logarithmic derivative of which is given by the series

$$-\frac{1}{m} \sum_{\mu=1}^{\infty} \text{tr } (e_\chi \circ F_r^\mu) t^{\mu-1},$$

where "tr" denotes the trace of the representation on $H_r(U)$.

If
$$P_\chi^{(r)} = \Pi \, (t - \beta_j^{(r)}(\chi))$$

is the characteristic polynomial of $e_\chi \circ F_r$, then one sees immediately that the logarithmic derivative of the polynomial $\Pi(1 - \beta_j^{(r)}(\chi)t)$ gives precisely the above series. Following classical terminology, we call this function an *L-series* and denote it by $L_F^{(r)}(t, \chi, U/V)$. This is of course the r-dimensional part of the full L-series, which can be obtained by taking the alternating product. We now have the following properties:

1. If χ and χ' are two characters of G, then

$$L_F^{(r)}(t, \chi + \chi', U/V) = L_F^{(r)}(t, \chi, U/V) \, L_F^{(r)}(t, \chi', U/V),$$

so that the L-series are multiplicative. The proof is obvious.

2. For the trivial covering $U = V$ and the identity character, the L-series is equal to the zeta function, i.e.,

$$L_F^{(r)}(t, 1, U/U) = Z_F^{(r)}(t).$$

The proof is again obvious.

3. Let f be the transformation induced on V by projection. Letting 1 denote the identity character for G, we have

$$L_F^{(r)}(t, 1, U/V) = Z_f^{(r)}(t),$$

i.e., the L-series is the zeta function of the *base space*. This formula is an immediate consequence of the first of our topological statements, namely:

Statement A: The characteristic polynomial of the transformation $(1/N) \times (\sum_{T \, \epsilon \, G} T_r) \circ F_r$ on $H_r(U)$ is equal to the characteristic polynomial of f_r on $H_r(V)$.

In the special case where F is the identity transformation (this being precisely the case used in Conner's theorem), Statement A takes on a particularly simple form, namely:

Statement B: The number of times that the identity occurs in the representation of G in $H_r(U)$ is equal to $B_r(V)$.

4. Let H be a subgroup of G, and let ψ be a character of H. Let W be the quotient space relative to H, and let χ_ψ be the induced character (or "transferred character," as it is sometimes called). Then

$$L_F^{(r)}(t, \psi, U/W) = L_F^{(r)}(t, \chi_\psi, U/V).$$

The proof here is formal and depends only on a trivial juggling of characters. Hence there is no need of reproducing it.

5. Assume that H is normal in G, and let $K = G/H$. Then we can consider W/V as a covering with group K. Let ψ be a character of K. Then ψ extends to

a character χ of G having the value of ψ on the cosets of H. Let φ be the transformation on W induced by projection of F on W. Then

$$L_\varphi^{(r)}(t, \psi, W/V) = L_F^{(r)}(t, \chi, U/V).$$

The proof is again an immediate consequence of Statement A.

The above properties exhaust Artin's formalism. Consider the special case where F is the identity, H is the identity subgroup, and ψ is the identity in Property 4. Then the induced character χ_ψ is the character of the regular representation, which we write χ_{reg}. It contains the identity character χ_1 of G exactly once. Let $\chi_2 = \chi_{\text{reg}} - \chi_1$. Using properties 2, 3, and 4, we get

$$(1 - t)^{B_r(U)} = (1 - t)^{B_r(V)} L^{(r)}(t, \chi_2, U/V).$$

This gives an explicit relation between the Betti numbers of U and those of V, which can be reduced to computations in Abelian coverings by using Brauer's theorem on induced characters.[4]

We can recover Conner's theorem as a special case of the previous statements by applying the Lefschetz fixed-point formula and one more topological statement:

Statement C: Let U_T be the space of fixed points of T. Let Γ_T be the graph of T. Then the Euler characteristic of U_T is equal to $I(\Gamma_T \cdot \Delta)$, where Δ is the diagonal and I is the total intersection number of the two cycle classes of Γ_T and Δ.

Suppose, finally, that we deal with a Galois covering of algebraic varieties defined over a finite field with q elements. Consider the case $r = 1$. Let $H_1(U)$ now denote the l-adic representation space of G arising from the Albanese variety $A(U)$ of U. Its dimension $B_1(U)$ is equal to $2g$, where g is the irregularity of U (equal to $\dim A(U)$). We let F be the Frobenius transformation of U, so that f is the Frobenius transformation of V. Then both Statements A and B make sense with the new interpretation and are easily verified after noting that $(1/N) \sum_{T \epsilon G} T_1$ essentially gives the projection from $A(U)$ onto $A(V)$. In addition, one can conjecture that the L-series obtained by the above representation should give exactly the contribution in the circle $|t| < q^{-(n-1)}$ to the L-series defined in terms of the prime rational cycles of V in the obvious manner (i.e., following exactly Artin's definition in number fields), provided, of course, that we multiply the characteristic roots by q^{n-1} (where $n = \dim U = \dim V$). For the zeta function, this last statement has already been conjectured by Weil.[5]

[1] P. Conner, "Concerning the Action of a Finite Group," these PROCEEDINGS, **42**, No. 6. 349, 1956.

[2] E. Artin, "Zur Theorie der L-Reihen mit allgemeinen Gruppencharakteren," *Abhandl. Math. Sem. Hamburg Univ.*, pp. 229–306, 1930.

[3] A. Weil, *Sur les courbes algébriques et les variétés qui s'en deduisent* (Paris: Hermann & Cie, 1948), p. 81.

[4] R. Brauer and J. Tate, "On the Characters of Finite Groups," *Ann. Math.*, **62**, No. 1, 1–7, 1955.

[5] S. Lang and A. Weil, "Number of Points of Varieties in Finite Fields," *Am. J. Math.*, **76**, No. 4, 819–827, 1954.

Bull. Soc. Math. France,
84, 1956, p. 385 à 407.

SUR LES SÉRIES L D'UNE VARIÉTÉ ALGÉBRIQUE;

Par Serge Lang.

1. Introduction. — Dirichlet a démontré l'existence d'une infinité de nombres premiers p dans toute progression arithmétique $p \equiv a \,(\mathrm{mod}\, m)$, où a est premier à m.

Dans la théorie du corps de classes, on a une autre interprétation de ce théorème d'existence. Soient $\mathbf{Q}$ le corps des nombres rationnels et E le corps des racines m-ièmes de l'unité. Alors à chaque nombre premier p ne divisant pas m, correspond l'automorphisme du corps E qui envoie chaque racine m-ième de l'unité dans sa p-ième puissance. Cet automorphisme est noté $(p, E/\mathbf{Q})$, et le théorème de Dirichlet peut s'interpréter en disant qu'à chaque élément σ du groupe de Galois de $E/\mathbf{Q}$ il existe une infinité de p tels que $(p, E/\mathbf{Q}) = \sigma$. On peut préciser cet énoncé en attribuant une densité à ces nombres premiers.

Plus généralement on prend un corps de nombres K comme corps de base, et l'on sait classifier les extensions abéliennes au moyen de progressions arithmétiques généralisées. Dans le cas non ramifié on prend le groupe des classes d'idéaux ordinaires, c'est-à-dire les idéaux de K modulo les idéaux principaux. On sait qu'il existe une extension abélienne maximale non ramifiée, dont le groupe de Galois est isomorphe au groupe des classes d'idéaux, groupe qu'on peut noter C_K. Cet isomorphisme s'obtient au moyen de la loi de réciprocité d'Artin, comme suit : Soit $\mathcal{O}$ l'anneau des entiers de E, et $\mathfrak{p}$ un idéal premier de K. Alors, $\mathfrak{p}$ étant non ramifié, on a $\mathfrak{p}\mathcal{O} = \mathfrak{P}_1 \ldots \mathfrak{P}_r$, où les $\mathfrak{P}_i$ sont maximaux dans $\mathcal{O}$. On considère le sous-groupe D du groupe de Galois G de E/K composé des éléments σ de G laissant l'un des $\mathfrak{P}_i$ invariant. Du fait que G est abélien, c'est le même pour tous les $\mathfrak{P}_i$. Ce groupe D, nommé groupe de décomposition, est cyclique et possède un générateur canonique σ (l'automorphisme de Frobenius) caractérisé par la congruence

$$(1) \qquad \sigma\alpha \equiv \alpha^{N\mathfrak{p}} \quad (\mathrm{mod}\,\mathfrak{P}_i) \quad \text{pour tout } \alpha \in \mathcal{O},$$

$N\mathfrak{p}$ étant le nombre d'éléments du corps des résidus $\mathfrak{o}/\mathfrak{p}$.

Ce générateur σ est noté $(\mathfrak{p}, E/K)$, car il ne dépend que de $\mathfrak{p}$, et pas du $\mathfrak{P}_i$

112

dont on s'est servi, du fait que G est abélien. La loi de réciprocité d'Artin dit alors que le symbole $(\mathfrak{p}, E/K)$ ne dépend que de la classe de $\mathfrak{p}$ dans C_K. Le théorème de la progression arithmétique peut s'exprimer en disant que pour chaque élément σ de G il existe une infinité d'idéaux premiers $\mathfrak{p}$ tels que $(\mathfrak{p}, E/K) = \sigma$. Comme ARTIN l'a remarqué, on peut donc interpréter une progression arithmétique ou bien comme une classe d'idéaux premiers modulo les idéaux principaux, ou bien comme classe d'idéaux premiers ayant le même automorphisme de Frobenius dans un groupe de Galois. Le cas ramifié se traite de façon analogue. Si $\mathfrak{m}$ est le diviseur de ramification, on prend le groupe des diviseurs premiers à $\mathfrak{m}$ modulo le sous-groupe des diviseurs principaux, provenant des éléments f de K tels que $f \equiv 1 \pmod{\mathfrak{m}}$, la congruence étant prise au sens valuatif. Dans le cas des racines de l'unité, cette congruence se réduit précisément à la congruence ordinaire, ce qui justifie la terminologie.

Dans le cas d'une extension galoisienne non abélienne E/K, on ne connaît pas d'analogue aux classes d'idéaux, mais on a toujours l'automorphisme de Frobenius, qui est alors une classe de conjugaison, qu'on notera aussi $(\mathfrak{p}, E/K)$, définie encore par la congruence (1). On peut encore énoncer un théorème d'existence d'idéaux premiers dans une progression arithmétique, à savoir qu'à chaque classe de conjugaison dans G, il existe une infinité d'idéaux premiers $\mathfrak{p}$, tels que $(\mathfrak{p}, E/K)$ soit une classe donnée. De plus, on peut attribuer à ces idéaux une densité, que l'on trouvera calculée chez ARTIN [1]. La démonstration se fait au moyen de séries L. Dans le cas le plus simple, celui des entiers $\bmod m$, on prend un caractère χ du groupe multiplicatif des entiers $\bmod m$, et premiers à m, et l'on définit la série L par son logarithme

$$\log L(s, \chi) = \sum \frac{\chi(p^{\mu})}{\mu p^{\mu s}},$$

la somme étant prise pour tous les p premiers à m, et μ entier ≥ 1. Dans le cas d'une extension galoisienne quelconque, on prend un caractère de G. Si σ est n'importe quel élément de la classe de conjugaison $(\mathfrak{p}, E/K)$, $\mathfrak{p}$ étant non ramifié, on définit

$$\chi(\mathfrak{p}^{\mu}) = \chi(\sigma^{\mu}).$$

La série L devient alors

$$\log L(s, \chi, E/K) = \sum \frac{\chi(\mathfrak{p}^{\mu})}{\mu (N\mathfrak{p})^s},$$

la somme étant prise pour tous les idéaux premiers non ramifiés, et μ entier ≥ 1. ARTIN a montré [2] comment on pouvait compléter cette définition pour inclure les idéaux premiers ramifiés, mais il est inutile d'entrer ici dans ces considérations.

Bien entendu, dans le cas abélien, on peut définir directement la fonction

$L(s, \chi, E/K)$ par un produit eulérien, ou une somme, de la façon suivante :

$$\prod \frac{1}{1 - \dfrac{\chi(\mathfrak{p})}{(N\mathfrak{p})^s}} = \sum \frac{\chi(\mathfrak{a})}{(N\mathfrak{a})^s}.$$

Comme d'habitude, on note par $\mathfrak{a}$ un idéal, c'est-à-dire un élément du groupe abélien libre engendré par les $\mathfrak{p}$, et $\chi(\mathfrak{a})$ est défini à partir des $\mathfrak{p}$ par linéarité. La somme est prise pour les $\mathfrak{a}$ entiers.

Il est facile de voir que la densité des idéaux premiers dans les progressions arithmétiques dépend du comportement des séries L pour $Re(s) > 1/2$. Si l'on sait que pour $\chi \neq \chi_1$ (χ_1 désignant une fois pour toutes le caractère principal) $L(s, \chi)$ n'a pas de zéro ni de pôle pour $Re(s) > 1/2$ (conjecture d'Artin) et que $L(s, \chi_1)$ a un pôle simple en $s = 1$, et pas d'autres zéros ni pôles pour $Re(s) > 1/2$ (hypothèse de Riemann), alors on voit par un calcul analytique simple qu'on a la densité voulue d'idéaux premiers dans une progression arithmétique, avec un terme d'erreur aussi bon que possible. Dans le cas de la fonction ζ, par exemple, on pose $\Lambda(n) = 0$ si $n = 1$, $\log p$ si $n = p^\mu$ est une puissance de p, et 0 si n n'est pas une puissance d'un nombre premier p. On pose

$$\Psi(x) = \sum_{n \leq x} \Lambda(n)$$

et l'on trouve alors

$$\Psi(x) = x + O\left(x^{\frac{1}{2} + \varepsilon}\right).$$

D'autre part, on sait que toutes ces questions ont un analogue dans le cas d'un corps de fonctions d'une variable sur un corps fini à q éléments (SCHMIDT [12] à la suite des travaux d'Artin). On a aussi la notion d'idéaux premiers du corps, mais on jouit d'un avantage de plus : le nombre d'éléments $N\mathfrak{p}$ du corps des restes est une puissance q^d. L'entier d est appelé le degré de $\mathfrak{p}$ et noté $\deg(\mathfrak{p})$. De ce fait, on voit que les puissances $(N\mathfrak{p})^s$ peuvent s'exprimer comme puissances de q, et il est alors utile de changer de variable : on pose $t = q^{-s}$, et la série L devient une série de puissances

$$\log L(t, \chi, E/K) = \sum_{\mathfrak{p}^\mu} \frac{\chi(\mathfrak{p}^\mu)}{\mu} t^{\mu.\deg(\mathfrak{p})}.$$

Si $\chi = \chi_1$, alors la série L est la fonction ζ attachée au corps.

On sait que WEIL a introduit une fonction ζ pour les variétés algébriques de dimension quelconque [13]. Il était donc naturel de chercher à définir les séries L dans ce cas plus général. Les séries L d'Artin se généralisent dans tous les cas de la façon suivante. Soit K un corps de fonctions sur le corps de constantes k, autrement dit

$$K = k(x_1, \ldots, x_n) = k(x).$$

On peut considérer (x) comme point générique d'une variété V qui est appelée un modèle de K. Chaque point Q de V détermine un anneau local dans K, et deux points ont le même anneau local si et seulement s'ils sont conjugués sur k.

Nous suivrons la terminologie de Chevalley, et nous appellerons tout anneau local $\mathfrak{o}$ obtenu de cette manière une *localité* de K. [Si $\mathfrak{p}$ est l'idéal maximal de $\mathfrak{o}$ nous dirons aussi quelquefois que la paire $(\mathfrak{o}, \mathfrak{p})$, ou $\mathfrak{p}$ lui-même, sont des localités]. Une localité sera dite simple si le point Q est simple. Le corps k étant parfait, on sait que ceci équivaut à la condition que $\mathfrak{o}$ soit régulier. Les localités simples sont alors en correspondance biunivoque avec les cycles premiers rationnels de V sur k. (Par « cycle » nous entendrons toujours cycle de dimension zéro.)

Pour définir les séries L en dimension > 1, il suffit de prendre la définition donnée dans le cas de dimension 1, de remplacer la notion d'idéal premier par la notion de localité, et de prendre la somme pour toutes les localités $\mathfrak{p}$ d'un modèle de K. On sait par les travaux de Krull que les groupes de décomposition et d'inertie se généralisent, et l'on a donc l'automorphisme de Frobenius. On peut alors chercher à démontrer un théorème de densité. C'est ce qui sera fait ici.

On voit que l'on peut aussi définir les localités pour un corps de type fini en caractéristique zéro, et l'on aura aussi des séries L. Il est à peine besoin de dire que la découverte des lois de décomposition (lois de réciprocité) dans ce cas serait d'un intérêt considérable. Mais jusqu'ici, nous ne savons rien sur ces séries L, en dehors de résultats qui dépendent de façon immédiate de ceux qu'on connaît en réduisant modulo une localité $\mathfrak{p}_0$ du corps de nombres contenu dans le corps de type fini K. Par exemple, démontrons la surjectivité de l'application de réciprocité pour une extension abélienne E de K. Soit $\mathfrak{p}$ une localité simple de K, non ramifiée dans E, et $D_\mathfrak{p}$ le groupe de décomposition de $\mathfrak{p}$ dans le groupe de Galois G de E sur K. Soit G' le sous-groupe de G engendré par les $D_\mathfrak{p}$, et soit E' le sous-corps de E invariant par G'. Alors tous les $\mathfrak{p}$ de K se décomposent complètement dans E'. Si k est le corps de nombres algébriques contenu dans K, autrement dit la clôture algébrique de $\mathbf{Q}$ dans K, on déduit d'abord de la remarque précédente que E' ne peut contenir une extension algébrique propre de k, en employant un résultat connu de la théorie des nombres, à savoir qu'il existe une infinité de localités de k qui ne se décomposent pas complètement dans une telle extension. On est donc ramené à une extension dite géométrique, et donc au résultat analogue en caractéristique p, après avoir réduit l'extension E de K mod $\mathfrak{p}_0$ pour n'importe quelle localité $\mathfrak{p}_0$ de k en dehors d'un ensemble fini exceptionnel.

Dans la suite de ce travail, nous nous bornerons exclusivement au cas de caractéristique p.

Nos résultats seront purement birationnels, et nous montrerons que les séries L (pour un caractère $\neq \chi_1$) n'ont pas de zéro ni de pôle dans le cercle $|t| < q^{-\left(r-\frac{1}{2}\right)}$. Nous avons donné autre part [5] la conjecture relative au comportement sur le cercle $|t| = q^{-\left(r-\frac{1}{2}\right)}$, qui est aussi birationnellement invariant. Quant au comportement pour les cercles suivants, si les deux variétés U et V sont sans singularités et complètes, alors on peut conjecturer que les séries L sont comme des fonctions ζ, sauf que les termes $(1-t)$ et $(1-q^r t)$ sont absents.

Ce cas est relativement rare, car ABHYANKAR a démontré qu'en général on ne peut pas résoudre simultanément les singularités des deux corps K et E par un modèle V de K et sa normalisation U dans E. Néanmoins on sait dans le cas topologique que les séries L satisfont au même formalisme que celui d'Artin quand on prend l'espace quotient d'une variété par un groupe fini d'homéomorphismes. Ceci indique que la présence de singularités doit être considérée comme normale, et que les progrès que l'on pourra faire dans cette direction dépendent de l'étude plus précise de l'effet que les singularités peuvent avoir sur la fonction ζ d'une variété. Contrairement à ce que prétend WEIL dans sa « Footnote... » [15], il semble au contraire parfaitement approprié, et même essentiel, de mieux connaître comment de telles fonctions ζ se comportent, même si elles n'ont pas d'équation fonctionnelle.

D'ailleurs, pour revenir au cas topologique, si l'on prend une variété (topologique) U et un groupe fini d'homéomorphismes G de V (avec des hypothèses convenables) les nombres de Betti de l'espace quotient $V = U/G$ vérifient la dualité de Poincaré. Cela conduit à penser que dans le cas arithmétique on n'aurait besoin que d'un modèle U non singulier par exemple. La variété V aurait des singularités; mais on pourrait espérer que sa fonction ζ aurait encore une équation fonctionnelle.

Enfin, on peut chercher dans le cas abélien à montrer l'identité des séries L définies par le groupe de Galois à la manière d'Artin, avec des séries L définies par un caractère du groupe des cycles de dimension zéro. C'est ce que nous pouvons faire dans un grand nombre de cas : on se donne une application rationnelle $\alpha : V \to A$ de V dans une variété de groupe commutatif qui induit un homomorphisme du groupe des cycles sur V, rationnels sur k, dans le groupe des points rationnels A_k de A. On peut alors dire que deux cycles $\mathfrak{a}$ et $\mathfrak{b}$ sont dans la même progression arithmétique s'ils ont la même image dans A_k. Dans le cas des extensions non ramifiées, on prend pour A la variété d'Albanese de V, et l'on peut ainsi construire une théorie analogue à celle du corps de classes de HILBERT [6].

Dans le dernier numéro de ce travail, nous indiquerons brièvement comment on peut généraliser une partie de cette théorie en prenant un groupe commutatif sujet à la seule condition que l'image réciproque de ses revêtements (par des isogénies) soit biunivoque.

On pourrait aussi chercher à étendre cette théorie au cas non abélien en prenant un groupe algébrique quelconque, mais pour le moment on ne sait à peu près rien sur cette question.

Je ne voudrais pas terminer cette introduction sans exprimer ma reconnaissance à J.-P. SERRE, qui a bien voulu se charger de la corection des phautes d'orthografe.

2. Le théorème de densité.

— Soit V une variété abstraite, normale, définie sur un corps fini k a q éléments. Soit K/k un corps de fonctions de V/k, et E/K une extension galoisienne de K, de degré n et de groupe G. Nous ne nous occuperons que du cas géométrique, et nous supposerons donc que k est le corps de constantes de E. Alors la normalisation U de V dans E est une variété abstraite, projective si V l'est, et l'on a une application rationnelle $f : U \to V$ partout définie sur U et telle que l'image inverse d'un point Q de V contienne au plus n points de U. Cela se voit facilement en considérant la décomposition de l'anneau local de Q dans E (*voir* KRULL [4]). On dira que $f : U \to V$ est un *revêtement de degré n*. Si Q est simple sur V et s'il y a exactement n points P_i sur U au-dessus de Q, alors tous les P_i sont simples aussi. C'est immédiat, à partir du fait que l'idéal maximal de Q se remonte à l'idéal maximal de chacun des P_i (NAGATA [9]). Pour la convenance du lecteur, nous allons reproduire ici un argument géométrique dû à MATSUSAKA. La question étant locale, nous pouvons nous borner aux variétés affines.

PROPOSITION 1. — *Soit $f : U \to V$ une application rationnelle de degré n, séparable, partout définie, et telle que l'image inverse de chaque point de V soit un ensemble fini de points de U. Soit Q un point simple de V, et supposons que $\Gamma_f \cap (U \times Q)$ contienne exactement n points distincts $(P_i \times Q)$. Alors chaque P_i est simple sur U.*

Démonstration. — Soient S^N et S^M les espaces affines ambiants des variétés U et V respectivement. Soit L une variété linéaire de S^M telle que $V.L$ soit défini, et contienne Q avec multiplicité 1. L'intersection $\Gamma_f.(S^N \times L)$ est alors définie, sur $S^N \times S^M$, et est égale à $\sum m_i(P_i \times Q) + X$, où X est un cycle positif sur $S^N \times S^M$, ne contenant aucun des $P_i \times Q$. En projetant sur S^M, et en appliquant le théorème de projection, on voit qu'on obtient $nQ + X'$, où X' ne contient pas Q et, de ceci, on conclut que $\sum m_i = n$. Mais l'hypothèse montre que chaque $m_i = 1$. Par le critère de multiplicité 1, on en déduit que chaque $P_i \times Q$ est simple sur le graphe Γ_f, et donc que P_i est simple sur U.

Soit $\mathfrak{p}$ un cycle premier rationnel de V/k, de degré d. Alors on peut écrire $\mathfrak{p} = \sum Q^{(q^i)}$. Les cycles premiers rationnels de V/k sont en correspondance biunivoque avec les anneaux locaux réguliers de K déterminés par n'importe quel point Q de $\mathfrak{p}$. Nous avons convenu dans l'introduction de les appeler des localités simples. Une telle localité est non ramifiée (au sens de KRULL [4]) si et seulement s'il existe n points distincts P_i dans U au-dessus de Q.

Si x est un point générique de V/k, nous identifions souvent $k(x)$ et K. On peut alors écrire $E = k(y)$, y étant un point générique de U/k tel que $f(y) = x$. Si σ est un automorphisme de E/K, alors (y, y^σ) est un point générique d'une correspondance birationnelle T_σ de U, et le groupe de Galois de E/K peut être identifié avec ce groupe de correspondances birationnelles. Pour chaque T_σ on a $f \circ T_\sigma = T_\sigma$. Ceci exprime géométriquement le fait que chaque σ laisse K invariant.

On sait d'après les travaux de KRULL [4] que les groupes de décomposition et d'inertie d'une localité se définissent comme dans le cas des corps de nombres algébriques. Nous allons répéter ici ces définitions pour le cas géométrique et arithmétique qui nous occupe.

Soit $\mathfrak{p}$ un cycle premier rationnel de V sur k, de degré d. Soit Q un point de $\mathfrak{p}$ et P n'importe quel point dans $f^{-1}(Q)$. On suppose que Q (donc $\mathfrak{p}$) est non ramifié. Alors il existe une transformation T de G uniquement déterminée par la condition $T(P) = P^{(q^d)}$, à une conjugaison près. Si l'on choisit un autre point Q_1 dans $\mathfrak{p}$ et un point P_1 dans $f^{-1}(Q_1)$, on voit immédiatement que la transformation T_1 telle que $T_1(P_1) = P_1^{(q^d)}$ est conjuguée à T dans G. Donc la classe de conjugaison de T est bien déterminée à partir de $\mathfrak{p}$. On la notera $(\mathfrak{p}, U/V)$ [ou $(\mathfrak{p}, E/K)$ si nous considérons les automorphismes du corps E/K plutôt que les transformations birationnelles]. Si G est abélien, cette classe de conjugaison se réduit à un seul élément.

Comme nous l'avons déjà dit plus haut, si χ est un caractère de G, μ un entier positif et T un élément de $(\mathfrak{p}, U/V)$, nous poserons

$$\chi(\mathfrak{p}^\mu) = \chi(T^\mu).$$

Nous allons maintenant profiter de la situation géométrique avec laquelle nous travaillons pour transformer la série L définie à partir des localités en une série L définie à partir des points.

Soit Q un point de V rationnel sur le corps k_m, unique extension de k de degré m. Si Q est simple sur V et non ramifié, on notera par $C_Q^{(m)}$ la classe de conjugaison associée à Q dans G comme on l'a fait plus haut, mais en prenant maintenant k_m comme corps de constantes. Autrement dit, $C_Q^{(m)}$ est l'unique classe contenant un élément T de G tel que pour P dans $f^{-1}(Q)$ on ait $T(P) = P^{(q^m)}$. On choisira un représentant de cette classe qu'on notera $T_Q^{(m)}$. Si $\mathfrak{p}$ est un cycle premier rationnel de V/k, de degré d divisant m, on a évidemment

$$(\mathfrak{p}, U/V)^{m/\deg(\mathfrak{p})} = T_Q^{(m)}$$

pour n'importe quel Q dans $\mathfrak{p}$.

Comme dans le présent travail nous nous occupons des séries L seulement du point de vue birationnel, nous allons adopter les conventions suivantes, quand il s'agira de points singuliers ou ramifiés : Soit Z un diviseur sur V, rationnel sur k, et contenant tous les points singuliers de V et tous les points ramifiés. Le corps k étant parfait, on peut supposer que toutes les composantes de Z ont la multiplicité 1. Un point Q de V appartient à Z si et seulement si tous ses conjugués (sur k) appartiennent aussi à Z. Si tel est le cas, on prendra $T_Q^{(m)}$ égal à l'élément neutre de G, et l'on posera $\chi(Q)=1$.

Nous nous servirons aussi de la définition de la série L à partir de sa dérivée logarithmique au lieu de son logarithme. Dans un cas comme celui-ci, on conviendra toujours de normer la fonction en prescrivant qu'elle prenne la valeur 1 pour $t=0$. Nous pouvons alors écrire

$$\frac{d}{dt}\log L(t,\chi,E/K)=\sum_{m=1}^{\infty}\left(\sum_{\deg(\mathfrak{p})\mid m}\chi(\mathfrak{p}^{m/\deg(\mathfrak{p})})\deg(\mathfrak{p})\right)t^{m-1}.$$

Compte tenu de nos remarques précédentes, on voit que le coefficient de t^{m-1} dans cette somme est égal à

$$\sum_{Q\in\Gamma_m}\chi(T_Q^{(m)}),$$

la somme étant prise pour tous les points Q de V rationnels sur k_m (ce ensemble étant noté V_m).

Nous allons maintenant énoncer le théorème principal que nous avons en vue.

THÉORÈME 1. — *Soit $f:U^r\to V^r$ un revêtement galoisien de groupe G, défini sur un corps fini k à q éléments. Soit Z un diviseur sur V, rationnel sur k, contenant tous les points singuliers et les points ramifiés de V. Soit χ un caractère simple de G. Alors il existe un nombre A dépendant de f, U, V, Z (mais pas de q) tel que*

$$\left|\sum\chi(T_Q)\right|\leq Aq^{\left(r-\frac{1}{2}\right)}\qquad si\quad\chi\neq\chi_1$$

et

$$\left|\sum\chi(T_Q)-q^r\right|\leq Aq^{\left(r-\frac{1}{2}\right)}\qquad si\quad\chi=\chi_1.$$

Le cas où $\chi=\chi_1$ est celui de la fonction ζ, et a déjà été démontré dans [7] et [11].

Dans le théorème, on a pris $T_Q=T_Q^{(1)}$, et les sommes sont prises pour Q dans $V_1=V_k$, l'ensemble des points rationnels de V dans k. On a $\chi(T_Q)=\chi(Q)$, en considérant Q comme point rationnel.

Le théorème sera démontré dans les n°s 3 et 4. Nous allons en tirer ici quelques corollaires.

Pour commencer, on peut appliquer l'argument formel d'ARTIN [1] pour en tirer la densité des localités dans une progression arithmétique. Soit C une classe de conjugaison de G, et T un élément de C. Pour $m \to \infty$, le théorème donne

$$\sum_{Q \in \Gamma_m} \chi(T_Q^{(m)}) = \mathrm{O}\left(q^{m\left(r-\frac{1}{2}\right)}\right) \qquad \text{si} \quad \chi \neq \chi_1$$

et

$$\sum_{Q \in \Gamma_m} \chi(T_Q^{(m)}) = q^{rm} + \mathrm{O}\left(q^{m\left(r-\frac{1}{2}\right)}\right) \qquad \text{si} \quad \chi = \chi_1.$$

On multiplie par $\chi(T^{-1})$ et l'on fait la somme. Si l'on note $N(C, m)$ le nombre de points Q rationnels sur k_m ayant leur $T_Q^{(m)}$ dans la classe C donnée, et h le nombre d'éléments de C, alors on trouve (en employant les relations d'orthogonalité)

$$\frac{n}{h} N(C, m) = q^{rm} + \mathrm{O}\left(q^{m\left(r-\frac{1}{2}\right)}\right).$$

Cette formule donne la densité des points, et l'on retrouve trivialement celle des localités. En effet, si un point Q dans V_m est de degré $d < m$, donc $d \mid m$, on a $d \leq m/2$. Une estimation grossière montre que la contribution de ces points à $N(C, m)$ peut être absorbée dans le terme d'erreur d'ordre $q^{m\left(r-\frac{1}{2}\right)}$. Si le point Q est de degré m exactement, alors il détermine une localité $\mathfrak{p}$, et il y a exactement m points dans $\mathfrak{p}$. Donc si l'on note par $N_0(C, m)$ le nombre de localités de degré m telles que $(\mathfrak{p}, U/V)$ soit égal à la classe C donnée, on trouve

$$\frac{n}{h} N_0(C, m) = \frac{1}{m} q^{rm} + \mathrm{O}\left(q^{m\left(r-\frac{1}{2}\right)}\right).$$

En particulier, quand m tend vers l'infini, le nombre de localités de degré m dans une progression arithmétique donnée tend aussi vers l'infini. Notons également que le théorème est formulé pour une variété abstraite, et que le diviseur peut être pris arbitrairement. Donc on peut toujours trouver les localités en dehors d'un ensemble algébrique $F \subset V$, $F \neq V$. Si deux variétés sont en correspondance birationnelle sur k, la correspondance est biholomorphe en dehors d'un tel ensemble algébrique. On voit donc que le contenu de notre théorème est bien birationnel.

Remarquons enfin que pour $\chi \neq \chi_1$, la série qui définit la dérivée logarithmique converge dans le cercle $|t| < q^{-\left(r-\frac{1}{2}\right)}$ (conséquence immédiate de nos inégalités) et donc que $L(t, \chi, U/V)$ ne peut avoir de zéros ou de pôles dans ce cercle. En prenant la représentation l-adique de G dans la variété d'Albanese on a fait ailleurs la conjecture relative au comportement sur le cercle $|t| = q^{-\left(r - \frac{1}{2}\right)}$. Au-delà de ce cercle, on ne sait quelles représentations il faut prendre.

La démonstration du théorème 1 se fera par réduction directe au cas des courbes, en prenant un système algébrique de revêtements, et en employant les résultats connus pour les courbes (démontrés par WEIL [14]).

Pour la convenance du lecteur nous allons rappeler ici ces résultats, et inclure quelques remarques sur l'uniformité qu'on obtient dans l'estimation de la somme $\sum \chi(T_Q)$.

Supposons d'abord que nous ayons un revêtement $f : W \to C$ où les deux courbes W et C sont non singulières, mais où il peut y avoir des ramifications.

Alors la série L est un polynome $\prod (1 - \alpha_i t)$, où les α_i sont des nombres algébriques de valeur absolue $q^{\frac{1}{2}}$. Il est essentiel aussi de remarquer que le degré de ce polynome est donné par la formule

$$\frac{1}{[W:C]} \sum_{T \in G} \chi(T^{-1}) \operatorname{Tr}(T_*),$$

où $\operatorname{Tr}(T_*)$ est la trace de la représentation l-adique de T, dérivée de la jacobienne de W. On voit donc que le degré du polynome est borné, en fonction du genre de la courbe W. Soit d ce degré. Alors en prenant la dérivée logarithmique de $\prod (1 - \alpha_i t)$, on trouve

$$\sum \chi(T_Q) = - \sum_{i=1}^{d} \alpha_i.$$

Ceci montre que pour les courbes qui nous occupent, notre somme est bornée par $A q^{\frac{1}{2}}$, pour un nombre A ne dépendant que du degré du caractère χ, du genre de W, et du degré $[W:C]$.

Naturellement, la série L considérée ci-dessus par WEIL est la série L complète, où l'on a tenu compte des points de ramification. Dans notre définition, où nous avons convenu de poser $\chi(T_Q) = 1$ si Q est ramifié, on voit que notre somme diffère de celle obtenue pour une courbe non singulière par une constante dépendant du nombre de points ramifiés.

Dans notre réduction au cas des courbes, nous obtiendrons des courbes spécialisées d'une courbe générique, et le nombre de points de ramification, ou de points singuliers sur chaque courbe sera uniformément borné. C'est ce qui nous permettra de démontrer notre théorème.

Nous supposerons désormais que V et U sont des variétés projectives et normales, et dans le numéro suivant, nous allons rappeler quelques propriétés de la courbe générique. Ensuite nous spécialiserons dans les corps finis.

3. La courbe générique. — Nous employons principalement comme références le début des *Critères d'équivalence* de Weil [16], et l'article de Chow [3]. On sait que la courbe générique C_u s'obtient en coupant V par des hyperplans génériques sur k, H_{u_1}, ..., $H_{u_{r-1}}$, où les u_i sont des points génériques indépendants de l'espace projectif $\mathbf{P}'$ dual de l'espace $\mathbf{P}$ dans lequel V est plongée. Le produit de $\mathbf{P}'$ avec lui-même $(r-1)$ fois est une variété qui sera notée Γ, et dont le point générique est $u = (u_1, ..., u_{r-1})$, produit des u_i. La courbe C_u est définie sur $k(u) = k_u$, et tout point générique x^* de C_u sur k_u est un point générique de V sur k. On a

$$C_u = V.H_{u_1}...H_{u_{r-1}} = V.L_u,$$

si L_u est la variété linéaire intersection des H_{u_i}. Nous allons faire une liste des propriétés de C_u qui nous serviront pour les calculs que nous avons en vue.

1° La courbe C_u est non singulière, et tout point de C_u est simple sur V

Regardons maintenant l'image inverse de C_u par f. Je dis que c'est une courbe W_u et que f induit une application rationnelle $f_u : W_u \to C_u$ de façon naturelle, et de même degré que f. Il suffit de démontrer ceci pour l'intersection de V avec un hyperplan générique H_u, et c'est alors un résultat à peu près équivalent au théorème de Bertini (*voir*, par exemple, Zariski [18] et Matsusaka [8]). Plus précisément, rappelons le lemme suivant :

Lemme. — *Soit $k(x)$ une extension régulière d'un corps infini k de caractéristique $p > 0$, et y, z deux éléments de $k(x)$ indépendants sur k et tels que z ne soit pas une puissance p-ième dans $k(x)$. Alors pour tous les éléments c de k sauf un nombre fini, $k(x)$ est une extension régulière de $k(y + cz)$.*

Si (x) est un point générique de V/k, indépendant de u, et (y) est un point générique de U/k tel que $f(y) = (x)$, alors on construit $V.H_u$ de la façon suivante. Supposons que (x) soit un point affine, $(x) = (x_1, ..., x_N)$. Si $u_{11}, ..., u_{1N}$ sont indépendants de (x) sur k, on pose

$$u_{1, N+1} = u_{11} x_1 + ... + u_{1N} x_N.$$

Alors la variété $V.H_u$ est le lieu du point générique (x) sur $k(u_{11}, ..., u_{1N+1})$. Étant donné que $k(y)$ est séparable sur $k(x)$, on voit par le lemme que l'extension $k_u(y)$ sur $k(u)$ est régulière [bien que le corps k soit fini, $k(u_{11})$ est infini, par exemple]. On peut monter ainsi jusqu'à la courbe C_u, et l'on voit que W est le lieu de (y) sur $k(u)$. L'application rationnelle f_u est définie par $f_u(y) = (x)$. En résumé, on a :

2" L'image inverse $f^{-1}(C_u)$ est une courbe W_u, simple sur U. On a un diagramme commutatif

$$
\begin{array}{ccc}
W_u & \xrightarrow{\ i\ } & U \\
f_u\downarrow & & \downarrow f \\
C_u & \xrightarrow{\ i\ } & V
\end{array}
$$

où i est l'inclusion, et $f_u : W_u \to C_u$ est un revêtement galoisien de même degré que f (¹).

(A vrai dire, nous nous sommes borné dans nos définitions d'un revêtement à des variétés normales. Dans le cas présent, W_u peut ne pas être normale. Pour être strict, il faudrait donc dire que f_u est une application rationnelle, partout définie, et telle que l'image inverse d'un point de C_u ne contienne qu'un nombre fini de points. Nous ferons un abus de langage, et dirons toujours que f_u est un revêtement.)

Le groupe de transformations birationnelles de U sur V peut être identifié à celui de W_u sur C_u, les conjugués de (y) sur $k_u(x)$ étant précisément les mêmes que sur $k(x)$. On peut aussi mettre cette condition sous forme de diagramme commutatif, à savoir

$$
\begin{array}{ccc}
W_u & \xrightarrow{\ i\ } & U \\
T_u\downarrow & & \downarrow T \\
W_u & \xrightarrow{\ i\ } & U
\end{array}
$$

Autrement dit, on a :

3° Chaque T de G induit une transformation birationnelle T_u de W_u, telle que $f_u T_u = T_u$, et l'application $T \to T_u$ est un isomorphisme de G sur le groupe de W_u sur C_u.

La courbe W_u peut avoir des singularités. Mais le nombre de ces singularités est borné. En fait, on peut donner une borne supérieure à ces singularités.

(¹) On peut retrouver un critère d'équivalence classique par la méthode des revêtements, en procédant comme suit : soit X un diviseur sur V, rationnel sur le corps k (supposé algébriquement clos); supposons que l'ordre de la classe d'équivalence linéaire de X soit égal à m premier à la caractéristique, et soit φ une fonction telle que $(\varphi) = mX$. Alors, on voit que l'extension $K(\varphi^{1/m})/K$ conserve son degré par restriction à la courbe générique C_u; comme $X.C_u$ est le diviseur de la fonction induite par φ sur C_u, l'ordre de la classe de $X.C_u$ sur C_u est exactement m.

4° L'intersection $Z.C_u$ est définie, et est un cycle de dimension o sur C_u et sur V. (C'est l'intersection de Z avec L_u.) Si Q est un point de C_u qui n'est pas dans $Z.C_u$, alors il existe exactement n points distincts de W_u au-dessus de Q, qui sont tous simples sur U et sur W_u.

Notre dernière assertion découle de la proposition 1, n° 2.

Les seules singularités de W_u se trouvent par conséquent au-dessus des points de $Z.C_u$. Soit d leur nombre. Il y a au plus nd points singuliers sur W_u.

Nous observons maintenant que les propriétés formulées ci-dessus restent valables pour presque toutes les spécialisations de u dans la clôture algébrique de k. Il existe un sous-ensemble k-fermé F de Γ, $F \neq \Gamma$, tel que pour une spécialisation a de u avec $a \notin F$, et a rationnel sur $\bar{k}$, les propriétés restent valables si u est remplacé par a. Par exemple, L_u se spécialise en L_a, $C_u = V.L_u$ se spécialise en $C_a = V.L_a$, et $C_u.Z$ se spécialise en un cycle de même degré $C_a.Z$. L'image inverse $f^{-1}(C_a)$ est une courbe W_a, et f induit une application rationnelle séparable $f_a : W_a \to C_a$. partout définie, de même degré que f. Chaque T de G étant partout défini sur U, induit une transformation birationnelle de W_a/C_a. Si Q est un point de C_a non contenu dans $C_a.Z$, alors les n points de U au-dessus de Q sont dans W_a. Si T et T' sont deux éléments de G, et si l'on a $T_a = T'_a$, alors pour n'importe lequel des P tels que $f(P) = Q$, on a :

$$T(P) = T_a(P) = T'_a(P) = T'(P), \qquad \text{donc} \quad T = T'.$$

Le groupe de Galois de U/V peut donc être identifié avec celui de W_a/C_a.

4. Démonstration du théorème 1. — Par $\varkappa_{n+1}$ nous désignerons le nombre des points de l'espace projectif $\mathbf{P}$ qui sont rationnels sur k. On a

$$\varkappa_{n+1} = \frac{q^{n+1} - \mathrm{I}}{q - \mathrm{I}}.$$

Si Γ est comme précédemment la variété produit de $r - \mathrm{I}$ espaces projectifs $\mathbf{P}'$ (dual de $\mathbf{P}$) alors le nombre de points rationnels de Γ dans k est évidemment égal à $(\varkappa_{n+1})^{r-\mathrm{I}}$.

Le nombre d'hyperplans dans $\mathbf{P}'$ rationnels sur k passant par un point rationnel Q donné est égal à $\varkappa_n$. Chaque point a de Γ détermine une variété linéaire, notée L_a, et le nombre des points rationnels sur k tels que L_a passe par le point donné Q est donc égal à $(\varkappa_n)^{r-\mathrm{I}}$. (La variété linéaire L_a peut dégénérer, c'est-à-dire avoir une dimension plus grande que celle de L_u, mais ceci ne pourra se passer que sur le sous-ensemble algébrique propre de Γ.)

On a

$$(\mathrm{I}) \qquad \frac{\varkappa_{n+1}}{\varkappa_n} = q + \frac{\mathrm{I}}{\varkappa_n},$$

$$(2) \qquad q^{n-1} \leq \varkappa_n.$$

Par conséquent,

$$(3) \qquad \left| \left(\frac{\mathsf{x}_{n+1}}{\mathsf{x}_n} \right)^{r-1} - q^{r-1} \right| \leq A_1 q^{r-2},$$

$$(4) \qquad q^{(n-1)(r-1)} \leq \mathsf{x}_n^{r-1},$$

A_1 étant une constante dépendant seulement de n et de r.

Soit, comme ci-dessus, Z un diviseur sur V, rationnel sur k, contenant tous les points ramifiés et tous les points singuliers, et ayant toutes ses composantes avec multiplicité 1. Soit F le sous-ensemble k-fermé de Γ défini au numéro précédent. On notera comme d'habitude par V_k (ou V_1) les points de V rationnels sur k. Si χ est un caractère de G, la relation suivante est alors évidente [nous écrivons $\chi(Q)$ au lieu de $\chi(T_Q)$]

$$\sum_{Q \in V_k} \chi(Q) = \frac{1}{\mathsf{x}_n^{r-1}} \sum_{a \in \Gamma_k} \sum_{Q \in \Gamma \cap L_a} \chi(Q).$$

Si X est un ensemble algébrique k-fermé, nous noterons $N(X)$ le nombre de points rationnels de X dans k. Il existe une constante A_2 telle que

$$N(F) \leq A_2 \, q^{n(r-1)-1} \leq A_2 \mathsf{x}_n^{r-1} \, q^{r-2}.$$

On trouve ceci en notant que la dimension de Γ est $n(r-1)$, $\dim F < \dim \Gamma$, et en appliquant les résultats de [7] et l'inégalité (4).

On peut séparer la somme $\sum \chi(Q)$ selon que le point a est bon ou mauvais, c'est-à-dire selon qu'il est dans F ou non. Si $a \notin F$, alors $V \cap L_a = V . L_a$ est une courbe définie sur k, et l'on récrit la somme comme suit :

$$\frac{1}{\mathsf{x}_n^{r-1}} \sum_{a \notin F} \sum_{Q \in V . L_a} \chi(Q) + \frac{1}{\mathsf{x}_n^{r-1}} \sum_{a \in F} \sum_{Q \in \Gamma \cap L_a} \chi(Q).$$

Considérons d'abord le cas où $\chi = \chi_1$ est le caractère principal. Les deux sommes deviennent

$$\frac{1}{\mathsf{x}_n^{r-1}} \sum_{a \notin F} N(V . L_a) + \frac{1}{\mathsf{x}_n^{r-1}} \sum_{a \in F} N(V \cap L_a).$$

On va voir que la méthode que nous sommes en train d'employer pour compter les points de V donne le même résultat que celle de [7]. Considérons la première somme. D'après la théorie des courbes, on a

$$(5) \qquad | N(V . L_a) - q | \leq A_3 q^{\frac{1}{2}}.$$

Compte tenu de l'inégalité pour $N(F)$ obtenue plus haut, et de l'inégalité (3), on trouve

$$\left| \frac{1}{\mathsf{x}_n^{r-1}} \sum_{a \notin F} N(V . L_a) - q^r \right| \leq A_4 q^{\left(r - \frac{1}{2}\right)}.$$

Compte tenu du résultat connu concernant le nombre de points [7], ceci montre que la somme prise pour les bons points a donne déjà le nombre de points $N(V)$, à un terme d'ordre $q^{r-\frac{1}{2}}$ près. Donc la somme prise sur les mauvais points satisfait à une inégalité du type

$$\left| \frac{1}{\chi_n^{r-1}} \sum_{a \in F} N(V \cap L_a) \right| \leq A_3 q^{r-\frac{1}{2}}.$$

On a donc montré que si l'on compte par induction avec une famille d'hyperplans (comme dans [7]) ou si l'on compte comme ici par une famille de courbes, on trouve le même résultat.

Nous revenons maintenant à un caractère simple arbitraire χ. Alors $\chi(Q)$ a une valeur absolue uniformément bornée. Donc la seconde somme, prise pour les mauvais points, satisfait à la même estimation que celle que nous venons de trouver pour le caractère principal, c'est-à-dire est d'un ordre de grandeur $q^{r-\frac{1}{2}}$. Tout ce qui reste à démontrer est que la première satisfait à cette estimation quand $\chi \neq \chi_1$.

Pour $a \notin F$, si Q n'est pas dans $C_a.Z$, alors Q n'est pas ramifié dans W_a : les n points distincts de U au-dessus de Q sont précisément les n points distincts de W_a au-dessus de Q. D'après la proposition 1, ils sont donc simples sur W_a. Soit W_a' la courbe non singulière birationnellement équivalente à W_a sur k. Le nombre de points où la transformation entre W_a et W_a' n'est pas biholomorphe ne peut dépasser

$$[U : V] \deg(C_a.Z) = [W_a : C_a] \deg(C_a.Z),$$

autrement dit il est uniformément borné.

D'autre part, le genre de C_a est égal au genre de C_u pour $u \notin F$ (cf. [10]) et le nombre de points ramifiés de C_a dans W_a' n'excède pas $\deg(C_a.Z)$. Quant à la courbe W_a', son genre est aussi borné par celui de W_u. En effet, si l'on prend une extension inséparable de $k(u)$, on a une transformation birationnelle entre W_u et une courbe W_u' non singulière. [J'ignore si la $k(u)$-normalisation de W_u est déjà non singulière.] Comme nous spécialisons dans un corps fini (qui est parfait), la spécialisation de W_u' [défini sur une extension purement inséparable de $k(u)$] sera néanmoins définie sur $k(a)$. Le genre de W_a' est donc uniformément borné.

Les groupes de Galois de W_a'/C_a et de W_a/C_a peuvent alors être identifiés, en tant que groupes de transformations sur les points de W_a' et W_a respectivement, en dehors des points où la transformation birationnelle n'est pas biholomorphe.

De plus, si Q est un point de $C_a = V.L_a$ qui n'est pas dans $C_a.Z$, les n points P_i au-dessus de Q dans V coïncident avec les n points au-dessus de Q dans W_a. Si la transformation T de G induit T_a sur W_a, alors la commutativité montre que $(Q,\ W_a/C_a) = (Q,\ U/V)_a$. De ces remarques, il résulte l'inégalité

$$\left| \sum_{Q \in C_a} \chi(Q) \right| \leq A_6\, q^{\frac{1}{2}}$$

le nombre A_6 ne dépendant que des genres de W_a (plutôt W'_a), de C_a, et du nombre de points de ramification, tous bornés.

Mais le nombre de termes dans la somme prise pour les points $a \notin F$ est égal à $N(\Gamma) - N(F)$.

De plus, on a l'inégalité

$$\left| \frac{N(\Gamma) - N(F)}{\varkappa_n^{r-1}} \right| \leq A_7\, q^{r-1}$$

qu'on déduit de (3) et de celle de $N(F)$ obtenue plus haut. En la combinant avec celle que nous venons de trouver pour les courbes, on trouve le théorème qu'il s'agissait de démontrer.

5. **L'image réciproque d'un revêtement.** — Soit V une variété abstraite, définie sur un corps k qu'on peut supposer parfait pour commencer. Soit $\alpha : V \to A$ une application rationnelle de V dans une variété de groupe commutatif A, α et A étant également définis sur k. (On pourrait s'occuper du cas plus général où A est une variété arbitraire, ou un groupe non commutatif, mais les applications que nous ferons porteront uniquement sur le cas d'un groupe commutatif. Nous nous plaçons donc immédiatement dans ce cas pour alléger l'exposition.) Nous supposerons de plus que V est sans singularités, et que α est défini en tout point de V. (Il ne s'agit pas de résoudre les singularités ; on enlève tout simplement les points singuliers, et les points où α n'est pas défini. Il reste toujours une variété abstraite.)

Soit $\lambda : B \to A$ une isogénie séparable, c'est-à-dire un homomorphisme séparable avec noyau fini. On suppose λ et B définis sur k. Si le noyau de λ est contenu dans les points rationnels B_k, on dira que k est *complet pour* λ. Nous supposerons ici que c'est bien le cas.

Le revêtement $\lambda : B \to A$ est non ramifié, ce qui veut dire qu'il existe exactement n points distincts de B dans l'image réciproque d'un point de A. Nous allons maintenant voir comment on peut construire l'image réciproque de B/A, pour obtenir un revêtement U de V.

Soit (y) un point générique de B/k et $(x) = \lambda(y)$. L'extension $k(y)$ sur $k(x)$ est abélienne, et son groupe de Galois $\mathfrak{g}$ est le groupe des translations sur B par un sous-groupe de B_k. Chaque automorphisme peut se représenter par une translation

$$y \to y + b_i, \qquad (b_i \in \mathfrak{g}).$$

Soit v un point générique de V/k, et posons $K = k(v)$. Soit $\xi = \alpha(v)$.

Alors $x \to \xi$ est une spécialisation. Soit η n'importe quelle spécialisation de y sur $x \to \xi$. Le corps $k(\eta)$ ne dépend pas de la spécialisation η, car n'importe quelle autre est de type $\eta + b_i$, et l'on a

$$k(\eta) = k(\eta + b_i).$$

De plus, $k(\eta)$ contient $k(\xi)$. Le cycle $\sum (\eta + b_i)$ est l'unique spécialisation du cycle $\sum (y + b_i)$ sur $x \to \xi$.

Il se peut que k ne soit pas le corps des constantes de $k(\eta)$, ou que k ne soit pas le corps des constantes du corps composé $k(v, \eta)$. Nous supposerons qu'il l'est c'est-à-dire que k est algébriquement clos dans $k(v, \eta)$. On peut alors définir la normalisation U de V dans $E = k(v, \eta)$. Ce sera une variété normale, définie sur k (supposé parfait). On obtient ainsi un revêtement $f : U \to V$ qui sera dit *dérivé* du revêtement $\lambda : B \to A$, ou *image réciproque* de $\lambda : B \to A$.

On peut écrire $k(v, \eta) = k(u)$ où u est un point générique de U/k, et $f(u) = v$. On a une application rationnelle $\beta : U \to B$ définie par $\beta(u) = \eta$. Si

$$[U : V] = [B : A],$$

autrement dit si $[k(u) : k(v)] = [k(y) : k(x)]$, on dira que l'image réciproque est *non dégénérée*. Si k est algébriquement clos et si l'image réciproque de tout revêtement de type $\lambda : B \to A$ donnée par une isogénie séparable λ est non dégénérée, on dira que V et A sont *bien adaptés par* α, ou simplement bien adaptés.

Considérons de plus près le cas non dégénéré.

La variété U peut encore se décrire d'une autre façon. Soit U_1 la variété dont le point générique sur k est (v, η). Elle est birationnellement équivalente à U, et nous verrons dans un instant qu'en fait elle lui est biholomorphe.

Si Q est un point de V, alors la spécialisation $v \to Q$ détermine une spécialisation $\xi \to \xi'$, car nous avons supposé α partout défini. Si $\lambda^{-1}(\xi') = \sum \eta'_i$, où les η'_i sont les n points distincts au-dessus de ξ', alors la spécialisation $(v, \xi) \to (Q, \xi')$ se prolonge de manière unique en $\sum (\eta_i) \to \sum (\eta'_i)$. On voit donc qu'au-dessus de chaque point Q de V il y a exactement n points de U_1. D'après la proposition 1, ils sont simples sur U_1 (car nous avons supposé V non singulière). De ceci on déduit immédiatement que U et U_1 sont biholomorphes, donc que U est non singulière. Ces remarques peuvent être résumées dans la proposition suivante :

PROPOSITION 2. — *Soit V une variété algébrique abstraite, non singulière.*

26.

Soient $\alpha : V \to A$ une application rationnelle, partout définie, et $\lambda : B \to A$ une isogénie séparable. Supposons l'image réciproque $f : U \to V$ non dégénérée. Alors tous les points de V sont non ramifiés dans U, qui est non singulière.

Si l'image réciproque est dégénérée, on voit facilement qu'on peut la réduire au cas non dégénéré en passant à une extension du corps de constantes, et en prenant l'image réciproque d'un revêtement C intermédiaire à B/A. On voit aussi que nous n'avons pas employé l'hypothèse que A et B sont des groupes. Mais ici, nous nous intéressons particulièrement au cas où le corps de base k est fini, avec q éléments. Comme on sait, on peut alors définir un revêtement $\rho : A \to A$ par la formule

$$\rho(y) = y^{(q)} - y,$$

qui sera appelé le *q-revêtement de A*. On a alors les exemples suivants :

A est le groupe multiplicatif G_m. Les images réciproques par applications dans G_m donnent les extensions kummériennes cycliques de K.

A est un groupe de Witt W_μ [17]. On obtient alors les extensions cycliques de degré p^μ.

(On peut prendre aussi des produits de G_m et des produits de W_μ pour donner les extensions abéliennes).

A est la variété d'Albanese. On obtient alors le corps de classe de Hilbert [6].

On remarquera que toute isogénie séparable $\lambda : B \to A$ définie sur la clôture algébrique $\bar{k}$ d'un corps fini k peut toujours s'obtenir comme revêtement intermédiaire d'un q-revêtement. En effet, on prend un corps de définition fini k, complet pour λ. Alors la commutativité de λ et ρ montre que λ est intermédiaire. Cette construction arithmétique remplace avantageusement la division des périodes dans le cas d'une variété abélienne. Elle ne peut se faire naturellement que sur la clôture algébrique d'un corps fini, mais dans ce cas elle donne bien une idée générale de la structure de la tour d'isogénies séparables au-dessus de A.

Considérons de nouveau le cas d'une image réciproque non dégénérée d'un revêtement $\lambda : B \to A$. Le revêtement U/V est alors abélien, et le groupe d'automorphismes de $k(u)$ sur $k(v)$ (ou de transformations birationnelles de U sur V) s'identifie avec celui de $k(y)$ sur $k(x)$ (ou de B/A). En effet, le groupe de translations b_i s'interprète comme groupe de translations de $k(\eta)$ sur $k(\xi)$, chaque translation τ appliquant η sur $\eta + b_i$. De plus une transformation T dans le groupe G de U/V est déterminée de façon unique par la relation

$$\beta T = \tau \beta.$$

Si T et τ se correspondent de cette façon, nous noterons cette relation par $T \leftrightarrow \tau$.

On a alors une loi de réciprocité (loi de décomposition) pour les cycles premiers rationnels de V dans U, relative à (α, A). Si $\mathfrak{a}$ est un cycle sur A, $\pi(\mathfrak{a})$ désignera la somme des points de $\mathfrak{a}$. Si $\mathfrak{a}$ est rationnel sur k, alors $\pi(\mathfrak{a})$ est un point rationnel de A dans k. Pour un cycle $\mathfrak{a}$ de V, nous écrirons $\pi(\mathfrak{a})$ au lieu de $\pi(\alpha(\mathfrak{a}))$, pour simplifier. L'opération π dépend de α bien entendu, et devrait être notée π_α, mais comme il s'agira toujours de la même application α, nous écrirons tout simplement π.

Si $\mathfrak{p}$ est un cycle premier rationnel de V, et $\pi(\mathfrak{p}) = a$, on voit trivialement que

$$(\mathfrak{p}, U/V) \leftrightarrow (a, B/A).$$

En particulier, si $\pi(\mathfrak{a}) = 0$, alors $(\mathfrak{a}, U/V)$ est l'identité.

On peut alors interpréter l'existence de localités dans une progression arithmétique de la façon suivante :

THÉORÈME 2. — *On suppose V^r abstraite, non singulière, définie sur le corps fini k. Soit $\alpha : V \to A$ une application rationnelle, partout définie, et supposons que l'image réciproque par α du q-revêtement $\rho(y) = y^{(q)} - y$ de A soit non dégénérée. Soit a un point rationnel dans A_k, et soit $N_0(a, \mu)$ le nombre des cycles premiers rationnels de V/k tels que $\deg(\mathfrak{p}) = \mu$ et que $\pi(\mathfrak{p}) = a$. Désignons par h le nombre de points de A_k. Alors*

$$N_0(a, \mu) = \frac{1}{h\mu} q^{\mu r} + O\left(q^{\mu\left(r - \frac{1}{2}\right)} \right)$$

pour $\mu \to \infty$.

Le théorème est une traduction du résultat du paragraphe 2, compte tenu de l'identification entre le groupe de Galois et le groupe des points rationnels A_k.

En particulier, nous retrouverons ici les propositions qui servent dans la théorie du corps de classes et démontrées en Appendice dans [6], car on sait que l'image réciproque d'une isogénie séparable de la variété d'Albanese est non dégénérée. De plus, nous les retrouvons généralisées à des groupes commutatifs non complets, pourvu seulement que l'image réciproque soit non dégénérée. Ceci sera essentiel pour l'élaboration de la théorie du corps de classes des extensions ramifiées. Dans ce qui suit, nous allons montrer comment la partie arithmétique de [6] se généralise.

La partie algébrique, qui a pour but de montrer quels revêtements de V on obtient par image réciproque d'une isogénie reste à faire. Pour les revêtement non ramifiés, on sait qu'on les obtient tous par la variété d'Albanese, sauf ceux dus à la torsion, ou à certaines p-extensions.

6. **Théorie du corps de classes.** — Soit V une variété abstraite, et $\alpha : V \to A$ une application rationnelle de V dans une variété de groupe commutatif. On

suppose tout ceci défini sur le corps fini k. De plus, on supposera V sans singularités, et α défini en tout point de V. (Il suffira d'enlever les points offensants).

On peut alors définir les classes de cycles relatives à (α, A) de la façon suivante. Soit $Z(V, k)$ le groupe des cycles de V rationnels sur k. Soit $Z_0(V, k)$ le sous-groupe des cycles de degré o. Alors l'application π induit un homomorphisme de $Z(V, k)$ dans A_k, et de $Z_0(V, k)$ dans A_k. Le noyau de π dans $Z_0(V, k)$ sera noté $Z_\alpha(V, k)$. On l'appellera le *noyau de* (α, A). Le Le groupe facteur Z/Z_α sera appelé le groupe des *classes de cycles* $C_K(V)$; on pourrait démontrer que c'est un invariant birationnel, ne dépendant que de K et de α. Nous ne le démontrerons que dans le cas où l'image réciproque du q-revêtement de A est non dégénérée. C'est alors une conséquence immédiate de l'identification de A_k avec le groupe de Galois de U/V, au moyen de l'application de réciprocité. En effet, si l'on considère la décomposition d'une localité de V dans le corps E/K, E étant le corps de fonctions de U, on sait qu'il existe toujours un cycle $\mathfrak{a}$ de degré 1, rationnel sur k, tel que $(\mathfrak{a}, U/V) = 1$. L'application de réciprocité étant surjective, on voit immédiatement que si V_1 est un autre modèle de K/k, on peut prendre un cycle $\mathfrak{a}_1$ sur V_1 de degré 1 tel que $(\mathfrak{a}_1, U_1/V_1) = 1$. L'application des classes de cycles qui envoie $\mathfrak{a}$ sur $\mathfrak{a}_1$ est un isomorphisme.

Dans tout ce qui suit, on supposera que V et A sont bien adaptés par α. Alors π applique $Z_0(V, k)$ sur A_k. Soit E une extension finie arbitraire de K. On dira que E est de *type* (α, A) si la normalisation U de V dans le corps $E\bar{k}$ (définie sur $\bar{k}$) s'obtient comme image réciproque d'une isogénie séparable $\lambda : B \to A$, λ et B étant définis également sur $\bar{k}$. L'extension E/K peut ne pas être galoisienne, mais sur $\bar{k}$, elle devient abélienne, bien entendu.

Supposons que E soit galoisienne. Si E est de type (α, A), alors on sait par la proposition 2 que tous les points de V sont non ramifiés dans U, qui est non singulière. Soit k' le corps des constantes de E. Alors U est définie sur k'. On a des applications rationnelles

$$U \xrightarrow{f} V \xrightarrow{\alpha} A,$$

f étant défini sur k', et α sur k. On voit que αf est partout défini. U étant non singulière, on peut alors définir les classes de cycles $Z(U, k')$ comme ci-dessus, et l'on a un homomorphisme

$$S_K^E : C_E(U) \to C_K(V)$$

des classes de cycles de E dans celles de K, au moyen de la trace.

Toutes les notions employées pour énoncer la théorie du corps de classes sont maintenant définies, et les énoncés de [6] (dans le § 5 et le théorème 5 du § 9) ont un sens, si l'on remplace partout l'expression « de type Albanese » par « de type (α, A) ». Sauf pour le théorème de limitation qui demande quelques remarques, les démonstrations données dans [6] s'appliquent sans changement. Il est donc inutile de les reproduire ici. (*Voir*, § 6 à 10 de [6]).

Remarquons seulement qu'en ce qui concerne la surjectivité de l'application de réciprocité, on peut la démontrer très simplement et directement dans une extension galoisienne E/K pouvant même admettre une extension du corps des constantes (et pas seulement géométrique comme nous l'avons fait plus haut). En effet, supposons E/K abélien pour simplifier. A chaque $\mathfrak{p}$ de V on associe le groupe de décomposition $D_\mathfrak{p}$ et l'on prend le corps K' fixe pour le sous-groupe engendré par les $D_\mathfrak{p}$. Il s'agit de montrer que $K' = K$. Tout $\mathfrak{p}$ se décompose complètement dans K' (on suppose naturellement $\mathfrak{p}$ non ramifié). S'il y avait une extension du corps de constantes dans K', on voit tout de suite que ce serait impossible. Nous sommes alors ramenés au cas géométrique, que nous connaissons.

Quant au théorème de limitation, supposons que F/K soit une extension galoisienne arbitraire. On rétrécit alors V en enlevant un sous-ensemble k-fermé X tel que tout point de $V - X$ soit non ramifié dans F. On peut de nouveau définir les classes de cycles. Elles ne changent pas pour $V - X$, et l'on en obtient pour F. De plus on a le lemme habituel :

LEMME. — *Soit F une extension galoisienne de K, et E une extension abélienne. Soit V un modèle abstrait de K, non singulier, et tel que toute localité soit non ramifiée dans EF. Soit $\mathfrak{q}$ une localité de F dans la normalisation de V dans F. Alors la restriction de $(\mathfrak{q}, EF/F)$ à E est égale à $(S_K^F(\mathfrak{q}), E/K)$.*

D'après le théorème d'existence, on prend pour E le corps de classes de type (α, A) appartenant au groupe de trace $S_K^F C_F$. Alors pour tout $\mathfrak{q}$ dans F, on doit avoir $(\mathfrak{q}, EF/F) = 1$. Ceci est impossible, car on sait qu'il y a une infinité de localités de F qui ne se décomposent pas complètement dans E si $E \neq F$.

La seule différence entre la démonstration donnée ici et celle de [6] est que nous sommes forcés de définir les classes de cycles à partir de α, tandis que si l'on travaille avec la variété d'Albanese, on peut définir les classes de cycles à chaque étage indépendamment. Pour les définir dans le cas présent, il faut de plus passer par un rétrécissement, qui se trouve en fin de compte être sans importance.

Enfin, pour terminer, montrons comment on peut développer la théorie du corps de classes pour une courbe C dans le cas ramifié. Supposons pour commencer que C soit définie sur un corps k arbitraire. Grâce à un théorème de Rosenlicht (encore inédit) on sait que ses jacobiennes généralisées (que nous appellerons aussi jacobiennes de Rosenlicht) satisfont à la propriété d'application universelle pour les applications de C dans les variétés de groupe commutatif. On peut d'ailleurs se restreindre aux jacobiennes provenant d'un anneau semi-local du type $k + \mathfrak{c}$ ou $\mathfrak{c}$ est un diviseur positif sur la courbe. Vu la remarque faite au paragraphe 5, que tous les revêtements abéliens d'une variété proviennent d'images réciproques d'isogénies de groupes algébriques commutatifs (sur le domaine universel, bien entendu) on peut conclure du théorème de Rosenlicht qu'ils proviennent tous d'images réciproques d'isogénies de jacobiennes généralisées du type ci-dessus. Comme Serre l'a remarqué, il s'ensuit immédiatement que l'application canonique $\alpha : C \to J$ de la courbe dans une jacobienne de Rosenlicht satisfait à la condition que nous avons énoncée plus haut, à savoir que l'image réciproque d'une isogénie de J conserve son degré, c'est-à-dire que C et (α, J) sont bien adaptées. Nous voyons donc que, pour les courbes, on obtient la classification géométrique qui nous fait défaut pour les variétés de dimension supérieure.

D'autre part, supposons C définie sur le corps fini k, et le diviseur $\mathfrak{c}$ rationnel sur k. D'après les résultats de Weil sur les groupes de transformations et les espaces homogènes, on sait qu'il existe une application canonique de C dans un espace principal homogène H de J, tous ces objets étant définis sur k. Comme k est fini, l'espace H possède un point rationnel [$cf.$ S. LANG, *Algebraic groups over finite fields* (*Amer. J. Math.*, July 1956, th. 2)], et l'on peut donc trouver une application canonique de C dans J définie sur k. La courbe et sa jacobienne étant bien adaptées, nous obtenons donc la théorie du corps de classes comme nous l'avons indiqué au début de ce numéro.

De plus, on peut retrouver la théorie exprimée en langage d'idèles (et donc la théorie cohomologique avec le cocycle fondamental) par un raisonnement facile, comme le fit Chevalley lorsqu'il introduisit les idèles pour la première fois. On voit alors que si $\mathfrak{c}$ est le conducteur au sens classique d'une extension abélienne géométrique du corps de fonctions de la courbe C, cette extension est une image réciproque d'une isogénie de la jacobienne définie par l'anneau $k + \mathfrak{c}$.

BIBLIOGRAPHIE.

[1] E. ARTIN, *Ueber eine neue art von L-Reihen* (*Abh. math. Sem. Hamburg Univ.*, Bd. 2, 1924).

[2] E. ARTIN, *Zur theorie der L-Reihen mit allgemeinen gruppencharakteren* (*Abh. math. Sem. Hamburg Univ.*, Bd. 8, 1930).

[3] W. L. CHOW, *Abstract theory of the Picard and Albanese varieties* (à paraître dans les *Annals of Mathematics*).

[4] W. KRULL, *Galoische theorie der ganz abgeschlossene Stellenringe* (*Sitzungsberichte der phys.-med. Soz. zu Erlangen*, Bd. 67-68, 1935-1936).

[5] S. LANG, *L-series of a covering* (à paraître aux *Proceedings of the National Academy*, U. S. A., 1956).

[6] S. LANG, *Unramified class field theory* (à paraître dans les *Annals of Mathematics*, 1956).

[7] S. Lang et A. Weil, *Number of points of varieties in finite fields* (*Amer. J. Math.*, vol. 72, No 4, octobre 1954, p. 818-827).

[8] T. Matsusaka, *The theorem of Bertini on linear systems in modular fields* (*Mem. College of Science, Univ. of Kyoto*, Series A, Math., vol. 26, No 1, juillet 1950).

[9] M. Nagata, *On the theory of Henselian Rings* (*Nagoya math. J.* vol. 5, 1953, p. 45-57, et vol. 7, 1954, p. 1-19).

[10] Y. Nakai, *On the genus of algebraic curves* (*Mem. College of Science, Univ. of Kyoto*, Series A, Math., No 2, 1952).

[11] L. B. Nisnevic, *Ueber die Anzahl der Punkte einer algebraische Mannigfaltigkeit in einem endlichen Primkörper* (en russe *Doklady Akad. Nauk U. S. S. R.*, 1954).

[12] F. K. Schmidt, *Die theorie der Klassenkörper..*, (*Sitzungsberichten der phys. med. Soz. zu Erlangen*, Bd. 62, 1930, p. 267-284).

[13] A. Weil, *Number of solutions of equations in finite fields* (*Bull. Amer. Math. Soc.*, vol. 55, No 5, 1949, p. 497-508).

[14] A. Weil, *Sur les courbes algébriques et les variétés qui s'en déduisent* (Paris, Hermann, 1948).

[15] A. Weil, *Footnote to a recent paper* (*Amer. J. Math.*, vol. 76, No 2, 1954).

[16] A. Weil, *Critères d'équivalence* (*Math. Annalen*, Bd. 128, 1954).

[17] E. Witt, *Zyklische Körper und algebren der characteristik p vom Grad p^n* (*J. reine angew. Math.*, Bd. 176, 1936).

[18] O. Zariski, *Pencils on an algebraic variety and a new proof of a theorem of Bertini* (*Trans. Amer. math. Soc.*, 1941).

(Manuscrit reçu en avril 1956.)

Am. J. Math., LXXVIII, 1956

ALGEBRAIC GROUPS OVER FINITE FIELDS.*

By Serge Lang.

---

1. Introduction. Let k be a finite field with q elements. Let G be an algebraic group defined over k. (For the foundations of the theory of algebraic groups and homogeneous spaces, see Weil [8], [9].) If x is a point of G, we denote by $x^{(q)}$ the point obtained by raising all coordinates of x to the q-th power, i.e. by applying to x the Frobenius automorphism of the universal domain leaving k fixed. The mapping $f(x) = x^{-1}x^{(q)}$ is a rational map of G into itself. It will be shown that it is in fact surjective, and that it gives a Galois, in general non abelian unramified covering of G over itself, the Galois group being that of the left rational translations. This sort of covering is of course impossible in characteristic 0.

More generally, it will be shown that G acts on itself as a homogeneous space, under the operation $F(x, y) = x \cdot y = x^{(q)}yx^{-1}$. Using this fact we shall show that every homogeneous space H of G defined over k has a rational point. If it is principal homogeneous, then it must be biregularly equivalent to G over k, and in case G is commutative, this means that the Weil group is trivial. We also use this result to give a new proof of a result due to Chatelet [4], that if a variety V/k becomes biregularly equivalent to projective space over the algebraic closure of k, then it is already so over k itself.

Finally we consider the class field theory for our covering defined by the map $z \to z^{-1}z^{(q)}$, get a partial non abelian reciprocity law, and prove that the Artin L-series are trivial. This is used to prove the following result: Let $\mathfrak{g}$ be a subgroup of the rational points of G over k. Let H be the homogeneous space of cosets of $G \bmod \mathfrak{g}$. Then G and H have the same number of rational points.

In case the group G is commutative, then one can get a complete reciprocity law, and one can use it to derive the class field theory over a variety V having a rational map into G by means of which the abelian coverings of G by commutative groups can be pulled back in a one-one manner.

---

* Received March 9, 1956.

555

This abelian class field theory is carried out in detail elsewhere, and in this paper, we have concentrated exclusively on the non abelian aspects of the covering $z^{-1}z^{(q)}$.

2. The map $f(z) = z^{-1}z^{(q)}$. The map $f(z)$ being as above we contend that if z is a generic point of G/k then $f(z)$ is also a generic point of G/k, and in fact the extension $k(z)$ over $k(f(z))$ is separable algebraic. Namely, putting $x = f(z)$ we have $k(z) = k(x,z) = k(x, z^{(q)})$. Our contention now follows from the following elementary and well known result of field theory:

PROPOSITION 1. *Let F be a field and E/F a finitely generated extension. We assume the characteristic p is $\neq 0$. If E/F is separable algebraic, then $E^{p^\mu}F = E$ for all powers p^μ, and conversely, if $E^{p^\mu}F = E$ for some power p^μ, then E/F is separably algebraic.*

(By E^{p^μ} we denote the field obtained by raising all elements of E to the p^μ-th power.)

We have trivially for z, w in G:

$$(1) \qquad (zw)^{(q)} = z^{(q)}w^{(q)}, \quad (z^{-1})^{(q)} = (z^{(q)})^{-1}.$$

Furthermore our rational map satisfies the following formalism:

$$(2) \qquad f(zw) = f(z)^w f(w)$$

where y^w is defined to be $w^{-1}yw$.

The subgroup of G consisting of the elements rational over the field k_d (unique extension of k of degree d) will be denoted by G_d.

(3) An element a of G is in G_1 if and only if $f(a) = 0$. More generally, $f(z) = f(w)$ if and only if $z = aw$ for some a in G_1.

The first part of the statement is obvious. If $z = aw$, it is also obvious that $f(z) = f(w)$. Suppose $f(z) = f(w)$. Then $z^{-1}z^{(q)} = w^{-1}w^{(q)}$, whence $(zw^{-1}) = (zw^{-1})^{(q)}$. This means that zw^{-1} is rational over k, as desired.

We now see that if z is a generic point of G/k, then the extension $k(z)/k(f(z))$ is Galois, the distinct conjugates of z over $k(f(z))$ being az, with a rational over k.

The mapping $g(z) = z^{(q)}z^{-1}$ has analogous properties with respect to right translations, and we shall use them freely whenever needed.

PROPOSITION 2. *Let y be an arbitrary point of G and x a generic point over $k(y)$. Then $x^{(q)}yx^{-1}$ is a generic point of G over $k(y)$.*

Proof. Let $w = x^{(q)}yx^{-1}$. Put $K = k(y)$. Then $K(w, x) = K(w, x^{(q)})$. This implies that $K(x)$ is separable algebraic over $K(w)$, and that w has the same dimension as x over K. Hence it is a generic point of G/K.

Given two arbitrary points x and y of G, we denote by $x \cdot y$ the point $x^{(q)}yx^{-1}$. We obviously have for x, y, z arbitrary,

$$(4) \qquad (xy) \cdot z = x \cdot (y \cdot z) \quad \text{and} \quad e \cdot x = x$$

To prove that G is a homogeneous space over itself with the above defined external law of composition, we need only prove the following statement:

THEOREM 1. *Given two points y and w in G, there exists a point x such that $x \cdot y = w$.*

Proof. In fact, using the associativity, it suffices to prove: Given y in G, there exists x such that $x \cdot y = e$. Let t be a generic point of G over $K = k(y)$. Then $t \cdot y$ is a generic point of G/K by Proposition 2. There is an isomorphism σ which is identity on K and maps $t^{(q)}t^{-1}$ on $t \cdot y = t^{(q)}yt^{-1}$. We can extend σ to the field $K(t)$. Let $u = t^{\sigma}$. Then

$$g(u) = g(t^{\sigma}) = g(t)^{\sigma} = t^{(q)}yt^{-1}.$$

If we put $x = u^{-1}t$, we have what we want.

COROLLARY. *The map $z \to z^{-1}z^{(q)}$ is surjective, i.e. given any y in G, there exists z such that $y = z^{-1}z^{(q)}$.*

Proof. According to the theorem, there exists z in G such that $z^{(q)}y^{-1}z^{-1} = e$. This element z does what is required.

From this corollary we see that we have indeed an unramified covering of G over itself. Given any point Q in G, there exists n points P such that $Q = P^{-1}P^{(q)}$, where n is the order of G_1. Given any one of them, all the others are simply the left rational translations of this one.

As another application of Theorem 1, we prove

THEOREM 2. *Let H/k be a homogeneous space over G. Then H has a rational point.*

Proof. We must show that there is some point u in H such that $u^{(q)} = u$. Let v be any point of H. Since H is defined over k, then $v^{(q)}$ is also in H. Since H is a homogeneous space, there exists an element x in G such that $xv^{(q)} = v$. By the corollary to Theorem 1, we can write $x = y^{-1}y^{(q)}$. From this we see that $(yv)^{(q)} = (yv)$ and hence $u = yv$ is the element we are looking for.

We would like to point out here that Theorem 3 of [6] is a special case of the preceding result. Indeed, if a variety V/k becomes biregularly equivalent to an abelian variety over the algebraic closure of k, then V can be viewed as a principal homogeneous space over its Albanese variety A, which is known (by Chow's work) to be defined over k. (See Weil [9], Prop. 4.) However, because of the completeness of the group, we could give a direct proof, without using the Albanese variety. The proof can in fact be further simplified as follows:

We wish to prove that if a variety V over a finite field k becomes bi-regularly equivalent to a complete group variety over the algebraic closure of k, and hence over a finite extension k' of k, then V has a rational point over k. Over k' we can put a law of composition on V which makes it into a complete group variety (we don't even need to know it is abelian). There is a unit element e, rational over k', but not necessarily over k. Let z be a generic point of V over k. With respect to the composition law over k', we consider the point $f(z) = z^{-1}z^{(q)}$. It is a generic point of V over k' (by Proposition 1). Since V is complete, we can extend a specialization of $f(z)$ on e to a specialization of z to a point ξ on V, and then $\xi^{-1}\xi^{(q)} = e$. This point ξ satisfies $\xi = \xi^{(q)}$, and is therefore a rational point.

The statement made in our above mentioned paper that there is a rational place is a consequence of the following well known fact: *Let k be any field, and V/k any variety (say affine). Let Q be a simple point of V, rational over k. Let v be a generic point of V/k. Then there exists a place ϕ of $k(v)$ over k such that $\phi(v) = Q$ and such that ϕ is rational over k (i. e. k-valued).* One often says that Q is at the center of ϕ. Here of course, we have the additional property that the place can be chosen in such a way that the residue class field is canonically isomorphic to k itself: No irrationalities are needed in extending the specialization $v \to Q$ to a place of the function field. There exist many proofs of the above fact, and one of them runs along the following lines. The completion of the local ring of Q in $k(v)$ is isomorphic to a power series ring in r variables ($r = \dim V$) with coefficients in k, because Q is simple. There is therefore a canonical isomorphism of $k(v)$ in the quotient field of this power series ring. This quotient field itself can be embedded in the repeated power series field, which has obviously a k-valued place mapping all the variables on 0, one after the other. The restriction of this place to $k(v)$ is the one we are looking for.

We now return to the arbitrary algebraic group G over the finite field k.

Let $\sigma, \tau, \cdots$ range over the group of automorphisms of the extension k_d of k. This group (which is cyclic) operates on G_d in an obvious fashion. Referring to this operation, we show that the 1-cocycles split:

PROPOSITION 3. *Let $\{x_\sigma\}$ be a set of elements of G_d such that $x_\tau{}^\sigma x_\sigma = x_{\sigma\tau}$. Then there exists y in G_d such that $x_\sigma = y^\sigma y^{-1}$.*

Proof. We can change our indices from σ to integers mod d. According to the corollary of Theorem 2, we can write $x_1 = y^{(q)} y^{-1}$ for some y in G (we do not know yet whether it is in G_d). Then by the cocycle property, we get $x_i = y^{(q^i)} y^{-1}$. Finally, taking $i = d$ we must have

$$e = x_0 = x_d = y^{(q^d)} y^{-1}.$$

This shows that y is rational over k, because $y^{(q^d)} = y$.

As an application, we prove Chatelet's theorem:

THEOREM 3. *Let V be an abstract variety defined over k, which becomes biregularly equivalent to projective space S over the algebraic closure of k. Then V is biregularly equivalent to S over k.*

Proof. Let $T: V \to S$ be the correspondence, defined over some k_d. With Chatelet ([3], [4]) we take $x_\sigma = T^\sigma T^{-1}$, which is a birational biregular correspondence between S and itself. It is therefore projective, and is an element of the projective group G, rational over k_d. It satisfies the condition of Proposition 3, and if we let y be the projective transformation as in Proposition 3, we consider $T_1 = y^{-1} T$. Then T_1 is obviously fixed under every automorphism σ of k_d over k, and hence T_1 is defined over k.

For a proof that the only biregular correspondences of S with itself are projective, see Chow [5]. In that paper, other varieties are proved to have that property, and our theorem applies to them as well.

3. Class field theory. We shall now investigate the covering $f(x) = x^{-1} x^{(q)}$ from the point of view of class field theory. By a cycle, we shall always mean a cycle of dimension 0. Let $\mathfrak{p}$ be a prime rational cycle of G over k, and let Q be any point in it. Let P be any point such that $f(P) = Q$. All other points are of type aP, where a is rational over k. If $\mathfrak{p}$ is or degree d, then $f(P^{(q^d)}) = f(P)^{(q^d)} = Q$. Hence $P^{(q^d)}$ also lies in the inverse image of Q under f, and hence there exists a rational point a in G_1 such that $aP = P^{(q^d)}$.

We define $\pi_d(Q)$ to be the product $QQ^{(q)} \cdots Q^{(q^{d-1})}$. Then we have

$$(6) \qquad\qquad aP = P^{(q^d)} = P\pi_d(Q).$$

It is clear that the point a in G_1 is completely well defined by the prime $\mathfrak{p}$, up to conjugacy in G_1. Furthermore we have

PROPOSITION 4. *Given two points* Q_1 *and* Q_2 *in* G_1, *let* a_1 *and* a_2 *in* G_1 *be any points determined by the condition*

$$a_1 P_1 = P_1^{(q)} = P_1 Q_1, \qquad a_2 P_2 = P_2^{(q)} = P_2 Q_2.$$

Then a_1 *is conjugate to* a_2 *in* G_1 *if and only if* Q_1 *is conjugate to* Q_2 *in* G_1.

The proof is trivial and formal, and we leave it to the reader.

We have a mapping from the primes to the conjugacy classes of G_1 defined by (6), and if the prime is of degree 1, then we obtain a 1-1 mapping of the conjugacy classes of G_1 (viewed as a set of primes of degree one) onto the conjugacy classes of G_1 (viewed as Galois group of left translations). The period of the point a is clearly equal to the period of $\pi_d(Q)$, and thus we can tell the period of the Frobenius class associated with a prime. Furthermore, we see that $\mathfrak{p}$ splits completely if and only if $\pi_d(Q) = e$.

I have not been able to determine whether the conjugacy class of a is always equal to that of Q (when Q is in G_1) and more generally to determine if a and $\pi_d(Q)$ are conjugate in G_d. In order to obtain the complete decomposition laws, we would need to know that any rational point b in G_1, conjugate to $\pi_d(Q)$ in G_d is conjugate to a in G_1. This would imply that the Frobenius class associated with $\mathfrak{p}$ in G_1 can be determined rationally. (This is the case for the full linear group.)

We now consider the L-series.

Let z be a generic points of G/k. Let

$$f_n(z) = z^{-1}z^{(q^n)}) \quad \text{and} \quad \pi_n(z) = zz^{(q)} \cdots z^{(q^{n-1})}$$

Then $\pi_n(f_1(z)) = f_n(z)$.

Let $E = k(z)$, $F = k(f_1(z))$, and $K = k(f_n(z))$. We have inclusions $E \supset F \supset K$, and E/F is Galois (it is the extension discussed previously). Let E_n, F_n, and K_n be the constant field extensions by k_n. Then E_n/K_n becomes Galois, with group G_n. The group G is a model of each one of our function fields, but of course in a different way each time.

Referring to these models, we shall prove that the L-series are trivial for a character not containing the identity.

Let χ be a character of G_1. Let $\mathfrak{p}$ be a prime cycle of G/k. Let $T_\mathfrak{p}$ be any one of the translations of G_1 in the Frobenius class associated with $\mathfrak{p}$ in G_1. Then the L-series are defined by

$$t\frac{d}{dt}\operatorname{Log} L(t, \chi, E/F) = \sum \chi(T_\mathfrak{p}{}^\mu)\deg(\mathfrak{p}) t^{\mu \deg(\mathfrak{p})}$$

the sum being taken over all primes and all $\mu \geq 1$.

If we look at the coefficient of t^n we see that we must prove that

$$\sum_{\deg(\mathfrak{p})\,|\,n} \chi(T_\mathfrak{p}{}^{n/\deg(\mathfrak{p})})\deg(\mathfrak{p}) = 0.$$

This sum can be rewritten in terms of points in G_n as follows: To each point Q in G_n, we can associate a translation $T_Q{}^{(n)}$ (well defined up to conjugacy) in G_1, such that for any P in $f_1^{-1}(Q)$ we have $T_Q{}^{(n)}(P) = P^{(q^n)}$. It is then clear that the above sum is equal to

$$(7) \qquad\qquad \sum_{Q\,\varepsilon\,G_n} \chi(T_Q{}^{(n)}).$$

For $n = 1$ we know by Proposition 4 that each class will have a representative $T_Q{}^{(1)}$ for some Q in G_1, and that this representative will occur in our sum as many times as there are element in that class. Consequently the value of our sum is the same as the value of the character taken over the sum of the conjugacy classes in the group ring of G_1. If χ does not contain the identity character, then this sum must be 0.

For arbitrary n, we note that the coefficient of t^n in our L-series $L(t, \chi, E/F)$ is by definition and formula (7) the same as the coefficient of t in $L(t, \chi, E_n/F_n)$. We have thus reduced the computation of the n-th term to the computation of the first term of an L-series over a bigger constant field. By one of the main theorems on L-series, we know that

$$L(t, \chi, E_n/F_n) = L(t, \chi^*, E_n/K_n)$$

where χ^* is the induced character. If χ does not contain the identity, then neither does χ^*. The extension E_n/K_n is now of a type analogous to that considered above for $n = 1$ (i.e. belonging to a rational map f_n). Hence the first coefficient of the L-series is equal to zero, as desired.

Let $\lambda: G \rightarrow H$ be a homomorphism of an algebraic group G onto an algebraic group H, defined over k, and with finite kernel. Let z be a generic point of G/k. As usual, $f_G(z) = z^{-1}z^{(q)}$. We also have an f-mapping on H, denoted by f_H. Then obviously $f_H\lambda = \lambda f_G$. Taking the degree of both sides, we see that the degree of f_H must equal that of f_G, i.e. that *G and H have* *the same number of rational points* (hence the same zeta function). If kernel of λ is contained in G_1, and λ is separable, then G is Galois over H, and this again suggests that the L-series for the covering are trivial. This is indeed the case, and can be proved as follows: Put $y = \lambda(z)$. Then $E = k(z) \supset k(y) = M \supset k(z^{-1}z^{(q)}) = F$. If χ is a character for the Galois group of E/M, and does not contain the identity, then the induced character χ^* to the Galois group of E/F does not contain the identity either, and by what has been proved before, the L-series belonging to it must be trivial.

More generally, if the identity for E/M occurs with some multiplicity in χ, then the identity for E/F occurs with the same multiplicity in χ^*. (See for instance Brauer-Tate [2], where we put $\Theta =$ identity in formula (5).) From this remark we can deduce the following result.

Let H be the homogeneous space obtained from G by the cosets of a subgroup of G_1. Then we have a rational map $\lambda : G \to H$ (not necessarily a homomorphism), and the same type of field inclusion as before: $k(z) \supset k(y) \supset k(x)$. The zeta function of H, denoted by $Z_H(t)$, can be written

$$Z_H(t) = L(t, 1, E/M) = L(t, 1^*, E/F)$$

where 1 stands for the identity character on the Galois group of E/M. The identity for E/F occurs exactly once in 1^*, and hence the above L-series is equal to $Z_G(t) \cdot L(t, \chi, E/F)$, where χ is some character for E/F, which does not contain the identity, and $Z_G(t)$ is the zeta function of G. This shows that the zeta function of G and H coincide, and hence that G and H have the same number of rational points.

Knowing that the L-series are trivial, we can of course apply the formal argument given by Artin (Satz 4 of [1]) to get the density of primes in a given arithmetic progression. Let χ_i be the simple characters of G_1. Then we know that

$$(8) \qquad \sum_{Q \, \varepsilon \, G_n} \chi_i(T_Q^{(n)}) = \begin{cases} 0 & i \neq 1 \\ q^{nr} + O(q^{n(r-1/2)}) & i = 1 \end{cases}$$

because for $i = 1$, we deal with the zeta function and can use the results of [7]. Let C_j be a fixed class in G_1 and T any element of C_j. Let h be the order of G_1 and h_j the number of elements in C_j. Multiplying (8) by $\chi_i(T^{-1})$ and summing, we use the orthogonality relations (see formula (3) of [1]) to get

$$\frac{h}{h_j} N(n, C_j) = q^{nr} + O(q^{n(r-1/2)}),$$

where $N(n, C_j)$ is the number of points in G_n having their $T_Q^{(n)}$ in the given class C_j. One sees trivially that to get an estimate for the number of primes, one has to divide by n.

COLUMBIA UNIVERSITY.

———

REFERENCES.

———

[1] E. Artin, "über eine neue Art von L-Reihen," *Abhandlungen aus dem Mathematischen Seminar der Hamburgischen Universität*, vol. 3 (1923), pp. 89-108.

[2] R. Brauer and J. Tate, "On the characters of finite groups," *Annals of Mathematics*, vol. 62 (1955), pp. 1-7.

[3] F. Chatelet, "Variations sur un thème de Poincaré," *Annales de l'Ecole normale superieure*, vol. 61 (1944), pp. 249-300.

[4] ———, "Les courbes de genre 1 dans un champ de Galois," *Comptes rendus des seances de l'Academie des Sciences*, vol. 224 (1947), pp. 1616-1618.

[5] W. L. Chow, "On the geometry of algebraic homogeneous spaces," *Annals of Mathematics*, vol. 50 (1949), pp. 32-67.

[6] S. Lang, "Abelian varieties over finite fields," *Proceedings of the National Academy of Sciences*, vol. 41 (1955), pp. 174-176.

[7] ——— and A. Weil, "Number of points of varieties in finite fields," *American Journal of Mathematics*, vol. 76 (1954), pp. 819-827.

[8] A. Weil, "Algebraic groups of transformations," *American Journal of Mathematics*, vol. 77 (1955), pp. 355-391.

[9] ———, "On algebraic groups and homogeneous spaces," *American Journal of Mathematics*, vol. 77 (July 1955), pp. 493-512.

SUR LES REVÊTEMENTS NON RAMIFIÉS DES VARIÉTÉS ALGÉBRIQUES.*

par Serge Lang et Jean-Pierre Serre.

Introduction. Soit V une variété algébrique normale définie sur le corps C des nombres complexes. Munissons V de la topologie "usuelle," déduite de celle de C. Tout revêtement algébrique non ramifié de V peut être considéré comme un revêtement topologique, et correspond donc à un sous-groupe d'indice fini du groupe fondamental $\pi_1(V)$ de V. Des propriétés connues du groupe fondamental résulte alors immédiatement:

a) Si V et W sont deux variétés, tout revêtement de $V \times W$ est quotient d'un revêtement de la forme $V' \times W'$, où V' et W' sont des revêtements de V et de W.

(C'est une conséquence de l'égalité $\pi_1(V \times W) = \pi_1(V) \times \pi_1(W)$).

b) Le nombre des revêtements de V de degré donné est fini.

(C'est une conséquence du fait que $\pi_1(V)$ peut être engendré par un nombre fini d'éléments).

Dans ce qui suit, nous nous proposons de donner des démonstrations *algébriques* de a) et de b), valables en toute caractéristique (cf. th. 1 et th. 4); nous devons toutefois faire l'hypothèse supplémentaire que V est une variété *complète*. Des contre-exemples simples (dus essentiellement à la présence de "ramification supérieure") montrent d'ailleurs que a) et b) ne sont pas vrais sans restriction en caractéristique $p > 0$.

Comme application de a), nous démontrons que tout revêtement non ramifié d'une variété abélienne est une isogénie (th. 2), généralisant ainsi un résultat bien connu pour les courbes elliptiques.

1. Conventions et notations. Nous adopterons la définition des *revêtements* donnée dans [4] § 1, à cela près que nous ne nous occuperons que de revêtements *géométriques*. Par définition, un revêtement est donc une application rationnelle partout définie

$$f: U \to V,$$

---

[1] Received December 21, 1956.

319

vérifiant les conditions suivantes:

α) U et V sont des variétés *normales* (absolument irréductibles) de même dimension.

β) Pour tout $v \in V$, l'ensemble $f^{-1}(v)$ est *fini*.

γ) La variété U est *complète* au-dessus de tout point de V.

δ) Si K et L désignent les corps des fonctions rationnelles sur V et sur U, l'extension L/K est *séparable*.

Si U, V et f sont définis sur un corps k, on dira plus brièvement que le revêtement est défini sur k.

Les propriétés ci-dessus entraînent que U est la normalisée de V dans l'extension L/K (pour tout ce qui concerne l'operation de normalisation, voir par exemple [6], Chap. II, §5); inversement, si l'on se donne une variété normale V ayant K pour corps de fonctions, et si l'on se donne une extension finie séparable L de K, la normalisée U de V dans L/K est un revêtement de V. Ainsi, V étant donnée, il y a correspondance bijective entre revêtements de V et extensions finies séparables de K. Le *degré* de l'extension L/K sera appelé le degré du revêtement $f: U \to V$, et sera noté $[U:V]$. De même, un revêtement sera dit *galoisien*, de groupe de Galois G, si l'extension L/K correspondante est galoisienne et de groupe de Galois G; on a alors $V = U/G$, avec les notations de [7], no. 13.

Soit $f: U \to V$ un revêtement de degré n, défini sur un corps algébriquement clos k. Si v est un point de V, l'ensemble $f^{-1}(v)$ a au plus n éléments; s'il en a exactement n, on dira que U est *non ramifié en v*. L'ensemble des points de ramification est un sous-ensemble k-fermé de V; s'il est vide, on dira que U est *non ramifié*. En fait, pour qu'une application rationnelle partout définie $f: U \to V$ soit un revêtement non ramifié, il suffit, d'après Krull ([2], voir aussi [4], p. 292), que V soit normale, que les conditions β) et γ) soient remplies, et que pour tout point $v \in L$ l'ensemble $f^{-1}(v)$ ait exactement n points (n désignant comme précédemment le degré de l'extension L/K); la variété U est alors automatiquement normale, et l'extension L/K séparable.

Revenons au cas d'un revêtement quelconque $f: U \to V$, et soit V' une sous-variété de V. Soit $f^{-1}(V')$ l'image réciproque de V' dans U, et soient U_i' les composantes irréductibles de $f^{-1}(V')$. En appliquant les théorèmes du type Cohen-Seidenberg, on voit tout de suite que les U_i' ont même dimension que V'; de plus, si l'on note $[U_i':V']$ le degré de la projection $U_i' \to V'$, on a:

$$\Sigma[U_i':V'] \leqq [U:V], \quad \text{cf. [2] ainsi que [4], } loc.\ cit.\,.$$

Lorsque tous les degrés $[U_i' : V']$ sont égaux à 1, on dira que le revêtement U *se décompose complètement* sur V'.

En général, les U_i' ne sont pas des variétés normales, et les extensions de corps correspondant aux projections $U_i' \to V'$ ne sont pas séparables. On a toutefois :

LEMME 1. *Supposons que le revêtement* $f : U \to V$ *soit non ramifié, et soit* V' *une sous-variété normale de* V. *Alors les composantes* U_i' *de* $f^{-1}(V')$ *sont des variétés normales et disjointes. Chacune d'elles est un revêtement non ramifié de* V', *et l'on a l'égalité :*

$$\Sigma[U_i' : V'] = [U : V].$$

Posons $n = [U : V]$ et $n_i = [U_i' : V']$; d'après ce qui a été dit plus haut, on a en tout cas $\Sigma n_i \leqq n$. D'autre part, soit $v \in V'$, et soit m_i le norbre des points de U_i' se projetant en v ; du fait que V' est normale, on a $m_i \leqq n_i$. Mais d'autre part, puisque $f : U \to V$ est non ramifié, il a exactement n points de U qui se projettent en v. On a donc $\Sigma m_i \geqq n$, d'où $m_i = n_i$ et $\Sigma n_i = n$. D'après le résultat de Krull déjà cité, l'égalité $m_i = n_i$ entraîne que U_i' est un revêtement non ramifié de V', et en particulier que U_i' est une variété normale ; enfin, si U_i' et U_j' avait en commun un point u, de projection v, l'ensemble des points de U se projetant en v aurait au plus $\Sigma m_i - 1 = n - 1$ éléments, ce qui est impossible ; les variétés U_i' sont donc bien disjointes, ce qui achève la démonstration.

(Notons que, si $f : V \to V$ est galoisien, il en est de même de chacun des revêtements $U_i' \to V'$; leurs groupes de Galois sont des sous-groupes de celui de U—les classiques "groupes de décomposition").

Plus généralement, soit $\alpha : V' \to V$ une application partout définie d'une variété V' dans V ; supposons encore que V' soit normale. Alors $V' \times U$ forme de façon naturelle un revêtement de $V' \times V$; si l'on plonge V' dans $V' \times V$ au moyen du graphe de α, on peut encore définir $f^{-1}(V') \subset V' \times U$, et l'on se retrouve dans la situation envisagée ci-dessus ; dans la suite, nous noterons $\alpha^*(U)$ l'ensemble algébrique $f^{-1}(V')$; par définition, c'est l'ensemble des couples $(v', u) \in V' \times U$ tels que $\alpha(v') = f(u)$. Lorsque U est non ramifié, le lemme 1 montre que les composantes irréductibles de $\alpha^*(U)$ sont des revêtements non ramifiés de V' (les "pull-back" de [4], §4).

2. Revêtements d'un produit de variétés. Soient V et W deux variétés normales. Dans ce no., nous aurons à considérer des revêtements du produit $V \times W$. Si $f : U \to V \times W$ est un tel revêtement, et si w est un

9

point de W, nous noterons U_w l'image réciproque de $V \times \{w\}$ dans U; si l'on note α_w l'injection de V dans $V \times W$ définie par la formule $\alpha_w(v) = (v, w)$, l'ensemble algébrique U_w n'est autre que $\alpha_w{}^*(U)$. Lorsque U est non ramifié, les composantes irréductibles de U_w sont des revêtements non ramifiés de V, d'après le lemme 1.

LEMME 2. *Soit k un corps de définition d'un revêtement $f : U \to V \times W$, et soit w un point générique de W sur k. Supposons que U soit complètement décomposé sur $V \times \{w\}$. Alors U est isomorphe à un revêtement $V \times W'$, où W' est un revêtement de W.*

Soit v un point générique de V, indépendant de w, et soit $u \in U$ tel que $f(u) = (v, w)$; le point u est alors un point générique de U sur k. On voit tout de suite que le lieu de u sur le corps $k(w)$ n'est autre que U_w, qui est donc irréductible sur $k(w)$. Soit L la fermeture algébrique de $k(w)$ dans $k(u)$; par hypothèse, $k(u)/k(v, w)$ est séparable, et il en est évidemment de même de $k(v, w)/k(w)$; donc $k(u)/k(w)$ est séparable, et l'extension $k(u)/L$ est régulière. Il s'ensuit que le lieu de u sur L est l'une des composantes irréductibles de U_w; puisque on a supposé U complètement décomposé sur $V \times \{w\}$, on a donc $k(u) = L(v)$. Si l'on désigne alors par W' la normalisée de W dans l'extension $L/k(w)$, la variété $V \times W'$ est la normalisée de $V \times W$ dans $k(u)/k(v, w)$, donc est isomorphe à U, cqfd.

Remarque. Le lemme 2 ne s'étend pas aux "revêtements inséparables" (c'est-à-dire ceux où l'on supprime l'hypothèse δ) du no. 1). Pour le voir, il suffit de poser $k(w) = k(x, y)$ et $k(u) = k(v, x, y, \theta)$, avec $\theta^p = x + v^p y$, k désignant un corps algébriquement clos de caractéristique p.

LEMME 3. *Soit $f : U \to V \times W$ un revêtement non ramifié, V étant un variété complète. Supposons qu'il existe $a \in W$ tel que U se décompose complètement sur $V \times \{a\}$. Alors U est isomorphe à un revêtement $V \times W'$, où W' est un revêtement non ramifié de W.*

Supposons d'abord que V soit une variété *projective*. Soit k un corps de définition du revêtement, et soit w un point générique de W par rapport à k. D'après le théorème de dégénérescence de Zariski ([8]), le nombre de composantes connexes de U_w est au moins égal à celui de sa spécialisation U_a; mais, d'après le lemme 1, ce dernier nombre est égal à celui des composantes irréductibles de U_a, lequel est égal à $[U : V]$ par hypothèse. Il s'ensuit que le nombre de composantes connexes de U_w est aussi égal à $[U : V]$, ce qui signifie que U se décompose complètement sur $V \times \{w\}$;

d'après le lemme 2, il s'ensuit que $U = V \times W'$, et on voit immédiatement que W' est non ramifié.

L'hypothèse que V est projective n'a été appliquée que pour pouvoir utiliser le théorème de dégénérescence de Zariski. Le cas général se ramène au cas projectif: d'après un résultat de Chow (cf. [1]), il existe une variété projective normale V_1 et une application birationnelle partout définie

$$\phi: V_1 \to V.$$

L'application ϕ définit une application $\psi: V_1 \times W \to V \times W$; posons $U_1 = \psi^*(U)$; en appliquant la première partie de la démonstration à U_1, on voit que $U_1 = V_1 \times W'$. Le corps $k(U_1)$ des fonction rationnelles sur U_1 est donc un composé $k(V_1) \cdot k(W') = k(V) \cdot k(W')$; il s'ensuit que $k(U) = k(V) \cdot k(W')$, d'où $U = V \times W'$, cqfd.

Remarque. Le lemme 3 est inexact lorsque l'on ne suppose plus que V est complète: on le voit en prenant pour V et W des droites affines de points génériques v et w, et en définissant U par l'équation $w^p - u = vw$ (si l'on fait $v = 0$ ou $w = 0$, l'équation se décompose, ce qui serait incompatible avec le lemme 3).

THÉORÈME 1. *Soit $f: U \to V \times W$ un revêtement non ramifié, V étant une variété complète. Il existe alors deux revêtements non ramifiés V' de V et W' de W tels que U soit un quotient du revêtement produit $V' \times W'$.*

Nous pouvons évidemment supposer que le revêtement U est galoisien; soit G son groupe de Galois. Choisissons alors un point $a \in W$, et soit $U_a = f^{-1}(V \times \{a\})$; d'après le lemme 1, les composantes irréductibles de U_a sont des revêtements non ramifiés de V; comme on a supposé U galoisien, il en sera de même pour ces composantes (leurs groupes de Galois étant des sous-groupes de G, les "groupes de décomposition"). Choisissons l'une de ces composantes, soit V', et soit $G' \subset G$ son groupe de Galois. La projection $V' \to V$ définit une projection

$$h: V' \times W \to V \times W,$$

qui fait de $V' \times W$ un revêtement de $V \times W$ de groupe de Galois G'. Posons $U = h^*(U)$, sous-ensemble algébrique de $V' \times W \times U$; comme précédemment pour U_a, les composantes irréductibles de U_1 sont des revêtements non ramifiés de $V' \times W$, galoisiens et de groupes de Galois des sous-groupes de G. Soit $b \in V'$ et soit U' la composante de U_1 qui contient le point $(b, a, b) \in U_1$; comme U' est un revêtement non ramifié de $V' \times W$, c'est aussi un revêtement non ramifié de $V \times W$; de plus, puisque la projection de U' sur $V \times W$ se factorise en $U' \to U \to V \times W$, le revêtement U est un quotient de U', et il nous suffira donc de prouver que U' est de la forme $V' \times W'$. Mais U' se décompose complètement sur $V' \times \{a\}$: en effet, U_1 contient la sous-variété de $V' \times W \times U$ formée des points (v', a, v') où v' parcourt V'; comme cette sous-variété a le point (b, a, b) en commun avec U', elle est contenue dans U', et constitue une composante de U_a' se projetant sur $V' \times \{a\}$ avec degré 1; du fait que U' est galoisien, ceci suffit à assurer que U' se décompose complètement sur $V' \times \{a\}$. En appliquant alors le lemme 3, on en déduit bien que U' est isomorphe à un revêtement de la forme $V' \times W'$, ce qui achève la démonstration.

COROLLAIRE 1. *Soit G un groupe abélien fini. Les notations étant celles de [7], no. 14, on a $\pi^1(V \times W, G) = \pi^1(V, G) \times \pi^1(W, G)$.*

C'est immédiat.

COROLLAIRE 2. *Les hypothèses étant celles du théorème 1, soient a et b deux points de W. Il existe alors un isomorphisme de l'ensemble algébrique U_a sur l'ensemble algébrique U_b compatible avec les projections $U_a \to V$ et $U_b \to V$.*

(On a un énoncé analogue en permutant les rôles de V et de W).

En effet, d'après le théorème 1, le revêtement U est quotient d'un revêtement $g : V' \times W' \to V \times W$; on peut supposer V' et W' galoisiens de groupes de Galois G_1 et G_2 respectivement; alors U est isomorphe à $(V' \times W')/H$, où H est un sous-groupe de $G_1 \times G_2$. Posons:

$$U_a' = g^{-1}(V \times \{a\}) \quad \text{et} \quad U_b' = g^{-1}(V \times \{b\}).$$

Il est clair que U_a' et U_b' sont tous deux isomorphes à $V' \times G_2$, ces isomorphismes commutant avec les opérations de $G_1 \times G_2$; comme U_a est isomorphe à U_a'/H, il est isomorphe à $(V' \times G_2)/H$, et de même pour U_b, ce que démontre le corollaire.

Lorsque U est galoisien, de groupe de Galois G, on a $G = (G_1 \times G_2)/H$; de plus, on peut supposer (quitte à changer V' et W') que $G_1 \cap H = \{e\}$ et $G_2 \cap H = \{e\}$. Les groupes G_1 et G_2 s'identifient ainsi à des sous-groupes de G, invariants, commutant entre eux, et engendrant G. De plus, le raisonnement fait ci-dessus montre que, pour tout $a \in W$, l'ensemble algébrique U_a se décompose en composantes irréductibles qui sont isomorphes à V', le groupe G_1 s'identifiant au groupe de décomposition correspondant; on a un résultat analogue pour U_v si $v \in V$. En définitive, on a obtenu:

COROLLAIRE 3. *Outre les hypothèses du théorème 1, supposons que U soit galoisien de groupe de Galois G. Soit $(v, w) \in V \times W$, et soit V' (resp. W') une composante de U_w (resp. U_v); soit G_1 (resp. G_2) son groupe de Galois. Les groupes G_1 et G_2 sont alors des sous-groupes invariants de G, qui commutent entre eux et engendrent G; si H désigne le noyau de l'homomorphisme $G_1 \times G_2 \to G$, le revêtement U est isomorphe à $(V' \times W')/H$.*

3. Revêtements non ramifiés des variétés abéliennes. Nous nous proposons de démontrer le résultat suivant:

THÉORÈME 2. *Toute revêtement non ramifié d'une variété abélienne est une isogénie (séparable).*

(Cela montre en particulier qu'un tel revêtement est *abélien*).

Soit donc V une variété abélienne, et soit $f\colon U \to V$ un revêtement non ramifié de V. On peut évidemment supposer que U est galoisien; soit G son groupe de Galois. Il nous faut montrer que l'on peut définir sur U une structure de variété abélienne (auquel cas f sera automatiquement une isogénie).

Soit $g\colon V \times V \to V$ l'application qui définit la loi de composition de V, i.e. $g(v, v') = v + v'$. Soit $U' = g^*(U)$, sous-ensemble algébrique de $V \times V \times U$ formé des triples (v, v', u) tels que $f(u) = v + v'$. Le groupe G permute les composantes irréductibles de U'; mais si l'on pose

$$U_1' = U' \cap (\{0\} \times V \times U),$$

l'ensemble U_1' s'identifie à U, donc est irréductible; *a fortiori*, il en est de même de U'. Nous noterons $f'\colon U' \to V \times V$ la projection de U' sur les deux premiers facteurs du produit $V \times V \times U$. On a évidemment $U_1' = f'^{-1}(\{0\} \times V)$, et l'on vient de voir que U_1' s'indentifie à U; de même $U_2' = f'^{-1}(V \times \{0\})$ s'identifie à U. Appliquant alors le corollaire 3 au théorème 1, on voit que. avec les notations de ce corollaire, on a $G_1 = G_2 = G$, ce qui montre d'abord que G est *abélien*, puis que U' s'identifie à $(U \times U)/H$, où H désigne le sous-groupe de $G \times G$ formé par les éléments (g, g^{-1}), $g \in G$. Comme on a une application canonique $h\colon U' \to U$, on en déduit par composition une application $g'\colon U \times U \to U' \to U$ qui "relève" l'application $g\colon V \times V \to V$. Si l'on a choisi dans U un point e tel que $f(e) = 0$, on peut en outre s'arranger pour que $g'(e, e) = e$. Il ne nous reste plus qu'à démontrer que l'application g' vérifie les axiomes d'une loi de groupe, ce que ne présente pas de difficultés:

a) Le point $e \in U$ est élément neutre pour la loi de composition g'. En effet, soit $\sigma : U \to U$ l'application définie par la formule :

$$\sigma(u) = g'(e, u).$$

On a $f \circ \sigma = f$, ce qui montre que σ est un automorphisme du revêtement U, i. e. un élément du groupe de Galois G. Mais on a $\sigma(e) = e$, et comme G opère sans points fixes, cela montre que σ est l'identité; un raisonnement analogue montre que $g'(u, e) = u$ pour tout $u \in U$.

b) La loi de composition g' est associative.

Soient k_1 et k_2 les applications de $U \times U \times U$ dans U définies par les formules :

$$k_1(v, v', v'') = g'(v, g'(v', v'')) \quad \text{et} \quad k_2(v, v', v'') = g'(g'(v, v'), v'').$$

Du fait que g' relève g, les projections de $k_1(v, v', v'')$ et de $k_2(v, v', v'')$ dans U sont les mêmes. On a donc $k_2(v, v', v'') = \sigma \cdot k_1(v, v', v'')$, avec $\sigma \in G$. Appliquons ceci à trois points génériques indépendants v, v', v''; en spécialisant v, v' et v'' en e, on voit que $\sigma = 1$, d'où $k_1 = k_2$ et g' est bien une loi de composition associative.

c) Il existe un inverse $u \to u^{-1}$ qui est un automorphisme de U.

Désignons par f'' l'application $f \times f : U \times U \to V \times V$; soit d'autre part $\phi : V \times V \to V \times V$ l'application définie par $(v, v') \to (v, v + v')$, et soit de même $\psi : U \times U \to U \times U$ l'application définie par

$$\psi(u, u') = (u, g'(u, u')).$$

On a évidemment la relation de commutation $f'' \circ \psi = \phi \circ f''$. En prenant les degrés des deux membres, et en tenant compte du fait que ϕ est bi-rationnelle, on voit qu'il en est de même de ψ. De plus, ψ est partout définie puisque ϕ l'est. L'existence d'un inverse, et le fait que $u \to u^{-1}$ est un automorphisme résultent aussitôt de là, ce qui achève la démonstration.

Remarques.

1) La démonstration précédente est calquée sur la démonstration classique prouvant que tout revêtement d'un groupe topologique est un groupe topologique (démonstration qui repose elle-même sur la formule $\pi_1(X \times Y) = \pi_1(X) \times \pi_1(Y)$).

2) Le théorème 2 *ne s'étend pas* aux groupes algébriques non complets. On trouvera des contre-exemples dans [3], relatifs au cas où V est un groupe

non commutatif, défini sur un corps fini. En fait, ou peut même construire des contre-exemples où V est le groupe additif G_a: si U' est une courbe elliptique d'invariant de Hasse nul, une différentielle de première espèce de U' peut s'écrire $\omega = df$, où f est une fonction rationnelle; soit U l'ouvert de U' obtenu en retirant l'ensemble des pôles de f; il est clair que $f: U \to G_a$ est un revêtement non ramifié qui n'est pas une isogénie.

4. Finitude du nombre des classes de revêtements non ramifiés d'une variété complète. Nous démontrerons d'abord le résultat suivant:

THÉORÈME 3. *Soit V une variété normale, complète, définie sur un corps algébriquement clos k. Alors tout revêtement non ramifié de V est isomorphe à un revêtement défini sur k.*

(Autrement dit, il n'y a pas de "systèmes algébriques" de revêtements non ramifiés).

Soit $f_1: U_1 \to V$ un revêtement non ramifié de V, défini sur une extension L/k que l'on peut supposer de type fini. Soient v un point générique de V sur L et u un point de U_1 tel que $f_1(u) = v$. Soit d'autre part W' une variété normale, définie sur k, telle que, si w est un point générique de W' sur k, on ait $L = k(w)$. On a $L(v) = k(v, w)$, corps des fonctions rationnelles sur $V \times W'$, et le corps $L(u)$ est une extension finie de $k(v, w)$. Soit U' la normalisée de $V \times W'$ dans l'extension $L(u)/k(v, w)$, et soit f la projection canonique de U' sur $V \times W'$. It est clair que $U_w' = f^{-1}(V \times \{w\})$ est une variété $k(w)$-normale, et en correspondance birationnelle, sur $k(w)$, avec le revêtement donné U_1; comme U_1 est normale, ceci montre que U_w' et U_1 sont isomorphes. Soit $Z \subset V \times W'$ l'ensemble des points au-dessus desquels le revêtement U' est ramifié; c'est un ensemble algébrique k-fermé, qui ne rencontre pas $V \times \{w\}$ puisque par hypothèse $U_1 = U_w'$ est non ramifié. Il est donc contenu dans un produit $V \times Z'$, où Z' est un sous-ensemble algébrique k-fermé de W'; posons $W = W' - Z'$, et soit U l'image réciproque de $V \times W$ dans U'. Le revêtement U est non ramifié, et l'on a $U_w = U_w' = U_1$; choisissons alors dans W un point a, rationnel sur k. D'après le corollaire 2 au théorème 1 les ensembles algébriques U_w et U_a sont isomorphes. Il s'ensuit que U_a est irréductible, et que c'est un revêtement de V isomorphe à U_1, cfqd.

THÉORÈME 4. *Une variété normale et complète ne possède qu'un nombre fini de revêtements non ramifiés de degré donné.*

Soit V une telle variété. D'après Chow [1], V est image d'une variété projective V_1, que l'on peut supposer normale, et qui est birationnellement équivalente à V. Tout revêtement non ramifié de V en définit un de V_1, et l'on est ainsi ramené à démontrer le théorème pour V_1. Autrement dit, nous pouvons supposer que V est une variété *projective*. Soit k un corps de définition de V, et soit C_u une *courbe générique* de V (sur k), au sens défini dans [5], §3; on sait (*loc. cit.*) que si U est un revêtement de V, défini sur k, la restriction de U à C_u est un revêtement irréductible de C_u. Il s'ensuit, par un raisonnement standard, que deux revêtements non isomorphes de V, tous deux définis sur k, ont des restrictions à C_u qui sont également non isomorphes. En prenant pour k un "domaine universel," on voit donc qu'il suffira de démontrer le théorème 4 pour C_u.

Nous supposerons donc, à partir de maintenant, que V est une *courbe* C, non singulière, complète, de genre g. Nous la supposerons plongée dans un espace projectif S_1, et nous désignerons son degré projectif par d. Si $f: U \to C$ est un revêtement non ramifié de C de degré n, le genre g_U de la courbe U est donné par la formule classique:

$$g_U - 1 = n(g-1),$$

et ne dépend donc pas de U. D'après le théorème de Riemann-Roch, on peut plonger birégulièrement U dans un espace projectif S_2 de dimension g_U (donc fixe), le degré projectif de ce plongement étant $d_2 = 2g_U + 1$. Soit $\Gamma \subset S_1 \times S_2$ le graphe de f. Si H_1 et H_2 sont des hyperplans de S_1 et S_2, posons:

$$H = S_1 \times H_2 + H_1 \times S_2;$$

un calcul immédiat montre que $\deg(\Gamma \cdot H) = d_2 + nd$, donc que ce degré ne dépend pas de U. Cela signifie que, si l'on plonge $S_1 \times S_2$ dans un espace projectif S_3 à la manière habituelle (i. e. de telle sorte que H devienne une section hyperplane dans ce nouveau plongement), le degré de Γ dans S_3 sera $d_2 + nd$, donc sera fixe.

Or, d'après la théorie des coordonnées de Chow, il n'existe qu'un nombre *fini* de familles algébriques irréductibles de cycles de $C \times S_2$ ayant un degré donné (dans S_3). Il nous suffit donc de prouver que, si deux revêtements U_1 et U_2 ont des graphes Γ_1 et Γ_2 qui appartiennent à la même famille algébrique F, ces deux revêtements sont isomorphes.

Pour cela, soit Γ un élément générique de la famille F; nous allons d'abord montrer que $\Gamma \to C$ est un revêtement non ramifié de degré n. Tout d'abord, puisque Γ admet pour spécialisation Γ_1, c'est une variété; comme la projection sur C est compatible avec les spécialisations, la relation

$\mathrm{pr}_1(\Gamma_1) = nC$ entraîne $\mathrm{pr}_1(\Gamma) = nC$, donc l'application $\Gamma \to C$ est de degré n. De plus, si Q est un point de C, on peut prolonger la spécialisation $\Gamma \to \Gamma_1$ en une spécialisation $Q \to Q_1$, et $Q \times S_2$ se spécialise en $Q_1 \times S_2$. Comme le cycle $\Gamma_1 \cdot (Q_1 \times S_2)$ est défini, et se compose de n points distincts, il en est de même du cycle $\Gamma \cdot Q \times S_2)$. D'après ce qu'on a vu au no. 1, il s'ensuit bien que Γ est un revêtement non ramifié de degré n de la courbe C.

Nous allons maintenant montrer que les revêtements Γ_1 et Γ sont isomorphes; comme le même argument montrera que Γ_2 et Γ sont isomorphes, il en résultera bien que Γ_1 et Γ_2 sont isomorphes, ce qui achèvera la démonstration, en vertu de ce qui a été dit plus haut.

D'après le théorème 3, le revêtement Γ est isomorphe à un revêtement Γ' défini sur le corps de base; on peut supposer Γ' plongé dans $C \times S_2$ comme ci-dessus. Désignons par T le graphe de l'isomorphisme $\Gamma \to \Gamma'$; c'est une sous-variété de $\Gamma \times \Gamma'$, lui-même contenu dans $C \times S_2 \times C \times S_2$. On a évidemment $\mathrm{pr}_{13}(T) = n\Delta$, où Δ est la diagonale de $C \times C$. Etendons maintenant la spécialisation $\Gamma \to \Gamma_1$ en une spécialisation $T \to T_1$; la variété Γ' reste fixe durant cette spécialisation, puisqu'elle est définie sur le corps de base. On a $\mathrm{pr}_{12}(T_1) = \Gamma_1$, $\mathrm{pr}_{34}(T_1) = \Gamma'$ et $\mathrm{pr}_{13}(T_1) = n\Delta$; les deux premières formules montrent que, ou bien T_1 est le graphe d'une correspondance birationnelle entre Γ_1 et Γ', ou bien T_1 est "dégénérée," i.e. égale à $\Gamma_1 \times \{\gamma'\} + \{\gamma_1\} \times \Gamma'$; la troisième formule exclut cette seconde possibilité. On a donc obtenu une correspondance birationnelle entre les revêtements Γ_1 et Γ', commutant avec les projections sur C. Il s'ensuit que Γ_1 et Γ' sont isomorphes, donc aussi Γ_1 et Γ, cqfd.

Remarques.

1) La démonstration du théorème 4 que nous venons de donner utilise de façon essentielle les résultats du no. 2. Inversement, si l'on pouvait démontrer directement le théorème 4 (dans le cas d'une courbe), on retrouverait facilement ces résultats comme corollaires.

2) Soit C une courbe, et soient $P_1, \cdots, P_r$ des points de C en nombre fini. Il est vraisemblable qu'il n'existe qu'un nombre fini de revêtements de C d'un degré n donné, ramifiés seulement aux points P_i, et n'y admettant pas de ramification supérieure (revêtements "tamely ramified"). Une démonstration de ce fait permettrait d'étendre notablement nos résultats.

3) Dans le cas des revêtements abéliens, le théorème 4 peut aussi se démontrer en utilisant le théorème de Néron-Severi (pour les revêtements d'ordre premier à la caractéristique), et la finitude du premier groupe de

cohomologie (pour les revêtements d'ordre une puissance de la caractéristique).
Cf. [7], nos. 15 et 16.

Columbia University
Institute for Advanced Study.

BIBLIOGRAPHIE.

[1] W-L. Chow, " On, the projective embedding of homogeneous varieties," *Symposium in honor of S. Lefschetz*, Princeton, 1957.

[2] W. Krull, " Der allgemeine Diskriminantensatz. Unverzweigte Ringerweiterungen," *Mathematische Zeitschrift*, vol. 45 (1939), pp. 1-19.

[3] S. Lang, "Algebraic groups over finite fields," *American Journal of Mathematics*, vol. 78 (1956), pp. 555-563.

[4] ———, " Unramified class field theory over function fields in several variables," *Annals of Mathematics*, vol. 64 (1956), pp. 285-325.

[5] ———, " Sur les séries L d'une variété algébrique," *Bulletin de la Société Mathématique de France*, vol. 84 (1956), pp. 385-407.

[6] M. Nagata, "A general theory of algebraic geometry over Dedekind domains. I. The notion of models," *American Journal of Mathematics*, vol. 78 (1956), pp. 78-116.

[7] J-P. Serre, " Sur la topologie des variétés algébriques en caractéristique p," *Symposium de topologie algébrique*, Mexico, 1956.

[8] O. Zariski, " Theory and applications of holomorphic functions on algebraic varieties over arbitrary ground fields," *Memoirs of the American Mathematical Society*, No. 5, 1951.

Am. J. Math., LXXIX No. 3, 1957

ON THE BIRATIONAL EQUIVALENCE OF CURVES UNDER SPECIALIZATION.*

By WEI-LIANG CHOW [1] and SERGE LANG.

Let k be a field with a discrete valuation, let p be the maximal prime ideal in the valuation ring on k, and let $\bar{k}$ be the residue field of k. Let S^n be the projective space of dimension n in the algebraic geometry over a universal domain containing k, and denote by $\bar{S}^n$ the projective space of dimension n over a universal domain containing $\bar{k}$; if Z is any cycle in S^n, rational over k, we shall denote by $\bar{Z}$ the cycle obtained from Z by reduction modulo p, in the sense of Shimura [4], and we shall call $\bar{Z}$ the specialization of Z (with respect to the given valuation of k). This definition applies also in case Z is a variety, which is then to be considered as a prime cycle; the the specialization $\bar{Z}$ is then a positive cycle, but in general not necessarily a variety. According to the generalization of the Principle of Degeneration recently proved by Chow, the specialization of a variety is always connected, but we shall not need this fact here.

Consider now two non-singular curves C_1 and C_2 in S^n, both defined over k, and assume that they are birationally equivalent over k. A theorem of Deuring ([2], Satz 3) can be expressed geometrically by saying that if C_1 and C_2 have genus 1 and if the cycles $\bar{C}_1$ and $\bar{C}_2$ are also non-singular curves of genus 1, then $\bar{C}_1$ and $\bar{C}_2$ are birationally equivalent $\bar{k}$. In this note we shall generalize this result to non-singular curves of arbitrary genus and also to abelian varieties; in fact, we shall first prove the result for abelian varieties, and then deduce the result for curves by embedding them in their Jacobians. Our method is quite different from Deuring's, and is based on the "compatibility" of the Chow construction of the Jacobian and the canonical mapping.

Let A be an abelian variety in S^n, defined over k. We shall say that the specialization $\bar{A}$ of A is *non-degenerate* if $\bar{A}$ is an abelian variety defined over $\bar{k}$ and if the law of composition in $\bar{A}$ is the specialization of the law of

---

* Received January 24, 1957.

[1] Work of W. L. Chow was supported in part by a National Science Foundation Grant, NSF-3022.

649

composition in A. Specifically, this means the following. Let G be the graph of the law of composition in A, so that G is a subvariety in $A \times A \times A$; then our condition can be expressed by saying that both $\bar{A}$ and $\bar{G}$ are varieties (in $\bar{S}^n$ and $\bar{S}^n \times \bar{S}^n \times \bar{S}^n$ respectively) and that $\bar{G}$ defines an abelian law of composition in $\bar{A}$. We observe that if e is the unit element in A, then $\bar{e}$ is the unit element in $\bar{A}$; in fact, e is characterized by the condition that $\mathrm{pr}_{32}(e \times A \times A))$ is contained in the diagonal in $A \times A$, and this condition is evidently preserved under the specialization.

THEOREM 1. *Let A and B be abelian varieties in S^n, and let T be a birational isomorphism between A and B; we assume that A, B, and T are all defined over k. If the specializations $\bar{A}$ and $\bar{B}$ are non-degenerate, then the specialization $\bar{T}$ is a variety and defines a birational isomorphism between $\bar{A}$ and $\bar{B}$.*

Proof. It is clear that $\bar{A} \times \bar{B}$ is the specialization of $A \times B$ and is non-degenerate. If we denote the graph of T by the same symbol, then the support $|\bar{T}|$ of $\bar{T}$ is an algebraic subgroup in $\bar{A} \times \bar{B}$; this follows from the fact that T is an abelian subvariety in $A \times B$, and the fact that the law of composition in $\bar{A} \times \bar{B}$ is the specialization of the law of composition in $A \times B$. Let $\bar{T}_0$ be a component in $\bar{T}$ which contains the unit element; then $\bar{T}_0$ is an abelian subvariety in $\bar{A} \times \bar{B}$, and every component in $|\bar{T}|$ and hence also every component in $\bar{T}$ is obtained from $\bar{T}_0$ by a translation. By assumption, we have $\mathrm{pr}_1 T = A$ and $\mathrm{pr}_2 T = B$; hence, by specialization, we have $\mathrm{pr}_1 \bar{T} = \bar{A}$ and $\mathrm{pr}_2 \bar{T} = \bar{B}$. Since all cycles obtained from $\bar{T}_0$ by translations must have the same projections on $\bar{A}$ and $\bar{B}$, we conclude readily that $\mathrm{pr}_1 \bar{T}_0 = \bar{A}$ and $\mathrm{pr}_2 \bar{T} = \bar{B}$, and that $\bar{T}_0$ is the only component in $\bar{T}$ and has the coefficient 1. This shows that $\bar{T} = \bar{T}_0$ defines a birational transformation between $\bar{A}$ and $\bar{B}$, which, since it preserves the unit elements, must be an isomorphism.

We turn now to the case of the curves C_1 and C_2 considered at the beginning of this note. Let C be a non-singular curve of genus $g(C)$ in S^n, defined over k; we shall say that the specialization $\bar{C}$ is *non-degenerate* if $\bar{C}$ is a non-singular curve and hence has the same genus $g(C)$ as C. We assume that there exists a birational correspondence F between C_1 and C_2, defined over k, and that the specialization $\bar{C}_1$ and $\bar{C}_2$ are both non-degenerate. Since the conditions $\mathrm{pr}_1 F = C_1$ and $\mathrm{pr}_2 F = C_2$ imply by specialization the conditions $\mathrm{pr}_1 \bar{F} = \bar{C}_1$ and $\mathrm{pr}_2 \bar{F} = \bar{C}_2$, we conclude that $\bar{F}$ is either a birational correspondence between $\bar{C}_1$ and $\bar{C}_2$, or has the form $x \times \bar{C}_2 + \bar{C}_1 \times y$, for some points x and y in $\bar{C}_1$ and $\bar{C}_2$ respectively. In case $g(C_1) = g(C_2) = 0$, it can be easily seen by examples that both possibilities can occur; however, in this case the assertion of Deuring is trivially true. In the next theorem we shall show that, except in this trivial case, the second possibility cannot occur; this implies that the specialization $\bar{F}$ is itself already such a birational correspondence.

THEOREM 2. *Let C_1 and C_2 be non-singular curves of genus > 0 in S^n, and let F be a birational correspondence between C_1 and C_2; we assume that C_1, C_2, and F are all defined over k. If the specializations $\bar{C}_1$ and $\bar{C}_2$ are non-degenerate, then the specialization $\bar{F}$ is a birational correspondance between $\bar{C}_1$ and $\bar{C}_2$.*

Proof. The theorem being "geometric," we may assume that the curves C_1 and C_2 have rational points over k, and hence the curves $\bar{C}_1$ and $\bar{C}_2$ have rational points over $\bar{k}$. In fact, we can enlarge k by a finite algebraic extension if necessary, and extend the discrete valuation in k to a discrete valuation of this extension; it is easily seen that the validity of our theorem over this extension will imply its validity in the original form.

For $i = 1, 2$, we consider the Jacobian J_i of C_i and the canonical mapping f_i of C_i into J_i with reference to a rational point p_i in C_i (i.e., $f_i(p_i) = e_i$, e_i being the unit element in J_i), both J_i and f_i being constructed as in Chow [1]. It is essential for our purpose that J_i and f_i are constructed in this particular way, which we shall call the Chow construction, not just some arbitrarily chosen projective models of J_i and the corresponding mapping f_i. The reason lies in the fact that the Chow construction is "compatible" with specializations, at least when they are non-degenerate. Specifically, this means that if the specialization $\bar{C}_i$ is non-degenerate, then the specialization $\bar{J}_i$ (which is a cycle in $\bar{S}^{t_i}$ if S^{t_i} is the ambient projective space of J_i) of the abelian variety J_i is also non-degenerate, and $\bar{J}_i$ and $\bar{f}_i$ are respectively the Jacobian and the canonical mapping of $\bar{C}_i$. This fact is a special case of a more general "compatibility" theorem of Igusa [3] (Igusa treats only the algebro-geometric specializations, but it is easily seen that his proof is also valid for the more general specializations considered here at least in the non-degenerate case). For the sake of convenience, we shall assume that $p_2 = F(p_1)$; then there exists a birational isomorphism T between J_1 and J_2, defined over k, such that $T \circ f_1 = f_2 \circ F$. If we denote by f the mapping $f_1 \times f_2$ of $C_1 \times C_2$ into $J_1 \times J_2$, and if we denote the graphs of F and T by the same symbols, then the above relation implies that $f(F)$ is contained in T. Since the specialization $\bar{f} = \bar{f}_1 \times \bar{f}_2$ is a rational mapping of $\bar{C}_1 \times \bar{C}_2$ into $\bar{J}_1 \times \bar{J}_2$, it follows that the cycle $\bar{f}(\bar{F})$ is defined and is contained in $|\bar{T}|$.

By Theorem 1, $\bar{T}$ is an abelian subvariety in $\bar{J}_1 \times \bar{J}_2$ (it is in fact sufficient to know that $|\bar{T}|$ is an algebraic group in $\bar{J}_1 \times \bar{J}_2$) and is a proper subset in $\bar{J}_1 \times \bar{J}_2$. If $\bar{F}$ is not a birational correspondence between $\bar{C}_1$ and $\bar{C}_2$, then by a previous remark we must have $\bar{F} = \bar{p}_1 \times \bar{C}_2 + \bar{C}_1 \times \bar{p}_2$ (we observe that $|\bar{F}|$ must contain the point $\bar{p}_1 \times \bar{p}_2$) and hence

$$\bar{f}(\bar{F}) = \bar{e}_1 \times \bar{f}_2(\bar{C}_2) + \bar{f}_1(\bar{C}_1) \times \bar{e}_2.$$

This shows that $|\bar{f}(\bar{F})|$ generates the entire variety $\bar{J}_1 \times \bar{J}_2$, in contradiction to the fact that $\bar{f}(\bar{F})$ is contained in $\bar{T}$. This proves our theorem.

THE JOHNS HOPKINS UNIVERSITY,
COLUMBIA UNIVERSITY.

<hr>

REFERENCES.

[1] W. L. Chow, "The Jacobian variety of an algebraic curve," *American Journal of Mathematics*, vol. 76 (1954), pp. 453-476.

[2] M. Deuring, "Die Zetafunktion einer algebraischen Kurve vom Geschlechte Eins (II)," *Nachrichten der Akademie der Wissenschaften in Göttingen*, 1955, pp. 13-42.

[3] J. Igusa, "Fibre systems of Jacobian varieties," *American Journal of Mathematics*, vol. 78 (1956), pp. 171-199.

[4] G. Shimura, "Reduction of algebraic varieties with respect to a discrete valuation of the basic field," *ibid.*, vol. 77 (1955), pp. 131-176.

DIVISORS AND ENDOMORPHISMS ON AN ABELIAN VARIETY.*

By Serge Lang.

We consider here the fundamental theorem on the algebra of endomorphisms of an abelian variety dealing with the existence of a positive definite quadratic form defined by means of the trace. We shall give a direct proof on abelian varieties for Weil's formula $\sigma(\xi\xi') > 0$, which will exhibit the positivity as a formal consequence of the definitions. Weil's method is to show that the trace can be expressed as an intersection number (Proposition 20 of [6], No. 47). However, even on a Jacobian, his proof that $\sigma(\xi)$ coincides with the penultimate coefficient of the characteristic polynomial is somewhat indirect (Theorem 34 of [6], No. 66). In the present exposition, this will be straightforward, and we shall obtain a simple representation of the trace as an intersection number canonically determined by a given projective embedding of the abelian variety A, or as Weil would say, a polarization of A [8].

More generally, if α_i $(i = 1, \cdots, d)$ are endomorphisms of A and m_i $(i = 1, \cdots, d)$ are integers, then we give an expression for the coefficients of the homogeneous polynomial $\nu(m_1\alpha_1 + \cdots + m_d\alpha_d)$ in terms of intersection numbers canonically determined by the α_i and the given polarization.

Our arguments are based on only two properties of abelian varieties: The theorem of the square (see below), and the fact that a divisor which is $\equiv 0$ is numerically equivalent to 0. (Weil [6], No. 62 and Morikawa [5], Lemma 7.) For the convenience of the reader, we reproduce in an appendix a simple proof of Weil for this fact.

We shall begin by defining a bilinear mapping of the endomorphisms of A into the Neron-Severi group of divisors $N(A)$, given by the formula $D_X(\alpha, \beta) \equiv (\alpha + \beta)^{-1}(X) - \alpha^{-1}(X) - \beta^{-1}(X)$. The properties of the trace are then reflected from analogous properties of our divisorial pairing. We shall take X to be a positive non-degenerate divisor. We call a class of divisors in $N(A)$ positive if a suitable positive integral multiple of it contains a positive divisor. The formula $\operatorname{tr}(\alpha\alpha') > 0$ then comes from the fact that $D_X(\alpha, \alpha)$ is positive.

We then discuss a pairing of endomorphisms into the group $N_1(A)$ of

---

* Received February 15, 1957.

4

1-cycles modulo those which are numerically equivalent to 0. It is dual to the above, and is defined by the formula $Z_C(\alpha, \beta) = (\alpha + \beta)(C) - \alpha(C) - \beta(C)$. Properties of $D_X(\alpha, \beta)$ are easily transferred to properties of $Z_C(\alpha, \beta)$. We introduce the Pontrjagin product between cycles of A, which is dual to the intersection product, and are then able to recover all the numerical statements contained in Weil [6] (for instance, the Lefschetz fixed point formula for curves). Some questions concerning the Pontrjagin product remain un-answered, principally those concerning the more precise duality which exists between this product and the ordinary intersection product.

The above section dealing with 1-cycles is not used in the section following it, where we analyse the positivity of divisor classes in $N(A)$. We generalize results of Morikawa [5], except that we work only in the tensor product of $N(A)$ with the rational numbers, and do not consider integrality questions related to the positivity. For instance, let Y be a divisor and suppose that there exists an integer $m > 0$ and a divisor $Y_1 > 0$ such that $mY \equiv Y_1$. Does there exist a divisor $Y_2 > 0$ such that $Y \equiv Y_2$?

Finally, we have pointed out how one can see directly on an abelian variety that the characteristic roots of the Frobenius endomorphism all have absolute value $q^{\frac{1}{2}}$ without having to go through the theory of curves. Together with the Lefschetz fixed point formula, this gives a more natural proof for the Riemann hypothesis for curves.

1. Applications of the theorem of the square.

For the rest of this paper, A, B will denote abelian varieties and $X, Y, Z, \cdots$ will be divisors on A or B. We denote by $H(A, B)$ the module of homomorphisms of A into B, and by $H(A)$ the ring of endomorphisms of A. The tensor products with Q will be denoted by $H_Q(A, B)$ and $H_Q(A)$ respectively. If $\alpha \in H_Q(A, B)$ and $\nu(\alpha) \neq 0$, then there is an element β of $H_Q(B, A)$ such that $\beta\alpha$ is the identity. We write $\beta = \alpha^{-1}$.

We denote by $\hat{A}$ the dual variety (Picard variety) of A. If X is a divisor on A, algebraically equivalent to 0, then $\text{Cl}(X)$ denotes the point on $\hat{A}$ associated with the linear equivalence class of X. The theorem of the square states that the mapping

$$\varphi_X : u \to \text{Cl}(X_u - X)$$

is a homomorphism of A into $\hat{A}$. By definition, $X \equiv 0$ if and only if $\varphi_X = 0$.

We let $D(A)$ be the group of divisors on A, and $D_l(A)$ the subgroup of divisors linearly equivalent to 0. Let $\alpha \in H(A, B)$. Let X be a divisor on B. Then $\alpha^{-1}(X)$ may not be defined, but if t is a generic point of B, then $\alpha^{-1}(X_t)$ is always defined. Furthermore, if X is algebraically equivalent to 0 on B, then $X_t \sim X$, and if $X \sim 0$ on B and $\alpha^{-1}(X)$ is defined, then $\alpha^{-1}(X) \sim 0$ on A. If V is a variety, we denote by $D_a(V)$ the group of divisors algebraically equivalent to 0 on V. Then the above remarks show that α induces a linear map α^{-1} of $D_a(B)/D_l(B)$ into $D_a(A)/D_l(A)$. (Later we shall use the terminology of the Picard variety, and we call this map the transpose of α.)

We denote by $N(A)$ the factor group of $D(A)$ by the subgroup of divisors of A which are $\equiv 0$. Then α will also induce an inverse mapping α^{-1} of $N(B)$ into $N(A)$ as follows. We contend that if t and u are generic points of B then $\alpha^{-1}(X_t) \equiv \alpha^{-1}(X_u)$. To prove this, we first state a lemma explicitly.

LEMMA. *Let* $\lambda : A \to B$ *be a homomorphism,* X *a divisor on* B, v *a point of* A *such that* $\lambda^{-1}(X_{\lambda v})$ *is defined. Assume also that* $\lambda^{-1}(X)$ *is defined. Then* $(\lambda^{-1}(X))_v = \lambda^{-1}(X_{\lambda v})$.

Proof. The translation $(v, \lambda v)$ on $A \times B$ leaves the graph of λ fixed. It moves $\lambda^{-1}(X)$ on its translate by v, and moves X on its translate by λv. The lemma is then obvious.

Returning to the proof of our contention, we let v be a generic point of A, independent of t and u. Then $\alpha^{-1}(X_{t+\alpha v})$ and $\alpha^{-1}(X_{u+\alpha v})$ are defined, and a trivial computation starting with the definitions, together with the theorem of the square proves our contention.

Given a divisor X on B, and $\alpha \in H(A, B)$, then from the theorem of the square, we know that $X_t \equiv X$, and hence the class $\alpha^{-1}(X_t)$ in $N(A)$ depends only on X and α and not on the auxiliary generic point t. By an abuse of language, we shall denote this class by $\alpha^{-1}(X)$. The context will always make clear whether we deal with a class in $N(A)$, or whether we deal with a divisor X in $D_a(B)$, in which case $\alpha^{-1}(X)$ will mean the linear equivalence class in $D_a(A)/D_l(A)$.

The following theorem and its proof are directly inspired from Weil's Proposition 31, [6], No. 73. For the special case of elliptic curves, see Hasse [4].

THEOREM 1. *Let* $\alpha, \beta \in H(A, B)$. *Define*

$$D_X(\alpha, \beta) \equiv (\alpha + \beta)^{-1}(X) - \alpha^{-1}(X) - \beta^{-1}(X),$$

this being an element of $N(A)$. *Let* m, n *be integers. Then*

$$(m\alpha + n\beta)^{-1}(X) \equiv m^2 \alpha^{-1}(X) + mn D_X(\alpha, \beta) + n^2 \beta^{-1}(X).$$

Proof. The proof will be carried out in three steps. Throughout, we let $s_n : B \times \cdots \times B \to B$ be the sum on the product of B n-times with itself, and we let $p_i : B \times \cdots \times B \to B$ be the projection on the i-th factor.

1.) $X \equiv 0$ if and only if $s_n^{-1}(X) \sim \sum p_i^{-1}(X)$.

For simplicity take $n = 2$. Then for u generic on B, we have

$$s_2^{-1}(X) \cdot (B \times u) = X_{-u} \times u.$$

Hence $[s_2^{-1}(X) - (X \times B)] \cdot (B \times u) = (X_{-u} - X) \times u$. If $X \equiv 0$, then this is linearly equivalent to 0, and hence we can write

$$s_2^{-1}(X) \sim (X \times B) + (B \times Y)$$

for some divisor Y on B. By symmetry, we must have $Y \sim X$. The converse is equally clear.

2.) If $X \equiv 0$ then

$$(\alpha + \beta)^{-1}(X) \sim \alpha^{-1}(X) + \beta^{-1}(X).$$

Let $\lambda : A \to B \times B$ be the composite map (α, β), so we have $\lambda u = (\alpha u, \beta u)$ for u in A. Then $s_2 \lambda = \alpha + \beta$, and $p_i \lambda = \alpha$ or β according as $i = 1$ or 2. Using the formula $(fg)^{-1} = g^{-1} f^{-1}$ which is applicable in the present case, together with step 1, we get what we want.

3.) We finish the proof in this step. Let X be arbitrary. Using the above lemma, we get

$$((m\alpha + n\beta)^{-1}(X))_u - (m\alpha + n\beta)^{-1}(X) = (m\alpha + n\beta)^{-1}(X_{(m\alpha+n\beta)u} - X).$$

For any $v \in B$, $X_v - X \equiv 0$. Using step 2 and making repeated use of the theorem of the square, we see that this expression is linearly equivalent to

$$(1) \qquad m^2 \alpha^{-1}(X_{\alpha u} - X) + mn\alpha^{-1}(X_{\beta u} - X)$$
$$+ mn\beta^{-1}(X_{\alpha u} - X) + n^2 \beta^{-1}(X_{\beta u} - X).$$

Let $Z = \alpha^{-1}(X_{\beta u} - X) + \beta^{-1}(X_{\alpha u} - X)$. To prove our theorem, it suffices to shrow that $Z \sim D_X(\alpha, \beta)_u - D_X(\alpha, \beta)$. One sees this by putting $m = n = 1$ in formula (1), and using the lemma.

To go further, it is convenient to use the language and formalism of the Picard variety. (At present, a complete exposition is not yet published. Some of it will be in Chow [3], and the commutative diagrams below are due to Weil. See also Morikawa [5].) Given a homomorphism $\alpha \in H(A, B)$, then one has an induced contravariant mapping on the Picard varieties,

$^t\alpha : \hat{B} \to \hat{A}$, which we shall call the *transpose* of α. We shall use constantly the linearity property

$$(2) \qquad {}^t(\alpha + \beta) = {}^t\alpha + {}^t\beta,$$

which is essentially step 2 in the proof of Theorem 1. If $\theta_A : A \to \hat{\hat{A}}$ denotes the canonical map of A onto its double dual, and X is a divisor on A, then the following diagram is commutative:

$$(3)$$

If $\alpha \in H(A, B)$ and if Y is a divisor on B, then the following diagram is commutative:

$$(4)$$

i. e., we have $\varphi_{\alpha^{-1}(Y)} = {}^t\alpha \circ \varphi_Y \circ \alpha$.

It follows just as in Theorem 1 that

$$(5) \qquad \varphi_{D_X(\alpha, \beta)} = {}^t\alpha \circ \varphi_X \circ \beta + {}^t\beta \circ \varphi_X \circ \alpha$$

for any divisor X on B. Conversely, one might also use the present formalism to deduce the following generalization. If $\alpha_1, \cdots, \alpha_d$ are homomorphisms of A into B, and $m_1, \cdots, m_d$ are integers, then

$$(6) \qquad (m_1\alpha_1 + \cdots + m_d\alpha_d)^{-1}(X) = \tfrac{1}{2} \sum m_i m_j D_X(\alpha_i, \alpha_j),$$

the sum being taken for $i, j = 1, \cdots, d$, so that each term appears twice and we therefore divide by 2.

In addition, the linearity property of the transpose gives us immediately a bilinear map of $H(A, B)$ into $N(A)$, namely the class of $D_X(\alpha, \beta)$ obeys the rules:

D1. $\qquad D_X(\alpha + \beta, \gamma) = D_X(\alpha, \gamma) + D_X(\beta, \gamma).$

D2. $\qquad D_X(\alpha, \beta) = D_X(\beta, \alpha).$

D3. $$D_X(m\alpha, \beta) \equiv mD_X(\alpha, \beta).$$

D4. $$D_X(\alpha, \alpha) \equiv 2\alpha^{-1}(X).$$

Condition D4 is due to the commutative diagram (4). The others are clear.

For the rest of this paper, we shall take $B = A$. Following Morikawa [5], we say that a divisor X on A is *non-degenerate* if the kernel of φ_X is finite. As Weil has shown [7], it is practically obvious that if X is positive and non-degenerate, then a suitable positive multiple of X is ample, i. e., becomes a hyperplane section in a projective embedding of A. From now on, *X will denote a positive non-degenerate divisor on A, whose class in $N(A)$ will remain fixed throughout.* It then follows that for any $\alpha \neq 0$ in $H(A)$, there exists an integer $m > 0$ and a divisor $Y > 0$ such that $m\alpha^{-1}(X) \equiv Y$. Condition D4 can thus be interpreted as a condition of positive definiteness.

Relative to our divisor X, we define an involution on $H_Q(A)$ by letting $\alpha' = \varphi_X^{-1} \circ {}^t\alpha \circ \varphi_X$. The commutativity relations (3) and (4) show that $\alpha \to \alpha'$ is an anti-automorphism, i. e., that

$$(\alpha + \beta)' = \alpha' + \beta', \qquad (\alpha')' = \alpha, \qquad (\alpha\beta)' = \beta'\alpha', \qquad \delta' = \delta.$$

Our pairing $D_X(\alpha, \beta)$ in $N(A)$ can then be extended to a bilinear map of $H_Q(A)$ into $N_Q(A)$. We can rewrite formula (5) in the form

$$(7) \qquad\qquad \varphi_{D_X(\alpha,\beta)} = \varphi_X(\alpha'\beta + \beta'\alpha).$$

It will be convenient to introduce still another symbol. We define

$$D_X(\alpha) \equiv D_X(\alpha, \delta) \equiv (\alpha + \delta)^{-1}(X) - \alpha^{-1}(X) - X.$$

If $\alpha \in H(A)$, then $\alpha \to D_X(\alpha)$ gives a linear map of $H(A)$ into $N(A)$ which can be extended to a Q-linear map of $H_Q(A)$ into $N_Q(A)$.

We can supplement D1-D4 by more formulas which give the behavior of our pairing under multiplication, namely, for any divisor Y, we have

D5. $$D_X(\varphi_X^{-1}\, {}^t\alpha\varphi_Y\beta) \equiv D_Y(\alpha, \beta).$$

D6. $$D_X(\lambda\alpha, \lambda\beta) \equiv D_{\lambda^{-1}(X)}(\alpha, \beta).$$

D7. $$D_X(\alpha\lambda, \beta\lambda) \equiv \lambda^{-1}D_X(\alpha, \beta).$$

These formulas are direct consequences of the definitions and commutativity.

From D5 we deduce immediately formulas for $D_X(\alpha)$, namely,

D8. $$D_X(\alpha, \beta) \equiv D_X(\alpha'\beta).$$

D9. $$D_X(\alpha\beta, \gamma) \equiv D_X(\beta, \alpha'\gamma).$$

D10. $$D_X(\beta'\alpha\beta) \equiv \beta^{-1}D_X(\alpha).$$

D11. $$D_X(\alpha) \equiv D_X(\alpha').$$

The preceding formulas exhaust the formal behavior of our divisorial pairing, and we shall now turn to its application to the trace.

2. The trace. In this section, we use as our basic fact that if $X \equiv 0$, then X is numerically equivalent to 0. In an appendix, we shall reproduce a simple proof due to Weil, based only on the theorem of the square.

We consider divisors up to numerical equivalence, since we are concerned here with numerical statements. If Y is a divisor on A, then the intersection of Y with itself s times is understood to be up to numerical equivalence. If $t_1, \cdots, t_s$ are independent generic points, then $Y_{t_1} \cdot Y_{t_2} \cdots Y_{t_s}$ is an element of that class, for instance. We let r be the dimension of A. If C is any positive cycle of dimension 1 on A, then $\deg(X \cdot C) > 0$ because we have assumed X positive non-degenerate, hence essentially a hyperplane section of A.

In what follows, we shall also use the formula

$$\alpha^{-1}(Y_1 \cdot Y_2) = \alpha^{-1}(Y_1) \cdot \alpha^{-1}(Y_2),$$

which holds whenever both sides are defined. Generic translations will always insure that this is the case. Note also if $\mathfrak{a}$ is a 0-cycle, and $\nu(\alpha) \neq 0$, then

$$\deg(\alpha^{-1}(\mathfrak{a})) = \nu(\alpha)\deg(\mathfrak{a}).$$

From these remarks, we first note that if we intersect both sides of (6) with themselves r times ($r = \dim A$), then we recover immediately Weil's theorem that $\nu(m_1\alpha_1 + \cdots + m_d\alpha_d)$ as a function of $(m_1, \cdots, m_d)$ is a homogeneous polynomial of degree $2r$. Moreover, its coefficients are easily determined explicitly in terms of intersections of the $D_X(\alpha_i, \alpha_j)$.

We are particularly interested in the special case which yields the characteristic polynomial of an endomorphism. In Theorem 1 we take $m = 1$ and $\beta = \delta$. We get:

$$(\alpha + n\delta)^{-1}(X) \equiv n^2 X + nD_X(\alpha) + \alpha^{-1}(X).$$

Again raising this equation to the r-th power, we get

$$\nu(\alpha + n\delta)\deg(X^{(r)}) = c_{2r}n^{2r} + c_{2r-1}n^{2r-1} + \cdots + c_0,$$

where $X^{(r)}$ denotes the class up to numerical equivalence of the intersection of X with itself r times, and the c_i are integers which are easily determined explicitly in terms of intersections of $\alpha^{-1}(X)$, X, and $D_X(\alpha)$. For instance, we have

$$c_{2r} = \deg X^{(r)}, \qquad c_{2r-1} = r \cdot \deg(X^{(r-1)} \cdot D_X(\alpha)), \qquad c_0 = \nu(\alpha) \deg X^{(r)}.$$

This shows that $\nu(\alpha + n\delta)$ is a polynomial in n with rational coefficients. In [6], No. 67-68, Weil gives a simple and direct argument showing that the characteristic polynomial is actually that obtained from an l-adic representation.

The trace of α, which we denote by $\mathrm{tr}(\alpha)$, is defined by the formula

$$\mathrm{tr}(\alpha) = r/\deg(X^{(r)}) \cdot \deg(X^{(r-1)} \cdot D_X(\alpha)).$$

From D5 we immediately get

$$\mathrm{tr}(\alpha'\beta) = r/\deg(X^{(r)}) \cdot \deg(X^{(r-1)} \cdot D_X(\alpha, \beta)).$$

The main properties of the trace are then directly traceable to the properties of the divisor class $D_X(\alpha, \beta)$.

On the algebra $H_Q(A)$ over the rational numbers, we can define a scalar product (taking its value in Q) by putting $(\alpha, \beta) = \mathrm{tr}(\alpha'\beta)$. We then have the usual properties of a bilinear form, whose quadratic form is positive definite, and in addition we have the multiplicative property $(\alpha\beta, \gamma) = (\beta, \alpha'\gamma)$. This follows immediately from the analogous properties of $D_X(\alpha, \beta)$. The strict positivity comes from the fact that for some integer $m > 0$, mX is a hyperplane section of A, and hence the class of $m\alpha^{-1}(X)$ in $N(A)$ contains a divisor $Y > 0$. As we have noted before, we must then have $\deg(X^{(r-1)} \cdot \alpha^{-1}(X)) > 0$. From D4 and the definition of (α, α) we conclude that $(\alpha, \alpha) > 0$.

We define the norm $\| \alpha \|$ as usual by $(\alpha, \alpha)^{\frac{1}{2}}$. Aside from the usual properties of a metric, we have in addition the following statements:

N1. $\quad \| \alpha \| = \| \alpha' \|$

N2. $\quad \| \alpha'\alpha \| = \| \alpha\alpha' \|$

N3. $\quad$ If $\alpha \neq 0$ and $\beta \neq 0$, then $\| \alpha\beta \| < \| \alpha \| \, \| \beta \|$.

The first two are clear from the commutativity of the trace. As to the third, we have

$$\| \alpha\beta \|^2 = \mathrm{tr}(\beta'\alpha'\alpha\beta) = \mathrm{tr}(\alpha'\alpha\beta\beta') = (\alpha'\alpha, \beta\beta') \leqq \| \alpha'\alpha \| \, \| \beta\beta' \|.$$

this last inequality being the Schwartz inequality. We are therefore reduced to proving the inequality $\| \alpha'\alpha \| < \| \alpha \|^2$. By definition we have $\| \alpha'\alpha \|^2 = \mathrm{tr}(\alpha'\alpha\alpha'\alpha) = \mathrm{tr}((\alpha\alpha')^2)$. Let ω_j be the characteristic roots of $\alpha\alpha'$. Then

ω_j^2 are the characteristic roots of $(\alpha\alpha')^2$. It is known, and we shall recall the proof in the next section, that the ω_j are totally real totally positive algebraic numbers. Our inequality is equivalent to $\sum \omega_j^2 < (\sum \omega_j)^2$ which is true in view of this fact.

As an application, we consider an abelian variety A defined over a finite field k with q elements, and assume that X is rational over k. Then one verifies trivially that if π is the Frobenius endomorphism relative to k, then $\pi\pi' = \pi'\pi = q\delta$. Hence in that case, $\|\pi\| = (2rq)^{\frac{1}{2}}$. Furthermore, $\mathrm{tr}(\pi) = \mathrm{tr}(\pi\delta) = (\pi, \delta)$ and by the Schwartz inequality we get

$$| \mathrm{tr}(\pi)| \leqq \|\pi\| \|\delta\| = 2rq^{\frac{1}{2}}.$$

For any integer n, we get $|\mathrm{tr}(\pi^n)| \leqq 2rq^{n/2}$. If ω_j are the characteristic roots of π, then ω_j^n are those of π^n. From this one sees easily that the absolute value of a characteristic root can be at most $q^{\frac{1}{2}}$. Since we have $\nu(\pi) = q^r$, the product of the characteristic roots of π must be equal to $\pm q^r$. Hence the absolute value of each one of the characteristic roots is exactly $q^{\frac{1}{2}}$.

3. A dual pairing.* We introduce a pairing of the endomorphisms of A into the group of 1-cycles on A modulo numerical equivalence which is dual to the divisorial pairing of Section 1. The symbol $\equiv$ will now denote numerical equivalence, and $N_i(A)$ denotes the factor group of i-cycles by those which are numerically equivalent to 0.

Let Z be a cycle on A. Let $\alpha \in H(A)$. As usual, we mean by $\alpha(Z)$ the intersection $\mathrm{pr}_2[\Gamma_\alpha \cdot (Z \times A)]$.

We shall be particularly interested in 1-cycles. Let C be a 1-cycle. If Y is a divisor on A, then we have the *transposition formula*

$$(8) \qquad \deg(\alpha(C) \cdot Y) = \deg(C \cdot \check{\alpha}^{-1}(Y)).$$

The easy proof runs as follows.

$$\deg(\alpha(C) \cdot Y) = \deg\{\mathrm{pr}_2[(\Gamma_\alpha \cdot (C \times A)) \cdot (A \times Y)]\}$$
$$= \deg\{((C \times A) \cdot \Gamma_\alpha) \cdot (A \times Y)\},$$

because the degree of a 0-cycle is preserved under projection. All that remains to be done is to use associativity, and then unwind the above expression with respect to Y, just as we have done with respect to C.

We can transfer to the 1-cycles properties of divisors by transposition. For $\alpha, \beta \in H(A)$ we define

$$Z_C(\alpha, \beta) \equiv (\alpha + \beta)(C) - \check{\alpha}(C) - \beta(C), \qquad Z_C(\alpha) \equiv Z_C(\alpha, \delta),$$

---

* This section was added March 22, 1957.

these being elements of $N_1(A)$. We get trivially

(9) $$\deg[Z_C(\alpha, \beta) \cdot Y] = \deg[C \cdot D_Y(\alpha, \beta)].$$

This shows that $Z_C(\alpha, \beta)$ gives a bilinear map of $H(A)$ into $N_1(A)$, and D4 becomes $Z_C(\alpha, \alpha) \equiv 2\alpha(C)$. In addition, we obtain

(10) $$(m\alpha + n\beta)(C) \equiv m^2\alpha(C) + mnZ_C(\alpha, \beta) + n^2\beta(C)$$

(11) $$(\alpha + n\delta)(C) \equiv n^2 C + nZ_C(\alpha) + \alpha(C).$$

In order to apply a numerical calculus dual to the intersection calculus of cycles used in Section 2, we must define the Pontrjagin product of cycles on A as follows. Let V, W be two subvarieties of A. By $V \oplus W$ we shall denote the set theoretic sum of V and W taken in A. It is the variety consisting of all points $v + w$ with $v \in V$ and $w \in W$. There is a rational map $F: V \times W \to V \oplus W$ which may or may not be of finite degree. By $d(V, W)$ we denote the degree of F if it is finite, and 0 otherwise. Similarly, we define $d(V_1, \cdots, V_m)$ for any finite number of subvarieties of A to be the degree of the rational map from their product to their set theoretic sum if it is finite, and 0 otherwise. We denote by $V * W$ the cycle $d(V, W)(V \oplus W)$, and one sees immediately that $V * W = \mathrm{pr}_3[S \cdot (V \times W \times A)]$, where S is the graph of the sum $s_2: A \times A \to A$. Note that $V * W = 0$ if and only if the dimension of $V \oplus W$ is smaller than $\dim V + \dim W$. From the definitions, one also sees that the Pontrjagin product is associative, and that $V_1 * V_1 * \cdots * V_m = d(V_1, \cdots, V_m)(V_1 \oplus V_2 \oplus \cdots \oplus V_m)$. The product extends naturally to cycles by linearity, and if Z_1, Z_1 are two cycles, then

$$Z_1 * Z_2 = \mathrm{pr}_3[S \cdot (Z_1 \times Z_2 \times A)].$$

From the fact that the expression on the right is built up from the standard operations of intersection theory, we see that the Pontrjagin product defines a commutative ring structure on the graded module $\sum N_i(A)$.

Let $\alpha \in H(A)$. Then α induces an endomorphism of our ring, i.e., we have

(12) $$\alpha(V_1 * V_2 * \cdots * V_m) = \alpha(V_1) * \alpha(V_2) * \cdots * \alpha(V_m).$$

This is easily seen by applying the definitions. We are now in a position to use the same method as in Section 2 to get an expression for the characteristic polynomial $\nu(\alpha + n\delta)$, except that we use the Pontrjagin product instead of the intersection product. We are mostly interested in the trace. Raising both sides of (11) to the r-th power, using (12), we get

(13) $$\mathrm{tr}(\alpha) = r/d(C, \cdots, C) \cdot \deg[C^{*(r-1)} * Z_C(\alpha)].$$

provided we take for C a curve which generates A. Here, $C^{*(r-1)}$ means the

Pontrjagin product of C with itself $(r-1)$ times, and the degree is defined to be the coefficient of A appearing in the expression on the right. It is convenient to transform this expression into an intersection number. Whenever V and W have complementary dimesion, we have

$$(14) \qquad d(V, W) = \deg(V \cdot W^-) = \deg(V^- \cdot W),$$

denoting by V^- the transform of V by the map sending each point into its inverse. This is a restatement of an elementary result concerning algebraic groups (Weil [6], No. 13, Cor. 2 of Th. 4). If V is a divisor, then $V^- \equiv V$, say, by taking $\alpha = -\delta$ in the commutative diagram (4), and hence the expression for the trace becomes

$$(15) \qquad \operatorname{tr}(\alpha) = r/d(C, \cdots, C) \cdot \deg(C^{*(r-1)} \cdot Z_C(\alpha)).$$

We have thus recovered all the numerical results contained in Weil's treatise on Abelian varieties. Consider the special case where $A = J$ is the Jacobian of a curve C. Then $C^{*(r-1)} = (g-1)!\Theta$, and $d(C, \cdots, C) = g!$. Using the transposition formula we get

$$(16) \qquad \operatorname{tr}(\alpha) = \deg(D_\Theta(\alpha) \cdot C).$$

We remark that if we had tried to obtain this directly from the expression of the trace obtained by the ordinary intersection product, we would have needed the extra information that

$$\Theta^{(g-1)} \equiv (g-1)!C \quad \text{and} \quad \deg(\Theta^{(g)}) = g!.$$

Remarkably enough, there does not seem to be an easy proof available for this fact. Such a proof would require a closer analysis of the relations which exist between the ordinary and the Pontrjagin products of cycles.

Formula (16) allows us to recover the Lefschetz fixed point formula (and hence the Riemann hypothesis for curves), for we can now use the proof given by Weil ([6], No. 47) which is quite transparent. For the convenience of the reader, we reproduce this proof. If T is a correspondence on C (i.e., a 1-cycle on $C \times C$), we let τ be the associated endomorphism of J. The fixed point formula states that

$$\deg(T \cdot \Delta) = d(T) - \operatorname{tr}(\tau) + d'(T).$$

We let s_2 be the sum on $J \times J$, and consider the following diagram:

$$\begin{array}{ccccccc}
C \times C & \xrightarrow{f_2} & J \times J & \xrightarrow{(\tau, \delta)} & J \times J & \xrightarrow{s_2} & J \\
\uparrow{\scriptstyle h_i} & & \uparrow{\scriptstyle H_i} & & & & \\
C & \xrightarrow{\quad f \quad} & J & & & &
\end{array}$$

Here, f is the canonical mapping, f_2 is the product of f with itself, and h_i, H_i $(i = 1, 2, 3)$ are described as follows: For $i = 1$ we take the diagonal mapping. For the others, we may assume that we have a point P on C such that $f(P) = 0$ is the origin on J. Then h_2 maps C on $P \times C$ and h_3 maps C on $C \times P$. Similarly, H_2 maps J on $0 \times J$ and H_3 maps J on $J \times 0$. Then the square is commutative. Put $\lambda = (\tau, \delta)$. Then $f_2^{-1} \lambda^{-1} s_2^{-1}(-\Theta)$ is a correspondence T on $C \times C$ whose associated endomorphism is precisely τ. (Of course, in taking our inverse mappings, we first make a generic translation on Θ.) Furthermore, we have

$$s_2 \lambda H_1 = \tau + \delta, \qquad s_2 \lambda H_2 = \delta, \qquad s_2 \lambda H_3 = \tau.$$

From the commutativity, it is then obvious that

$$\deg[(\tau + \delta)^{-1}(-\Theta) \cdot C] = \deg(T \cdot \Delta), \qquad \deg[\tau^{-1}(-\Theta) \cdot C] = d(T),$$
$$\deg(-\Theta \cdot C) = d'(T).$$

This concludes the proof.

4. Positive endomorphisms. We shall say that an element α in $H_Q(A)$ is *symmetric* if $\alpha = \alpha'$. These elements form a subspace $S_Q(A)$ of $H_Q(A)$ over the rational numbers. We have a Q-linear map $D_X \colon S_Q(A) \to N_Q(A)$, defined by $\alpha \to D_X(\alpha)$. It is in fact an isomorphism, and we shall define its inverse $\Phi_X \colon N_Q(A) \to S_Q(A)$ as follows. If Y is a divisor representing a class in $N(A)$ then we put $\Phi_X(Y) = \frac{1}{2}\varphi_X^{-1}\varphi_Y$. The map $Y \to \Phi_X(Y)$ clearly induces a Q-linear map of $N_Q(A)$ into $H_Q(A)$. The commutativity relations (3) and (4) show that it is into $S_Q(A)$, i.e., that $\varphi_X^{-1}\varphi_Y$ is symmetric. From (7), we see that $D_X\Phi_X$ and $\Phi_X D_X$ are both equal to the identity, and hence are isomorphisms.

If Y is a divisor representing a class in $N(A)$, we shall write $\alpha \leftrightarrow Y$ to mean that $\alpha = \Phi_X(Y)$.

PROPOSITION 1. *Let α be a symmetric element of $H_Q(A)$, and suppose $\alpha \leftrightarrow Y$. Let λ be any element of $H_Q(A)$. Then $\lambda'\alpha\lambda \leftrightarrow \lambda^{-1}(Y)$. In particular, $\lambda'\lambda \leftrightarrow \lambda^{-1}(X)$.*

Proof. This is a reformulation of D10.

We shall say that an element of $H_Q(A)$ is *positive* (relative to the divisor X) if it is symmetric, and if there exists an integer $m > 0$ such that the class of $D_X(m\alpha)$ in $N(A)$ contains a divisor $Y > 0$. We then write $\alpha > 0$.

PROPOSITION 2. *Let α be symmetric.*

a.) *If $\alpha \geq 0$, and λ is any element of $H_Q(A)$ then $\lambda'\alpha\lambda \geq 0$.*

b.) *If $\lambda \in H_Q(A)$ and $\lambda \neq 0$, then $\lambda'\lambda > 0$.*

c.) *If $\alpha > 0$, then $\operatorname{tr}(\alpha) > 0$.*

Proof. Our statements are immediate from Proposition 1 and the definitions.

As a corollary to Theorem 2, we shall see that if $\alpha > 0$ and $\lambda'\alpha \neq 0$, then actually $\lambda'\alpha\lambda > 0$.

We shall give below some equivalent conditions for an endomorphism to be positive. Referring to Albert [1] and Morikawa [5] we recall first an abstract theorem for the convenience of the reader.

THEOREM 2. *Let R be an algebra over the rationals Q, with an involution $x \to x'$ satisfying $(xy)' = y'x'$ and $(x')' = x$, and a Q-linear functional $\sigma: R \to Q$ which satisfies $\sigma(xy) = \sigma(yx)$, and such that for $x \neq 0$ we have $\sigma(xx') > 0$. Then:*

1.) *R is semisimple.*

2.) *If $x = x' \neq 0$, then $Q[x]$ is a direct sum of totally real fields.*

3.) *If $x = x' \neq 0$, then $\sigma(yxy) \geq 0$ for all $y \in Q[x]$ if and only if x is a sum of squares in $Q[x]$.*

Proof of 1. We first prove that if $x = x' \neq 0$, then x cannot be nilpotent. If it were, then we would have $x^{2^m} = 0$ but $x^{2^{m-1}} \neq 0$, and hence $x^{2^m} = (x^{2^{m-1}})^2 = 0$. As $x^{2^{m-1}}$ is symmetric, this contradicts the strict positivity of σ. Now if $x \neq 0$ is an element of a nilpotent right ideal, then xx' is symmetric, is in the ideal, and hence is nilpotent, thereby contradicting again the positivity of σ.

Proof of 2. Consider $Q[x]$ with $x = x' \neq 0$. It is a commutative subalgebra of R, and the involution and trace on R induce on it an involution and trace with similar properties. Being semisimple, it is a direct sum of fields. On each such field F we have again induced an involution and a trace. We may therefore assume that $R = F$. But any Q-linear function on F is of type $\sigma(\xi) = S(\alpha\xi)$ for some α in F. Here S denotes the ordinary trace from F to Q. We have by assumption $S(\alpha\xi^2) = \sigma(\xi^2) > 0$, if $\xi \neq 0$. If some conjugate of α is not real, then the corresponding conjugate of F is dense in the complex numbers, and we choose $\xi \in F$ such that $\alpha\xi^2$ is very large negative at this conjugate, and very close to 0 everywhere else. This gives a contradiction, and α is totally positive. Similarly, one sees that R is totally real.

Proof of 3. We may again assume that $R = F$ is a totally real field. Then $\sigma(x\xi^2) > 0$ for all $x \neq 0$ in F, and as we have seen above, this implies that x is totally positive. Hence it is a sum of squares. The converse is equally clear. This concludes the proof of our theorem.

COROLLARY 1. *Let R me as in the theorem, and $\xi \in R$ be such that $\xi = \xi'$, and $\sigma(\lambda\xi\lambda) \geqq 0$ for all $\lambda \in \boldsymbol{Q}[\xi]$. Then given $\beta \in R$ such that $\beta'\xi\beta \neq 0$, we have $\sigma(\beta'\xi\beta) > 0$.*

Proof. We can write $\xi = \sum y_i^2$ with $y_i \in \boldsymbol{Q}[\xi]$. If we put $\mu_i = y_i\beta$, then at least one element $\mu_i'\mu_i$ is not 0, and

$$\sigma(\beta'\xi\beta) = \sum \sigma(\mu_i'\mu_i) > 0.$$

COROLLARY 2. *Let R be as in the theorem, let $\alpha = \alpha'$ and assume that α is a sum of squares in $\boldsymbol{Q}[\alpha]$. If $\lambda \in R$ and if $\lambda'\alpha \neq 0$, then $\lambda'\alpha\lambda \neq 0$.*

Proof. Writing $\alpha = \sum \beta_i^2$ with $\beta_i \in \boldsymbol{Q}[\alpha]$, we must have $\lambda'\beta_i \neq 0$ for some i. Hence $0 \neq (\lambda'\beta_i)(\lambda'\beta_i)' = \lambda'\beta_i^2\lambda$. Since $\sigma(\lambda'\beta_i^2\lambda) \geqq 0$ for each i, and for at least one i is > 0, this shows that $\lambda'\alpha\lambda$ cannot be 0.

THEOREM 3. *Let $\alpha \in H_{\boldsymbol{Q}}(A)$ and assume $\alpha = \alpha' \neq 0$. Then the following conditions are equivalent:*

1.) *α is positive.*

2.) *All the characteristic roots of α are totally real and totally positive.*

3.) *α is a sum of squares in $\boldsymbol{Q}[\alpha]$.*

Proof. The equivalence between 2.) and 3.) is well known. If α is positive, and $\beta \in \boldsymbol{Q}[\alpha]$ then by D10 and the definition of the trace, we see that $\operatorname{tr}(\beta\alpha\beta) \geqq 0$. Hence α is a sum of squares by Theorem 2. This proves that the first condition implies the third. If we can write $\alpha = \sum \beta_i\beta_i' = \sum \beta_i^2$ with $\beta_i \in \boldsymbol{Q}[\alpha]$, then there is an integer $e > 0$ and there are positive divisors Y_i such that $(e\beta_i)(e\beta_i)' = \varphi_X^{-1}\varphi_{Y_i}$ (Proposition 1). Putting $Y = \sum Y_i$, we see that $e^2\alpha = \varphi_X^{-1}\varphi_Y$, and hence that α is positive.

COROLLARY 1. *If α and β are two symmetric elements of $H_{\boldsymbol{Q}}(A)$ such that $\alpha > 0$, $\beta > 0$ and $\alpha\beta \neq 0$, then $\operatorname{tr}(\alpha\beta) > 0$.*

Proof. We write $\alpha = \sum \lambda_i^2$ with $\lambda_i \in \boldsymbol{Q}[\alpha]$, and we have therefore $\operatorname{tr}(\alpha\beta) = \sum \operatorname{tr}(\lambda_i^2\beta) = \sum \operatorname{tr}(\lambda_i\beta\lambda_i) \geqq 0$. Actually, for some i, we have

$\lambda_i\beta \neq 0$, and hence by Corollary 2 of Theorem 2, $\lambda_i\beta\lambda_i \neq 0$. Hence the strict inequality holds, as desired.

COROLLARY 2. *Let $Y > 0$ be a divisor on A, and $Z^* > 0$ a divisor on $\hat{A}$. If $\alpha = \theta^{-1}\varphi_Z \cdot \varphi_Y$ and $\alpha \neq 0$, then $\operatorname{tr}(\alpha) > 0$.*

Proof. Let $\lambda = \varphi_X^{-1}\varphi_W$ with $W = \varphi_X^{-1}(Z^*)$ and let $\beta = \varphi_X^{-1}\varphi_Y$. We use the commutativity relations (3) and (4) to conclude that $\alpha = \lambda\beta$, and we can then apply Corollary 1.

Finally, note that if the abelian variety is defined over a finite field k with q elements, if X is rational over k, and if π is the Frobenius endomorphism, then from the relation $\pi\pi' = q\delta$, we can conclude from the above considerations without taking powers of π that the characteristic roots have absolute value $q^{\frac{1}{2}}$. Indeed, the algebra generated over $\mathbf{Q}$ by π and π' is commutative, and, in view of Theorem 2, is a direct sum of fields which are either totally real, or totally imaginary such that $x \to x'$ is the complex conjugation. If e_i is the idempotent of one of those fields, and if we put $\pi_i = \pi e_i$, then we see directly that the absolute value of π_i is $q^{\frac{1}{2}}$.

Appendix.

We shall now reproduce Weil's proof that if X is a divisor, $X \equiv 0$, then X is numerically equivalent to 0. The only property of abelian varieties that is used is the theorem of the square. To begin with, we have:

THEOREM. *Let $f: C \to A$ be a rational map of a complete non-singular curve into A, such that $f(C)$ generates A (i.e., $A = f(C) + \cdots + f(C)$ n times, where $n = \dim A$). Let $W = \sum_{1}^{n-1} f(C)$. Let $X \equiv 0$ be a divisor on A, and let*

$$f^{-1}(X) = \sum m_j(P_j).$$

(If necessary, make a generic translation on X so that $f^{-1}(X)$ is defined.) Let $M_1, \cdots, M_n$ be independent generic points of C over a common field of definition k for f, C, and A, and let

$$d = [k(M_1, \cdots, M_n) : k(\sum_{i=1}^{n} f(M_i))], \quad d_0 = [k(M_1, \cdots, M_{n-1}) : k(\sum_{i=1}^{n-1} f(M_i))].$$

Then $dX \sim nd_0 \sum m_j W_{f(P_j)}$.

Proof. Let $f_n: C \times \cdots \times C \to A \times \cdots \times A$ be the product of f with itself n times, and s_n the sum on A. Let $F = s_n f_n$. Then d is the degree of F. A standard computation yields $F \circ F^{-1}(X) = dX$. On the other hand, using step 1 in Theorem 1, we get

$$F^{-1}(X) \sim \sum_i f_n^{-1} p_i^{-1}(X), \quad \text{which is} \quad \sum_j m_j \sum_{i=1}^n (C \times \cdots \times P_j \times \cdots \times C).$$

We shall take F of this, and see that it gives the desired result. Let G be the graph of F, and let P be a generic point of C. Then by F-VII$_6$ Th. 12, we first take $G(P)$ on the product $C \times \cdots \times \hat{C} \times \cdots \times C \times A$ (where C is omitted in the i-th place), i.e., we take the projection on this variety of

$$G \cdot (C \times \cdots \times P \times \cdots \times C \times A).$$

This is a variety, whose set theoretic projection on A is $W_{f(P)}$. Taking now its projection on A in the sense of intersection theory, we obviously get $d_0 W_{f(P)}$. If we work with a special point P_j which is a specialization of P, then we use the compatibility of intersection and projection with specializations, together with the fact that $W_{f(P)}$ has the uniquely determined specialization $W_{f(P_j)}$, to conclude the proof.

Now by the theorem of the square, we get

$$n d_0 \sum m_j (W_{f(P_j)} - W) \sim W_u - W \equiv 0,$$

where $u = n d_0 \sum m_j f(P_j)$. Consequently, we get $0 \equiv dX \equiv n d_0 (\sum m_j) W$. If we can prove that $\sum m_j = 0$, this will show immediately that $dX \sim W_u - W$ and will prove the numerical equivalence of X to 0. We are therefore reduced to proving the following lemma.

LEMMA. *Let Y be a divisor, $Y > 0$. Then Y cannot be $\equiv 0$.*

Proof. If $Y \equiv 0$, then for t generic on A, we have $Y_t \sim Y$. This would yield a representation of A into the complete linear system of Y, and thus a representation of A into the projective linear group. This representation must be trivial, and hence $Y_t = Y$, which is an absurdity.

It should be noted that Morikawa's Lemma 7 [5] is an immediate consequence of the above result, which, in fact, shows that the divisor W is non-degenerate. Cf. also Weil's arguments concerning the projective embedding of abelian varieties [7].

COLUMBIA UNIVERSITY.

REFERENCES.

[1] A. Albert, "On involutorial algebras," *Proceedings of the National Academy of Sciences*, vol. 41 (1955), pp. 480-482.

[2] I. Barsotti, "Il teorema di dualita per le varieta abeliane ed altri risultati," *Rendiconti di Matematica e delle sue applicazioni*, vol. 13 (1954), pp. 1-17.

[3] W. L. Chow, "Abstract theory of the Picard and Albanese varieties," to appear.

[4] H. Hasse, "Über die Riemannsche Vermutung in Funktionenkörpern," *Comptes Rendus du Congres International des Mathématicians*, Oslo, 1936, pp. 189-206.

[5] H. Morikawa, "On abelian varieties," *Nagoya Mathematical Journal*, vol. 6 (1953), pp. 151-170.

[6] A. Weil, *Variétés abéliennes et courbes algébriques*, Hermann et Cie., Paris, 1948.

[7] ———, On the projective embedding of abelian varieties, in the volume in honor of S. Lefschetz, Princeton, 1957.

[8] ———, "On the theory of complex multiplication," *Proceedings of the International Symposium on Algebraic Number Theory*, Tokyo-Nikko. 1955. pp. 9-22.

Additional Comment, 1999. I learned of Castelnuovo's fundamental contributions to the subject matter of the above paper from Kani [Kan 84]. In the complex case, Castelnuovo [Cas 21] establishes the relation between his equivalence defect and the trace in the complex representation of endomorphisms on the Jacobian of a curve. Furthermore, he defines the characteristic polynomial, expresses the coefficients as intersections, and gives intersection formulas for the sum of the curve with itself r times and the theta divisor, as well as powers of the theta divisor. In the fifties, I learned such results from Weil's book and lectures on abelian varieties, and I referred to Weil's work at the beginning of my paper. In my book on abelian varieties, I also stated: "Weil…was the first to recognize that Castelnuovo's theorem on the equivalence defect of correspondences on a curve could be expressed as a theorem on abelian varieties." As Kani pointed out in 1984, I was wrong. But there are no references to Castelnuovo on these matters in Weil's works, nor were there in his courses. What Weil did in the forties was to algebraicize Castelnuovo's theory, and extend it, following Hasse's discovery that the key to the Riemann Hypothesis in function fields lay with the theory of correspondences and endomorphisms of the Jacobian (1934, 1936).

[Cas 21] G. CASTELNUOVO, Sulle funzioni abeliane, *Rend. Accad. Lincei* XXX (1921), Four Notes, Reproduced in his collected papers, pp. 529–549.

[Has 34] H. HASSE, Abstrakte Begrundung der komplexen Multiplikation und Riemannsche Vermutung in Funktionenkörpern, *Abh. Math. Sem. Univ. Hamburg* **10** (1934) pp. 325–348.

[Kan 84] E. KANI, On Castelnuovo's equivalence defect, *J. reine angew. Math.* (1984) pp. 24–70.

Am. J. Math., LXXX No. 2, 1958

RECIPROCITY AND CORRESPONDENCES.*

By Serge Lang.

To Artin on his 60th birthday

The reciprocity theorem $f((g)) = g((f))$ on a curve arose first in Weil's investigations of Artin's reciprocity law on curves [4]. It has recently been used to deal with various questions of duality, for instance by Igusa [1]. Furthermore, Tate has just given a pairing of certain Galois cohomology groups of abelian varieties [3]. To prove the pairing well defined and bilinear on Jacobians, he used the above theorem.

In this manner, we are therefore led in a natural way to formulate it on abelian varieties, and by pull back on varieties. Aside from its application to the Tate pairing, we get an obvious complement to Kummer theory arising from the theory of divisorial correspondences.

1. Notation. Let U, V be two complete varieties, non-singular in co-dimension 1. This insures that a divisor which is linearly equivalent to 0 has this property over a given field of rationality and, also, that such a divisor determines uniquely the function of which it is the divisor, up to a multiplicative constant.

A cycle of codimension 1 on a variety will always be called a divisor. A cycle of dimension 0 will be called simply a cycle. Let $\mathfrak{a} = \sum n_i(P_i)$ be a cycle on U. Let ϕ be a function on U such that ϕ is defined at every point of $\mathfrak{a}$ and does not take on the value 0. Then we define

$$\phi(\mathfrak{a}) = \prod \phi(P_i)^{n_i}.$$

If $\mathfrak{a}$ is of degree 0, and ψ is a function such that $\psi = c\phi$ where c is constant, then $\psi(\mathfrak{a}) = \phi(\mathfrak{a})$.

Let D be a divisor on the product $U \times V$. Let $\mathfrak{a}$, $\mathfrak{b}$ be two cycles of degree 0 on U and V respectively. Assume that $D(\mathfrak{a})$ is defined and that it is the divisor of a function f on V. If in addition $f(\mathfrak{b})$ is defined, then we put $D(\mathfrak{a}, \mathfrak{b}) = f(\mathfrak{b})$ and say that $D(\mathfrak{a}, \mathfrak{b})$ is defined.

Suppose that, for every point P in $\mathfrak{a}$ and every point Q in $\mathfrak{b}$, the point

---

* Received January 6, 1958.

431

(P, Q) does not lie in the support of D (or as we shall say more briefly, in D). Then first, $D(\mathfrak{a})$ is defined, and Q does not lie in the divisor of the function f. Hence $f(Q)$ is defined and not equal to 0, and consequently, $D(\mathfrak{a}, \mathfrak{b})$ is defined. This sufficient condition will be used constantly in the sequel.

2. The theorem of the square. The proof of our main reciprocity theorem (Theorem 4) is based on the theorem of the square, which runs as follows.

Let U, V, W be three varieties and assume that W is complete and non-singular in codimension 1. Let D be a divisor on $U \times V \times W$, and let u_i, v_j $(i, j = 0, 1)$ be simple points of U, V respectively, such that $D(u_i, v_j)$ is defined. Then $\sum_{i,j} (-1)^{i+j} D(u_i, v_j)$ is linearly equivalent to 0 on W.

This is an immediate consequence of the existence of the Picard variety for W. Indeed, if a, b are two simple points of U, V respectively, such that $D(a, b)$ is defined, then for u, v generic we have a rational map

$$(u, v) \to \mathrm{Cl}[D(u, v) - D(a, b)]$$

of $U \times V$ into the Picard variety of W. (As usual, Cl denotes the point associated with the linear equivalent class on the Picard variety.) A rational map of a product into an abelian variety splits into a sum of mappings of the factors. It is then clear that in our alternating sum, the constants will cancel, and the alternating sum will map on 0 in the Picard variety. It is therefore linearly equivalent to 0.

Conversely, as Weil has shown, one can give a direct proof for the theorem of the square, and one can base on it the construction of the Picard variety. For this as well as all further information concerning abelian varieties, the reader is referred to [2].

3. The theorem of the hypercube. In order to avoid too many indices, we restrict ourselves to four-fold products, which will be all that is needed for the applications we have in mind to abelian varieties. The notation we now describe will remain fixed throughout this section.

Let U, V, W, T be four varieties, defined over a field k. We assume them complete and non-singular in codimension 1.

Let D be a divisor on the product $U \times V \times W \times T$, rational over k. Let i, j, k, l range over 0 and 1, and take two copies of each one of our four varieties, U_i, V_j, W_k, T_l. On the double product

$$U_0 \times U_1 \times V_0 \times V_1 \times W_0 \times W_1 \times T_0 \times T_1 = U^{(2)} \times V^{(2)} \times W^{(2)} \times T'^{(2)},$$

consider the divisor D_{ijkl} consisting of D on the partial product

$$U_i \times V_j \times W_k \times T_l$$

taken with the full varieties on the others. (In other words, it is the inverse image of D under the projection of the eight-fold product to the four-fold product.) We can take a coboundary of D, by setting

$$E = \Sigma \, (-1)^{i+j+k+l} D_{ijkl},$$

and observe that this formula could have been generalized to arbitrary products.

Let
$$P = (u_0, u_1, v_0, v_1) \quad \text{and} \quad Q = (w_0, w_1, t_0, t_1)$$

be two independent generic points of $U^{(2)} \times V^{(2)}$ and $W^{(2)} \times T^{(2)}$ over k. By the theorem of the square, there exists a function f_P on $W \times T$, defined over $k(P)$, and a function g_Q on $U \times V$, defined over $k(Q)$ such that

$$(1) \qquad (f_P) = \Sigma \, (-1)^{i+j} D(u_i, v_j) \quad \text{and} \quad (g_Q) = \Sigma \, (-1)^{k+l} \, {}^t D(w_k, t_l).$$

Furthermore, it is obvious from the definitions that we have

$$(2) \qquad E(P) = \Sigma \, (-1)^{k+l} (f_P)_{kl} = \sum_{i,j,k,l} (-1)^{i+j+k+l} D(u_i, v_j)_{kl}$$

$$(3) \qquad {}^t E(Q) = \Sigma \, (-1)^{i+j} (g_Q)_{ij} = \sum_{i,j,k,l} (-1)^{i+j+k+l} \, {}^t D(w_k, t_l)_{ij}.$$

All further reciprocity theorems derived from the correspondence D between $U \times V$ and $W \times T$ arise from the symmetry of our divisor E.

THEOREM 1. *The divisor E above is linearly equivalent to 0.*

Proof. There exists a function f^* on $U^{(2)} \times V^{(2)} \times W \times T$ defined over k such that

$$f^*(P, w, t) = f_P(w, t),$$

and we have
$$P \times (f_P) = (f^*) \cdot (P \times W \times T).$$

There exists a function F on the eight-fold product such that

$$(4) \qquad F(u_0, u_1, v_0, v_1, w_0, w_1, t_0, t_1) = \Pi \, f^*(P, w_k, t_l)^{(-1)^{k+l}},$$

and from the definitions, we see that $(F)(P) = E(P)$. Hence (F) and E differ by a divisor which is degenerate on $U^{(2)} \times V^{(2)}$. Inducing (F) and E on the variety $U^{(2)} \times V^{(2)} \times W^{(2)} \times \Delta_T$, we obtain 0. Hence the degenerate component is equal to 0, and we have $(F) = E$, as desired.

13

Note that the function F such that $E = (F)$ is defined at the point obtained by setting $t_0 = t_1$, and takes the value 1. This gives us a way of normalizing it, since two functions representing E differ by a multiplicative constant.

On the other hand, from the symmetry of F, we see that if we go from right to left in the correspondence given by D, then we obtain the following reciprocity formula, from (4).

THEOREM 2. *Let F be a function on the eight-fold product, defined over k, such that $E = (F)$, and normalized as above. Then*

$$F(P, Q) = \prod f_P(w_k, t_l)^{(-1)^{k+l}} = \prod g_Q(u_i, v_j)^{(-1)^{i+j}}.$$

As usual, such a formula at generic points has its counterpart for special points.

THEOREM 3. *Let (u_0', u_1', v_0', v_1') and (w_0', w_1', t_0', t_1') be two simple points of $U^{(2)} \times V^{(2)}$ and $W^{(2)} \times T^{(2)}$ respectively. Put*

$$\mathfrak{a} = \sum (-1)^{i+j} (u_i', v_j') \qquad\qquad \mathfrak{b} = \sum (-1)^{k+l} (w_k', t_l').$$

Assume that none of the points (u_i', v_j', w_k', t_l') lies in the support of D. Then $D(\mathfrak{a}, \mathfrak{b}) = {}^t D(\mathfrak{b}, \mathfrak{a})$.

Proof. It is immediate from our hypothesis that (P', Q') does not lie in $(F) = E$, and hence $F(P', Q')$ is defined. Our equality is now obvious.

4. Application to abelian varieties. Let A, B be two abelian varieties defined over k, and let D be a divisor on $A \times B$, rational over k. Let $\mathfrak{a}$ be a cycle on A of degree 0, such that $D(\mathfrak{a})$ is defined. If $\mathfrak{a}$ is in the kernel of Albanese, i. e., if $S(\mathfrak{a}) = 0$, then $D(\mathfrak{a})$ is linearly equivalent to 0 on B, say $D(\mathfrak{a}) = (f)$. If now $\mathfrak{b}$ is a cycle of degree 0 on B, also in the kernel of Albanese on B, such that no point of $\mathfrak{a} \times \mathfrak{b}$ is contained in D, then we may form symmetrically either $D(\mathfrak{a}, \mathfrak{b})$ or ${}^t D(\mathfrak{b}, \mathfrak{a})$.

THEOREM 4. *Let A, B be two abelian varieties. Let D be a divisor on $A \times B$. Let $\mathfrak{a}$, $\mathfrak{b}$ be cycles on A, B respectively, of degree 0, such that $S(\mathfrak{a}) = 0$ and $S(\mathfrak{b}) = 0$. Assume no point of $\mathfrak{a} \times \mathfrak{b}$ is contained in D. Then $D(\mathfrak{a}, \mathfrak{b})$ and ${}^t D(\mathfrak{b}, \mathfrak{a})$ are defined, and they are equal.*

Proof. Suppose first that D is the divisor of a function ϕ on $A \times B$. Our hypotheses imply that ϕ is defined at every point of $\mathfrak{a} \times \mathfrak{b}$, and our theorem simply asserts $\phi(\mathfrak{a}, \mathfrak{b}) = \phi(\mathfrak{a}, \mathfrak{b})$. Taking into account an obvious linearity, this remark allows us to change D by suitable linear equivalence.

We shall reduce our theorem to the case where $\mathfrak{a}$ and $\mathfrak{b}$ consist of four points each. This will then allow us to use Theorem 3.

Suppose that $\mathfrak{a}$ is written

$$\mathfrak{a} = (a_1) - (a_1') + (a_2) - (a_2') + \cdots + (a_n) - (a_n').$$

Let x be any point of A. We can write

$$\mathfrak{a} = (a_1) - (a_1') + (x) - (x + a_1 - a_1')$$
$$+ (a_2) - (a_2') + (x + a_1 - a_1') - (x + a_2 - a_2' + a_1 - a_1') + \cdots,$$

so that we correct each successive step to cancel the preceding term. We must come back at the end with $-(x)$ using the fact that $S(\mathfrak{a}) = 0$. Thus our cycle can be written as a sum of cycles of degree 0, in the kernel of Albanese, and consisting of four points.

From the theory of linear equivalence, one knows that it is possible to find a function ϕ on $A \times B$ such that none of the points of a given finite set of points on $A \times B$ lies in the support of $D + (\phi)$. (We reproduce a proof of this fact in an appendix.) We apply this to the product of the set of points entering in the expression of $\mathfrak{a}$ obtained above, with a similar set for $\mathfrak{b}$. In view of our previous remark, and of an obvious linearity when all necessary expressions are defined, we see that it indeed suffices to prove our theorem when $\mathfrak{a}$ and $\mathfrak{b}$ consist of four points.

Let us therefore write

$$\mathfrak{a} = (a_0) - (a_1) + (a_2) - (a_3),$$
$$\mathfrak{b} = (b_0) - (b_1) + (b_2) - (b_3).$$

Let $\lambda_A : A \times A \to A$ be the modified law of composition, such that $\lambda_A(u, v) = u - v$, and let $\lambda = (\lambda_A, \lambda_B)$. Let $D^* = \lambda^{-1}(D)$. It is a divisor on the product $A \times A \times B \times B$. Let u be a generic point of A, and let

$$u_0' = u + a_0, \quad u_1' = u + a_3, \quad v_0' = u, \quad v_1' = u + a_0 - a_1.$$

Do a similar construction for the points of $\mathfrak{b}$, to obtain points w_k' and t_l'. By hypothesis, no point of $\mathfrak{a} \times \mathfrak{b}$ lies in D, and hence none of the points (u_i', v_j', w_k', t_l') lies in D^*. We have

$$\sum (-1)^{i+j} D^*(u_i', v_j') = \lambda_B^{-1}\left[\sum (-1)^\nu D(a_\nu)\right].$$

If we put $\mathfrak{a}^* = \sum (-1)^{i+j}(u_i', v_j')$, we can also write

$$D^*(\mathfrak{a}^*) = \lambda_B^{-1} D(\mathfrak{a}).$$

This comes from commutativity in the diagram

$$(x, y) \times B \times B \xrightarrow{\;\lambda\;} (x - y) \times B$$

$$\text{inj} \Big\downarrow \qquad\qquad\qquad \Big\downarrow \text{inj}$$

$$A \times A \times B \times B \xrightarrow{\;\lambda\;} A \times B$$

where inj is the injection, together with the formalism of the inverse mappings of cycles applied to the divisor D (cf. the appendix of [2]).

Similarly, we have a cycle $\mathfrak{b}^*$ such that

$$^tD^*(\mathfrak{b}^*) = \lambda_A^{-1}\, {}^tD(\mathfrak{b}).$$

If $(f) = D(\mathfrak{a})$, we have a function $f^* = f\lambda_B$, and if $(g) = {}^tD(\mathfrak{b})$ we have a function $g^* = g\lambda_A$. Then

$$(f^*) = \lambda_B^{-1}(f) = \lambda_B^{-1} D(\mathfrak{a}) \quad \text{and} \quad (g^*) = \lambda_A^{-1}(g) = \lambda_A^{-1}\, {}^tD(\mathfrak{b}).$$

By Theorem 3, $f^*(\mathfrak{b}^*) = g^*(\mathfrak{a}^*)$ and hence clearly, $f(\mathfrak{b}) = g(\mathfrak{a})$. This proves our theorem.

By pull back, we now obtain the following result for varieties.

THEOREM 5. *Let V, W be two complete varieties, non-singular in co-dimension 1, and such that any finite set of points on them can be represented on an affine open subset. Let $\phi : V \to A$ and $\psi : W \to B$ be two canonical maps into their Albanese varieties. Let $\mathfrak{a}$, $\mathfrak{b}$ be two cycles on V, W respectively, of degree 0, and in the kernel of Albanese. Let D be a divisor on $V \times W$ such that no point of $\mathfrak{a} \times \mathfrak{b}$ lies in D. Then $D(\mathfrak{a}, \mathfrak{b})$ and $^tD(\mathfrak{b}, \mathfrak{a})$ are defined and they are equal.*

Proof. If the theorem is true for one divisor D in a correspondence class on $V \times W$, then it is true for every divisor in that class not containing any point of $\mathfrak{a} \times \mathfrak{b}$. This is first checked for linear equivalence, and then for degenerate divisors. A representative divisor in a correspondence class can always be obtained as an inverse image of a divisor on $A \times B$, and our result is now an immediate consequence of the formalism of inverse mappings of cycles, applicable in the present case to the commutative diagram

$$\mathfrak{a} \times W \xrightarrow{\;(\phi, \psi)\;} \phi(\mathfrak{a}) \times B$$

$$\text{inj} \Big\downarrow \qquad\qquad\qquad \Big\downarrow \text{inj}$$

$$V \times W \xrightarrow{\;(\phi, \psi)\;} A \times B.$$

COROLLARY. *Let C be a complete non-singular curve, and let f, g be* *two functions on C whose divisors have no point in common. Then $f((g)) = g((f))$.*

Proof. Take the diagonal on $C \times C$. Our hypothesis means that no point of $(f) \times (g)$ lies in it, and we can apply the theorem.

5. The Tate pairing. Let A, B be two abelian varieites, defined over the field k. Let D be a divisor on $A \times B$, rational over k, and let K be a finite Galois extension of k, with group G. If M is a G-module, we have cohomology groups $H^r(G, M)$. Since G will be fixed throughout, we shall also write $H^r(M)$.

The groups A_K and B_K of rational points of A and B respectively, in K are such G-modules. We wish to define a pairing of $H^1(A_K)$ and B_k into $H^2(K^*)$, where K^* is the multiplicative group of K. This is done as follows.

Let (a_σ) be a 1-cocycle in A_K and let $b \in B_k$. Denote by $Z_0(V, K)$ the group of zero cycles on a variety V, of degree 0, rational over K. Let $\mathfrak{a}_\sigma$ be in $Z_0(A, K)$, and such that $S(\mathfrak{a}_\sigma) = a_\sigma$. We can form the coboundary

$$\mathfrak{a} = (\delta \mathfrak{a})_{\sigma, \tau} = \sigma \mathfrak{a}_\tau - \mathfrak{a}_{\sigma\tau} + \mathfrak{a}_\sigma,$$

and we have $S(\mathfrak{a}) = 0$. Select $\mathfrak{b}$ in $Z_0(B, k)$ such that $S(\mathfrak{b}) = b$. We put

$$\alpha_{\sigma, \tau} = D(\mathfrak{a}, \mathfrak{b})$$

after changing D by the divisor of a function over k if necessary, to make this expression defined.

It is clear that $(\alpha_{\sigma, \tau})$ is a 2-cocycle of G in K^*. We contend that its cohomology class depends only on the linear equivalence class of D, and, in fact, on its correspondence class, and that the pairing

$$[(a_\sigma), b] \to (\alpha_{\sigma, \tau})$$

induces a well defined bilinear map of $H^1(A_K)$ and B_k into $H^2(K^*)$. In addition, if an element of B_k is a trace of an element of B_k, it is orthogonal to $H^1(A_K)$ so that we may replace B_k by $H^0(B_K)$. This will be proved in the following sequence of assertions.

a.) If D and D' differ by a trivial divisor, of type $X \times B + A \times Y + (\phi)$, and if $\alpha_{\sigma, \tau}$ and $\alpha_{\sigma, \tau}'$ are obtained, respectively, from D and D' as above, then they are cohomologous.

Proof. To begin with, we must insure that if D and D' differ by the divisor of a function (ϕ), $\phi \in k(A \times B)$, then our cocycles are cohomologous.

Indeed, $\alpha_{\sigma,\tau}$ and $\alpha_{\sigma,\tau}'$ differ by the coboundary of $\phi(a_\sigma, \mathfrak{b})$. The rest is now trivially verified.

b.) If a_σ is changed by a cycle of type $a_\sigma' + (\sigma - 1)\mathfrak{c}$ with $a_\sigma' \in Z_0(A, K)$ and $S(a_\sigma') = 0$, then $\alpha_{\sigma,\tau}$ changes by a coboundary.

Proof. In fact, that coboundary is that of $D(a_\sigma', \mathfrak{b})$.

c.) If we change $\mathfrak{b}$ by a cycle $\mathfrak{c} \in Z_0(B, k)$ such that $S(\mathfrak{c}) = 0$, then $\alpha_{\sigma,\tau}$ changes by a coboundary.

Proof. Using the reciprocity theorem, one sees that the coboundary is that of ${}^tD(\mathfrak{c}, a_\sigma)$.

The above statements show that our pairing is well defined. That it is bilinear now follows from the bilinearity of $D(a, \mathfrak{b})$.

d.) If b is a trace, $b = \sum \rho b'$, then $\alpha_{\sigma,\tau}$ is a coboundary.

Proof. We can choose $\mathfrak{b} = \sum (\rho' b') - n(0)$. Then the coboundary is that of

$$c_\sigma = \prod_\rho f_{\sigma,\rho}(\sigma\rho b') / f_{\sigma,\rho}(0).$$

6. Kummer theory. Let A, B be two abelian varieties and D a divisor on the product. Let a, $\mathfrak{b}$ be two cycles on A and B respectively, of degree 0. We write $a \sim 0$ if $S(a) = 0$. Assume that there is an integer n such that $na \sim 0$ and $n\mathfrak{b} \sim 0$. Let D_1 be linearly equivalent to D, and such that no point of $a \times \mathfrak{b}$ lies in D_1. Then one verifies immediately that ${}^tD_1(n\mathfrak{b}, a)/D_1(na, \mathfrak{b})$ is an n-th root of unity which is independent of the auxiliary divisor D_1 selected in the linear equivalence class of D. We shall denote it by $\varepsilon_{n,D}(a, \mathfrak{b})$. Furthermore, if $a_1 \sim a$ and $\mathfrak{b}_1 \sim \mathfrak{b}$, then $\varepsilon_{n,D}(a, \mathfrak{b}) = \varepsilon_{n,D}(a_1, \mathfrak{b}_1)$. Thus our root of unity defines a bilinear pairing of the points of order n on A and B respectively. If $a = S(a)$ and $b = S(\mathfrak{b})$, so that $na = 0$ and $nb = 0$, we can also write $\varepsilon_{n,D}(a, b)$ instead of $\varepsilon_{n,D}(a, \mathfrak{b})$. Note finally that it depends only on the correspondence class of D.

On the other hand, we can obviously generalize the root of unity defined by Weil which is usually denoted by $e_n(a, b)$ in the special case of divisors on an abelian variety, i. e., when $B = \hat{A}$ and D is a Poincaré divisor. Namely, it is clear that $(n\delta)^{-1}\, {}^tD(\mathfrak{b})$ is linearly equivalent to 0 on A if $n\mathfrak{b} \sim 0$. Say $(\omega) = (n\delta)^{-1}\, {}^tD(\mathfrak{b})$. Let u be a generic point of A. Then $\omega(u + a)/\omega(u)$ is by definition, the n-th root of unity $e_{n,D}(a, b)$. (See [2], Ch. 7.)

THEOREM 6. *Let D be a divisor on a product of abelian varieties $A \times B$.*

Let $\mathfrak{a}$, $\mathfrak{b}$ be two cycles on A and B respectively, of degree 0, such that $n\mathfrak{a} \sim 0$ and $n\mathfrak{b} \sim 0$. Let $a = S(\mathfrak{a})$ and $b = S(\mathfrak{b})$. Then $\varepsilon_{n,D}(a, b) = e_{n,D}(a, b)$.

Proof. Consider the rational map $n\delta: A \to A$ followed by the divisorial correspondence D on $A \times B$. Just as in the appendix of [2], we form the composed divisor on $A \times B$, which we denote by $E = D \circ (n\delta)$. Then it is obvious that E and nD are in the same correspondence class on $A \times B$, and thus that

$$E = nD + (F) + X \times B + A \times Y$$

for some function F on $A \times B$, and divisors X on A and Y on B. Let k be a field of rationality for D, E, F, X and Y, and let u, w be independent generic points of A, B over k. We put

$$\mathfrak{a} = (u + a) - (u) \quad \text{and} \quad \mathfrak{b} = (w + b) - (w).$$

Then

$$e_{n,D}(\mathfrak{a}, \mathfrak{b}) = {}^t E(\mathfrak{b}, \mathfrak{a}) = {}^t D(n\mathfrak{b}, \mathfrak{a}) F(\mathfrak{a}, \mathfrak{b}).$$

On the other hand, $E(\mathfrak{a}) = D(nu + na) - D(nu) = 0$, and hence $E(\mathfrak{a}, \mathfrak{b}) = 1$. Thus $D(n\mathfrak{a}, \mathfrak{b})^{-1} = F(\mathfrak{a}, \mathfrak{b})$. This proves our theorem.

One should observe that, just as we proved Theorem 5 for arbitrary varieties, we can generalize our root of unity to arbitrary varieties by pull back. The details are obvious and are left to the reader.

Appendix.

From the theory of linear equivalence, one knows that, given two cycles on a projective non-singular variety, it is possible to move one of them by linear equivalence so that the intersection is defined. Samuel has pointed out to me that this can be done rationally over a given field of rationality for all objects involved (at least over an infinite field). We need the result here only for divisors, and I shall reproduce below a proof due to Chevalley. We first deal with the local problem.

THEOREM. *Let V be an affine variety defined over a field k. Let D be a divisor on V, rational over k. Let S be a finite set of simple points of V. Then there exists a function ϕ on V, defined over k, such that no point of S lies in the support of $D + (\phi)$.*

Proof. We may obviously assume that D is positive and, in fact, a prime rational cycle. Furthermore, we may assume that the points of S are algebraic over k, because D contains the specializations over k of all of its points.

Let $R = k[v]$ be a coordinate ring for V over k and let I be the ideal of R consisting of those functions $f \in R$ such that $(f) \geq D$. This ideal has a finite basis, $I = (f_1, \cdots, f_n)$. Let $\mathfrak{p}$ be a maximal ideal of R, and $\mathfrak{o} = R_{\mathfrak{p}}$ its local ring with maximal ideal $\mathfrak{m}$. We assume that $\mathfrak{p}$ belongs to a simple point, so that $\mathfrak{o}$ is a unique factorization domain. Then there is a function t in $\mathfrak{o}$ which represents D locally at $\mathfrak{p}$. We can write $t = \sum a_i f_i$ with $a_i \in \mathfrak{o}$. I contend that if $b_i \in \mathfrak{o}$ is such that $b_i \equiv a_i \pmod{\mathfrak{m}}$ then $\sum b_i f_i$ differs from t by a unit in $\mathfrak{o}$. Indeed,

$$\sum b_i f_i = \sum (b_i - a_i) f_i + t.$$

Since $(f_i) \geq D$, we can write $f_i = t g_i$ with $g_i \in \mathfrak{o}$. Hence $\sum b_i f_i \in t[1 + \mathfrak{m}]$, thereby proving our contention.

Now let $\mathfrak{p}_j$ be a finite number of maximal ideals of R, belonging to simple points. Let $\mathfrak{o}_j = R_{\mathfrak{p}_j}$ be their local rings with maximal ideals $\mathfrak{m}_j$. Given elements x_j in $\mathfrak{o}_j$. it is possible to find $x \in R$ such that $x \equiv x_j \pmod{\mathfrak{m}_j}$, by the well known Chinese remainder theorem, applicable since $\mathfrak{p}_j + \mathfrak{p}_{j'} = R$ for $j \neq j'$. We now see that it suffices to approximate at each $\mathfrak{o}_j$ by an element of R, the coefficients of an element t, representing D at $\mathfrak{o}_j$ in terms of the f_i. This proves our theorem.

If we wish to apply the local result to an abstract variety V, then we must assume that the given finite set of points (algebraic over k) can be represented on an affine k-open subset of V. This is the case on a projective variety, although, if the ground field is finite, we may have to change the projective embedding (for instance, by finding first a hypersurface section of V which does not pass through any of the given points, and then dehomogenizing V at this hypersurface, in the projective embedding in which this hypersurface becomes a hyperplane).

Paris.

<hr>

REFERENCES.

[1] J. Igusa, " Fibre systems of Jacobian varieties, II," *American Journal of Mathematics*, vol. 78 (1956), pp. 745-760.

[2] S. Lang, *Abelian varieties*, Interscience Publishers, New York, 1958.

[3] J. Tate, " Galois cohomology of Abelian varieties over p-adic fields," to appear.

[4] A. Weil, " Sur les fonctions algébriques à corps de constantes fini," *Comptes Rendus. de l'Académie des Sciences*, vol. 210 (1940), pp. 592-594.

Am. J. Math., LXXX No. 3, 1958

PRINCIPAL HOMOGENEOUS SPACES OVER ABELIAN VARIETIES.*

By Serge Lang and John Tate.[1]

Let A be a commutative group variety defined over a field k. If K/k is a Galois extension, the group A_K of points of A rational over K is a module for the Galois group $G(K/k)$, and we denote the associated cohomology groups simply by $H^r(K/k, A)$ or by $H^r(k, A)$ in case K is the separable closure of k. In case K/k is infinite, we mean of course the cohomology groups constructed with cochains of finite type, i.e. coming by inflation from finite extensions.

In § 1, we have carried over to the infinite case the basic propositions of Galois cohomolgy which are familiar in the finite case [2]. This generalization is essentially trivial, but it furnishes a good review for the non-expert, it fixes our notation, and mainly, it is urgently called for when one studies algebraic groups for the following reason: If $A \to B \to C$ is an exact sequence of (separable) homomorphisms defined over k, then $A_K \to B_K \to C_K$ is not necessarily exact but *is* exact if K is the (separable) algebraic closure of k. Actually, in the remainder of the paper, we require the cohomological results only for dimension 1, and most of our applications are based on the special "Kummer sequence" discussed at the end of § 1.

In § 2 we discuss systematically the representation of principal homogeneous spaces for A over k by elements of the 1-dimensional cohomology group (a set if A is non-commutative) $H^1(k, A)$. One establishes an injection of the classes of k-isomorphic spaces into $H^1(k, A)$, and as Serre has remarked, using Weil's theorems concerning the field of definition of a variety, one sees immediately that one actually gets a bijection. Since this representation has been carried out in special cases by F. Chatelet, we call $H^1(k, A)$ the Chatelet group.

Although the rest of the paper is essentially independent of the principal homogeneous space interpretation, it is mainly this interpretation and the consequent relation of the cohomology to diophantine problems which motivates our study of $H^1(k, A)$. For example, as Chatelet has shown, an

---

* Received January 7, 1958.

[1] Fulbright and Sloan Fellows respectively.

659

elliptic curve Γ is a homogeneous space over its Jacobian A, and the question whether Γ has a rational point in an extension field K is the same as whether its cohomology class $\alpha(\Gamma) \in H^1(k, A)$ is split by K.

The essential content of this paper is contained in the theorems of the last two sections, which concern the structure of $H^1(k, A)$ when A is an abelian variety.

We first consider the case where k is a local field. After recalling in §3 some basic facts concerning reduction mod $\mathfrak{p}$, we treat in §4 the case when A has a non-degenerate reduction. We show (Theorem 1) how the study of elements of order prime to p in $H^1(k, A)$ can be reduced to a study of the reduced variety. There results (Theorem 2) a complete description of the part of $H^1(k, A)$ prime to p when k is a $\mathfrak{p}$-adic number field. In particular, it is a finite group.

In §5, we deal with global fields, essentially number fields, function fields over algebraically closed fields, or fields of finite type, and obtain various qualitative results. Theorem 3 gives a new variant of the proof that A_k/mA_k is finite. (For another variant, cf. Roquette [10].) From it, we deduce the finiteness of $H^r(K/k, A)$ for finite K/k and all $r > 0$, in Theorem 4. We then consider the subgroup of $H^1(k, A)$ consisting of those elements which split at all primes. Although it is known that this group is not necessarily trivial, we can show that for each integer m (prime to the characteristic of k), its subgroup of elements of period m is finite.

Theorems 6 and 7, which are independent of the preceding five, show on the other hand that $H^1(k, A)$ is large, in different senses. From Theorem 7, it is a corollary that given any positive integer m one can construct a function field in one variable of genus 1, over a suitable algebraic number field k, the degrees of whose divisors rational over k are exactly the multiples of m. One still does not have such examples when $k = \mathbf{Q}$ is the field of rational numbers.

1. Galois cohomology. Let k be a field and Ω an algebraically closed field containing k. In Galois cohomology, one deals with a functor A which attaches to each field K between k and Ω a group $A(K)$ and to each k-isomorphism $\phi: K \to K^\phi$, an isomorphism $A(\phi): A(K) \to A(K^\phi)$. The functor is subjected to three axioms:

(A1) For each K we have $A(K) = \bigcup A(F)$, the union being taken over all finitely generated subextensions F/k of K/k.

Notice that (A1) implies that all of our groups $A(K)$ are subgroups of one big group, namely $A(\Omega)$, and that we have $A(K) \subset A(L)$ whenever $K \subset L$.

(A2) If ψ prolongs ϕ, then $A(\psi)$ prolongs $A(\phi)$. If ψ and ϕ can be composed, then $A(\psi \circ \phi) = A(\psi) \circ A(\phi)$. $A(\text{identity}) = \text{identity}$.

From (A2) it follows that any group of k-automorphisms of a field L operates on $A(L)$.

(A3) If L/K is a Galois extension, then $A(K)$ is the set of fixed points for the operation of the Galois group on $A(L)$.

In most of our applications, A will be (the functor derived from) an algebraic group defined over k, for which $A(K) = A_K$ is the group of points of A which are rational over K, i. e. whose coordinates lie in K, and $A(\phi)$ is the map obtained by applying ϕ to the coordinates.

From now on in this section, we assume that our functor A is commutative, that is, each $A(K)$ is a commutative group. Let K/k be a (possibly infinite) Galois extension. Then $A(K)$ is a module for the Galois group $G(K/k)$, and we can consider the standard r-cochains $a = a(\sigma_1, \cdots, \sigma_r)$ of $G(K/k)$ with values in $A(K)$. We shall say that such a cochain is of *finite type* if there exists a finite sub-extension F such that $a(\sigma_1, \cdots, \sigma_r)$ depends only on the effects of the automorphisms σ_i on F. The cochains of finite type, which are the only ones arising from the usual algebraic processes, form a subcomplex of the standard cochain complex. We shall denote this subcomplex simply by $C(K/k, A)$ and its cohomology groups by $H^r(K/k, A)$. These latter are what we mean by Galois cohomology groups. Of course, if K/k is a finite extension then every cochain is of finite type, and we have achieved nothing but a simplification of notation: $H^r(K/k, A) = H^r(G(K/k), A(K))$. Also, even for infinite K, we have $H^0(K/k, A) = H^0(G(K/k), A(K))$ and this group is just $A(k)$ because of (A3). However, for $r > 0$ and K/k infinite, our groups may differ from the usual ones. For example, if $A(k) = A(K)$, that is, if $G(K/k)$ operates trivially on $A(K)$, then $H^1(K/k, A)$ is the group of continuous homomorphisms of $G(K/k)$ (Krull topology) into $A(K)$ (discrete topology), whereas $H^1(G(K/k), A(K))$ is the group of all homomorphisms.

Let K'/k' be another Galois extension, such that $K' \supset K$ and $k' \supset k$. Then each cochain $a \in C(K/k, A)$ determines a cochain $a' \in C(K'/k', A)$ by the rule $a'(\cdots, \sigma', \cdots) = a(\cdots, \sigma'_K, \cdots)$, where σ'_K denotes the effect on K of $\sigma' \in G(K'/k')$. The cochain map $a \to a'$ induces a homomorphism

$H^r(K/k, A) \to H^r(K'/k', A)$ which we simply call the *canonical* homomorphism. However, in the extreme case $k = k'$, it goes by the name of *inflation* (inf), and in the other extreme case $K = K'$, it is known as *restriction* (res).

PROPOSITION 1. *Let K/k be Galois and let k' be an arbitrary subfield. Then*

$$H^r(K/k', A) = \lim_{F} \{H^r(F/F \cap k', A)\},$$

the limit being taken inductively with respect to the canonical homomorphisms, as F runs over the finite subextensions of K.

Since passage to (co)homology commutes with inductive limits (cf. [2], Ch. V, Prop. 9.3*), our proposition will follow if we can show that the corresponding formula holds for the cochain groups. Each of the canonical maps $C^r(F/F \cap k', A) \to C^r(K/k', A)$ is an injection, because the inclusion $A(F \cap k') \to A(k')$ is injective and the natural map $G(K/k') \to G(F/F \cap k')$ is surjective. Thus we have only to show that each element $a \in C^r(K/k', A)$ is in the image of $C^r(F/F \cap k', A)$ for some F depending on a. Since a is of finite type, it has only a finite set of distinct values and by Axiom (A1), it follows that we can find a finitely generated, hence finite, sub-extension F/k of K/k such that all valuesof a lie in $A(F)$. Enlarging F, if necessary, so that $a(\sigma_1, \cdots, \sigma_r)$ depends only on the effects of the σ_i on Fk', we can well-define a cochain $b \in C^r(F/F \cap k', A)$ whose image is a by putting $b(\cdots, \sigma_F, \cdots) = a(\cdots, \sigma, \cdots)$, where σ_F denotes the effect of σ on F.

Taking the special case $k' = k$, we find as a corollary

$$H^r(K/k, A) = \lim_{F} \{H^r(F/k, A)\} = \lim_{F} \{H^r(G(F/k), A(F))\}.$$

Thus our cohomology groups are just the inductive limits, under inflation, of the ordinary cohomology groups of the finite subextensions.

Let L/K be a Galois extension. A k-isomorphism $\phi: L \to L^\phi$ induces an isomorphism $C(L/K, A) \to C(L^\phi/K^\phi, A)$ by the rule $a^\phi(\cdots, \sigma, \cdots) = \phi a(\cdots, \phi^{-1}\sigma\phi, \cdots)$, and it is a fact that the induced cohomology map $H^r(L/K, A) \to H^r(L^\phi/K^\phi, A)$, called *conjugation*, depends only on the effect of ϕ on K. In particular, if L and K are Galois extensions of k, then the Galois group $G(K/k)$ operates on $H^r(L/K, A)$.

Although there is no doubt that the whole spectral sequence of Hochschild-Serre carries over to the case of infinite Galois extensions we are content here to discuss only a small corner of it, namely:

PROPOSITION 2. *Let $K \supset k'$ be Galois extensions of k. Then there exists* *a canonical transgression homomorphism* (tg) *such that the following sequence is exact:*

$$0 \to H^1(k'/k, A) \xrightarrow{\text{inf}} H^1(K/k, A) \xrightarrow{\text{res}} H^1(K/k', A)^{G(k'/k)}$$

$$\xrightarrow{\text{tg}} H^2(k'/k, A) \xrightarrow{\text{inf}} H^2(K/k, A).$$

Indeed, this proposition is known in the case of ordinary cohomology groups (Cf. Hochschild, G. and Serre, J-P., "Cohomology of Group Extensions," *Trans. A. M. S.*, Vol. 74, 1953). Consequently, for each finite Galois subfield F of K, we have, exactly, with obvious abbreviations of notation:

$$0 \to H^1(F \cap k'/k) \to H^1(F/k) \to H^1(F/F \cap k')^{G(F \cap k'/k)}$$

$$\xrightarrow{\text{tg}} H^2(F \cap k'/k) \to H^2(F/k).$$

The commutativities required for passing to the inductive limit over F are satisfied and, by Proposition 1, the limit sequence is the one we are looking for. The superscript G on the middle term carries over to the limit because each $(F \cap k'/k)$ is finite.

It is easy to see that our limiting transgression map cap be characterized in the same way as the ordinary one, namely, we have $\alpha = \text{tg}\,\beta$ if and only if there is a cochain $b \in C^1(K/k, A)$ whose restriction to $C^1(K/k', A)$ is a cocycle representing β and whose coboundary δb is the inflation to $C^2(K/k, A)$ of a cocycle $a \in C^2(k'/k, A)$ representing α.

If K/k is Galois and E a finite (not necessarily Galois) subextension, there exists a *transfer* map

$$\text{tr}: H^r(K/E, A) \to H^r(K/k, A)$$

going in the opposite direction from restriction, whose definition we recall here, although we have little use for it in the sequel. Let Φ be the set of $[E:k]$ distinct k-isomorphisms of E. For each $\phi \in \Phi$, let $\phi_* \in G(K/k)$ be a chosen prolongation of ϕ to K. Then for $\sigma \in G(K/k)$, both $\sigma\phi_*$ and $(\sigma\phi)_*$ have the same effect on E, and, consequently, the automorphism $(\sigma\phi)_*^{-1}\sigma\phi_*$, which we shall denote by (σ, ϕ) leaves E elementwise fixed, i.e. belongs to $G(K/E)$. The cohomology transfer is that induced by the cochain transfer $a \to \text{tr}(a)$ defined by

$$(\text{tr}(a))(\sigma_1, \cdots, \sigma_r)$$

$$= \sum_{\phi \in \Phi} (\sigma_1\sigma_2 \cdots \sigma_r\phi)_* a((\sigma_1, \sigma_2 \cdots \sigma_r\phi), (\sigma_2, \sigma_3 \cdots \sigma_r\phi), \cdots, (\sigma_r, \phi)).$$

Although this cochain transfer depends on the choice of prolongations, the induced cohomology map does not. The main relation between transfer and restriction is the rule $\mathrm{tr} \circ \mathrm{res} = [E:k]$. This and all other such relations can be guessed from the case of dimension 0, where the transfer $A(E) \to A(k)$ is just the *trace*: $a \to \sum \phi a$, and the restriction $A(k) \to A(E)$ is just the inclusion map.

The relation $\mathrm{tr} \circ \mathrm{res} = [E:k]$ shows that if an element $\alpha \in H^r(K/k)$ restricts to 0 in $H^r(K/E)$, then $[E:k]\alpha = 0$. Applying this remark in case $E = K$ is a finite Galois extension of k, we see that $H^r(E/k)$ is a torsion group of exponent $[E:k]$ for $r > 0$, because $H^r(E/E, A) = 0$ for positive r. From Proposition 1, it now follows that the Galois cohomology groups in positive dimension are torsion groups for arbitrary K/k.

We conclude this section by considering the cohomology maps induced by a *homomorphism* $f: A \to B$ of one of our functors into another. By this we mean a collection of homomorphisms $f(K): A(K) \to B(K)$ satisfying

(f1)   If $L \supset K$, then $f(K)$ is the restriction of $f(L)$ to $A(K)$.

(f2)   If $\phi: K \to K^\phi$ is a k-isomorphism, then

$$f(K^\phi) \circ A(\phi) = B(\phi) \circ f(K).$$

For example, suppose A and B are algebraic groups defined over k. Then a k-homomorphism $f: A \to B$ (that is, an everywhere defined rational map over k which is a group homomorphism) yields a homomorphism of the functor A into the functor B. We simply let $f(K)$ denote the restriction of f to $A(K)$. Condition (f2) is satisfied because f is defined over k.

A functor homomorphism $f: A \to B$ gives rise to cohomology homomorphisms

$$H^r(K/k, A) \to H^r(K/k, B),$$

namely, those induced by the cochain map $a \to f \circ a$, and these induced cohomology maps obviously commute with canonical maps, conjugations, and transfers. A sequence of functor homomorphisms $A' \to A \to A''$ is said to be *K-exact* if the sequence $A'(K) \to A(K) \to A''(K)$ is exact in the ordinary sense.

PROPOSITION 3.   *Let K/k be Galois. A K-exact sequence*

$$0 \to A' \to A \to A'' \to 0$$

gives rise to an infinite exact cohomology sequence

$$\cdots \to H^r(K/k, A) \to H^r(K/k, A'') \xrightarrow{\;\delta\;} H^{r+1}(K/k, A') \to H^{r+1}(K/k, A) \to \cdots$$

This proposition follows as usual from the lemma: *If $A' \to A \to A''$ is K-exact, then the cochain sequence*

$$C(K/k, A') \to C(K/k, A) \to C(K/k, A'')$$

is exact. If a cochain $a \in C(K/k, A)$ goes to 0 in $C(K/k, A'')$, then each of its values goes to 0 in $A''(K)$. Hence, by K-exactness, we can pick a pre-image in $A'(K)$ for each of these values and thereby define a cochain a' whose image is a. Moreover, a' will be of finite type, provided we select one single pre-image for each of the distinct values of a, because of Axiom (A1).

In many applications, we shall deal with the case where K is the separable closure k_s of k, and where the sequence of Proposition 3 is k_s-exact. The resulting cohomology sequence involves the cohomology groups $H^r(k_s/k, A)$ which we shall simply write $H^r(k, A)$ since they occur so frequently.

Such a sequence arises from commutative algebraic groups. As a matter of notation, if X is an abelian group and m a natural number, we shall denote by X_m the kernel of the map $X \xrightarrow{\;m\;} X$. Let A be a commutative group variety defined over k. If m is prime to the characteristic, the sequence

$$0 \to A_m \to A \xrightarrow{\;m\;} A \to 0$$

is k_s-exact. Writing A_K instead of $A(K)$, our cohomology sequence becomes

$$0 \to A_m \cap A_k \to A_k \xrightarrow{\;m\;} A_k \to H^1(k, A_m) \to H^1(k, A) \xrightarrow{\;m\;} H^1(k, A) \to \cdots.$$

A portion of this exact sequence may be written more simply as follows.

$$0 \to A_k/mA_k \to H^1(k, A_m) \to H^1(k, A)_m \to 0.$$

The group A_k/mA_k is well known to be of interest in arithmetical questions, and we shall investigate it from this point of view in §5. For the moment, we assume further that $A_m \subset A_k$, and let G_k be the Galois group of k_s over k, i.e. $G_k = G(k_s/k)$. Then G_k operates trivially on A_m, and our sequence becomes

$$0 \to A_k/mA_k \to \operatorname{Hom}(G_k, A_m) \to H^1(k, A)_m \to 0,$$

where Hom means, of course, the continuous homomorphisms as always. We shall have many applications for this sequence, and we shall call it the *Kummer* sequence, because, in case A is the multiplicative group, we have

8

$H^1(k, A) = 0$ by Hilbert's Theorem 90, so that we find the familiar Kummer duality

$$k^*/k^{*m} \approx \mathrm{Hom}(G_k, A_m),$$

always provided that the group of m-th roots of unity A_m is contained in $A_k = k^*$.

2. Principal homogeneous spaces.

Let A be a functor of the type described in the first paragraph of § 1. If the groups $A(K)$ are not commutative, we cannot define cohomology groups $H^r(K/k, A)$ for Galois extensions K/k. The best we can do is define cohomology *sets* in dimension *one*, as follows. We consider 1-cochains a of finite type and we call 1-cocycles those satisfying the identity $a_\sigma a_\tau^\sigma = a_{\sigma\tau}$. Here we write $A(K)$ multiplicatively and write a_σ instead of $a(\sigma)$ for $\sigma \in G(K/k)$. We do not attempt to multiply 1-cocycles, but we explain when two 1-cocycles a and a' are cohomologous, namely, when there exists an element $b \in A(K)$ such that $a'_\sigma = b^{-1} a_\sigma b^\sigma$. The cohomology thus defined is an equivalence relation, and we denote the set of equivalence classes by $H^1(K/k, A)$. This set is not a group but it does have a distinguished element, namely the class of coboundaries of the form $a_\sigma = b^{-1} b^\sigma$, which we call the trivial class.

The reader will easily verify that the canonical maps, in particular, restriction and inflation, the conjugation maps, and the induced maps make sense for this non-commutative cohomology.

Proposition 1 holds in dimension 1.

Proposition 2 holds in a weaker form, namely, in the sequence

$$0 \to H^1(k'/k, A) \xrightarrow{\ \mathrm{inf}\ } H^1(K/k, A) \xrightarrow{\ \mathrm{res}\ } H^1(K/k', A)^{G(k'/k)}$$

inf is an injection whose image is the inverse image of the trivial class under restriction.

Proposition 3 collapses to the exactness of the sequence

$$0 \to A(k) \to B(k) \to C(k) \xrightarrow{\ \delta\ } H^1(K/k, A) \to H^1(K/k, B) \to H^1(K/k, C),$$

where we now deal with a K-exact sequence

$$0 \to A \xrightarrow{\ f\ } B \xrightarrow{\ g\ } C \to 0.$$

Here exactness means that the image of each map is the inverse image of the trivial element under the succeeding map, and there is the additional feature that δ injects the right coset space $C(k)/gB(k)$ into $H^1(K/k, A)$.

Suppose now that A is a group variety defined over k. We wish to explain the connection between the 1-dimensional cohomology sets and the right principal homogeneous spaces for A. To recall the definition of this latter object, a right phs for A over k is a variety V defined over k on which A operates simply and transitively, in such a way that the map $(v, w) \to v^{-1}w$ of $V \times V$ into A is an everywhere defined rational map over k. Here we use the symbol $v^{-1}w$ to denote the uniquely determined element of A which carries the point v into the point w, so that the formula $v(v^{-1}w) = w$ becomes an identity. Two phs's V and V' are said to be *k-isomorphic* if there exists a birational biholomorphic transformation $f: V \to V'$ such that $f(va) = f(v)a$ for $v \in V$ and $a \in A$. Among the k-isomorphism classes, there is a distinguished class, namely, the class of spaces which are k-isomorphic to A itself, viewed as phs under right multiplication. A space is in this class if and only if it has a rational point in k. In case A is commutative, Weil [13] has defined a geometric law of composition which makes the classes of phs's into a commutative group.

The following proposition has been proved by F. Chatelet [3], [4] for various special groups.

PROPOSITION 4. *Let K/k be a Galois extension. There is a canonical bijection between the first cohomology set $H^1(K/k, A)$ and the set of k-isomorphism classes of principal homogeneous spaces for A over k which have rational points in K. This bijection is a group isomorphism when A is commutative.*

Proof. Let V be a phs with a rational point v in K. Each such point gives rise to a 1-cocycle $a_\sigma = v^{-1}v^\sigma$. The different choices of v lead to cocycles filling out a cohomology class. Denote this class by $\alpha(V)$. If V and V' are k-isomorphic, then points v and v' which correspond under a k-isomorphism yield the same cocycle; hence $\alpha(V) = \alpha(V')$. Conversely, if V and V' are given such that $\alpha(V) = \alpha(V')$, then there do exist points v and v' yielding the same cocycle. When this is the case, the map $x \to v'(v^{-1}x)$ is a k-isomorphism of V onto V'. Indeed, it is a K-isomorphism, and a simple computation shows that it is invariant under all $\sigma \in G(K/k)$. We have thus obtained an injection of the classes of spaces into the cohomology set.

As Serre has observed, it follows immediately from Weil's theorems concerning the field of definition of a variety that every cocycle comes from a space. Indeed, consider the transformations $f_\sigma : A \to A$ defined by $f_\sigma(x) = a_\sigma x$. They satisfy the identity $f_\sigma \circ f_\tau{}^\sigma = f_{\sigma\tau}$. Since the cocycle a is of finite type, it follows from [14] that there exists a variety V defined over k and a birational biholomorphic transformation $F : A \to V$ defined over K such that $f_\sigma = F^{-1} \circ F^\sigma$. The operation $vb = F(F^{-1}(v)b)$ for $v \in V$ and $b \in A$ makes V into a phs for A over k, not only over K, because a simple computation shows that $(v^{-1}w)^\sigma = ((v^\sigma)^{-1}w^\sigma)$. Finally, one sees that the cocycle a with which we started is that arising from the point $v = F(1)$ on V.

If A is commutative, let U, V, W be phs's and suppose $f : U \times V \to W$ is an everywhere defined rational map over k which exhibits the fact that the class of W is the product of the classes of U and V (cf. [13], prop. 5). Select $u \in U$, $v \in V$ rational over K and put $w = f(u, v)$. Using the characteristic property of f, namely $f(ua, vb) = f(u, v)ab$, one finds that the cocycle derived from w is the product of those derived from u and v. This concludes the proof.

Since every variety defined over k has a point whose coordinates lie in the separable algebraic closure k_s of k, it follows from Proposition 4 that the set (group) of all principal homogeneous spaces for A over k is isomorphic to $H^1(k_s/k, A)$ which we have agreed to write $H^1(k, A)$, and which we shall call the *Chatelet set (group)* for A over k.

If k' is an extension field of k, it is obvious that the canonical cohomology map $H^1(k, A) \to H^1(k', A)$ reflects the homogeneous space operation of extending the ground field from k to k'. The cohomology classes in the kernel of this map are said to be *split* by k', and the same terminology is applied to the corresponding homogeneous spaces. Thus a phs for A over k is *split* by an extension k' if and only if it is k'-isomorphic to A, or what is the same, if and only if it has a rational point in k'. In case k'/k is a Galois extension, the exactness of the sequence

$$0 \to H^1(k'/k, A) \xrightarrow{\text{inf}} H^1(k, A) \xrightarrow{\text{res}} H^1(k', A)$$

is obvious from the point of view of homogeneous spaces, for the cohomology inflation map simply reflects the inclusion of the set of phs's split by k' in the set of all phs's for A over k.

Let X be a variety defined over k. We define the *index* of X over k (ind X) to be the greatest common divisor of the degrees of the 0-cycles on X which are prime rational over k. The degree of a prime rational 0-cycle is equal to the degree of the extension generated over k by the coordinates of one of its points. Thus, in the case of a principal homogeneous space V

for A over k, we see that ind V is the greatest common divisor of the degrees of the finite extensions of k which split V, and is consequently subject to a cohomological analysis.

From now on, we assume that A is commutative. Defining the *separable index* of V (ind$_s$ V) as the greatest common divisor of the degrees of the finite separable splitting extensions, and the *period* of V (per V) as the period of V in the group of classes of principal homogeneous spaces for A over k, we have

PROPOSITION 5. *Let V be a principal homogeneous space for A over k. Then* per V *divides* ind V *divides* ind$_s$ V, *and all three numbers have the same prime factors.*

Proof. It is trivial that the index divides the separable index. Let v_0 be a rational point of V in some Galois extension of k. Let $f: V \to A$ be the map $v \to v_0^{-1}v$, and let $a_\sigma = v_0^{-1}v^\sigma$ be the cocycle determined by v_0. Then for any point $v \in V$, we have $a_\sigma(f(v))^\sigma = f(v^\sigma)$, and consequently, by linearity, we have

$$a_\sigma^{\deg(\mathfrak{v})}(f(\mathfrak{v}))^\sigma = f(\mathfrak{v}^\sigma)$$

for any 0-cycle $\mathfrak{v}$ on V. If $\mathfrak{v}$ is rational over k and of degree d, then $\mathfrak{v}^\sigma = \mathfrak{v}$, and the d-th power of the cocycle a_σ is split by the coboundary of $f(\mathfrak{v})$. This proves that the period divides the index.

Finally, let p be a prime number not dividing the period. We must show that p does not divide the separable index, that is, we must construct a finite separable splitting extension E_p whose degree is not divisible by p. To this effect, we simply take any finite Galois splitting extension K/k and let E_p be the subextension cut out by a p-Sylow subgroup of $G(K/k)$. If $\alpha \in H^1(K/k, A)$ is the cohomology class of V, then the period of res $\alpha \in H^1(K/E, A)$ divides the p-power $[K:E]$ on the one hand, and divides per V on the other. Consequently, res $\alpha = 0$, i.e. E splits V.

The reader will have noticed the analogy between the Chatelet group of classes of principal homogeneous spaces for A over k and Brauer's group of classes of central simple algebras over k. We have in fact taken over Proposition 5 and its proof almost word for word from the theory of algebras. In the case of algebras, the period is not in general equal to the index as a counterexample of Albert [1] shows. We have found a similar counterexample in the case of principal homogeneous spaces (end of §4). Nevertheless, our counterexample, like Albert's, involves a comparatively complicated ground field k, and over number fields and their completions, all examples which we have satisfy the condition period = index.

The theory of algebras suggests other questions. Since every central division algebra contains a separable splitting field, it is true for algebras that (1) the index equals the separable index, and (2) the index is not only the greatest common divisor, but is actually the minimum of the degrees of the finite splitting fields. We have not investigated the corresponding statements for homogeneous spaces, except to notice that (2) is true in case A is an elliptic curve, for in that case, the Riemann-Roch theorem shows that every divisor class of positive degree contains positive diivsors.

The divisibilities of Proposition 5 imply

COROLLARY. *Suppose $x \in H^1(k, A)$ is of period m and is split by an extension E/k of degree m. Then the corresponding homogeneous space has index m.*

This corollary will enable us to construct, with abelian varieties A, over various ground fields k, examples of spaces whose index is a preassigned integer m (cf. Theorem 7, § 5). These examples are of particular interest in case A is an elliptic curve. Then V, which becomes birationally equivalent to A over an extended ground field, is also a curve of genus 1. The point is that the index of a curve of genus $g \neq 1$ is bounded by, and in fact divides, $2g - 2$, for the Riemann-Roch theorem shows the existence of divisors rational over k of degree $2g - 2$. On the other hand, our examples show that the index is completely arbitrary for $g = 1$, a fact conjectured by Artin, but unknown until now.

3. Non-degenerate reduction. Throughout this section, we assume that we have a field k and a place of k onto a field k' such that any two extensions of it to the algebraic closure $\bar{k}$ of k are conjugate over k'. In terms of valuations, this means that the extension of the valuation is unique, and this condition is certainly satisfied if k is complete, and the valuation discrete.

We denote by θ a definite extension of our place to $\bar{k}$. An automorphism σ of $G(\bar{k}/k)$ induces an automorphism σ' of the algebraic closure of k', characterized by the relation $\theta\sigma = \sigma'\theta$. The map $\sigma \to \sigma'$ is a homomorphism of $G(\bar{k}/k)$ onto $G(\bar{k}'/k')$.

We shall also assume that the valuation is discrete, although this can presumably be dispensed with. If Z is a cycle rational over an algebraic extension K of k (say in some projective space), then it has a reduction, or, as we shall also say, a specialization Z' in the reduced projective space determined by θ. and rational over K' (Shimura [12]). The only facts concerning the specialization $Z \to Z'$ which we shall use in the sequel is that it commutes with projections and intersections, i.e. under suitable hypotheses, $(\mathrm{pr}\,Z)' = \mathrm{pr}\,(Z')$, and for two positive cycles X, Y, we have $(X \cdot Y)' = X' \cdot Y'$. We refer to [12] for the proof.

Our place θ induces a mapping of points P of our projective space, algebraic over k, onto points P', algebraic over k'. We shall also write $P' = \theta P$. We have for any automorphism σ of $\bar{k}/k$, $(\sigma P)' = \sigma' P'$. If Z is a positive cycle, then $\mathrm{supp}\,(Z') = (\mathrm{supp}\,Z)'$.

We shall now list a few properties of reduction of cycles, and especially varieties in non-degenerate cases.

To begin with, we have a result which will serve as Hensel's lemma for points on varieties.

LEMMA. *Let $\mathfrak{a}$ be a positive 0-cycle rational over k. Let P' be a point of $\mathfrak{a}'$ which occurs in $\mathfrak{a}'$ with multiplicity 1. Then there exists a unique point P of $\mathfrak{a}$ specializing to P', and P is rational over k.*

Proof. Note that P' is rational over k' since $\mathfrak{a}'$ is rational over k'. Let P be in $\mathfrak{a}$, specializing to P'. Then obviously P occurs with multiplicity 1, so that P is separable over k. If σ is any automorphism of $\bar{k}$ over k, then $(\sigma P)' = (\sigma' P') = P'$, and hence σP also specializes to P'. Since $\mathfrak{a}$ is rational over k, it follows that σP also occurs in $\mathfrak{a}$, and since P' has multiplicity 1, we conclude that $\sigma P = P$, and hence that P is rational over k.

We now have the surjectivity of θ, whose proof is due to Chow.

PROPOSITION 6. *Let the cycle Z be a variety V, and assume that Z' is also a variety V', i.e. consists of one component with multiplicity 1. Let P' be a simple point of V', rational over k'. Then there exists a point P of V rational over k which specializes to P'.*

Proof. Our statement being local, we may assume that V' is affine. There exists a linear variety L' defined over k' such that $V' \cdot L'$ is defined and contains P' with multiplicity 1. Lift L' to a linear variety L over k, such that $V \cdot L$ is defined. Then $V \cdot L$ contains a point P specializing to P'. Since $V \cdot L$ is rational over k, we need merely apply the lemma and the compatibility of specializations with intersections to conclude the proof.

For the applications, we are principally interested in abelian varieties. Let A be an abelian variety defined over k. Its specialization A' is then a cycle. If this cycle has one component with multiplicity 1, which is an abelian variety whose law of composition is obtained by reduction of that of A, then we shall say that the specialization $A \to A'$ is *non-degenerate*. (The uniqueness of non-degenerate specializations has been proved in [5].) In that case, if a, b are two points of A algebraic over k, then $(a + b)' = a' + b'$,

the first $+$ referring to addition on A and the second on A'. In view of
Proposition 6, we have an exact sequence

$$0 \rightarrow S_k \rightarrow A_k \rightarrow A'_{k'} \rightarrow 0,$$

where S_k is simply the kernel of the specialization homomorphism θ. If K
is a finite Galois extension of k with group G then each one of the groups
S_K, A_K, $A'_{K'}$ is a G-module, in view of the relation $(\sigma x)' = \sigma' x'$. It is to
the exact sequence

$$0 \rightarrow S_K \rightarrow A_K \rightarrow A'_{K'} \rightarrow 0$$

that we shall apply the cohomology theory in the next section.

Suppose for a moment that k is a p-adic field $\mathbf{Q}_p$ and A is an elliptic
curve of the type considered by Lutz [8]. Then A_k contains a subgroup
isomorphic to the integers of $\mathbf{Q}_p$, which is none other than our kernel S_k
when A' is non-degenerate. More generally, if k is an arbitrary $\mathfrak{p}$-adic field,
and A an abelian variety of dimension r, then Mattuck [9] has shown that
A_k contains a subgroup isomorphic to r copies of the integers of k. We can
always choose such a subgroup M_k of Lutz-Mattuck such that $M_k \subset S_k$. We
shall prove below that S_k is uniquely divisible by an integer m prime to p.
From this one sees that $(S_k : M_k)$ is a power of p.

We return to an arbitrary field k subject to the conditions stated at the
beginning of this section. Let m be an integer not divisible by the characteristic p of k'. Let a be a point of A rational over k. The cycle

$$(m\delta)^{-1}(a) = \mathrm{pr}_1\{\Gamma \cdot (A \times (a))\},$$

where Γ is the graph of $m\delta$, consists of the points x on A such that $mx = a$,
each one occurring with multiplicity 1. Specializing, we see that

$$[(m\delta)^{-1}(a)]' = (m\delta')^{-1}(a'),$$

where δ' is the identity on A'. Taking $a = 0$, this shows in particular that
θ induces an isomorphism of the points of finite order prime to p on A onto
those points on A', i.e. of A_m onto A'_m. Actually, from Hensel's lemma,
we get

PROPOSITION 7. *Assume that the specialization $A \rightarrow A'$ is non-degenerate.
Let m be an integer prime to p. Then the homomorphism θ induces an
isomorphism of $A_m \cap A_k$ onto $A'_m \cap A'_{k'}$.*

Furthermore, we also get information concerning the kernel S_k in our
exact sequence above.

PROPOSITION 8. *Assume that the specialization $A \rightarrow A'$ is non-degenerate.
Let m be an integer prime to p. Then S_k is uniquely divisible by m.*

Proof. Let a be in S_k, so that $a' = 0'$. The cycle $(m\delta)^{-1}(a)$ specializes to the cycle $(m\delta')^{-1}(a') = (m\delta')^{-1}(0')$ which consists of the points of order m on A'. Since $(0')$ occurs with multiplicity 1 in the reduced cycle, it follows by Hensel's lemma that there is exactly one point on $(m\delta)^{-1}(a)$ which specializes to $0'$ and that this point is rational over k.

PROPOSITION 9. *Assume that the specialization $A \to A'$ is non-degenerate. Let m be an integer prime to p. Let a be a point of A_k, and let $K = k(1/m \cdot a)$ be the field obtained by adjoining all points x such that $mx = a$. Then K is unramified over k, and $K' = k'(1/m \cdot a')$.*

Proof. This is again an immediate consequence of Hensel's lemma. Taking into account the fact that the points in $(m\delta')^{-1}(a')$ are all separable over k', we see that K is in fact the uniquely determined unramified extension of k such that $K' = k'(1/m \cdot a')$.

We conclude this section by a remark on the reduction of principal homogeneous spaces. Let V be such a space over A, defined over k. Suppose that V' has one component with multiplicity 1. Then V' has a simple rational point in some separable extension of k', and hence, by Proposition 6, V has a point in, and is split by, an unramified extension of k.

4. Cohomology and non-degenerate reduction. Let k be a field complete with respect to a discrete valuation and let A be an abelian variety defined over k which has a non-degenerate reduction A' modulo the prime in k. The only fields we consider are the algebraic extensions K of k and their residue class fields K'. From the preceding section we know the following facts:

(R1) The reduction map $\theta_K : A_K \to A'_{K'}$ is surjective for each K and its kernel S_K is uniquely divisible by an integer m prime to the residue class characteristic, p.

(R2) If K' is separably closed, then $A'_{K'}$ is divisible by m and contains A'_m for m prime to p.

The following cohomological analysis makes no reference to the details of the reduction process itself, nor even to abelian varieties. It applies to any pair of commutative functors A and A' of the fields K and K' (in the sense of §1) which are linked by a natural homomorphism $\theta : A \to A'$ so that conditions (R1) and (R2) are satisfied. Note that (R1) implies, trivially,

(R1') θ_K induces a bijection $A_m \cap A_K \to A'_m \cap A'_{K'}$ for each K, and m prime to p.

Let k_u denote the maximal unramified extension of k. Then the residue class field of k_u is the separable closure of the residue class field of k, or in symbols, $(k_u)' = (k')_s$.

For the rest of this section, we let m be an integer prime to p. Our goal is now Theorem 1 below.

LEMMA 1. *If $K \supset k_u$, then A_K is divisible by m and contains A_m.*

Proof. K' is separably closed because $K \supset k_u$. Hence, by (R2), we know the divisibility by m of the factor group $A_K/S_K \approx A'_{K'}$. Since from (R1), we know S_K is also divisible by m, we conclude that A_K is. We also know from (R2) that $A'_m \subset A'_{K'}$. This together with (R1') implies that $A_m \cap A_K$ is independent of K for $K \supset k_u$, and consequently, $A_m \subset A_K$.

The group Ω_m of m-th roots of unity is contained in k_u, and the group of units in k_u is divisible by m. Therefore the maximal abelian extension of k_u of exponent m is the field $k_u(\pi^{1/m})$ obtained by adjoining the m-th root of any prime element π of k_u. The Galois group of this extension is canonically isomorphic to Ω_m. Indeed, the map $\sigma \to \zeta_\sigma = (\pi^{1/m})^{\sigma-1}$ is a homomorphism of G_{k_u} onto Ω_m, whose kernel cuts out $k_u(\pi^{1/m})$, and this homomorphism is independent of π and its chosen m-th root.

LEMMA 2. *There is a canonical isomorphism*

$$\mathrm{Hom}(\Omega_m, A_m) \approx H^1(k_u, A)_m$$

which attaches to each homomorphism $\chi : \Omega_m \to A_m$ the cohomology class of the cocycle $\sigma \to \chi(\zeta_\sigma)$, $\sigma \in G(k_s/k_u)$.

Proof. Since A_{k_s} is divisible by m, the Kummer sequence at the end of § 1 is applicable, and since A_{k_u} is divisible by m, it yields an isomorphism

$$\mathrm{Hom}(G_{k_u}, A_m) \approx H^1(k_u, A)_m.$$

Since A_m is commutative of exponent m, any homomorphism $h : G_{k_u} \to A_m$ will have the property that $h(\sigma)$ depends only on the effect of σ on an abelian extension of exponent m. Consequently, according to the discussion preceding this lemma, we can factor h through $\sigma \to \zeta_\sigma$, i.e. we can write $h(\sigma) = \chi(\zeta_\sigma)$ for some $\chi : \Omega_m \to A_m$.

The Galois group G_k of k_s/k operates on $H^1(k_u, A)$ as explained in § 1. It also operates on Ω_m and on A_m, and consequently, on $\mathrm{Hom}(\Omega_m, A_m)$ in the usual manner, so that

$$(\chi^\phi)(\zeta^\phi) = \phi(\chi(\zeta))$$

for $\phi \in G_k$.

LEMMA 3. *The isomorphism of Lemma 2 is a G_k-isomorphism.*

Proof. Let $\chi \in \mathrm{Hom}(\Omega_m, A_m)$ and let $a_\sigma = \chi(\zeta_\sigma)$ be the corresponding cocycle. Let $\phi \in G_k$. Then for $\sigma \in G_{k_u}$, we have

$$(a^\phi)_\sigma = \phi a_{\phi^{-1}\sigma\phi} = \phi\chi(\zeta_{\phi^{-1}\sigma\phi}) = \chi^\phi(\zeta_{\phi^{-1}\sigma\phi}^\phi) = \chi^\phi(\zeta_\sigma)$$

because

$$\zeta_{\phi^{-1}\sigma\phi}^\phi = (\pi^{1/m})^{\sigma\phi-\phi} = ((\pi^\phi)^{1/m})^{\sigma-1} = \zeta_\sigma.$$

Note that because of the way G_k is defined to operate on $\mathrm{Hom}(\Omega_m, A_m)$, we have

$$\mathrm{Hom}(\Omega_m, A_m)^{G_k} = \mathrm{Hom}_{G_k}(\Omega_m, A_m);$$

in other words, a homomorphism is left fixed by the operation of G_k if and only if it is a G_k-isomorphism.

LEMMA 4. *The sequence*

$$0 \to H^1(k_u/k, A)_m \xrightarrow{\ \mathrm{inf}\ } H^1(k, A)_m \xrightarrow{\ \mathrm{res}\ } [H^1(k_u, A)_m]^{G_k} \to 0$$

is exact.

Proof. Of course, it is only necessary to prove that the restriction is surjective. By the preceding lemmas, an element $\alpha \in H^1(k_u, A)_m$ which is invariant under G_k will be represented by a cocycle of the form $a_\sigma = \chi((\pi^{1/m})^{\sigma-1})$, $\sigma \in G_{k_u}$, where $\chi: \Omega_m \to A_m$ is a G_k-homomorphism. Here π is an arbitrary prime in k_u. Choosing π in k, and choosing a definite m-th root, we see that that the expression for a_σ makes sense for all $\sigma \in G_k$, not only for $\sigma \in G_{k_u}$. Now

$$\sigma \to (\pi^{1/m})^{\sigma-1}$$

is a cocycle in Ω_m (it is a coboundary in Ω), and since $\chi: \Omega_m \to A_m$ is a G_k-homomorphism, it follows that a_σ is still a cocycle. By construction, its class $\beta \in H^1(k, A)_m$ restricts to α.

Since k_u/k is unramified, its Galois group may be identified with the Galois group of its residue class extension. We shall make this identification from now on, and will designate both groups by the simpler symbol $G_{k'}$. Of course, $\theta: A_{k_u} \to A'_{k'_s}$ is a G_k-homomorphism.

LEMMA 5. θ *induces an isomorphism* $H^1(k_u/k, A)_m \approx H^1(k', A')_m$.

Proof. Since A_{k_u} and A'_{k_s} are divisible by m, we can apply the Kummer sequence to the extensions k_u/k and k'_s/k', obtaining the exact horizontal rows in the diagram

$$
\begin{array}{ccccccccc}
0 \to & A_k/mA_k & \to & H^1(k_u/k, A_m) & \to & H^1(k_u/k, A)_m & \to 0 \\
 & \downarrow \theta & & \downarrow \theta & & \downarrow \theta & \\
0 \to & A'_{k'}/mA'_{k'} & \to & H^1(k', A'_m) & \to & H^1(k', A')_m & \to 0
\end{array}
$$

The left vertical arrow is surjective by (R1). The middle is bijective because A_m is $G_{k'}$-isomorphic to A'_m (recall that $G_k = G(k_u/k)$). Consequently, the right vertical arrow is bijective as contended.

Putting our results together, we see that we have proved a good part of

THEOREM 1. *Let k be a field complete with respect to a discrete valuation with residue class field k' of characteristic $p \geqq 0$. Let A and A' be two commutative functors as in §1, satisfying (R1) and (R2). Then for each natural number m not divisible by p, there exists a canonical exact sequence*

$$0 \to H^1(k', A')_m \to H^1(k, A)_m \to \mathrm{Hom}_{G_{k'}}(\Omega'_m, A'_m) \to 0,$$

where Ω'_m is the group of m-th roots of unity in the residue field. Furthermore, if K is a finite extension of k with ramification index e, the following diagram is commutative:

$$
\begin{array}{ccccc}
H^1(k', A')_m & \to & H^1(k, A)_m & \to & \mathrm{Hom}_{G_{k'}}(\Omega'_m, A'_m) \\
\downarrow{\scriptstyle \mathrm{res}} & & \downarrow{\scriptstyle \mathrm{res}} & & \downarrow{\scriptstyle e} \\
H^1(K', A')_m & \to & H^1(K, A)_m & \to & \mathrm{Hom}_{G_K}(\Omega'_m, A'_m).
\end{array}
$$

Proof. The indicated exact sequence results from the preceding two lemmas and the trivial replacement of Ω_m and A_m by the isomorphic Ω'_m and A'_m. The commutativity of the left square is obvious. That of the right follows from the relationship

$$\chi\big((\pi^{1/m})^{\sigma-1}\big) = \chi^e\big((\pi_1^{1/m})^{\sigma-1}\big)$$

for $\sigma \in G_{K_u}$, where π_1 is a prime in K_u and $\pi \sim \pi_1^e$ is a prime in k_u.

Naturally, we are interested here in the case where the functors are given by an abelian variety A defined over k, and a non-degenerate reduction A'. Theorem 1 gives especially precise information when $H^1(k', A') = 0$. This is the case, for example, if k' is algebraically closed, or finite [7]. Then we obtain simply an isomorphism

$$H^1(k, A)_m \approx \mathrm{Hom}_{G_{k'}}(\Omega'_m, A'_m).$$

From the commutative diagram of the theorem, we obtain the following corollaries.

COROLLARY 1. *Assume k' is either finite or algebraically closed. Let A be an abelian variety defined over k, with a non-degenerate reduction A'. Then a finite extension K/k splits an element $\alpha \in H^1(k, A)_m$ if and only if the ramification index $e(K/k)$ is divisible by the period of α.*

In terms of homogeneous spaces, this means that the homogeneous space

V corresponding to α has a rational point in K if and only if $e(K/k)$ is divisible by its period.

COROLLARY 2. *Let A, A' be as in Corollary 1. Let V be a principal homogeneous space of A defined over k, of period prime to p. Then the index of V is equal to its period.*

Proof. There exist ramified extensions, e. g. $k(\pi^{1/e})$, with preassigned ramification index, whose degree is equal to that index.

To describe $\mathrm{Hom}(\Omega'_m, A'_m)$ more concretely is, of course, easy. We choose a generator ζ for Ω'_m. Then a homomorphism χ is determined by the image $\chi(\zeta)$ of ζ, and this may be any element of A'_m. In order that χ be a $G_{k'}$-homomorphism, it is necessary and sufficient that $\chi(\zeta^\sigma) = \sigma\chi(\zeta)$ for each $\sigma \in G_{k'}$. Defining the integer $\nu_\sigma \pmod m$ by $\zeta^\sigma = \zeta^{\nu_\sigma}$, we see that the condition becomes $\nu_\sigma\chi(\zeta) = \sigma\chi(\zeta)$. Thus, $\mathrm{Hom}_{G_k}(\Omega'_m, A'_m)$ is isomorphic to the group of solutions $x \in A'_m$ of the equations $\sigma x = \nu_\sigma x$, $\sigma \in G_{k'}$. The isomorphism is not canonical, but depends on the choice of a primitive m-th root of unity ζ.

In particular, suppose that k' is a finite field with q elements. Then $G_{k'}$ has a canonical generator, namely the Frobenius automorphism $\xi \to \xi^q$. Its effect on an element x of A' is denoted by $x^{(q)}$. We may therefore express our result in this case as follows.

THEOREM 2. *Let A be an abelian variety defined over k with a non-degenerate reduction A', and suppose k' is a finite field with q elements. Then the group of elements of order prime to p in $H^1(k, A)$ is isomorphic to the group of solutions of $x^{(q)} = qx$ on A', of order prime to p.*

We observe that $x \to x^{(q)} - qx$ is an endomorphism of A'. Furthermore, if π denotes the Frobenius endomorphism of A', then its transpose ${}^t\pi$ on the Picard variety is easily seen to be given by the formula

$$ {}^t\pi y = qy^{(1/q)}. $$

Consequently, the transpose of $\pi - \delta$ is given by the endomorphism $y \to qy^{(1/q)} - y$ on $\hat{A}'$. Raising the kernel of this endomorphism to the q-th power, we see that the kernel is isomorphic to the subgroup of $\hat{A}'$ satisfying $y^{(q)} = qy$. In particular, we see that the group of solutions of $x^{(q)} = qx$ on A', of order prime to p, is dual to the group of rational points of $\hat{A}'$ in k', of order prime to p.

We close this section with an example of a homogeneous space whose index is not equal to its period. Suppose first of all that we could construct an abelian variety A over a field k such that $A_m \subset A_k$ and such that A_K is divisible by m for each algebraic extension K of k. Then the Kummer sequence gives isomorphisms

$$\operatorname{Hom}(G_k, A_m) \approx H^1(k, A)_m$$

and, moreover, for each K/k, a commutative diagram

$$
\begin{array}{ccc}
\operatorname{Hom}(G_k, A_m) & \approx & H^1(k, A)_m \\
\text{res} \downarrow & & \downarrow \text{res} \\
\operatorname{Hom}(G_K, A_m) & \approx & H^1(K, A)_m,
\end{array}
$$

where the left hand res means restriction of homomorphisms from G_k to G_K. If $f \colon G_K \to A_m$ is a homomorphism, we see that K splits the corresponding cohomology class if and only if G_K is in the kernel of f. Thus the index of the corresponding homogeneous space would be the *order* of the group $f(G_k)$, whereas its period would be the *exponent* of the group $f(G_k)$. In particular, if k has an abelian extension whose group is isomorphic to A_m, i.e. to the direct product of $2r$ cyclic groups of order m ($r = \dim A$), then we can construct a homogeneous space with index m^{2r} and period m.

An actual example can be constructed easily enough. Let A be defined over an algebraically closed field k_0. Define $k_i = k_{i-1}((t_i))$ (power series in one variable) for $1 \leqq i \leqq 2r$. One proves by induction that $A_{k_i} = mA_{k_i}$, and a similar divisibility statement for finite extensions. Then we can construct our example with $k = k_{2r}$ and the abelian extension $k(t_1^{1/m}, \cdots, t_{2r}^{1/m})$.

One could also construct a similar example over a purely transcendental function field in $2r$ variables. Incidentally, we note that the Kummer sequence shows, for any A and k, that if $A_m \subset A_k$ and $\operatorname{per}_k V = m$, then $\operatorname{ind}_k V$ divides m^{2r}. Thus our example is as bad as possible under the conditions $A_m \subset A_k$.

5. Global fields. Let k be a field with a fixed set of (inequivalent) discrete valuations $\mathfrak{p}$ which we shall call primes. By a prime in a finite algebraic extension K of k, we mean, of course, a valuation of K which extends a prime of k. Let m be a natural number not divisible by the characteristic of k. We shall say that k is an *m-global field* if each finite extension K of k has the following two properties.

(Gl1) If A is an abelian variety defined over K, then A has a non-degenerate reduction at all but a finite set of primes $\mathfrak{p}$ of K.

(Gl2, m) The set of primes of K dividing m is finite, and if S is any finite set of primes of K, there is only a finite number of abelian extensions of K of exponent m which are unramified outside S.

By well known elementary properties of number fields, one sees immediately that an algebraic number field of finite degree is m-global for every m, if we take as primes all the inequivalent discrete valuations.

The same remark applies to a field of algebraic functions of one variable over a finite constant field, and shows that it is m-global for m prime to its characteristic.

More generally, let k be an algebraic function field in n variables (the case $n = 0$ is not excluded) over a constant field k_0 such that $k_0*/(k_0*)^m$ is finite, for instance, a p-adic field, or an algebraically closed field. Choose as set of primes of k those arising from the prime divisors rational over k_0 of a projective normal model of k. Then it is easy to see (using results of algebraic geometry) that k is m-global.

Finally, if k is a function field over an algebraic number field of finite degree, then k is m-global for every m. The set of primes is then to be the set of prime divisors on a "mixed characteristic" model. This can easily be seen by reducing the proof to a geometric statement. As this, and the above fields are meant mostly to give concrete examples of m-global fields to the reader, we do not go in detail into the proofs that they are m-global, since they are essentially well known.

For the convenience of the reader, we indicate very briefly a sketch of the manner in which the above fields can be proved to satisfy our two global axioms.

(Gl 1) actually depends on the fact that an element of the field k has only a finite number of zeros and poles. To check that a reduction is non-degenerate, one checks that each property entering into the definition of an abelian variety has a non-degenerate reduction. For instance, we have: the absolute irreducibility (say of the Chow form if the variety is in projective space) ; the non-singularity (which depends on the non-vanishing of a determinant which is an element of k) ; the fact that some mappings are everywhere defined; associativity; etc.

As for (Gl 2), say for number fields, one may look at it either from the point of view that there is only a finite number of unramified extensions of degree m (abelian or non-abelian), a fact already stated in Hilbert's *Zahlbericht*, or from the point of view of Kummer theory: After adjoining the m-th roots of unity, the extension can be obtained by extracting radicals, and one uses the finiteness of class number and unit theorem. In the geometric case, one uses the analogous facts, which involve the Jacobian for the case of curves, and the Picard variety as well as the torsion part of the Neron-Severi group in higher dimension.

We shall now reprove for global fields the weak part of the Mordell-Weil theorem, i.e. Theorem 3 below.

PROPOSITION 10. *Let A be an abelian variety defined over a field k. Let m be prime to the characteristic of k and let $K = k(1/m \cdot A_k)$. Then $(A_k : mA_k)$ is finite if and only if $[K : k]$ is finite.*

Proof. It is trivial that if $(A_k : mA_k)$ is finite, then $[K : k]$ is finite. Conversely, consider the exact sequence

$$0 \to A_m \to A_K \to mA_K \to 0$$

Let $G = G(K/k)$. We have the cohomology sequence

$$0 \to A_m \cap A_k \to A_k \to mA_K \cap A_k \to H^1(G, A_m).$$

By assumption, $mA_K \cap A_k = A_k$, and thus we get the injection

$$0 \to A_k/mA_k \to H^1(G, A_m)$$

which shows that A_k/mA_k is finite.

COROLLARY. *Let A be an abelian variety defined over a field k, and let m be a natural number prime to the characteristic of k. Let L be a finite extension of k. If A_L/mA_L is finite then so is also A_k/mA_k.*

Proof. By the proposition, $L(1/m \cdot A_L)$ is finite over L. This implies a fortiori that $k(1/m \cdot A_k)$ is finite over k. Using the proposition once more, we get what we want.

THEOREM 3. *Let A be an abelian variety defined over an m-global field k. Then A_k/mA_k is finite.*

Proof. By the corollary, we may assume that $A_m \subset A_k$. Let $K = k(1/m \cdot A_k)$. By Proposition 9, we know that K/k is unramified at every prime $\mathfrak{p}$ not dividing m at which A has a non-degenerate reduction. From the definition of m-global, it follows that K is finite over k because it is abelian of exponent m over k. We now apply Proposition 10, to conclude the proof.

If K is a finite Galois extension of k, and if A is an abelian variety defined over k such that its group of rational points A_K in K is finitely generated (i.e. satisfies the strong Mordell-Weil theorem), then obviously $H^r(K/k, A)$ is finite for $r > 0$. However, Theorem 3 will suffice to prove the analogous result for m-global fields. For use in the statement of our next theorem, we recall that the m-primary part of an abelian group consists of those elements whose period divides a power of m.

THEOREM 4. *Let A be an abelian variety defined over an m-global* *field k. If K/k is a finite Galois extension, then the m-primary part of the group $H^r(K/k, A)$ is finite for each $r > 0$.*

Proof. This is an immediate consequence of the following general cohomological fact.

LEMMA. *Let G be a finite group and X a G-module such that X/mX and X_m are finite. Then the m-primary part of $H^r(G, X)$ is finite for each $r > 0$.*

Proof. Since $H^r(G, X)$ is a torsion group of bounded exponent $(G : 1)$, it is enough to prove $H^r(G, X)_m$ finite. The map $m : X \to X$ can be written as a product $m = g \circ h$ of the maps g and h in the exact sequences

$$0 \to X_m \to X \xrightarrow{\ h\ } mX \to 0$$

$$0 \to mX \xrightarrow{\ g\ } X \to X/mX \to 0.$$

For the induced cohomology maps in dimension r, we have $m_* = g_* h_*$. From the exact cohomology sequences derived from our two exact sequences, we see that the kernels of h_* and g_* are finite, being homomorphic images of $H^r(G, X_m)$ and $H^{r-1}(G, X/mX)$ respectively. Consequently, the kernel of m_* is finite, which is what we wanted to prove.

If A is an abelian variety defined over an m-global field k, we say that an element $\alpha \in H^1(k, A)$ *splits at a prime* $\mathfrak{p}$ of k if α is in the kernel of the canonical map $H^1(k, A) \to H^1(k_\mathfrak{p}, A)$, where $k_\mathfrak{p}$ denotes the completion of k at $\mathfrak{p}$. In other words, α splits at $\mathfrak{p}$ if the corresponding homogeneous space has a rational point in $k_\mathfrak{p}$.

Selmer [11] has given examples of principal homogeneous spaces for abelian varieties of dimension 1 over the rational number field $\boldsymbol{Q}$ which split at all primes p but do not split in $\boldsymbol{Q}$. The elliptic curve $3X^3 + 4Y^3 + 5Z^3 = 0$ is such a space over its Jacobian. On the other hand, we can show that the number of such spaces with given period is finite. More precisely, let us denote by $H^1(k, A, S)$ the subgroup of $H^1(k, A)$ consisting of the elements which split at all primes outside a finite set S. We have

THEOREM 5. *Let A be an abelian variety defined over an m-global field k. Then $H^1(k, A, S)_m$ is finite.*

Proof. Without loss of generality, we may assume $A_m \subset A_k$. Indeed, the restriction of $H^1(k, A)$ to $H^1(K, A)$, where $K = k(A_m)$, has finite kernel by Theorem 4, and a fortiori that of $H^1(k, A, S)_m$ to $H^1(K, A, S)_m$. Furthermore, we may enlarge S until it contains all primes $\mathfrak{p}$ dividing m

9

and all primes $\mathfrak{p}$ where A does not have a non-degenerate reduction. Now for $\mathfrak{p} \notin S$, consider the commutative diagram

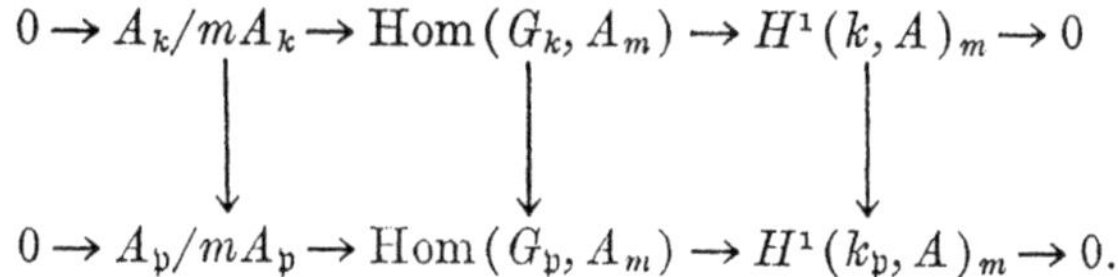

Here we have abbreviated by $A_\mathfrak{p}$ the group $A_{k_\mathfrak{p}}$ and by $G_\mathfrak{p}$ the decomposition group $G_{k_\mathfrak{p}}$ which we may view as a subgroup of G_k, uniquely determined up to a conjugation. The horizontal arrows are the exact Kummer sequences and the vertical arrows are the canonical maps, the middle one therefore denoting simply the operation of restricting to $G_\mathfrak{p}$ a homomorphism of G_k. What we shall actually prove is that the inverse image of $H^1(k, A, S)_m$ in $\mathrm{Hom}(G_k, A_m)$ is finite. To this effect, it will be enough to show that if $\chi: G_k \to A_m$ is a homomorphism whose cohomology class in $H^1(k, A)$ splits outside S, then the abelian extension K_χ/k of exponent m which is cut out by the kernel of χ is unramified outside S. By the commutativity of the right square of the above diagram we know that the restriction of χ to G_p is of the form $\chi(\sigma) = (\sigma - 1)(1/m \cdot a)$ for some $a \in A_\mathfrak{p}$. Thus we have $K_\chi \subset k_\mathfrak{p}(1/m \cdot a)$, and, by Proposition 9, it follows that K_χ is unramified at $\mathfrak{p}$. This completes the proof of Theorem 5.

It is natural to raise the question whether, given an element of $H^1(k_\mathfrak{p}, A)$, there exists an element of $H^1(k, A)$ which restricts to it. We can prove this in a special case.

THEOREM 6. *Let k be any field with a discrete valuation $\mathfrak{p}$ and let $k_\mathfrak{p}$ be its completion. Let m be prime to the characteristic of k and assume that the group of m-th roots of unity Ω_m lies in k. Let A be an abelian variety defined over k such that $A_m \subset A_k$. Then given $\alpha_\mathfrak{p} \in H^1(k_\mathfrak{p}, A)_m$, there exists an element $\alpha \in H^1(k, A)_m$ whose canonical image in $H^1(k_\mathfrak{p}, A)$ is $\alpha_\mathfrak{p}$.*

Proof. We use the Kummer sequence again, and the exact commutative diagram

$$\mathrm{Hom}(G_k, A_m) \to H^1(k, A)_m \to 0$$
$$\downarrow \qquad\qquad \downarrow$$
$$\mathrm{Hom}(G_\mathfrak{p}, A_m) \to H^1(k_\mathfrak{p}, A)_m \to 0.$$

We are trying to prove that the right hand vertical arrow is surjective. To do so, it is enough to prove the left hand vertical arrow is surjective. Since

A_m is isomorphic as an abstract group to the direct sum of $2r$ copies of Ω_m ($r = \dim A$), we have

$$\text{Hom}(G, A_m) \approx (\text{Hom}(G, \Omega_m))^{2r},$$

and, consequently, it is enough if we show the surjectivity of

$$\text{Hom}(G_k, \Omega_m) \to \text{Hom}(G_\mathfrak{p}, \Omega_m).$$

Recall now that Hom means here continuous homomorphisms, so that by the usual Kummer duality, we have $\text{Hom}(G_k, \Omega_m) \approx k^*/(k^*)^m$, and similarly for $k_\mathfrak{p}$. Thus, we must simply show the surjectivity of $k^*/(k^*)^m \to k_\mathfrak{p}^*/(k_\mathfrak{p}^*)^m$. This is now obvious since $(k_\mathfrak{p}^*)^m$ is an open subgroup of $k_\mathfrak{p}^*$, m being prime to the characteristic.

Remark. Under the same hypotheses as in the theorem, one can deal with a finite number of discrete valuations, and use the approximation theorem to show that there is an α restricting simultaneously to a finite number of given $\alpha_{\mathfrak{p}_i}$.

In analogy with Grunwald's theorem in class field theory, one may conjecture that if k is an algebraic number field and $\mathfrak{p}$ a given prime, then given $\alpha_\mathfrak{p} \in H^1(k_\mathfrak{p}, A)$, there exists $\alpha \in H^1(k, A)$ restricting to $\alpha_\mathfrak{p}$.

Finally, we prove a result indicating that $H^1(k, A)$ is usually a large group when k is a global field.

THEOREM 7. *Let m be a natural number and let k be a field with an infinite number of abelian extensions of exponent exactly m. Let A be an abelian variety defined over k such that (1) A_k/mA_k is finite and (2) A_k contains at least one element a of exact period m. Then $H^1(k, A)$ contains an infinite number of elements of period m, and, in fact, an infinite number such that the corresponding homogeneous spaces have index m as well as period m.*

Proof. Let $L = k(1/m \cdot A_k)$. Since we have assumed A_k/mA_k finite, L is a finite extension of k. Consequently, by our hypothesis about the existence of abelian extensions of exponent m, there exists a Galois extension K of k linearly disjoint from L whose group $G = G(K/k)$ is the direct product of an arbitrarily large number of cyclic groups of order m. For each homomorphism χ of G into the cyclic group $\langle a \rangle$ of order m generated by a, let $\alpha_\chi \in H^1(K/k, A) \subset H^1(k, A)$ denote the cohomology class of the 1-cocycle $\sigma \to \chi(\sigma)$. We claim that the map $\chi \to \alpha_\chi$ is an injection. Indeed, if $\chi(\sigma) = (\sigma - 1)b$, $b \in A_k$, $\sigma \in G$, then $m\chi(\sigma) = (\sigma - 1)mb = 0$ for all σ; hence $mb \in A_k$. Since K is disjoint from L, it follows that $b \in A_k$ and $\chi = 0$.

Thus, $H^1(k, A)$ contains a subgroup isomorphic to $\mathrm{Hom}(G, \langle a \rangle)$, that is, to the direct product of arbitrarily many cyclic groups of order m. Moreover, for each element of this subgroup, the corresponding homogeneous space has index $=$ period. This follows from the Corollary of Proposition 5, § 2 because, since $\langle a \rangle$ is cyclic, the index in G of the kernel of the homomorphism $\chi: G \to \langle a \rangle$ is equal to the period of χ, and the subfield of K cut out by this kernel splits α.

PARIS.

REFERENCES.

[1] A. Albert, "On primary normal division algebras of degree 8," *Bulletin of the American Mathematical Society*, vol. 39 (1933), pp. 265-272.

[2] H. Cartan and S. Eilenberg, *Homological algebra*, Princeton University Press, 1956.

[3] F. Chatelet, "Méthode Galoisienne et courbes de genre 1," *Annales de l'Universite de Lyon*, Section A., t. 89 (1946), pp. 40-49.

[4] ———, "Variations sur un thème de Poincaré," *Annales de l'Ecole Normale superieure*, vol. 59 (1944), pp. 249-300.

[5] W. L. Chow and S. Lang, "On the birational equivalence of curves under specialization," *American Journal of Mathematics*, vol. 79, No. 3 (1957), pp. 649-652.

[6] J. Igusa, "Fibre systems of Jacobian varieties, II," *American Journal of Mathematics*, vol. 78 (1956), pp. 745-760.

[7] S. Lang, "Algebraic groups over finite fields," *American Journal of Mathematics*, vol. 78 (1956), pp. 555-563.

[8] E. Lutz, "Sur l'équation $y^2 = x^3 - Ax - B$ dans les corps p-adiques," *Journal für die reine und angewandte Mathematik*, Bd. 177 (1937), pp. 238-247.

[9] A. Mattuck, "Abelian varieties over p-adic ground fields," *Annals of Mathematics*, Series 2, vol. 62 (1955), pp. 92-119.

[10] P. Roquette, "Uber das Hassesche Klassenkorper-Zerlegungesetz," *Journal für die reine und angewandte Mathematik*, Bd. 197 (1957), pp. 49-67.

[11] E. Selmer, "The diophantine equation $ax^3 + by^3 + cz^3 = 0$," *Acta Mathematika*, vol. 85 (1951), pp. 203-362.

[12] G. Shimura, "Reduction of algebraic varieties with respect to a discrete valuation of the basic field," *American Journal of Mathematics*, vol. 77 (1955), pp. 134-176.

[13] A. Weil, "On algebraic groups and homogeneous spaces," *American Journal of Mathematics*, vol. 77 (1955), pp. 493-512.

[14] ———, "The field of definition of a variety," *American Journal of Mathematics*, vol. 78 (1956), pp. 509-524.

ALGEBRAIC GROUPS AND THE GALOIS THEORY OF DIFFERENTIAL FIELDS.*

By Ellis Kolchin[1] and Serge Lang.

The purpose of this note is to draw together morely closely the Galois theory of differential fields and the theory of algebraic groups and homogeneous spaces.

Given a strongly normal extension of a differential field of characteristic 0, it is known (see [2] or Matsumura [3]) that the Galois group is isomorphic to the group of points of an algebraic group which are rational over the field of constants. We show below that the extension itself has a model which is a principal homogeneous space for the algebraic group, such that the operation of the elements of the Galois group on the extension is induced by the operation of the corresponding elements of the algebraic group on the space. The Galois theory can then be deduced from known facts concerning principal homogeneous spaces (Rosenlicht [4]). This provides an analysis of the Galois theory into two parts, one depending on the foundations of differential algebra and the other on the foundations of the theory of algebraic groups.

There remains almost untouched a converse problem which, for the sake of simplicity, we state for connected algebraic groups, or as we shall also say, group varieties. Given a differential field F of characteristic 0 with algebraically closed field of constants C, a group variety G defined over C, and a principal homogeneous space V for G defined over F, to determine all extensions of the differential field structure on F to the function field $F(V)$ which makes $F(V)$ a strongly normal extension of F with automorphism group induced by the group G_C of points of G rational over C. Put another way, the problem is to describe all possible ways of extending the derivation operators of F to derivation operators on $F(V)$ which are invariant relative to G_C, which commute with each other, and which have precisely C as field of constants. We observe here merely that the first condition alone

---

* Received August 3, 1957.

[1] The work of this author was assisted by a grant from the National Science Foundation.

103

presents no difficulty, i. e. that it is easily shown that the derivations of F can be extended to invariant derivations of $F(V)$.

1. Algebraic groups. We recall here briefly the notions of algebraic group and principal homogeneous space. For the general theory, see Weil [5] and [6], and Rosenlicht [4].

An *algebraic set* V is a finite union of varieties, $V = \bigcup V_i$, which for the purposes of this paper we shall always assume disjoints; the V_i are the components of V. Let V, W be two algebraic sets. A rational map $f: V \to W$ is an algebraic set, having one component f_i for each component V_i of V such that f_i is a rational map in the usual sense of V_i into a component of W. We say that f is defined at a point $u \in U_i$ if f_i is defined at that point, and we write $f(u) = f_i(u)$.

An *algebraic group* is an algebraic set G together with an everywhere defined rational map of $G \times G$ into G, relative to which G is a group, having the further property that the inverse mapping $x \to x^{-1}$ is an everywhere defined rational map of G into G.

If G is an algebraic group, a *transformation space* for G is an algebraic set V together with an everywhere defined rational map of $V \times G$ into V (the operation of G on V) which, if we denote the image of (v, x) by vx, satisfies the conditions

$$v(xy) = (vx)y \text{ and } ve = v$$

for all $v \in V$, $x \in G$, $y \in G$ (e denoting the unity element of G.) A transformation space is a *homogeneous space* if, given any $v, w \in V$, there exists an $x \in G$ such that $vx = w$. The homogeneous space is *principal* if this point x is determined uniquely and rationally. By the latter we mean that there exists an everywhere defined rational map $\lambda: V \times V \to G$ such that $x = \lambda(v, w)$. If V is a principal homogeneous space for G then the number of components of V equals that of G and each component of V is a principal homogeneous space for the component G_0 of G containing e, G_0 being a normal subgroup of G the cosets of which are precisely the components of G.

We shall say that the algebraic group G is defined over a field k if all the components of G, of its group law, and of its inverse mappings are defined over k. Similarly, we shall say that the principal homogeneous space V for G is defined over k if G is defined over k, and if all the components of V, of the operation of G on V, and of λ are also defined over k. Although this notion of field of definition may present some drawbacks, it is adequate for our present purpose.

2. Differential-algebraic preliminaries. Let F be a differential field of characteristic 0. Denote the field of constants of F by C. We work in a fixed universal extension F^* of F in the sense of [1], and denote the field of constants of F^* by C^*. The facts stated in the following propositions were proved in [1].

PROPOSITION 1. *The fields F and C^* are linearly disjoint over C.*

By an extension of F, we shall always mean a differential field extension contained in the universal domain. Similarly, an isomorphism will always mean a differential field isomorphism. When we wish to consider only the field structure, we shall say field extension, and field isomorphism.

Let E be an extension of F, and suppose that as a field extension E is finitely generated. Let F_0 be the algebraic closure of F in E. We recall if σ and σ' are isomorphisms of E, σ' is called a *specialization* of σ if the family $(\sigma'\alpha)_{\alpha \in E}$ is a specialization of $(\sigma\alpha)_{\alpha \in E}$ over E in the usual sense of algebraic geometry. The specialization is called *generic* if also σ is a specialization of σ'. An isomorphism of E over F is called *isolated* if it is not a nongeneric specialization of any isomorphism of E over F.

PROPOSITION 2. (a) *There exists a finite set of isolated isomorphisms of E over F such that every isomorphism of E over F is a specialization of precisely one of these.*

(b) *An isomorphism σ of E over F is isolated if and only if the transcendence degree of the compositum $E \cdot \sigma E$ over E equals that of E over F.*

(c) *An alement of E invariant under an isolated isomorphism of E over F belongs to F_0.*

(d) *If σ is isolated then σ' is a specialization of σ if and only if their restriction to F_0 coincide. Every isomorphism of F_0 over F is the restriction to F_0 of an isolated isomorphism of E over F.*

Suppose further that E has the same field of constants C as F has, and that C is algebraically closed. We recall that an isomorphism σ of E is called *strong* if $\sigma E \subset E \cdot C^*$ and $E \subset \sigma E \cdot C^*$. Letting $C(\sigma)$ denote the field of constants of $E \cdot \sigma E$, we can see with the help of Proposition 1 that σ is strong if and only if

$$E \cdot C(\sigma) = E \cdot \sigma E = \sigma E \cdot C(\sigma).$$

Every automorphism of E is obviously a strong isomorphism of E.

PROPOSITION 3. (a) *Every strong isomorphism of E over F has a specialization which is an automorphism of E.*

(b) *An isomorphism of E over F is strong if and only if it is the restriction to E of an automorphism of $E \cdot C^*$ over $F \cdot C^*$.*

As the automorphism mentioned in (b) is evidently unique, we may identify the strong isomorphism with it. The set of strong isomorphisms of E over F is then identified with the group of automorphisms of $E \cdot C^*$ over $F \cdot C^*$. We denote this group by $\mathfrak{G}$.

Let $\sigma, \tau \in \mathfrak{G}$. As $C(\sigma^{-1})$ is the field of constants of $E \cdot \sigma^{-1}E$, $\sigma C(\sigma^{-1}) = C(\sigma^{-1})$ must be the field of constants of $\sigma(E \cdot \sigma^{-1}E) = E \cdot \sigma E$. Thus

$$C(\sigma^{-1}) = C(\sigma).$$

Writing $C(\sigma, \tau)$ to denote the compositum $C(\sigma) \cdot C(\tau)$, we see that

$$E \cdot C(\sigma, \sigma\tau) = E \cdot C(\sigma) \cdot C(\sigma\tau) = E \cdot \sigma E \cdot \sigma\tau E = E \cdot \sigma(E \cdot \tau E) = E \cdot \sigma(E \cdot C(\tau))$$

$$= E \cdot \sigma E \cdot C(\tau) = E \cdot C(\sigma) \cdot C(\tau) = E \cdot C(\sigma, \tau).$$

It follows with the help of Proposition 1 (applied to E instead of F) that $C(\sigma, \tau) = C(\sigma, \sigma\tau)$. Therefore $C(\tau^{-1}, \sigma^{-1}) = C(\tau^{-1}, \tau^{-1}\sigma^{-1}) = C(\tau, \sigma\tau)$. Thus

$$(1) \qquad C(\sigma, \sigma\tau) = C(\sigma, \tau) = C(\sigma\tau, \tau).$$

We observe also, using Propositions 2(b) and Proposition 1, that $\sigma \in \mathfrak{G}$ is isolated if and only if the transcendence degree of $C(\sigma)$ over C equals that of E over F.

We suppose, finally, that the extension E of F is *strongly normal*, that is, that in addition to all the hypotheses made above, every isomorphism of E over F is strong. $\mathfrak{G}$ is then called the *Galois group* of E over F.

3. The theorem and its proof. We temporarily impose the restriction that $F = F_0$. Then by Proposition 2(a) and (d), all isolated elements of $\mathfrak{G}$ are generic specializations of each other.

Let $\sigma_0 \in \mathfrak{G}$ be isolated, and let W be any model of the field $C(\sigma_0)$ over C. This means that W is a variety defined over C and there exists a generic point x of W over C such that $C(\sigma_0) = C(x)$. If σ is any isolated element of $\mathfrak{G}$ then σ is a generic specialization of σ_0 so that there exists an isomorphism $E \cdot \sigma_0 E \approx E \cdot \sigma E$ over E mapping $\sigma_0\alpha$ onto $\sigma\alpha$ ($\alpha \in E$). This isomorphism maps $C(\sigma_0)$ onto $C(\sigma)$ and therefore maps x onto a generic point of W over C, which we call x_σ (so that in particular $x_{\sigma_0} = x$) and which has the property that $C(\sigma) = C(x_\sigma)$. Because $E \cdot \sigma_0 E = E \cdot C(\sigma_0)$ the isomorphism $E \cdot \sigma_0 E \approx E \cdot \sigma E$ over E is determined by its restriction to $C(\sigma_0)$, so that σ is determined by x_σ.

Furthermore if y is any point of W which is rational over C^* and generic over C, there exists a unique field isomorphism $C(x) \approx C(y)$ over C mapping x onto y, and because E and C^* are linearly disjoint over C this field isomorphism can be extended to a field isomorphism ϕ over E of $E \cdot C(x)$ $= E \cdot C(\sigma_0) = E \cdot \sigma_0 E$ onto $E \cdot C(y)$. As ϕ maps $C(\sigma)$ into $C(y)$, ϕ commutes on $C(\sigma)$ with the derivation operators of the differential field structure, and as ϕ is the identity on E, ϕ commutes on E with these derivation operators, so that ϕ commutes on $E \cdot C(\sigma)$ with these operators. In other words, ϕ is a differential field isomorphism. Letting $\tau = \phi \sigma_0$, we see that $y = x_\tau$. We have thus proved that $\sigma \to x_\sigma$ is a bijective mapping of the set of isolated elements of G onto the set of points of W which are rational over C^* and generic over C.

Let $\sigma, \tau \in \mathfrak{G}$. If in addition to being isolated, σ and τ are independent, in the sense that $C(\sigma)$ and $C(\tau)$ are linearly disjoint over C, then (1) shows that $\sigma\tau$ is isolated. Furthermore, by Proposition 1, $E \cdot \sigma E = E \cdot C(\sigma)$ and $E \cdot \tau E = E \cdot C(\tau)$ are linearly disjoint over E. If σ' and τ' are also independent isolated elements of $\mathfrak{G}$, then the isomorphism $E \cdot \sigma E \approx E \cdot \sigma' E$ over E mapping $(\sigma\alpha)_{\alpha \in E}$ onto $(\sigma'\alpha)_{\alpha \in E}$ and the analogous isomorphism $E \cdot \tau E \approx E \cdot \tau' E$ have a unique common extension.

$$(2) \qquad E \cdot \sigma E \cdot \tau E \approx E \cdot \sigma' E \cdot \tau' E.$$

For any element $\alpha \in E$, $\tau\alpha$ can be expressed in the form $\tau\alpha = f(x_\tau)$, where f is a rational function on W defined over E. Hence we have $\tau'\alpha = f(x_{\tau'})$, $\sigma\tau\alpha = f^\sigma(x_\tau)$, and $\sigma'\tau'\alpha = f^{\sigma'}(x_{\tau'})$. From this it is clear that the isomorphism (2) maps $(\sigma\tau\alpha)_{\alpha \in E}$ onto $(\sigma'\tau'\alpha)_{\alpha \in E}$ and therefore maps $x_{\sigma\tau}$ onto $x_{\sigma'\tau'}$. Thus $(x_{\sigma'}, x_{\tau'}, x_{\sigma'\tau'})$ is a specialization of $(x_\sigma, x_\tau, x_{\sigma\tau})$ over C (and even over E).

It follows that we may define a law of composition on W by the formula $x_\sigma x_\tau = x_{\sigma\tau}$, this law of composition being defined over C. From (1) we have $C(x_\sigma, x_\sigma x_\tau) = C(x_\sigma, x_\tau) = C(x_\sigma x_\tau, x_\tau)$. Furthermore, if ω is an isolated element of $\mathfrak{G}$ such that ω, σ, τ are independent, i.e. such that $x_\omega, x_\sigma, x_\tau$ are independent generic points of W, then it is clear that $x_\omega(x_\sigma x_\tau) = (x_\omega x_\sigma)x_\tau$. Thus we have a normal law of composition on W, so that by the fundamental theorem of Wiel [5], W is birationally equivalent over C to a group variety defined over C. Thus we may suppose that W is itself a group variety, and we henceforth denote it by G.

Since we have assumed that $F = F_0$, E is a regular extension of F.

Therefore there exists a model V for E over F, i. e. a variety V defined over F such that E may be written $E = F(v)$ where v is a generic point of V over F. Letting σ be an isolated element of $\mathfrak{G}$, so that x_σ is a point of G which is rational over C^* and generic over C (and therefore over E), we define a rational map of $V \times G$ into V by the formula

$$(v, x_\sigma) \to v x_\sigma = \sigma v.$$

It follows from the above considerations that this rational map, which is defined over F, does not depend on which isolated element σ is used. It is also easy to see from what precedes that if σ and τ are independent isolated elements of $\mathfrak{G}$, then

$$F(v x_\sigma, x_\sigma) = F(v, x_\sigma) = F(v, v x_\sigma)$$

and

$$v(x_\sigma x_\tau) = (v x_\sigma) x_\tau.$$

It follows from another result of Weil ([6] Prop. 3) that V is birationally equivalent over F to a principal homogeneous space for G defined over F, and therefore we may assume that V itself is such a space. This principal homogeneous space is, moreover, uniquely determined up to an everywhere defined birational correspondence compatible with the operation of G.

This being the case, we see that every point $a \in G$ (generic or not) such that v is generic on V over $F(a)$ induces a field isomorphism σ_a of $E = F(v)$ over F mapping v onto va. If a is any point of the group G_{C^*} of points of G rational over C^*, then v is automatically generic on V over $F(a)$, because E and C^* are linearly disjoint over C, and hence E and $F \cdot C^*$ are linearly disjoint over F. In this case, σ_a is an isomorphism of the differential field structure. Indeed, for elements of G_{C^*} which are generic over C this has been proved above, and for an arbitrary $a \in G_{C^*}$, there exists an element $x \in G_{C^*}$ such that x^{-1} and xa are generic over C, and $\sigma_a = \sigma_{x^{-1}}\sigma_{xa}$.

The mapping $a \to \sigma_a$ ($a \in G_{C^*}$) is obviously an isomorphism of G_{C^*} into $\mathfrak{G}$. It is actually onto $\mathfrak{G}$, for given any $\tau \in \mathfrak{G}$ we can select an isolated $\tau' \in \mathfrak{G}$ so that τ and τ' are independent. Then $\tau\tau'$ and τ'^{-1} are isolated and $\tau = (\tau\tau')\tau'^{-1}$.

We have now proved our main result in the special case that $F = F_0$. We shall indicate below how to extend it to the general case, and we state here the general theorem which for the case $F = F_0$ summarizes the above discussion.

THEOREM. *Let F be a differential field of characteristic 0 with algebraically closed field of constants C, and let E be a strongly normal extension* *of F, all lying in a universal extension F^* with field of constants C^*. Let $\mathfrak{G}$ be the Galois group of E over F. Then there exists an algebraic group G defined over C, and a principal homogeneous spave V for G, defined over the algebraic closure F_0 of F in E, having the following properties.*

(a) V is a model of E over F, that is, any component of V has a point v generic over F_0 such that $E = F_0(v)$, and these components are all conjugates to each other over F.

(b) For each $\sigma \in \mathfrak{G}$ there exists a unique point $x_\sigma \in G_{C^*}$ (the group of points of G which are rational over C^*) such that $\sigma v = v x_\sigma$, and the mapping $\sigma \to x_\sigma$ ($\sigma \in \mathfrak{G}$) is a group isomorphism of $\mathfrak{G}$ onto G_{C^*}.

(c) This isomorphism is rational in the sense that $C(\sigma) = C(x_\sigma)$ and that σ' is a specialization of σ if and only if $x_{\sigma'}$ is a specialization of x_σ over C.

In order to extend the result proved for $F = F_0$ to the general case, we apply the proved result to the extension E of F_0. We thus obtain a connected algebraic group G_0 defined over C and a principal homogeneous space V_0 defined over F_0 such that V_0 has a point v generic over F_0 with $E = F_0(v)$, and the Galois group $\mathfrak{G}_0$ of E over F_0 is isomorphic with G_{0C^*} in the manner described in the statement of the theorem. Every $\sigma \in \mathfrak{G}$ maps F_0 onto the set of elements of σE which are algebraic over F, so that if σ is an automorphism of E then $\sigma F_0 = F_0$. Since by Proposition 3, σ always has a specialization which is an automorphism of E, which by Proposition 2(d) coincides with σ on F_0, we see that $\sigma F = F_0$ for every $\sigma \in \mathfrak{G}$. It now follows by Proposition 2(d) and (a) that the mapping

$$\sigma \to \text{restriction of } \sigma \text{ to } F_0$$

is a homomorphism of $\mathfrak{G}$ with kernel $\mathfrak{G}_0$ onto a finite group of automorphism of F_0 with fixed field F. If σ_i ($i = 1, \cdots, r$) are the elements of the finite set of itsolated isomorphisms mentioned in Proposition 2(a) (with say $\sigma_0 \in \mathfrak{G}_0$) and if $\mathfrak{G}_i$ is the set of specializations of σ_i, then the $\mathfrak{G}_i$ are cosets of $\mathfrak{G}_0$ in $\mathfrak{G}$.

Using the fact (Proposition 3(a)) that each $\mathfrak{G}_i$ contains an element σ_i' which is an automorphism of E, we follow [2] to extend the isomorphism $\mathfrak{G}_0 \approx G_0$ to an isomorphism $\mathfrak{G} \approx G$ (rational in the sense of (c)), where G is an algebraic group of which the component of the unity element is G_0. The varieties $V_i = V_0^{\sigma_i'}$ determined by the restrictions of the σ_i' to F_0 are precisely the conjugates of V_0 over F, and for each i the point $v_i = \sigma_i' v_0$ is a generic point of V_i over F_0. We now define the operation of G on the algebraic set $V = \bigcup V_i$ in the obvious manner, namely by setting $v_i x_{\sigma_j} = \sigma_j v_i$ for all i and j. We leave to the reader the verification that this makes V into a principal homogeneous space for G with the stated properties.

COLUMBIA UNIVERSITY.

REFERENCES.

[1] E. R. Kolchin, " Galois theory of differential fields," *American Journal of Mathematics*, vol. 75 (1953), pp. 753-824.

[2] ———, " On the Galois theory of differential fields," *American Journal of Mathematics*, vol. 77 (1955), pp. 868-894.

[3] H. Matsumura, "Automorphism-groups of differential fields and group varieties," *Memoirs of the College of Science, University of Kyoto, Series A*, vol. 28 (1954), pp. 283-292.

[4] M. Rosenlicht, " Some basic theorems on algebraic groups," *American Journal of Mathematics*, vol. 78 (1956), pp. 401-443.

[5] A. Weil, " On algebraic groups of transformation," *American Journal of Mathematics*, vol. 77 (1955), pp. 355-391.

[6] ———, " On algebraic groups and homogeneous spaces," *American Journal of Mathematics*, vol. 77 (1955), pp. 493-512.

RATIONAL POINTS OF ABELIAN VARIETIES OVER
FUNCTION FIELDS.*

S. Lang[1] and A. Néron.

We intend to give here a systematic and simplified exposition of the theorem of the base for divisors [5]. Since the appearance of [5], the theory of the Picard and Albanese varieties has become available, as well as Chow's theory of algebraic systems of abelian varieties (the K/k-trace and image). Using first an elementary equivalence criterion [9], we reduce the theorem of the base on a variety (projective, non-singular in codimension 1) to a theorem concerning the group of rational points of an abelian variety defined over a function field, by means of the above mentioned theories. We then prove the finiteness statement in two steps as usual: First, the so-called weak Mordell-Weil theorem, for which we reproduce the proof given in [3], and second the infinite descent.

We also take this opportunity of reproducing simultaneously a proof of the ordinary Mordell-Weil theorem concerning the group of rational points of an abelian variety defined over a number field, to the effect that this group is finitely generated. This proof is contained in § 3, § 5, §6, and § 7.

Aside from the above mentioned equivalence criterion, which is used only in § 1, we assume only that the reader is familiar with the basic theory of abelian varieties, as it is done for instance in [2]. We do not need the full theory of distributions, and reproduce the definition of the height of a point, together with its elementary functorial properties. This is all that is needed to carry out the infinite descent. We thus make this paper independent of [5], [6], [8].

1. The theorem of the base. If G denotes a set of geometric objects, we shall denote by G_k the subset of these objects which are rational over a field k.

Let V be a projective variety, non-singular in codimension 1, and defined over an algebraically closed field k which we may take to be a

---

* Received May 6, 1958.
[1] Fulbright Fellow.

95

universal domain. The group of divisors $D(V)_k$ contains the usual sub-groups of divisors which are algebraically equivalent to 0 and linearly equivalent to 0 respectively.

$$D(V)_k \supset D_a(V)_k \supset D_l(V)_k.$$

The Picard group $D_a(V)_k/D_l(V)_k$ can be given the structure of an abelian variety, the Picard variety. The theorem of the base asserts that *the factor group $D(V)_k$ modulo $D_a(V)_k$ is of finite type, i. e. finitely generated.* We are going to show here how this theorem can be reduced to a theorem concerning abelian varieties.

Let C_u be a generic curve on V over k, depending on generic linear parameters u. This means that there is a generic linear variety L_u over k such that $C_u = V \cdot L_u$. (Cf. [1], Ch. 7.) Let J be its Jacobian, defined over the field $k(u) = K$. Let $D_0(V)_k$ be the subgroup of $D(V)_k$ consisting of those divisors $X \in D(V)_k$ such that $X \cdot C_u$ is of degree 0. Then $D(V)_k/D_0(V)_k$ is infinite cyclic, and it will suffice to prove that $D_0(V)_k/D_a(V)_k$ is of finite type.

Let $\phi \colon C_u \to J$ be a canonical map of C_u into its Jacobian, defined over the algebraic closure of $k(u)$. Then

$$a \to S(\phi(a))$$

is the canonical map of $D_a(C_u) = D_0(C_u)$ into J. We have a homomorphism

$$h \colon D_0(V)_k \to J_{k(u)}$$

given by the formula

$$h(X) = S(\phi(X \cdot C_u)).$$

According to the elementary equivalence criterion, one knows that the kernel E of h, which contains $D_l(V)_k$, is of finite type modulo $D_l(V)_k$. Hence the factor group $[E + D_a(V)_k]/D_a(V)_k$ is of finite type. To prove the theorem of the base, it will therefore suffice to prove that $D_0(V)_k/[E + D_a(V)_k]$ is of finite type.

The inverse image $h^{-1}(h(D_a(V)_k))$ is precisely equal to $E + D_a(V)_k$. We have therefore an injection

$$0 \to D_0(V)_k/[E + D_a(V)_k] \to J_{k(u)}/hD_a(V)_k.$$

Now let $\psi \colon V \to A$ be a canonical map of V into its Albanese variety. For a suitable constant $b \in A$, we have a commutative diagram

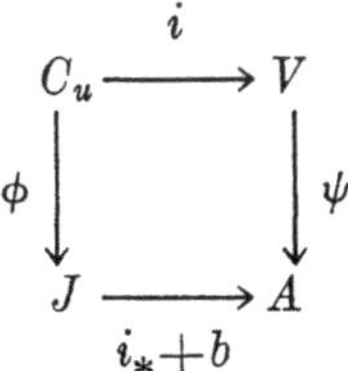

where i_* is the induced homomorphism.

Let us look at the inverse images of divisors in $D_a(A)_k$ under the composite maps $(i_* + b) \circ \phi$ and $\psi \circ i$. The formalism $(f \circ g)^{-1} = g^{-1} \circ f^{-1}$ can be applied to the first according to Appendix 1 of [2] since J and A are non-singular. A direct verification using intersection theory (associativity and the definition of $C_u = V \cdot L_u$) shows that it can also be applied to the second. According to the definition of the Picard variety, which one knows is obtained by pull back from the Albanese variety, we see that the map

$$h : D_a(V)_k \to J_{k(u)}$$

induces the rational homomorphism ${}^t i_*$ on $\hat{A}_k = D_a(V)_k / D_l(V)_k$, if we denote as usual by the upper index t the transpose of i_* on the Picard varieties. Consequently we have an injection

$$0 \to D_0(V)_k / [E + D_a(V)_k] \to J_{k(u)} / {}^t i_* \hat{A}_k.$$

By Chow's theory of the K/k-trace (whose definition is recalled below), one knows that $(\hat{A}, {}^t i_*)$ is a $k(u)/k$-trace of $\hat{J} = J$ ([2], Ch. 8, Th. 12). Consequently, to prove the theorem of the base, it will suffice to prove the following result.

THEOREM 1. *Let K be a finitely generated regular extension of a field k. Let A be an abelian variety defined over K, and let (B, τ) be its K/k-trace. Then $A_K / \tau B_k$ is of finite type.*

For the convenience of the reader, we recall that a couple (B, τ) consisting of an abelian variety B defined over k and an injective (i.e. purely inseparable) homomorphism $\tau : B \to A$ defined over K is said to be a K/k-trace of A if it satisfies the following universal mapping property: For any abelian variety C defined over an extension E of k which is free from K over k, and a homomorphism $\alpha : C \to A$ defined over KE, there exists a homomorphism $\alpha' : C \to B$ defined over E such that the following diagram is commutative.

$C \xrightarrow{\ \alpha\ } A$
$\alpha' \searrow \quad \nearrow \tau$
B

7

Essentially the same arguments which will be used to prove Theorem 1 will also give us

THEOREM 2. *Let K be an algebraic number field, of finite degree over the rationals. Let A be an abelian variety defined over K. Then A_K is finitely generated.*

From these two theorems we recover immediately the fact that if K is finitely generated over the prime field, then A_K is also finitely generated. It suffices to apply the two theorems to K and the algebraic closure of the prime field in K, viewed as a constant field.

2. Reduction steps. The propositions which we prove in this section will be used to reduce our main theorem (Theorem 1) to special cases which are technically easier to handle. Throughout this section, K will denote a finitely generated regular extension of a field k. By a regular extension, we shall always mean a finitely generated one.

We first show that to prove our main theorem, we may extend our function field by a finite separable extension.

PROPOSITION 1. *Let $L \supset K$ be a finite separable extension of K, also regular over k. Let A be an abelian variety defined over K. Let $(A^{K/k}, \tau_K)$ be its K/k-trace and $(A^{L/k}, \tau_L)$ its L/k-trace. Then the factor group*

$$[\tau_L(A^{L/k})_k \cap A_K]/\tau_K(A^{K/k})_k$$

is finite.

Proof. Let Γ_L be the graph of τ_L. It is defined over L. Let us take the intersection $H = \bigcap_\sigma \Gamma_L{}^\sigma$ of Γ_L and its conjugates over K, where σ ranges over the distinct isomorphisms of L over k. Then H is K-closed and is an algebraic subgroup of $A^{L/k} \times A$. Its connected component H_0 is therefore defined over K by Chow's theorem ([2], Ch. II).

$$\Gamma_L \cap [(A^{L/k})_k \times A_K] = H \cap [(A^{L/k})_k \times A_K]$$

and this group contains $H_0 \cap [(A^{L/k})_k \times A_K]$ as a subgroup of finite index, since H_0 is of finite index in H.

We have a surjective homomorphism

$$\Gamma_L \cap [(A^{L/k})_k \times A_K] \to \tau_L(A^{L/k})_k \cap A_K$$

by projection on the second factor. We contend that the inverse image of $\tau_K(A^{K/k})_k$ contains $H_0 \cap [(A^{L/k})_k \times A_K]$.

$$H_0 \cap [(A^{L/k})_k \times A_K] \to \tau_K(A^{K/k})_k$$

From this it will be clear that the desired factor group is finite.

Since H_0 is contained in the graph of a homomorphism of $A^{L/k}$ into A, it is itself the graph of a homomorphism β of its projection B on the first factor into A. Furthermore, B is contained in $A^{L/k}$, is defined over K, and hence over k by Chow's theorem. By the universal property of the K/k-trace, there exists a commutative diagram

$$\begin{array}{ccc} & \beta & \\ B & \longrightarrow & A \\ {}^{\beta'}\searrow & \nearrow^{\tau_K} & \\ & A^{K/k} & \end{array}$$

with β' defined over k, and hence $\beta(B_k)$ is contained in $\tau_K(A^{K/k})_k$. This proves our contention, and concludes the proof of our proposition.

COROLLARY. *If Theorem 1 is true for L/k, then it is true for K/k, i. e. if $A_L/\tau_L(A^{L/k})_k$ is finitely generated, so is $A_K/\tau_K(A^{K/k})_k$.*

Proof. This is immediate from the proposition and the fact that we have an injection

$$0 \to A_K/[\tau_L(A^{L/k})_k \cap A_K] \to A_L/\tau_L(A^{L/k})_k.$$

Next, we show that we may extend the constant field.

PROPOSITION 2. *Let k^* be any extension of k which is independent of K, and let $K^* = Kk^*$. Let A be an abelian variety defined over K and let (B, τ) be its K/k-trace. Then*

$$A_K \cap \tau B_{k^*} = \tau B_k.$$

Proof. From the definition of the trace, it is clear that (B, τ) is also a K^*/k^*-trace of A. Our assertion is obvious if k^* is separable over k. Hence it suffices to deal with a finite purely inseparable extension k^* of k. The inclusion $\tau B_k \subset A_K \cap \tau B_{k^*}$ is obvious. Conversely, let b be a point of B_{k^*} such that τb is rational over K. If we knew that τ is regular, i. e. that it is birational biholomorphic between B and its image in A, then in view of the fact that τ is defined over K, we would see immediately that $\tau^{-1}\tau b$ is rational over k. As we do not know that τ is regular, we use Chow's regularity theorem, according to the standard technique of [2], Ch. 9. We can write $K = k(u)$ for suitable parameters u, and $A = A_u$, $\tau = \tau_u$. Let $u_1, \cdots, u_m$ be independent generic specializations of u over k. For m large, the map

$$x \to (\tau_{u_1}x, \cdots, \tau_{u_m}x)$$

of B into $A_{u_1} \times A_{u_2} \times \cdots \times A_{u_m}$ is regular, and thus biholomorphic between

B and its image. If τb is rational over K, then $(\tau_{u_1} b, \cdots, \tau_{u_m} b)$ is rational over $k(u_1, \cdots, u_m)$, hence b is rational over $k(u_1, \cdots, u_m)$, hence b is rational over that field. Since b is purely inseparable over k, and since $k(u_1, \cdots, u_m)$ is regular over k, it follows that b is rational over k. This concludes the proof.

COROLLARY. *If Theorem 1 is true for K^*/k^*, then it is true for K/k.*

Proof. The factor group $A_K/\tau B_k$ can be identified with a subgroup of $A_{K^*}/\tau B_{k^*}$.

Finally, it will be convenient to deal with the case where K is a function field of dimension 1 over k, i.e. the function field of a curve, and the next proposition shows that the reduction to this case is trivial.

PROPOSITION 3. *Let $K \supset E \supset k$ be a tower of fields such that K is regular over E and E regular over k. If Theorem 1 is true for K/E and E/k, then it is true for K/k.*

Proof. Let A be as in Theorem 1, an abelian variety defined over K. Let $(A^{K/E}, \tau_{K/E})$ be a K/E-trace of A, and $(A^{E/k}, \tau_{E/k})$ a E/k-trace of $A^{K/E}$. By the universal mapping property, there is a purely inseparable homomorphism $\beta: A^{E/k} \to A^{K/k}$ defined over k such that the following diagram is commutative:

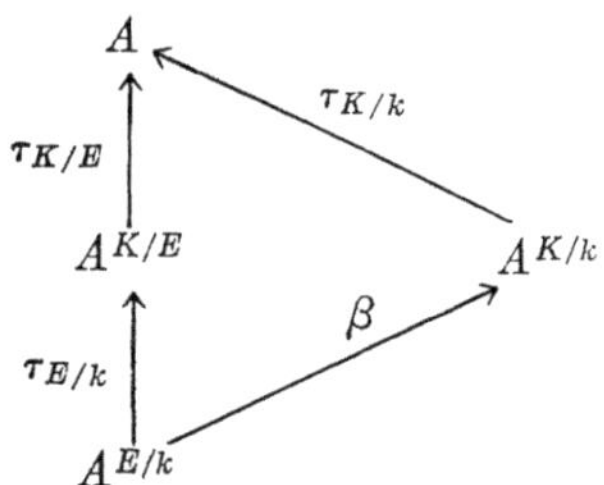

All the homomorphisms τ are purely inseparable, and thus injective. We have

$$A_K \supset \ \tau_{K/E}(A^{K/E})_E \supset \ \tau_{K/E}\tau_{E/k}(A^{E/k})_k = \tau_{K/k}\beta(A^{E/k})_k.$$

If we assume that $(A^{K/E})_E$ is of finite type modulo $\tau_{E/k}(A^{E/k})_k$, it follows that A_K modulo $\tau_{K/k}\beta(A^{E/k})_k$ is also of finite type. But this last group is contained in $\tau_{K/k}(A^{K/k})_k$. We see therefore that it suffices to prove Theorem 1 for each step K/E and E/k of our tower.

It is well known and easy to show that one can always construct a tower $E_0 = k \subset E_1 \subset \cdots \subset E_n = K$ such that each step E_i/E_{i-1} is regular and of transcendence degree 1. This comes from the lemma of Bertini's theorem, to the effect that if x, y are two elements of K, algebraically independent over k, and such that $y \notin K^p k$, then for all but a finite number of constants $c \in k$, K is regular over $k(x + cy)$. (Cf. [1], Ch. 6.) If k is finite, one needs a small additional argument. However, in view of Proposition 2, for our purposes, we may assume k infinite.

We have thus reduced the proof of Theorem 1 to the case where k is algebraically closed and K is of transcendence degree 1 over k, i.e. is the function field of a curve.

3. The weak Mordell-Weil Theorem.

By a *global field* we shall mean a field K which is either

a function field over an algebraically closed constant field k (i.e. a finitely generated regular extension of k)

or *an algebraic number field of finite degree over the rationals.*

We refer to these as the *function field case* and *number field case* respectively. We let $\dim K$ be the transcendence degree of K over k in the function field case. It is the dimension of a model of K over k. Our aim is to prove Theorems 1 and 2. The results of §2 (and eventually those of §8) will be used only to deal with the function field case. The case of number fields is thus somewhat simpler to deal with.

Let m be a natural number prime to the characteristic of K. Let A be an abelian variety defined over K, and let A_m denote the group of points of order m on A, i.e. the kernel of $m\delta$. To prove Theorem 1 or 2, we may assume that all point of A_m are rational over K, because we may deal with the finite separable extension $K(A_m)$ of K instead of K itself. This is obvious in the case of number fields, and follows from Proposition 1 in the case of function fields.

In this section, we prove the weak Mordell-Weil theorem, namely: *If A is an abelian variety defined over a global field K, such that $A_m \subset A_K$, and such that $\dim K = 1$ in the function field case, then the factor group A_K/mA_K is finite.* We reproduce essentially word for word the proof given in [3]. Although we could have limited ourselves to making a reference to that paper, we prefer to make our treatment here as self-contained as possible.

We shall use only two elementary and well-known properties of K. To state them, we denote by $\{\mathfrak{p}\}$ the set of all finite primes of K (i.e. in the case of function fields with $\dim K = 1$, the set of points of a model of K/k, rational over k, and in the case of number fields, the set of all finite primes). Our properties may then be stated as follows.

a.) Let S be a finite set of primes. Then there are only a finite number of abelian extensions of K of exponent m (i.e. such that $\sigma^m = 1$ for all automorphisms of the extension over K) which are unramified outside S.

b.) There exists a finite set S of primes such that for $\mathfrak{p} \notin S$, the abelian variety $A_\mathfrak{p}'$ obtained by reducing A modulo $\mathfrak{p}$ is non-degenerate.

For the reduction of varieties, we refer the reader to Shimura [7]. We recall that a reduction (or specialization) of an abelian variety is said to be non-degenerate if the specialized cycle has one component, with multiplicity 1, which is an abelian variety whose law of composition is obtained by specializing that of A. Property b.) may of course also be expressed by saying that in an algebraic system of varieties whose generic member is an abelian variety, almost all members of the system are also abelian varieties whose laws of composition are obtained by specializing that of the generic member. Here, the parameter variety is a model of K/k in the case of function fields, and is a so-called absolute curve in the case of number fields.

The following two lemmas achieve what we want. The first one ties up the factor group A_K/mA_K with the extension $K(1/m \cdot A_K)$ of K obtained by adjoining to K all points $y \in A$ such that $my \in A_K$. It is an abelian extension, whose automorphisms are induced by translations of A_m, in view of the fact that we assumed $A_m \subset A_K$. Indeed, if σ is in the Galois group G and $my = x \in A_K$ then $\sigma(my) = m\sigma y = x$, so that $\sigma y - y \in A_m$.

LEMMA 1. *The factor group A_K/mA_K is finite if and only if $K(1/m \cdot A_K)$ is a finite extension of K.*

Proof. One implication is obvious. Conversely, assume that $K(1/m \cdot A_K)$ is finite over K. We have a bilinear map

$$(A_K, G) \to A_m$$

obtained as follows. Let $x \in A_K$. Let y be such that $my = x$. We put $(x, \sigma) = \sigma y - y$. This element of A_m obviously does not depend on the y selected such that $my = x$. It is clear that (x, σ) is bilinear in x and σ. It is trivially verified that the right-hand kernel of our pairing is 1, while the left-hand kernel is precisely mA_K. The groups A_K/mA_K and G are therefore paired exactly into A_m. Since G and A_m are finite, it follows that A_K/mA_K is also finite.

To conclude the proof of the weak Mordell-Weil Theorem, there remains but to show that $K(1/m \cdot A_K)$ is finite over K. This follows from a.) and b.) and our second lemma.

LEMMA 2. *If $\mathfrak{p}$ is a prime such that $A_{\mathfrak{p}}'$ is non-degenerate, and $\mathfrak{p} \nmid m$, then $K(1/m \cdot A_K)$ is unramified over K at $\mathfrak{p}$.*

Proof. This being a local statement, we may go over to the completion $K_{\mathfrak{p}}$ of K with respect to $\mathfrak{p}$. In the function field case, $K_{\mathfrak{p}}$ is the power series field over k. Let $x \in A_k$. It suffices to show that $K_{\mathfrak{p}}(1/m \cdot x)$ is unramified over $K_{\mathfrak{p}}$. We use a prime to denote the reduction of objects modulo $\mathfrak{p}$. Let $\mathfrak{a} = (m\delta)^{-1}(x)$. Then $\mathfrak{a}' = (m\delta')^{-1}(x')$, since $A_{\mathfrak{p}}'$ is non-degenerate. All points in $(m\delta')^{-1}(x')$ occur with multiplicity 1, since $\mathfrak{p} \nmid m$, and are rational over a finite separable extension L' of $K_{\mathfrak{p}}'$. (Here, $K_{\mathfrak{p}}'$ is the residue class field, and in the function field case, we have $L' = K_{\mathfrak{p}}' = k$.) By a suitable form of Hensel's lemma [3], it follows that all points of $\mathfrak{a}$ are rational over the unramified extension of $K_{\mathfrak{p}}$ obtained by lifting L'. In the case of function fields, this unramified extension is of course $K_{\mathfrak{p}}$ itself.

The needed form of Hensel's lemma, as stated in [3], is the following. *Let L be complete under a discrete valuation, with residue class field L'. Let $\mathfrak{a}$ be a positive 0-cycle, in a projective space, say, rational over L, and let P' be a point of $\mathfrak{a}'$ which is rational over L', and of multiplicity 1 in $\mathfrak{a}'$. Then there is a unique point P in $\mathfrak{a}$ which specializes to P', and P is rational over L.* The proof is immediate, taking into account the uniqueness of the extension of the valuation to algebraic extensions of L.

4. Heights in function fields. In order to carry out the infinite descent in § 7, we need to be able to measure the size of a point rational over K, or as is customary to say, its height. It is convenient to give the definition of the height separately for the function field and number field case. We do this in this section and the next. The fundamental properties of heights are given in § 6, and there the statements and proofs can again be formulated uniformly for the two cases.

In this section, we assume that k is algebraically closed. Let W be a complete non-singular curve defined over k, and let $K = k(w)$ be a function field for W over k, (w) being a generic point of W over k. We follow [6] essentially without change.

We shall define the height of a point P in projective space $\boldsymbol{P}^n$, rational over K. Let $(y_0, \cdots, y_n)$ be a set of homogeneous coordinates for P, rational over K. Then the y_i may be viewed as functions on W, defined over k. We define the *height of P* to be

$$(1) \qquad h(P) = h(y) = - \deg \inf_i (y_i),$$

where (y_i) denotes as usual the divisor of y_i. Since the degree of the divisor of a function is equal to 0, we see on the one hand that $h(P)$ does not depend on the set of homogeneous coordinates representing P, and on the other hand that $h(P) \geqq 0$ (because we could take say $y_0 = 1$). Furthermore, $h(P) = 0$ if and only if P is rational over k, i. e. is constant.

One can give an alternate geometric definition of the height. Let $T(P)$, which we also write $T(y)$, be the locus of P over k. It is a curve, which may of course have singularities. We have a rational map $f: W \to \boldsymbol{P}^n$ such that that $f(w) = (y)$. It is induced by a surjective rational map of W onto $T(y)$. We contend that

$$(2) \qquad h(y) = (\deg f) \deg T(y),$$

where $\deg f$ is the degree of the rational map of W onto $T(y)$ (non-zero if f is not constant), and $\deg T(y)$ is the projective degree of the variety $T(y)$. To prove this, we may assume without loss of generality that none of the y_i is 0, and that (y) is not constant. Let (Y) be the variables of $\boldsymbol{P}^n$, and let $H = H(Y)$ be a hyperplane defined over k, such that $T(y) \cdot H$ is defined. Put $z = H(y)$. Then for each i, y_i/z is an element of K, and thus a function on W. Furthermore, if we denote by H_i the divisor of the hyperplane $Y_i = 0$, then $H_i - H$ is the divisor of a function on $\boldsymbol{P}^n$, which can also be viewed as a function on $W \times \boldsymbol{P}^n$. Let Γ_f denote the graph of f. It is biholomorphic to W under projection on the first factor. Under this identification, it is clear that the function on $W \times \boldsymbol{P}^n$ whose divisor is $W \times (H_i - H)$ induces y_i/z on Γ_f. Hence by [10], F-VIII$_3$, Th. 4, Cor. 2, we get

$$(y_i/z) = f^{-1}(H_i) - f^{-1}(H).$$

By definition, we have

$$h(y) = - \deg \inf_i (y_i/z) = - \deg \inf_i [f^{-1}(H_i) - f^{-1}(H)].$$

Since the hyperplanes H_i have no point in common, neither do the divisors $f^{-1}(H_i)$ on W, and consequently, $h(y) = - \deg(-f^{-1}(H)) = \deg(f^{-1}(H))$. We have

$$h(y) = \deg[\Gamma_f \cdot (W \times H)]$$

since the degree of a 0-cycle does not change under projection. Projecting on the right, we have

$$h(y) = (\deg f)(\deg T(y) \cdot H)$$

which proves our contention.

From definition (2) of he height, we see that for $h(y)$ bounded, the degree of $T(y)$ is also bounded. Hence by the theory of Chow coordinates, we get our first property of heights.

PROPERTY 1F. *In the function field case, with* $\dim K = 1$, *let* (y) *be a point of* $\boldsymbol{P}^n$ *rational over* K. *Let* $T(y)$ *be the locus of* (y) *over* k. *If* $h(y)$ *is bounded, then* $\deg T(y)$ *is bounded, and* $T(y)$ *can belong only to a finite number of algebraic families.*

The definition of heights can also be given when $\dim W > 1$ [6]. In fact, let W be a projective normal model of K over k. If $P = (y)$ is again a point in $\boldsymbol{P}^n$, rational over K, we can define $h(P)$ as in (1), with the understanding that deg now denotes the projective degree of the divisor $\inf_i(y_i)$, in the given projective embedding of W. Thus if $\dim W > 1$, the height of P depends on the choice of model for K over k, while it does not if $\dim W = 1$. As in the case where $\dim W = 1$, we have

PROPOSITION 4. *Let* W *be a projective normal model of* K *over the algebraically closed field* k. *Let* $(y_0, \cdots, y_n)$ *be projective coordinates of a point* P *in* $\boldsymbol{P}^n$, *with* $y_i \in k(W) = K$. *Let* $T(P)$ *be the locus of* P *over* k, *and* $f: W \to \boldsymbol{P}^n$ *the corresponding rational map of* W *into* $\boldsymbol{P}^n$. *Then for a generic hyperplane* H *of* $\boldsymbol{P}^n$, *we have*

$$h(P) = -\deg \inf_i(y_i) = \deg f^{-1}(H).$$

Proof. The arguments follow exactly those given above for curves. The degree is the projective degree in the given projective embedding of W. One sees immediately that the projection on W of any subvariety of codimension 1 of Γ_f must be of codimension 1 on W, and hence simple on Γ_f since f is defined at such a subvariety. Thus W and Γ_f are biholomorphic at such a subvariety, and the arguments given above hold (especially [10], F-VIII$_3$, Th. 4, Cor. 2).

The analogue of (2) if $\dim W > 1$ could also be given, but we omit it. We do not need it for the proof of Theorem 1. We note that the properties of heights proved in §6 depend only on definition (1).

When dealing simultaneously with the function field and number field case, it is often convenient to replace the height as we have defined it in (1) by an exponential of it. Indeed, in number fields, the valuations are usually written multiplicatively because of the archimedean ones. To make arguments run completely parallel and to avoid a repetition of formulas in additive and multplicative notation, we therefore define the multiplicative height in function fields as follows. Let c be a number, $0 < c < 1$. We put

$$(3) \qquad h^*(P) = c^{-h(P)}$$

and call this the *multiplicative height*, or simply height if it is used constantly throughout a section. Let $\dim W = 1$. For each point $\mathfrak{p}$ of W, rational over k, and a function $y \in K$, we define the absolute value $v_\mathfrak{p}$ as usual by

$$(4) \qquad v_\mathfrak{p}(y) = c^{\operatorname{ord}_\mathfrak{p} y},$$

where $\operatorname{ord}_\mathfrak{p} y$ is the order of the zero of y at $\mathfrak{p}$ (negative if y has a pole). Thus a function has a zero of high order at $\mathfrak{p}$ if $v_\mathfrak{p}(y)$ is close to 0, and a pole of high order at $\mathfrak{p}$ if $v_\mathfrak{p}(y)$ is large, close to $+\infty$.

In view of the — sign in (1) and the fact that $0 < c < 1$, we get by a trivial computation

$$(5) \qquad h^*(P) = \prod_\mathfrak{p} \sup_i v_\mathfrak{p}(y_i).$$

It is this expression which is used to define the height in number fields, as we shall see in the next section. We note that $h^*(P) \geq 1$.

If $\dim W > 1$, we shall say that $\mathfrak{p}$ is a *prime divisor of K* if it is a prime divisor of W, i. e. a subvariety of W of codimension 1, defined over k. Then $\deg \mathfrak{p}$ is the projective degree of $\mathfrak{p}$ in the given projective embedding of W, and $v_\mathfrak{p}(y)$ is defined by

$$v_\mathfrak{p}(y) = c^{(\deg \mathfrak{p})\operatorname{ord}_\mathfrak{p} y}.$$

Formula (5) is then clearly applicable without change.

5. Heights in number fields. Let K be a number field of finite degree over the rationals $\boldsymbol{Q}$. Let $\mathfrak{p}$ be a finite prime of K, corresponding to a prime ideal in its ring of integers. Let $N\mathfrak{p}$ be as usual the number of elements in the residue class field. For $y \in K$, we let

$$v_\mathfrak{p}(y) = (1/N\mathfrak{p})^{\operatorname{ord}_\mathfrak{p} y},$$

where $\operatorname{ord}_\mathfrak{p} y$ is the order of y at the discrete valuation of K determined by $\mathfrak{p}$.

If $\mathfrak{p}$ is an infinite prime, i. e. is a real or a pair of complex conjugate embeddings of K into the complex numbers, then we let $v_{\mathfrak{p}}(y)$ be the ordinary absolute value if the embedding is into the reals, and the square of the ordinary absolute value if the embedding is not real. The product formula states that for $y \in K$, $y \neq 0$, we have

$$\prod_{\mathfrak{p}} v_{\mathfrak{p}}(y) = 1.$$

Let $(y_0, \cdots, y_m)$ be a set of homogeneous coordinates for a point P in $\boldsymbol{P}^m$, with $y_i \in K$. We define the *height* $h(P)$ of P by the relation

$$(6) \qquad\qquad h(P)^{[K:\boldsymbol{Q}]} = \prod_{\mathfrak{p}} \sup_i v_{\mathfrak{p}}(y_i).$$

The product formula guarantees that this depends only on the point in projective space and not on the coordinates chosen. Since one of the y_i could have been chosen equal to 1, we see that $h(P) \geqq 1$. The extra term $[K:\boldsymbol{Q}]$ has been added to insure that $h(P)$ does not depend on the field K over which P is rational.

If σ is an automorphism of K over $\boldsymbol{Q}$, then by transport of structure, we have $h(P^\sigma) = h(P)$.

The (absolute) degree $d(P)$ of P is defined to be the degree $[K:\boldsymbol{Q}]$, where $K = \boldsymbol{Q}(P)$ is the field obtained by adjoining to $\boldsymbol{Q}$ a set of affine coordinates for P, say $y_1/y_0, \cdots, y_m/y_0$. We have the following strong version of Property 1F.

PROPERTY 1N. (Northcott) *Let h_0, d_0 be two fixed numbers. Then there is only a finite number of points P in $\boldsymbol{P}^m$ such that $d(P) \leqq d_0$ and $h(P) \leqq h_0$.*

Proof. Let us first consider the case where $d(P) = 1$, i.e. P is rational over $\boldsymbol{Q}$. Multiplying the y_i by a suitable integer, we may assume that all y_i are integers, and that their g. c. d. is equal to 1. For all finite primes p of $\boldsymbol{Q}$, we then get $\sup_i v_p(y_i) = 1$. The height of P is then determined by the ordinary absolute value, and the result is obvious.

We prove the property in general by reducing it to the preceding case.

Let $(T_0, \cdots, T_m)$ be variables. Let

$$F(T) = \sum x_\alpha \boldsymbol{M}_\alpha(T)$$

be a form in the T's, where the $\boldsymbol{M}_\alpha$ are monomials, whose coefficients x_α lie in a number field. Then (x) may be viewed as a point in some projective space $\boldsymbol{P}^N$. We define the *height of F* to be the height of (x), and write it $h(F)$.

LEMMA. *Let d_1, d_2 be two natural numbers. Then there exists a number s depending only on d_1, d_2 such that if F_1, F_2 are two forms of degrees d_1, d_2 respectively with coefficients in a number field K, then*

$$h(F_1F_2) \leqq sh(F_1)h(F_2).$$

Proof. Let $F_1 = \sum x_\alpha M_\alpha(T)$ and $F_2 = \sum y_\beta M_\beta(T)$. Put $F_1F_2 = \sum z_\lambda M_\lambda(T)$. Then

$$z_\lambda = \sum x_{\alpha(\lambda)} y_{\beta(\lambda)},$$

where $\alpha(\lambda)$, $\beta(\lambda)$ range over those indices α, β such that $M_{\alpha(\lambda)} M_{\beta(\lambda)} = M_\lambda$. If $\mathfrak{p}$ is a finite prime of K, then obviously

$$v_\mathfrak{p}(z_\lambda) \leqq \sup_\alpha v_\mathfrak{p}(x_\alpha) \sup_\beta v_\mathfrak{p}(y_\beta),$$

and hence we can add a $\sup_\lambda$ to the left hand side of this inequality.

If $\mathfrak{p}$ is archimedean, then there obviously exists an integer $s_\lambda(d_1, d_2)$ such that

$$v_\mathfrak{p}(z_\lambda) \leqq s_\lambda(d_1, d_2) \sup_\alpha v_\mathfrak{p}(x_\alpha) \sup_\beta v_\mathfrak{p}(y_\beta).$$

This $s_\lambda(d_1, d_2)$ is the number of terms in the sums expressing z_λ in terms of $x_{\alpha(\lambda)} y_{\beta(\lambda)}$, or its square if $\mathfrak{p}$ is complex. Hence there exists an integer $s(d_1, d_2)$ such that

$$\sup_\lambda v_\mathfrak{p}(z_\lambda) \leqq s(d_1, d_2) \sup_\alpha v_\mathfrak{p}(x_\alpha) \sup_\beta v_\mathfrak{p}(y_\beta).$$

Taking the product over all $\mathfrak{p}$, we get

$$h(z)^{[K:Q]} \leqq s(d_1, d_2)^r h(x)^{[K:Q]} h(y)^{[K:Q]},$$

with an integer $r \leqq [K:Q]$. This proves our lemma.

To conclude the proof of Property 1N, let $(y_0, \cdots, y_m)$ be a point of degree d_0, with say $y_0 = 1$. Let

$$F(T) = \prod_\sigma (y_0^\sigma T_0 + \cdots + y_m^\sigma T_m)$$

the product being taken over all conjugates of the field $Q(y)$, so that $F(T)$ is of degree d_0. We call $F(T)$ the *Chow form* of $P = (y)$. It is clear that two points have the same Chow form if and only if they are conjugate over Q, because $F(T)$ factorizes essentially uniquely. Hence by the lemma, we get

$$(7) \qquad h(F) \leqq s(d_0) h(P)^{d_0},$$

where $s(d_0)$ is an integer depending only on d_0, and not on P. Our property is obvious from this relation, the first paragraph of the proof, and the fact that $h(P) = h(P^\sigma)$.

6. Properties of heights. Let K be a global field, and let V be an abstract variety defined over K. Let λ, λ' be two real-valued functions on V_K (the rational points of V in K). We shall say that they are *equivalent*, and write $\lambda \sim \lambda'$, if there exist two real numbers $c_1, c_2 > 0$ such that for all $P \in V_K$, we have

$$c_1 \lambda(P) \leqq \lambda'(P) \leqq c_2 \lambda(P).$$

This equivalence relation will be applied to heights. In this section and the next, the height is taken to be multiplicative in the function field case, and we write h instead of h^*. (If it were taken additive, as in (1), then we would make the above relation read

$$c_1 + \lambda(P) \leqq \lambda'(P) \leqq c_2 + \lambda(P)$$

and we would omit the condition $c_1, c_2 > 0$.)

Actually, we remark (as in [8]) that in number fields, we can consider the stronger equivalence relation where λ, λ' are functions on all absolutely algebraic points of V, the constants working uniformly for all such points. All the statements which we shall make concerning the equivalence of heights in the sequel are valid under this stronger interpretation. If we adopted a device in function fields analogous to the one in number fields, i.e. once K is fixed, define the height for points rational over the algebrais closure $\bar{K}$ of K by taking suitable roots, then a similar remark would apply in function fields.

Let now V be a complete, abstract, normal variety defined over K. Let $\phi: V \to \boldsymbol{P}^m$ be an everywhere defined rational map of V into projective space, defined over K. Then for each point P of V_K, $\phi(P)$ is a point of $\boldsymbol{P}^m$, rational over K, and we can thus define its height, which we shall denote by $h_\phi(P)$. We shall give below conditions under which the real-valued functions h_ϕ are equivalent. As mentioned before, the properties of heights in the case of function fields depend only on definition (1) or (3), and for the proof of Theorem 1, are needed only in the case $\dim K = 1$.

It follows immediately from the definition of the height that if ϕ' is another rational map of V into $\boldsymbol{P}^m$ which differs from ϕ by a projective transformation defined over K, then $h_\phi \sim h_{\phi'}$. Indeed, the elements of K which enter into such a transformation can introduce only a uniformly bounded change in the height of a point because they are fixed once for all.

We shall obtain mappings $\phi: V \to \boldsymbol{P}^m$ by means of linear systems. Let $\mathfrak{L}$ be a linear system on V, defined over K. This means that we can find a divisor $X_0 \in \mathfrak{L}$, rational over K, such that the space of functions L_0 consisting of all functions on V whose divisors are of type $X - X_0$ with $X \in \mathfrak{L}$ has a basis defined over K. If $(f_0 = 1, \cdots, f_m)$ is such a basis, then it defines a rational map $\phi: V \to \boldsymbol{P}^m$. If $\mathfrak{L}$ is without fixed point, then ϕ is everywhere defined ([1], Ch. 6). In view of the remark in the preceding paragraph, we see that the equivalence class of h_ϕ actually depends only on the linear system.

Let $\mathfrak{M}$ be another linear system on V, also defined over K. Then we can define the sum $\mathfrak{L} + \mathfrak{M}$ as a linear system in the usuall manner. If $(f_0 = 1, \cdots, f_m)$ is a basis for L_0 over K, and $(g_0 = 1, \cdots, g_n)$ is a basis for M_0 over K obtained in a similar manner, then we get a vector space of functions N_0 generated by the products $f_i g_j$. We have $f_0 g_0 = 1$. The divisor of $f_i g_j$ is $(f_i g_j) = (f_i) + (g_j)$. If we write $(f_i) = X_i - X_0$ and $(g_j) = Y_j - Y_0$ then $(f_i g_j) = X_i + Y_j - (X_0 + Y_0)$. The space N_0 gives rise to a linear system $\mathfrak{N}$ called the sum of $\mathfrak{L}$ and $\mathfrak{M}$. If $\mathfrak{L}$ and $\mathfrak{M}$ are both without fixed points, so is $\mathfrak{L} + \mathfrak{M}$.

Let $\phi: V \to \boldsymbol{P}^m$ and $\psi: V \to \boldsymbol{P}^n$ be two rational maps derived from $\mathfrak{L}$ and $\mathfrak{M}$. The functions $\{f_i g_j\}$ give rise to a rational map η of V into $\boldsymbol{P}^{(m+1)(n+1)-1}$. If we denote by $\phi + \psi$ any one of the rational maps (defined over K) derived from $\mathfrak{L} + \mathfrak{M}$, and determined only up to a projective transformation, then obviously, $h_\eta = h_\phi h_\psi$ and $h_\eta \sim h_{\phi + \psi}$. Summarizing, we get

PROPERTY 2. *Let V be a complete, abstract, normal variety defined over K. Let $\mathfrak{L}$, $\mathfrak{M}$ be two linear systems on V, also defined over K, and without fixed points. Let ϕ, ψ be two rational maps of V into $\boldsymbol{P}^m$, $\boldsymbol{P}^n$ respectively, defined by these systems over K, and let $\phi + \psi$ be a rational map defined over K by $\mathfrak{L} + \mathfrak{M}$. Then $h_{\phi + \psi} \sim h_\phi h_\psi$.*

The next property also follows immediately from the definitions.

PROPERTY 3. *Let U, V be two complete, abstract varieties defined over K. Assume U normal and V non-singular. Let $\omega: U \to V$ be an everywhere defined, surjective rational map, defined over K. Let $\mathfrak{L}$ be a linear system on V, defined over K, and let $\omega^{-1}(\mathfrak{L})$ be the linear system on U consisting of all divisors $\omega^{-1}(X)$ as X ranges over $\mathfrak{L}$. Let ϕ be a rational map of V into $\boldsymbol{P}^m$ defined by $\mathfrak{L}$ over K. Then $\phi \circ \omega$ is a rational map associated with the linear system $\omega^{-1}(\mathfrak{L})$. Assume in addition that $\mathfrak{L}$ is without fixed points. Then so is $\omega^{-1}(\mathfrak{L})$, and we have*

$$h_{\phi \circ \omega} = h_\phi \circ \omega.$$

Proof. We have assumed V non-singular in order to insure that the inverse image of a divisor linearly equivalent to 0 is also linearly equivalent to 0 (cf. [2], App. 1). That $\omega^{-1}(\mathfrak{L})$ is then a linear system is obvious, and so is the rest of our assertions.

The preceding three properties of heights have been trivial consequences of the definitions. Our fourth and last property, although not difficult to prove, will require a slightly more elaborate argument. We have taken it and its proof from Weil's paper [8].

PROPERTY 4. *Let V, $\mathfrak{L}$, $\mathfrak{M}$, ϕ, ψ be as in Property 2. If the divisors of $\mathfrak{L}$ and $\mathfrak{M}$ are linearly equivalent to each other, then $h_\phi \sim h_\psi$.*

Proof. In the course of the proof we shall use frequently the fact that if a function $f \in K(V)$ is not defined at a point Q of V, then it has a pole passing through Q ([1], Ch. 6) and hence if there exists a place of $K(V)$ over K which maps f on ∞ (resp. on 0), then f has a pole (resp. a zero) passing through Q.

By transitivity, we may clearly assume that $\mathfrak{M}$ is the complete linear system containing $\mathfrak{L}$.

Let X_0 be as before a divisor of $\mathfrak{L}$, rational over K, such that the derived space of functions L_0 has a basis $(f_0, \cdots, f_m)$ defined over K. For each f_i we can write

$$(f_i) = X_i - X_0.$$

We denote by ϕ_i the rational map of V into the affine K-open subset of $\boldsymbol{P}^m$ determined by the functions $(f_0/f_i, \cdots, f_m/f_i)$. We let V_i be the K-open subset $V - \mathrm{supp}(X_i)$ of V. Then the V_i cover V, and ϕ_i is defined at every point of V_i.

In view of our assumption on $\mathfrak{M}$, we may take $Y_0 = X_0$ and $M_0 = L(X_0)$. Let $(g_0, \cdots, g_n)$ be a basis of $L(X_0)$ defined over K, and put

$$(g_j) = Y_j - X_0.$$

We may assume that $g_i = f_i$ $(i = 0, \cdots, m)$. We have a rational map ψ_j of V into the affine K-open subset of $\boldsymbol{P}^n$ determined by $(g_0/g_j, \cdots, g_n/g_j)$, and ψ_i is defined at every point of V_i for $i = 0, \cdots, m$. Thus to compute $h_\psi(P)$ for $P \in V_K$, we may restrict ourselves to $h_{\psi_i}(P)$ for $P \in V_i \cap V_K$.

Consider first the points of V_K which are in V_0. Let g be any one of the g_j $(j = 0, \cdots, n)$. We contend that the ideal generated by $(f_0/g, \cdots, f_m/g)$ in the ring $K[f_0/g, \cdots, f_m/g]$ is the unit ideal. Otherwise, it is contained in a maximal ideal, and there is a place of the function field $K(V)$ over K which maps all f_i/g on 0. This place induces a point Q on V, and hence the functions f_i/g must all have a zero passing through Q. Since (f_i/g) $= X_i - Y$ with $Y > 0$, this implies that $Q \in \mathrm{supp}(X_i)$ for all i, and contradicts our assumption that they have no point in common.

From the above, we can write

$$1 = \sum z_\nu \boldsymbol{M}_\nu (f_i/g)$$

with $z_\nu \in K$. Here, $\boldsymbol{M}_\nu$ stands for a monomial $(f_0/g)^{\nu_0} \cdots (f_m/g)^{\nu_m}$. Note that the z_ν do not depend on $P \in V$, and that $\deg \boldsymbol{M}_\nu \geqq 1$.

Put $x_i = f_i(P)$, where P is any point in V_0, and $y = g(P)$. Suppose first that $y \neq 0$. Then

$$1 = \sum z_\nu \boldsymbol{M}_\nu (x_i/y).$$

For every prime $\mathfrak{p}$ of K, it follows that there exists a number $c_\mathfrak{p} > 0$ which is equal to 1 for all but a finite number of $\mathfrak{p}$, such that

$$\sup_i v_\mathfrak{p} (x_i/y) \geqq c_\mathfrak{p}.$$

(In function fields, this means that there exists a finite set S of prime divisors of W such that the x_i/y cannot have a common zero for $\mathfrak{p} \notin S$, and that for $\mathfrak{p} \in S$, the order of such a common zero cannot be arbitrarily high. The set S can be taken to be the set where some z_ν has a pole.)

Since we selected g arbitrarily among the g_j, we get

$$\sup_i v_\mathfrak{p} (x_i) \geqq c_\mathfrak{p} \sup_j v_\mathfrak{p} (y_j),$$

where $y_j = g_j(P)$. We emphasize that $c_\mathfrak{p}$ depends only on the z_ν, and not on $P \in V_0 \cap V_K$. Furthermore, this formula clearly holds whether $y_j = 0$ or not, i.e. for all P in $V_0 \cap V_K$. Taking the product, we see that there exists a number $c_1 = c_1(V_0) > 0$ such that for all $P \in V_0 \cap V_K$ we have

$$c_1 h_\psi (P) \leqq h_\phi (P).$$

By symmetry, using the same arguments as above, there exists a constant $c_2 > 0$ such that for all $P \in V_0 \cap V_K$, we have

$$c_1 h_\psi (P) \leqq h_\phi (P) \leqq c_2 h_\psi (P).$$

Since V is covered by the finite number of K-open sets $V_0, \cdots, V_m$, we can repeat the above procedure for each one of them. We thus obtain numbers $c_1(V_i)$ and $c_2(V_i)$. In the statement of our property, we simply take the smallest of the former and the largest of the latter to conclude the proof.

7. The infinite descent. Let K be a global field, and let A be an abelian variety embedded in a projective space $\mathbf{P}^n$ over K. The height of a point $P \in A_K$ determined by the identity mapping of A into $\mathbf{P}^n$ is denoted by $h(P)$. In the function field case, we deal with the multiplicative height.

Let $m > 1$ be a natural number such that A_K/mA_K is finite, and let $a_1, \cdots, a_s$ be representatives of A_K/mA_K. Then any point P in A_K is congruent to some $a_i \bmod mA_K$. Denote by S a sequence $\{P_0, P_1, \cdots\}$ of points of A_K, constructed by starting with an arbitrary point P_0, and such that

$$mP_{\nu+1} = P_\nu - a_{i_\nu}.$$

We contend that there is a number $c_1 > 0$ independent of S, and an integer $\nu(S)$ depending on S such that for any sequence S, we have $h(P_\nu) \leqq c_1$ for $\nu > \nu(S)$. This contention is an obvious consequence of the fact, to be proved below, that there exists a number $c_2 > 0$ independent of S such that

$$(8) \qquad\qquad h(P_{\nu+1})^{m^2-1} \leqq c_2 h(P_\nu).$$

Actually, we shall prove

PROPOSITION 5. *Let K be a global field. Let A be an abelian variety embedded in projective space $\mathbf{P}^n$ over K. Let $m > 1$ be a natural number, and let $a \in A_K$. Then there exists a number $c_2(a, m)$ depending on a and on m, such that for any $P \in A_K$, we have*

$$h(P)^{m^2-1} \leqq c_2(a, m) h(mP + a).$$

To deduce (8) from Proposition 5, we let a range over the a_i, and let $c_2 = \sup_i c_2(a_i)$.

From the manner in which the sequences are constructed, we see that for each ν, there exist integers $n_1, \cdots, n_s$ such that

$$P_0 = m^\nu P_\nu + n_1 a_1 + \cdots + n_s a_s$$

and we obtain the pay-off of our infinite descent.

COROLLARY. *Let m be such that A_K/mA_K is finite, and let $a_1, \cdots, a_s$ be representatives in A_K of A_K/mA_K. There exist a number c_1 and a subset $\mathfrak{S}$ of A_K such that:*

(i) $h(P) \leqq c_1$ *for all $P \in \mathfrak{S}$, and*

(ii) *for any $P_0 \in A_K$, there exist integers $n_0, n_1, \cdots, n_s$ and a point $P \in \mathfrak{S}$ such that*

$$P_0 = n_0 P + n_1 a_1 + \cdots + n_s a_s.$$

8

In the case of number fields, we apply Property 1N to conclude the proof of Theorem 2. The case of function fields will require an additional argument, which will be supplied in the next and final section.

We observe that A_K/mA_K was proved finite in §3 only for a special case of K. A posteriori, once Theorems 1 and 2 are proved, one sees of course that A_K/mA_K is finite for all global fields.

Let us now prove Proposition 5. Let a be a point of A_K. Let $\omega : A \to A$ be the rational map $\omega u = mu + a$. Let X be a hyperplane section of A in its given projective embedding, rational over K. From now on we use constantly the results and notations of [2], Ch. 5 concerning divisors on abelian varieties. Aside from those, we use only Properties 2, 3, 4 of heights. To begin with, we have immediately from the definitions

$$\omega^{-1}(X) = (m\delta)^{-1}(X_{-a}) = (m\delta)^{-1}(X_{-a} - X) + (m\delta)^{-1}(X).$$

We also have

$$(m\delta)^{-1}(X) \equiv m^2 X,$$

this being deeper. One knows that the equivalence $\equiv$ is the same as algebraic equivalence. (This uses the fact that there is no torsion, not proved in [2]. Using only what is proved there, namely that it is the same as the torsion equivalence, we could dispense with this, at the cost of introducing in an obvious manner a multiple of the equivalences below.) Since X is a hyperplane section, the homomorphism $u \to \mathrm{Cl}(X_u - X)$ of A into $\hat{A}$ is surjective. Hence there is a point b of A such that

$$(m\delta)^{-1}(X) - m^2 X \sim X_b - X.$$

Furthermore, for the same reason, there exists a point $c \in A$ such that $(m\delta)^{-1}(X_{-a} - X) \sim X_c - X$. Thus we get finally

$$\omega^{-1}(X) \sim X_c - X + m^2 X + X_b - X$$
$$\sim m^2 X + X_d - X$$
$$\sim (m^2 - 1)X + X_d$$

with $d = b + c$. Now d is not necessarily rational over K, but since X is ample, so is X_d. Since $\omega^{-1}(X) - (m^2 - 1)X$ is linearly equivalent to an ample divisor over some field, and is itself rational over K, we conclude that there exists an ample positive divisor X' rational over K such that

$$(9) \qquad\qquad \omega^{-1}(X) \sim (m^2 - 1)X + X'.$$

The divisors in the complete linear system containing $\omega^{-1}(X)$ are linearly equivalent to the divisors of the sum of the complete linear systems containing $(m^2-1)X$ and X'. If $Y > 0$ is a divisor on A rational over K, whose complete linear system $\mathfrak{L}(Y)$ is without fixed point, we select a definite rational map ϕ_Y defined over K, associated with $\mathfrak{L}(Y)$, in the class of such mappings determined only up to a projective transformation. We shall write h_Y instead of h_{ϕ_Y}. From (9) and Properties 2 and 4 we get

$$h_{\omega^{-1}(X)} \sim h_X{}^{(m^2-1)}h_{X'}.$$

Also by Property 4, we have $h \sim h_X$. (The linear system of hyperplane sections in the given embedding of A in $\boldsymbol{P}^n$ might not be complete, so we cannot write an equality.) Hence by Property 3 and 4 we see that there exists a number $c_2 = c_2(a, m)$ such that

$$h(P)^{m^2-1}h_{X'}(P) \leqq c_2 h(\omega(P)).$$

From the definition of the height (multiplicative in function fields), we know that $h_{X'}(P) \geqq 1$ for all $P \in A_K$. Hence

$$h(P)^{m^2-1} \leqq c_2 h(mP + a).$$

This concludes the proof of Proposition 5, and of Theorem 2.

8. End of the proof of Theorem 1. We consider the function field case. We let $W \subset \boldsymbol{P}^r$ be a projective, normal variety defined over the algebraically closed field k. Let w be a generic point of W over k, and $K = k(w)$ a function field for W over k. Let A be an abelian variety defined over K, and embedded in projective space $\boldsymbol{P}^n$ over K. Let u be a generic point of A over K. We shall denote by $W \circ A$ the locus of (w, u) over k. Then $W \circ A$ is the graph of the algebraic family parametrized by W, of which A is a generic member. (If we put $Z = W \circ A$, then $A = Z(w) = \mathrm{pr}_2[Z \cdot (w \times \boldsymbol{P}^n)]$.)

Given a rational point P of A over K, we shall denote by W_P the locus of (w, P) over k. Then

$$W_P \subset W \circ A \subset W \times \boldsymbol{P}^n \subset \boldsymbol{P}^r \times \boldsymbol{P}^n.$$

Note that W_P and W are birationally equivalent under the projection pr_1 (biholomorphic if W is a curve).

There is a projective embedding $\phi: \boldsymbol{P}^r \times \boldsymbol{P}^n \to \boldsymbol{P}^{(r+1)(n+1)-1}$ which to each products of points with homogeneous coordinates $(x_0, \cdots, x_r) \times (y_0, \cdots, y_n)$ assigns the point with homogeneous coordinates $(x_i y_j)$. One verifies immediately that for any $P \in A_K$, we have

$$h_\phi(w, P) \leqq h(w) + h(P) = \deg W + h(P).$$

Assume $\dim K = 1$. If $\mathfrak{S}$ is a subset of A_K such that $h(P) \leqq c_3$ (the constant of the corollary to Proposition 5), then by Property 1F, there exists a constant c_4 such that $\deg \phi(W_P) \leqq c_4$, and hence W_P can belong only to a finite number of algebraic families on $W \circ A$ for $P \in \mathfrak{S}$. In view of the corollary to Proposition 5, we shall be through with the proof of Theorem 1 once we have proved

THEOREM 3. *Let W be a projective, normal variety defined over the algebraically closed field k. Let w be a generic point of W over k, and $K = k(w)$ a function field of W over k. Let A be an abelian variety defined over K, and let (B, τ) be its K/k-trace. For each point $P \in A_K$, let W_P be the locus of (w, P) over k. If $P, P' \in A_K$ are such that W_P and $W_{P'}$ are in the same algebraic family on $W \circ A$, then P and P' are congruent modulo τB_k (i.e. $P - P' \in \tau B_k$).*

Proof. We shall prove our theorm by going from W_P to $W_{P'}$ by passing through a generic member of the family. Our first step is to show that a generic member is of the same type as W_P.

LEMMA. *Let U be a subvariety of $W \circ A$, defined over an extension k^* of k, independent of K over k, and such that W_P is a specialization of U over k. Put $K^* = Kk^*$. Then there exists a rational point P^* of A in K^* such that $U = W_{P^*}$.*

Proof. We have $\mathrm{pr}_1 W_P = W$. Hence by the compatibility of projections with specializations [4], we must have $\mathrm{pr}_1 U = W$. But $\dim U = \dim W$, and hence a generic point of U over k^* is of type (w, Q) for some $Q \in \boldsymbol{P}^n$, algebraic over $k^*(w)$. We must show that Q is rational over $k^*(w)$, and that Q is in A.

On $W \times \boldsymbol{P}^n$ the intersection $U \cdot (w \times \boldsymbol{P}^n)$ is obviously defined, and in view of the above remarks, we have

$$U \cdot (w \times \boldsymbol{P}^n) = \sum_{i=1}^{d} (w, Q_i)$$

for suitable points Q_i. Taking the projection on W, we see immediately that $d = 1$. Since $U \cdot (w \times \boldsymbol{P}^n)$ is rational over $k^*(w)$, so is $Q_1 = P^*$. Finally, we have $P^* \in A$ because

$$w \times P^* = U \cdot (w \times \boldsymbol{P}^n) \subset (W \circ A) \cdot (w \times \boldsymbol{P}^n) = w \times A.$$

Note that in the lemma, we have made no use of the fact that A is an abelian variety: The same statement holds for any algebraic system of varieties.

It will suffice to prove Theorem 3 for W_P and W_{P^*}. Indeed, given W_P and $W_{P'}$ as in the theorem, there exists a subvariety W_{P^*} of $W \circ A$, defined over some k^*, such that both W_P and $W_{P'}$ are specializations of W_{P^*}. If $P - P^* \in \tau B_{k^*}$ and $P' - P^* \in \tau B_{k^*}$, then $P - P' \in \tau B_{k^*}$, and by Proposition 2, $P - P' \in \tau B_k$.

The field k^* may also be assumed to be of transcendence degree 1 over k, i. e. the parameter variety in the algebraic system joining W_P to W_{P^*} may be assumed to be a curve, by blowing up the point corresponding to W_P on the given parameter variety joining W_P and W_{P^*}. Algebraically speaking, there exists a field $k_1 \supset k$, a complete non-singular curve T defined over k_1, a generic point t^* of T over k_1, and a point t of T, rational over k_1, such that if we put $k^* = k_1(t^*)$, the variety W_{P^*} is defined over k^*, and W_P is the unique specialization of W_{P^*} over $t^* \to t$ (relative to k_1) [4]. Without loss of generality for the proof of Theorem 3, we may let $k = k_1$ (using once more Proposition 2). We have the following diagram of fields:

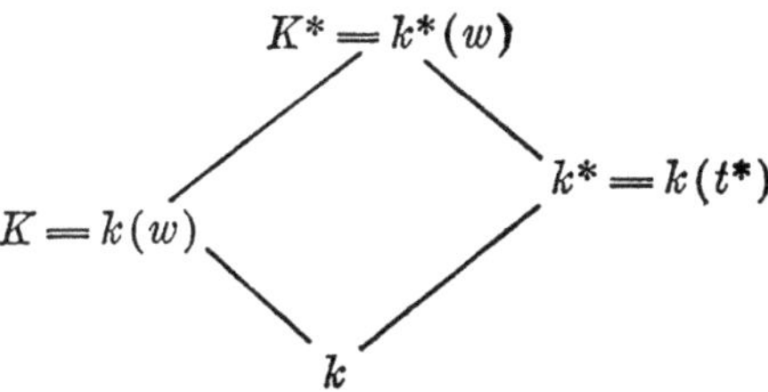

Let $\mathrm{Alb}(T)$ be the Albanese variety of T, defined over k, and let $f : T \to \mathrm{Alb}(T)$ be a canonical map, defined over k. Let $g : T \to A$ be the rational map defined over K by the expression $g(t^*) = P^*$. Then there is a homomorphism $\alpha : \mathrm{Alb}(T) \to A$ and a point $a \in A$ such that the following diagram is commutative.

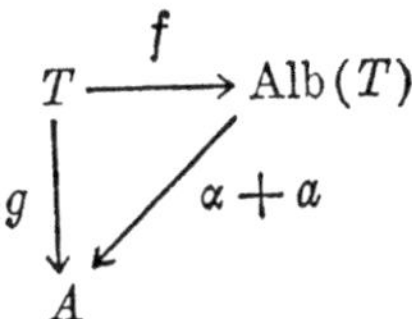

By the definition of the K/k-trace, there exists a homomorphism $\beta : \mathrm{Alb}(T) \to B$ defined over k such that the following diagram is commutative.

S. LANG AND A. NÉRON.

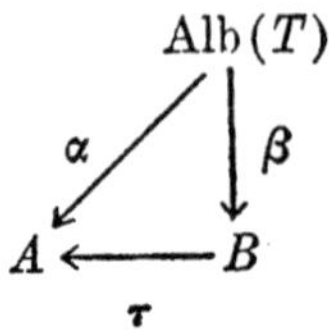

Since t is simple on T, f is defined at t. Furthermore, since W_P is the unique specialization of W_{P^*} over $t^* \to t$ (relative to k or K), it follows that P is the unique specialization of P^* over $t^* \to t$ (relative to K). Hence $g(t) = P$, and

$$P^* - P = g(t^*) - g(t) = \alpha f(t^*) + a - \alpha f(t) - a$$
$$= \tau \beta f(t^*) - \tau \beta f(t)$$
$$= \tau b$$

with $b = \beta[f(t^*) - f(t)] \in B_{k^*}$. QED.

Paris.

REFERENCES.

[1] S. Lang, *Introduction to algebraic geometry*, Interscience Publishers, New York, 1958.

[2] ———, *Abelian varieties*, Interscience Publishers, New York, 1958.

[3] S. Lang and J. Tate, "Principal homogeneous spaces over abelian varieties," *American Journal of Mathematics*, vol. 80 (1958), pp. 659-684.

[4] T. Matsusaka, "Specialization of cycles on a projective model," *Memoirs of the College of Science*, University of Kyoto, Series A, vol. 26, No. 2 (1950), pp. 167-173.

[5] A. Neron, "Problèmes arithmétiques et géometriques rattachés a la notion de rang courbe algébrique dans un corps," *Bulletin de la Societe Mathematique de France*, vol. 80 (1952), pp. 101-166.

[6] ———, "Arithmétique et classes de diviseurs sur les variétés algébriques," *Proceedings of the International Symposium on Algebraic number theory*, Tokyo-Nikko, 1955, pp. 139-154.

[7] G. Shimura, "Reduction of algebraic varieties with respect to a discrete valuation of the basic field," *American Journal of Mathematics*, vol. 77 (1955), pp. 134-176.

[8] A. Weil, "Arithmetic on algebraic varieties," *Annals of Mathematics*, vol. 53 (1951), pp. 412-444.

[9] ———, "Sur les criteres d'équivalence en géometrie algébrique," *Mathematische Annalen*, vol. 128 (1954), pp. 95-127.

[10] ———, Foundations of Algebraic Geometry. New York, 1946.

EXISTENCE OF INVARIANT BASES

BY

ELLIS KOLCHIN AND SERGE LANG

Reprinted from the
PROCEEDINGS OF THE AMERICAN MATHEMATICAL SOCIETY
Vol. 11, No. 1, February, 1960
pp. 140-148

EXISTENCE OF INVARIANT BASES

ELLIS KOLCHIN AND SERGE LANG[1]

Let K be a field, G a group of automorphisms of K, and M a vector space over K on which G acts in such a way that $\sigma(aD) = \sigma a \cdot \sigma D$ for $\sigma \in G$, $a \in K$, and $D \in M$. The problem arises to find whether M has a basis consisting of invariant elements under G. In other words, letting K_0 be the fixed field under G, and M_0 the set of fixed elements of M under G so that M_0 is a vector space over K_0, to find out whether M is isomorphic to the tensor product

$$M \approx K \otimes_{K_0} M_0$$

under the natural map. We shall see that this is so if and only if a certain cocycle of G in the full linear group is trivial.

In some applications, a rational structure is added to K and G, namely K is the function field of a principal homogeneous space over a group variety G. We shall show that the cocycle involved is then determined rationally. This leads us into a discussion of rational cocycles in §3, and of their comparison with the ordinary cocycles of Galois theory, i.e. where G is a finite Galois group. All cocycles involved with coefficients in the full linear group split, and in fact the Galois cohomology (in dimension 1) of the group variety of units in an algebra is trivial (Propositions 2 and 5).

1. **The invariant subspace.** Let K be a field, and G a group of automorphisms of K. By a (G, K)-*space* M we shall mean a vector space over K which is also a unitary G-module, such that

$$\sigma(aD) = \sigma a \cdot \sigma D$$

for $a \in K$, $D \in M$ and $\sigma \in G$. An element D of M is said to be invariant under G if $\sigma D = D$ for all $\sigma \in G$. A basis $(D) = (D_i)$ of M over K will be called *invariant* if $\sigma D_i = D_i$ for every $\sigma \in G$ and every i.

PROPOSITION 1. *Let M be a finite dimensional (G, K)-space. If $(D) = (D_1, \cdots, D_m)$ is any basis of M, and if*

$$A(\sigma) = (a_{ij}(\sigma)) \qquad (i, j = 1, \cdots, m)$$

is the matrix defined by

$$\sigma D_j = \sum_{r} a_{rj}(\sigma) D_r \qquad (1 \leqq j \leqq m),$$

Received by the editors April 1, 1959.

[1] This paper was prepared in connection with contract NONR 266(57).

140

then $A(\sigma) \cdot \sigma A(\tau) = A(\sigma\tau)$. A necessary and sufficient condition that M have an invariant basis is that there exist an invertible matrix B with coordinates in K such that $A(\sigma) = B^{-1}\sigma B$.

PROOF. That $A(\sigma)$ satisfies the cocycle relation is easy to see. Suppose $A(\sigma) = B^{-1}\sigma B$. Using matrix notation, we may write $\sigma(D) = {}^t A(\sigma)(D)$, where for any matrix X, we denote by ${}^t X$ the transpose of X. Let a new basis (D') be defined by $(D') = {}^t B^{-1}(D)$. Then

$$\sigma(D') = \sigma^t B^{-1}\sigma(D) = {}^t\sigma(B)^{-1} \cdot {}^t A(\sigma) \cdot (D) = {}^t B^{-1}(D) = (D')$$

and thus (D') is invariant. Conversely, if (D') is an invariant basis, define the matrix B by the relation $(D) = {}^t B(D')$. Then

$$ {}^t A(\sigma)(D) = \sigma(D) = {}^t\sigma B \cdot \sigma(D') = {}^t\sigma B \cdot (D') = {}^t\sigma B \cdot {}^t B^{-1}(D)$$

and $A(\sigma) = B^{-1}\sigma B$. This proves our proposition.

If we denote by M_0 the set of G-invariant elements of M and by K_0 the fixed field of K under G, then M_0 is a vector space over K_0. If M admits an invariant basis, then one sees immediately that M is isomorphic to the tensor product

$$K \otimes_{K_0} M_0 \approx M$$

under the natural map $a \otimes D \to aD$, $a \in K$, $D \in M_0$.

We observe that the set of K_0-linear transformations of K, denoted by $\mathrm{End}_{K_0}(K)$, forms a (G, K)-space in a natural fashion: If $D \in \mathrm{End}_{K_0}(K)$, and $\sigma \in G$, then one defines

$$(\sigma D)(a) = \sigma(D(\sigma^{-1}a)),$$

and verifies trivially that $\sigma(aD) = \sigma a \cdot \sigma D$.

In the applications, one frequently takes the subspace of $\mathrm{End}_{K_0}(K)$ consisting of the derivations of K over K_0, or a finite dimensional space of linear transformations over a subfield of K.

REMARK. Proposition 1 and its proof generalizes so that K can be a ring (with unity, not necessarily commutative) and M can be a unitary K-module with finite basis. In this situation, a matrix B over K with m rows and n columns is *invertible* if there exists a matrix C over K with n rows and m columns such that BC is the unity matrix of degree m and CB is the unity matrix of degree n. C is then unique and is denoted by B^{-1}. The generalized proposition states that if (D) is a basis of m elements, and $A(\sigma)$ is defined as above, then $A(\sigma) \cdot \sigma A(\tau) = A(\sigma\tau)$, and a necessary and sufficient condition that there exist an invariant basis of n elements is that there exist an invertible matrix B over K with n rows and m columns such that $A(\sigma) = B^{-1}\sigma B$.

2. **Galois cohomology in dimension 1.** As we have seen in the last section, it is useful to have a criterion to split a 1-cocycle. We shall give one in this section.

Let G be a group variety (i.e. a connected algebraic group) defined over a field k. Let K be a finite Galois extension of k with Galois group $\mathfrak{g}$. Then $\mathfrak{g}$ operates as a group of automorphisms of the subgroup of G consisting of the points of G which are rational over K. We denote this subgroup by G_K. Suppose we are given a family $(x_\sigma)_{\sigma \in \mathfrak{g}}$ of points of G_K satisfying

$$x_\sigma \cdot \sigma x_\tau = x_{\sigma\tau}, \qquad\qquad \sigma, \tau \in \mathfrak{g}.$$

Such a family is called a cocycle of $\mathfrak{g}$ in G_K. The set of these cocycles is denoted by $Z^1(\mathfrak{g}, G_K)$. We say that (x_σ) is *cohomologous* to (y_σ) and write $(x_\sigma) \sim (y_\sigma)$ if there exists an element $z \in G_K$ such that

$$y_\sigma = z^{-1} x_\sigma \sigma z$$

for each $\sigma \in \mathfrak{g}$. This is obviously an equivalence relation between cocycles. The set of equivalence classes is called the *first cohomology set* of $\mathfrak{g}$ in G_K and is denoted by $H^1(\mathfrak{g}, G_K)$. If $z \in G_K$ then $(z^{-1}\sigma z)$ is a cocycle, and such a cocycle is called a coboundary of $\mathfrak{g}$ in G_K. The set of all such coboundaries is denoted by $B^1(\mathfrak{g}, G_K)$ and is itself an equivalence class, i.e. an element of $H^1(\mathfrak{g}, G_K)$. (Of course, if G is commutative, these sets are groups, and $H^1(\mathfrak{g}, G_K)$ is a commutative group.)

If $L \supset K$ is another Galois extension of k, then there is a natural map of $H^1(\mathfrak{g}_{K/k}, G_K)$ into $H^1(\mathfrak{g}_{L/k}, G_L)$ obtained by inflation: A cocycle for $\mathfrak{g}_{K/k}$ determines one for $\mathfrak{g}_{L/k}$ simply by extending the function to cosets of the subgroup of $\mathfrak{g}_{L/k}$ of which $\mathfrak{g}_{K/k}$ is a factor group. It is trivially seen that this natural map is actually *injective* and we may take the injective limit of these cohomology sets as L becomes larger and larger. The limiting set, union of all $H^1(\mathfrak{g}_{L/k}, G_L)$, will be denoted by $H^1(k, G)$.

We are interested in a noncommutative group, namely the full linear group. More generally, let A_0 be a finite dimensional associative algebra with unity element over the field k, let Ω be a universal domain containing k, and let A be the algebra over Ω which is the tensor product of A_0 with Ω. Let $e_1, \cdots, e_m$ be a linear basis of A_0 over k, and therefore of A over Ω. Expressing elements of A in terms of this basis, $x = \sum \xi_i e_i$, one sees that there exists a polynomial $P(X_1, \cdots, X_m) \in k[X_1, \cdots, X_m]$ such that x is invertible if and only if $P(\xi_1, \cdots, \xi_m) \neq 0$, or as we shall abbreviate, $P(x) \neq 0$. These invertible elements therefore form a group variety defined over k (it is a k-open subset of affine m-space in the Zariski topology), and we shall denote it by $\Gamma(A)$ or simply Γ. The general linear group $GL(m)$ is an example of such a group variety defined over the prime field.

PROPOSITION 2. *Let $\Gamma = \Gamma(A)$ be as above the group variety of units in an algebra defined over k, and let K be a Galois extension of k of finite degree, with Galois group $\mathfrak{g}$. Then $H^1(\mathfrak{g}, \Gamma_K)$ is trivial, that is, every cocycle is a coboundary.*

PROOF. The case in which k is finite is a special case of the fact that $H^1(\mathfrak{g}, G_K)$ is trivial for any group variety defined over a finite field k (see [2]). In the infinite case, the theorem is proved in [1]. We reproduce the proof here for the convenience of the reader. Let (x_τ) be in $Z^1(\mathfrak{g}, \Gamma_K)$. If $(t_\tau)_{\tau\in\mathfrak{g}}$ is a family of elements of Ω, algebraically independent over K, then $P(\sum_\tau t_\tau x_\tau) \neq 0$ because this polynomial does not vanish when one t_τ is replaced by 1 and all the others by 0. It is well known (see, for instance, Bourbaki, *Algèbre*, Chapter V, §10, Theorem 4, p. 57) that this implies the existence of an element $\alpha \in K$ such that $P(\sum(\tau\alpha)x_\tau) \neq 0$. Writing $y = \sum(\tau\alpha)x_\tau$ we see that $y\in\Gamma_K$ and $\sigma y = \sum(\sigma\tau\alpha)\sigma x_\tau = \sum(\sigma\tau\alpha)x_\sigma^{-1}x_{\sigma\tau} = x_\sigma^{-1}y$, so that $(x_\sigma) = (y\sigma y^{-1})$ is in $B^1(\mathfrak{g}, \Gamma_K)$. This concludes the proof.

3. **Rational cohomology.** We recall the notion of a homogeneous space and use the terminology of Weil [4]. Let V be a variety and G a group variety. Suppose we are given a rational map

$$f\colon V \times G \to V$$

which is everywhere defined. Given $v\in V$ and $x\in G$, we write vx instead of $f(v, x)$. We say that V is a *transformation space* for G if

$$v(xy) = (vx)y \qquad \text{and} \qquad ve = v$$

for $x, y\in G$ and $v\in V$. Here, as usual, e denotes the unity element of G. We say that the transformation space V is defined over k if G, V and the rational map f are defined over k.

We observe that if Ω is the universal domain, then G can be viewed as a group of automorphisms of the function field $\Omega(V)$. Indeed, for each $x\in G$, we have the automorphism σ_x such that

$$(\sigma_x f)(v) = f(vx)$$

whenever f is a function in $\Omega(V)$, and v is a generic point of V over a field of definition for f over which x is rational. In the same manner, if the transformation space V is defined over k, then G_k is a group of automorphisms of $k(V)$.

A transformation space is said to be a *homogeneous space* if given two points $v, w\in V$, there exists $x\in G$ such that $vx = w$. We say that

V is a *principal* homogeneous space if the element x is uniquely and rationally determined. By this we mean that there exists an everywhere defined rational map $\mu\colon V\times V\to G$ such that $x=\mu(v, w)$. One may write symbolically $x=v^{-1}w$. The principal space V is said to be defined over k if, as a transformation space it is defined over k, and the rational map μ is defined over k.

Let now G, G' be group varieties, let V be a transformation space of G, all these being defined over k. A rational map

$$f\colon V\times G\to G'$$

of $V\times G$ into G', defined over k, is said to be a 1-*cocycle* if it satisfies the relation

$$f(v, x)f(vx, y) = f(v, xy)$$

whenever v, x, y are independent generic points of V, G, and G' over k. *It then follows that $f(v, x)$ is defined whenever x is any point and v is a generic point of V over $k(x)$,* because we can write

$$f(v, x) = f(v, xy)f(vx, y)^{-1}$$

with y generic over $k(v, x)$. The 1-cocycles form a set denoted by $Z_k^1(G, V, G')$. We say that two cocycles are *cohomologous* and write $f\sim g$, if there exists a rational map $\phi\colon V\to G'$ defined over k such that $f(v, x)=\phi(v)^{-1}g(v, x)\phi(vx)$. This establishes an equivalence relation among the cocycles, and the equivalence classes form the first cohomology set $H_k^1(G, V, G')$. The cocycles in the identity class, i.e. those of type $\phi(v)^{-1}\phi(vx)$ are called *coboundaries*, and form a set $B_k^1(G, V, G')$.

(If G' is commutative, and written additively, we can define cocycles in higher dimension, an r-cocycle being by definition a rational map

$$f\colon V\times G\times\cdots\times G\to G'$$

defined over k satisfying the coboundary formula

$$\begin{aligned}
0 = (\delta f)(v, x_1, \cdots, x_{r+1}) &= f(vx_1, x_2, \cdots, x_{r+1})\\
&+ \sum (-1)^r f(v, x_1, \cdots, x_\nu x_{\nu+1}, \cdots, x_{r+1})\\
&+ (-1)^{r+1} f(v, x_1, \cdots, x_r).
\end{aligned}$$

One then has an rth cohomology group for $r\geqq 0$.)

We return to the noncommutative case, and assume that V is a principal homogeneous space for the group variety G, all defined over k. Let $f\in Z_k^1(G, V, G')$. If $w_0\in V$ has the property that f is defined at $(u, u^{-1}w_0)$ for u generic on V over $k(w_0)$ and if we define

$\phi(v) = f(v, v^{-1}w_0)^{-1}$, so that ϕ is a rational map of V into G' defined over $k(w_0)$, then $f(v, x) = \phi(v)^{-1}\phi(vx)$. Thus f is trivial as an element of $Z^1_{k(w_0)}(G, V, G')$. It follows that if the points of V which are rational over k are dense in the Zariski topology, then every element of $Z^1_k(G, V, G')$ is a coboundary. (This is the case for instance if k is separably closed.)

Now we transform the above cocycles into a homogeneous form. Let $\lambda : V \times V \to V \times G$ be the canonical map such that $\lambda(u, v) = (u, u^{-1}v)$. Then the inverse of λ is a rational map sending (v, x) onto (v, vx). Given a cocycle $f \in Z^1_k(G, V, G')$, there exists therefore a rational map $F : V \times V \to G'$ which makes the following diagram commutative:

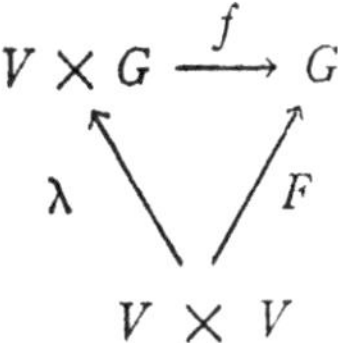

and the mapping F satisfies the relation

$$F(u, v)F(v, w) = F(u, w)$$

whenever u, v, w are independent generic points of V over k. We may of course start with an arbitrary variety V defined over k and a group variety G' with such a mapping. We may thus define homogeneous cocycles $Z^1_k(V, G')$, and coboundaries $B^1_k(V, G')$, these being rational maps of type $F(u, v) = \phi(u)^{-1}\phi(v)$ where $\phi : V \to G'$ is a rational map defined over k. This allows us to define $H^1_k(V, G')$ for any variety V defined over k.

PROPOSITION 3. *Let V be a principal homogeneous space for G, and let G' be another group variety defined over k. Then there is a bijective mapping between $H^1_k(G, V, G')$ and $H^1_k(V, G')$ given by*

$$f(v, x) = F(v, vx).$$

The proof is trivial.

As Serre has pointed out to us, we can inject $H^1_k(V, G')$ into the Galois cohomology set $H^1(k, G')$ as defined in §2. The way this is done is described in the following proposition.

PROPOSITION 4. *Let V be a principal homogeneous space for G, and let G' be another group variety, defined over k. For each $\overline{F} \in H^1_k(V, G')$ choose a representative cocycle F in $Z^1_k(V, G')$, and choose a finite Galois extension K of k with group denoted by $\mathfrak{g}$ such that V has a point v_0 rational over K for which $F(u, v_0)$ is defined when u is generic over K.*

For each $\sigma \in \mathfrak{g}$, let $x_\sigma = F(u, v_0)^{-1}F(u, \sigma v_0)$, where u, v_0 are chosen as above. Then $(x_\sigma)_{\sigma \in \mathfrak{g}}$ is a cocycle in $Z^1(\mathfrak{g}, G'_K)$, the corresponding element $\tilde{x}$ of $H^1(k, G')$ is independent of the choice of F, K, v_0, u and the mapping $\overline{F} \to \tilde{x}$ is an injection

$$H^1_k(V, G') \to H^1(k, G').$$

PROOF. From the coboundary relation one verifies immediately that for independent generic points u, v of V over K, we have

$$F(u, v_0)^{-1}F(u, \sigma v_0) = F(v, v_0)^{-1}F(v, \sigma v_0),$$

so that each x_σ is rational over K, and (x_σ) is a cocycle. The rest of the proof is straightforward and is left to the reader.

In particular, we get

COROLLARY. *Let V be a principal homogeneous space for G and let G' be another group variety, defined over k. If $H^1(k, G')$ is trivial, then so is $H^1_k(V, G')$.*

Examples of group varieties G' for which $H^1(k, G')$ is trivial are:

The group variety of units in a finite dimensional algebra as we showed in Proposition 2, and in particular the full linear groups $GL(m)$;

The additive and multiplicative groups of the universal domain, this being Hilbert's Theorem 90 in its multiplicative and additive forms;

The group varieties G' having a normal sequence $G' = G_0 \supset G_1 \supset \cdots \supset G_r = 1$ defined over k, with each G_{i-1}/G_i either the additive or multiplicative group as above.

One may ask how it is possible to characterize those elements of $H^1(k, G')$ which come from an element of $H^1_k(V, G')$. It is known [3, Proposition 4] that $H^1(k, G')$ is in bijective correspondence with the set of isomorphism classes of principal homogeneous spaces of G' over k, and the reader may easily verify that the image of $H^1_k(V, G')$ in $H^1(k, G')$ corresponds to those principal homogeneous spaces of G', defined over k, which have a rational point in a field $k(v)$ where v is a generic point of V over k. The reader will also note that a cocycle $F: V \times V \to G'$ determines a principal homogeneous space of G' through which it can be factored [4, Proposition 4] and that this space corresponds precisely to the one determined by the cocycle described in Proposition 4 (Serre).

4. Rational determination of the invariant subspace. Let V be a principal homogeneous space for the group variety G. For each point

x of G, there is a unique automorphism σ_x of the function field $\Omega(V)$ over the universal domain Ω, such that, if $f \in \Omega(V)$ then $\sigma_x f$ is defined at a point v of V whenever f is defined at vx, and $(\sigma_x f)(v) = f(vx)$. The mapping $x \to \sigma_x$ is a group isomorphism permitting us to identify G with a group of automorphisms of $\Omega(V)$ over Ω. If k is a field of definition of V, and if we make the assumption that x is rational over k, then σ_x maps $k(V)$ onto itself. Therefore without this assumption, σ_x maps $k(V)$ into $k(x)(V)$.

Let $f: V \times G \to W$ be a rational map of $V \times G$ into some variety W, and let x be a point of G. Assume that the following condition is satisfied. For some field of definition k for f (i.e. for V, G, W and the graph of f), and for some generic point v of V over $k(x)$, f is defined at (v, x). Then there exists a unique rational map $f_x: V \to W$ defined over $k(x)$, such that $f_x(v) = f(v, x)$. When the above condition is satisfied, we shall say that f_x is *meaningful*. Of course, if x is generic on G over k, then f_x must be meaningful.

Let M be a vector space over $\Omega(V)$, of finite dimension m, which is a $(G, \Omega(V))$-space. If $(D) = (D_1, \cdots, D_m)$ is a basis of M, then for each point x of G, every $\sigma_x D_j$ is a linear combination of the basis vectors D_i with coefficients in $\Omega(V)$. We shall call the basis *rational* if there exist rational functions f_{ij} on $V \times G$ such that, for some common field of definition k of V and the f_{ij}, and for every generic point x of G over k, we have

$$(1) \qquad\qquad \sigma_x D_j = \sum_i f_{ijx} D_i, \qquad\qquad (1 \leq j \leq m).$$

A simple computation shows that when this is the case then, for independent generic points x, y of G over k, we have

$$f(v, xy) = f(v, x)f(vx, y)$$

where we denote by f the matrix (f_{ij}). In other words, f is a 1-cocycle in $Z_k^1(G, V, GL(m))$. It follows (§3) that f_x (i.e. each f_{ijx}) is meaningful for every point x of G, and also, that Equation (1) holds for every point x of G since each point of G can be expressed as a product of generic points $(x = xy \cdot y^{-1})$. In particular, any common field of definition of V and the f_{ij} must enjoy the same properties that have been attributed to k above. Such a common field of definition will be called a *field of rationality* of the rational basis (D). It is obvious that every invariant basis of M is rational, and admits as field of rationality any field of definition of the principal homogeneous space V.

It is almost immediate that if one basis of a $(G, \Omega(V))$-space is rational, then so are all its bases. By a *rational $(G, \Omega(V))$-space* we shall mean a $(G, \Omega(V))$-space with rational bases. The nature of such spaces is completely described by the following theorem, the proof of which follows that of Proposition 1, making use of the results of §2 and §3.

THEOREM. *Let M be a finite dimensional $(G, \Omega(V))$-space. A necessary and sufficient condition that M be rational is that M have an invariant basis. If (D) is any rational basis of M, and k is a field of rationality of (D), then there exists an invertible matrix over $k(V)$ transforming (D) into an invariant basis of M.*

PROOF. Since invariant bases are rational, the sufficiency is clear. To prove the necessity and the final part of the theorem, let (D) be a rational basis of M with field of definition k. Denoting the corresponding cocycle in $Z_k^1(G, V, GL(m))$ by $f = (f_{ij})$, we conclude from §§2, 3 that there exist rational functions $\phi_{ij} \in k(V)$ such that the matrix $\phi = (\phi_{ij})$ is invertible and $f(v, x) = \phi(v)^{-1}\phi(vx)$ for $x \in G$ and v generic on V over $k(x)$. Setting $E_j = \sum_i \psi_{ij} D_i$ $(1 \leq j \leq m)$, where $(\psi_{ij}) = (\phi)^{-1}$, we conclude that (E) is an invariant basis of M.

In the applications of the above theorem, one is usually given a subset $M_{k(V)}$ of M which is a vector space over $k(V)$ such that we have an isomorphism

$$M \approx \Omega(V) \otimes_{k(V)} M_{k(V)}$$

under the natural map $f \otimes D \to fD$, and such that a basis of $M_{k(V)}$ over $k(V)$ is a rational basis of M, having k as field of rationality. One may then say that M is *rationally $k(V)$-extended*. In that case one may say that an element D of M is *defined* over a field $k' \supset k$ if D lies in $k'(V)M_{k(V)}$. If M_k^0 is the set of elements of M which are invariant and defined over k, then M_k^0 is a vector space over k, and our theorem shows that we have an isomorphism

$$k(V) \otimes_k M_k^0 \approx M_{k(V)}$$

again under the map $f \otimes D \to fD$.

BIBLIOGRAPHY

1. E. Kolchin, *On the Galois theory of differential fields*, Amer. J. Math. vol. 77 (1955) pp. 868–894.

2. S. Lang, *Algebraic groups over finite fields*, Amer. J. Math. vol. 78 (1956) pp. 555–563.

3. S. Lang and J. Tate, *Principal homogeneous spaces over abelian varieties*, ibid. vol. 80 (1958) pp. 659–684.

4. A. Weil, *Algebraic groups and homogeneous spaces*, ibid. vol. 77 (1955) pp. 493–512.

COLUMBIA UNIVERSITY

Pub. IHES, No. 6, 1960

INTEGRAL POINTS ON CURVES

by SERGE LANG [1]

Siegel has shown in [14] that an affine curve $f(x, y) = 0$ with coefficients in a number field and of genus ≥ 1 has only a finite number of points whose coordinates are integers of that field. Mahler [8] has conjectured that a similar statement holds for points having only a finite number of primes in their denominators, and proved this for curves of genus 1 over the rationals by his p-adic analogue of the Thue-Siegel theorem.

In view of Roth's recent result, and the progress which has been made in the theory of abelian varieties (especially the Jacobian) since Siegel and Mahler's papers appeared, it seemed worth while to reconsider the question, and I give below a modernized exposition of Siegel and Mahler's proof, which automatically carries with it a proof of Mahler's conjecture. The Jacobian is used in order to take a pull-back over the given curve of the standard covering given by $u \to mu + a$ where m is a large integer, and $a \in J$ is a suitable translation.

Aside from Roth's theorem (whose statement is reproduced in § 1) we use only the classical properties of heights and the weak Mordell-Weil theorem. This paper is thus a natural sequel of [7].

A proof of Mordell's conjecture [10] that a curve of genus ≥ 2 has only a finite number of *rational* points would of course supersede the Siegel-Mahler theorem for such curves, but I would conjecture that the latter holds in fact for abelian varieties : If A is an abelian variety defined over a number field K, if U is an open affine subset, and R a subring of K of finite type over **Z**, then there is only a finite number of points of U in R. The difficulty in trying to extend the proof to abelian varieties lies in the fact that there is a whole divisor at infinity, whereas for curves, there is only a finite number of points, which are all algebraic.

It is easy to see that if the conjecture is true, then it remains true if K is replaced by a field of finite type over **Q**, and R by a subring of finite type over **Z**. In § 7 we shall carry this out for curves. This could be applied to strengthen in a like manner Siegel's result on curves of genus 0 as on p. 47 of [14]. There is no point in carrying this out here, but it is worth while to go deeper into one of Siegel's arguments.

We observe that if G is a group variety, and Γ a subgroup of finite type, then there is a field K of finite type over the prime field over which G is defined, and over which all points of Γ are rational.

Now the argument of Siegel can be used to prove the following theorem.

---

[1] Sloan Fellow.

Let K *be a field of characteristic* 0, *and* Γ *a subgroup of finite type of its multiplicative group. Then the curve* $ax + by = 1$ *with* $a, b \in K$ *and* $ab \neq 0$ *has only a finite number of points with* $x, y \in \Gamma$.

PROOF. If there were infinitely many, let m be an integer ≥ 3. Then infinitely many x (resp. y) would lie in the same coset mod Γ^m, so that for such x and y we can write $x = a_1 \xi^m$ and $y = a_2 \eta^m$, and we get infinitely many points in $\Gamma \times \Gamma$ on the curve

$$aa_1 \xi^m + ba_2 \eta^m = 1$$

which has genus ≥ 1. Contradiction.

The straight line just considered should in fact be regarded as a subvariety of the product of the multiplicative group with itself. Recall that an *algebraic torus* (*torus* for short) is a group variety which is a finite product of multiplicative groups. Infinitely many rational points on the line give rise to integral points on a curve of genus ≥ 1, and using this same idea, we get more generally :

Let G *be a torus in characteristic* 0. *Let* C *be a subvariety of dimension* 1 *of* G. *Let* Γ *be a subgroup of finite type of* G. *If* C *intersects* Γ *in an infinite number of points, then* C *is the translation in* G *of a subtorus of dimension* 1.

PROOF. We can find a field K of finite type over $\mathbf{Q}$ over which G is written as an n-fold product of multiplicative groups, over which all points of Γ are rational, and over which C is defined. Let $(x_1, \ldots, x_n)$ be a generic point of C over K. Then C has genus 0, and we can write $x_i = \varphi_i(t)$ where φ_i is a rational function of a parameter t, defined over K. We proceed as above. Let us take m large and relatively prime to the orders of zeros and poles of the functions $\varphi_i(t)$. For suitable elements $a_1, \ldots, a_n \in K$, the curve whose generic point is $(\xi_1, \ldots, \xi_n)$ where $\xi_i^m = a_i x_i$ (over possibly a finite extension of K) must also have genus 0. Consider first a covering of the t-line given by the equation $\xi^m = a\varphi(t)$ with a, φ any one of the a_i, φ_i. We use the classical formula for the genus of a covering :

$$2\, g' - 2 = m(2\, g - 2) + \Sigma(e_\mathrm{P} - 1).$$

Then $\varphi(t)$ can have at most one zero and one pole. Indeed, the degree is m, and the ramification index above such a zero or pole is m also. We have $m(2\, g - 2) = -2\, m$, and if there were at least 3 distinct zeros and poles, then the term $\Sigma(e_\mathrm{P} - 1)$ would grow at least like $3(m - 1)$, so we would get a covering of genus ≥ 1, having infinitely many points with coordinates in Γ, which is impossible.

After a linear transformation, we can write say for $i = 1$,

$$x_1 = a_1 t^{r_1}$$

for some integer $r_1 \neq 0$. From the same argument, $x_i = a_i \varphi_i(t)$ where $\varphi_i(t)$ is a power of some linear transformation of t. In fact, $\varphi_i(t) = t^{r_i}$ because our covering has the intermediate covering defined by the equation

$$\xi^m = a_1 x_1 (a_i x_i)^{\pm 1} = at^{r_1} \varphi_i(t)^{\pm 1}$$

320

which would be of genus ≥ 1 unless $\varphi_i(t)$ is of the prescribed type. We thus conclude that C is the translation of a subtorus.

As Siegel already observed, the coset argument is formally the same as the one used to carry out the proof that curves of genus ≥ 1 have only a finite number of integral points, but using the Jacobian instead of the torus (cf. below Proposition 1).

The analogy between toruses and abelian varieties was again observed by Chabauty, who in two papers [2], [3] considers infinite intersections of a subvariety of a torus or an abelian variety with particular subgroups of finite type, namely subgroups of units and groups of rational points in a number field respectively. Thus one is led to generalize and reformulate a conjecture of Chabauty [2] in the following manner, which includes the Mordell conjecture.

Let G be torus (resp. an abelian variety) in characteristic 0. Let V be a subvariety of G, having an infinite intersection with a subgroup of finite type Γ of G. Then V contains a finite number of translations of group subvarieties of G which contain all but a finite number of points of $V \cap \Gamma$.

This statement has been proved above when V is of dimension 1 and G is a torus. When G is an abelian variety and again V is of dimension 1, this is Mordell's conjecture. Of course, one may ask whether such a statement would not be valid also for a commutative group extension of an abelian variety by a torus.

Returning to the question of integral points, we shall see that our theorems have analogues in function fields K (finitely generated regular extensions) over arbitrary constant fields k of characteristic 0, and the finiteness statement can be given a relative formulation : One proves that certain points have bounded heights (Theorems 2, 3). In view of Theorem 3 [7], one sees that the conjecture we made above concerning integral points on affine subsets of abelian varieties can also be formulated relatively : If A is defined over k, if (B, τ) is a K/k-trace, then integral points of A_k lie in a finite number of cosets of τB_k.

In this connection, the Mordell conjecture becomes a conjecture in algebraic geometry, and it is worth while to make further comments on it here. Let k be as above, $K = k(t)$ a function field over k, where t is the generic point of a variety T, and let C be a curve of genus ≥ 2, defined over K. Then $C = C_t$ can be viewed as the generic member of an algebraic family. *The conjecture then asserts that if C_t has infinitely many rational points in $k(t)$ (cross sections of the parameter variety T in the graph of the family), then C_t is birationally equivalent over $k(t)$ to a curve C_0 defined over k, and all but a finite number of these points arise from points of C_0 in k.*

Evidence for this comes from the special case where $C_t = C_0$ is already defined over k, and then one obtains a classical theorem of de Franchis, to the effect that *given a variety V and a curve C of genus ≥ 2 (in characteristic 0) there exists only a finite number of generically surjective rational maps of V on C.* We give a quick proof of this theorem. Taking a generic hyperplane section U of V and inducing the rational map on it, one reduces the theorem to the case where V is itself a curve.   Indeed, two distinct generically surjective rational maps f, f' : V→C induce distinct generically surjective maps on U, as one sees by taking the induced homomorphisms on the Albanese varieties, using Theorem 4 of [5], Chapter VIII, § 2.

Assuming now that V is a curve, we have the formula for the genus :

$$2 . g(V) - 2 = d[2 . g(C) - 2] + \lambda$$

where $\lambda \geq 0$.   Thus the degree of V over C is bounded.   Taking suitable projective embeddings, we see that the degree of the graph of our rational maps f must be bounded. Hence these graphs Γ_f lie on finitely many algebraic families on V × C.   On the other hand, a generic element of such families is likewise a generically surjective rational map of V onto C (as one sees by projecting on both factors).   Taking the induced homomorphisms on the Jacobians, and using the fact that an abelian variety has no algebraic family of abelian subvarieties, we see that all induced maps coming from the same family differ by translations.   We use now the fact that C is not equal to a non-zero translation of itself in its Jacobian.   (If it were, so would the divisor Θ, and it isn't, even up to linear equivalence by Th. 3 of Ch. VI, § 3, [5].)   We conclude that a graph Γ_f actually must constitute by itself a maximal algebraic family on V × C, and thus finally that there is only a finite number of such graphs, or maps f.   This concludes the proof.   (When V is a curve, we do not need characteristic o, only the assumption that the map f : V→C is separable, to be able to use the genus formula above.)

The Mordell conjecture thus gives rise to diophantine criteria for lowering fields of definition, and we can actually prove such a criterion in the context of integral points (Proposition 2).

One remark on notation to conclude this introduction : If O is a set of geometric objects, and K a field, we denote by O_K the subset of O consisting of those objects which are rational over K.   For example, if V is a variety defined over K, then V_K denotes the set of its rational points in K.

§ 1. **Diophantine approximations.**

Let K be a number field (by definition, a finite extension of the rationals **Q**) and let $N = [K : \mathbf{Q}]$ be its degree over **Q**.   For each prime p of K (finite or archimedean) let $N_p = [K_p : \mathbf{Q}_p]$ be the local degree of the completions.   If p is archimedean, then $\mathbf{Q}_p = \mathbf{R}$ is the field of real numbers.   Otherwise, it is the field of p-adic numbers where p is a prime number.   We denote by $|\xi|_p$ the absolute value on K corresponding to the prime p, which induces on **Q** the usual absolute value if p is archimedean, and otherwise the p-adic absolute value, so that $|p|_p = 1/p$.   We assume that this absolute value is extended to the algebraic closure $\overline{K}$ of K in some way.   This amounts to embedding $\overline{K}$ in $\overline{\mathbf{Q}}_p$ and taking the absolute value induced by that of $\overline{\mathbf{Q}}_p$.

If β is an element of K, we can define its height

$$H_K(\beta) = \prod_p \sup(1, |\beta|_p)^{N_p}$$

the product being taken over all primes of K. More generally, if P is a point in projective n-space, with coordinates $(\xi_0, \ldots, \xi_n)$ rational over K, then

$$H_K(P) = \prod_{\mathfrak{p}} \sup_i \left[|\xi_i|_{\mathfrak{p}}^{N\mathfrak{p}} \right].$$

The product formula [1] guarantees that this does not depend on the choice of coordinates. Thus $H_K(\beta)$ is the height of the point having $(1, \beta)$ as coordinates in $\mathbf{P}^1$, and if $\beta \neq 0$ we see that $H_K(\beta) = H_K(1/\beta)$. If $\beta = m/n$ is a rational number, with m, n relatively prime integers, then $H_{\mathbf{Q}}(\beta) = \sup(|m|, |n|)$.

If K is fixed, and the reference to a projective space is fixed throughout a discussion, then we write H instead of H_K.

We recall that one can define the absolute height

$$h(P) = H_K(P)^{1/[K : \mathbf{Q}]}$$

which is then independent of the field in which P is rational. Thus H_K is a function on points in projective space rational over K while h is a function on points in projective space rational over $\overline{K}$. Note that $h(P) \geq 1$ and $H_K(P) \geq 1$.

Two positive functions λ, λ' on a set of points are called *equivalent* (we write $\lambda \sim \lambda'$) if there exist two numbers $c_1, c_2 > 0$ such that

$$c_1 \lambda \leq \lambda' \leq c_2 \lambda.$$

It will also be convenient to define λ, λ' to be *quasi-equivalent* (we write $\lambda \approx \lambda'$) if given $\varepsilon > 0$, there exist two numbers $c_1, c_2 > 0$, depending on ε, such that for all points P in the set, we have

$$c_1 \lambda(P)^{1-\varepsilon} \leq \lambda'(P) \leq c_2 \lambda(P)^{1+\varepsilon}.$$

These relations are obviously equivalence relations (symmetric, reflexive, transitive).

On the set of elements $\alpha \in K$ such that $K = \mathbf{Q}(\alpha)$, our function H_K is equivalent to the height function used for instance by Roth, i.e. the maximum value of the coefficients in the irreducible equation satisfied by β over $\mathbf{Z}$, the integers. This is trivially verified. If E is a subfield of K, then on E we have $H_K = H_E^{[K : E]}$. From this one sees immediately that the set of elements of K of bounded height is finite (such elements can satisfy only a finite number of equations over $\mathbf{Z}$).

The Thue-Siegel-Mahler-Roth theorem (Roth's theorem for short) can be stated as follows. *Let α be algebraic over* K. *Let $\varkappa$ be a number > 2, and let S be a finite set of primes of* K. *Then the solutions β in* K *of the inequality*

$$\prod_{\mathfrak{p} \in S} \inf(1, |\alpha - \beta|_{\mathfrak{p}}) \leq \frac{1}{H(\beta)^{\varkappa}}$$

have bounded height.

For a proof, see for instance [12], which follows Roth closely, includes the Mahler version, and obviously generalizes to number fields. Of course, the set of elements of K with bounded height is finite, but we have stated the theorem in the above form so as to have a uniform terminology with the function field case (characteristic o). We discuss this below. The above statement can be slightly strengthened (as in Mahler) : If $\mathfrak{b}$ is the ideal which is the denominator in the ideal factorization of β in K, and if one defines $|\mathfrak{b}|_\mathfrak{p}$ for finite primes $\mathfrak{p}$ in the obvious manner, then one can replace $|\alpha-\beta|_\mathfrak{p}$ by $|\alpha-\beta|_\mathfrak{p}|\mathfrak{b}|_\mathfrak{p}$ in the above inequality, for the finite primes appearing in S.

Actually, for the sequel, we need only the approximation for one prime : *The solutions β in K of*

$$|\alpha-\beta|_\mathfrak{p} \leq \frac{1}{H(\beta)^\varkappa}$$

have bounded height. In fact, we shall need it in the following context, as in Mahler [8].

Let $G(Y)$ be a polynomial in $\overline{K}[Y]$, and assume that the multiplicity of its roots is at most r for some integer $r > 0$. Say G has leading coefficient 1, so $G(Y) = \prod(Y-\alpha_i)^{e_i}$. Let $c > 0$ be a number, and $\mathfrak{p}$ a prime of K. Then the solutions β in K of

$$|G(\beta)|_\mathfrak{p} \leq \frac{c}{H(\beta)^{\varkappa r}}$$

have bounded height if $\varkappa > 2$.

It is a trivial matter to get this from the preceding statement. Indeed, our absolute value comes from an embedding of $\overline{K}$ in $\overline{K}_\mathfrak{p}$. If β stays away from all the α_i, our statement is clear. If β comes close to one of them, then its distance from the others is greater than some fixed lower bound. Thus in evaluating $|G(\beta)|_\mathfrak{p}$ precisely one term $|\alpha_i-\beta|_\mathfrak{p}^{e_i}$ becomes small, and we get

$$|\alpha_i-\beta|_\mathfrak{p}^{e_i} \leq \frac{c'}{H(\beta)^{\varkappa r}}$$

for a suitable $c' > 0$, and a sequence of β's such that $H(\beta) \to \infty$. Since $e_i \leq r$, we can replace it by r, still preserving the inequality, and then take an r-th root. By making $\varkappa$ a little smaller, but still > 2, one can omit the constant c', and thus reduce our statement to the previous one.

Note that if we put $G(Y) = Y-\alpha$, we recover Roth's theorem in its original form.

We now discuss the function field case. Let K be a function field (of arbitrary dimension) over a constant field k of characteristic o. Let W be a projective model of K over k, non-singular in codimension 1. Let w be a generic point of W over k, so that we can write $K = k(w)$. As in [7] we use W to compute heights. If $\mathfrak{p}$ is a prime rational divisor of W over k, then $\deg(\mathfrak{p})$ denotes its projective degree. We then have the absolute value

$$|\xi|_\mathfrak{p} = \gamma^{\deg(\mathfrak{p})\mathrm{ord}_\mathfrak{p}\,\xi}$$

where $\mathrm{ord}_\mathfrak{p}\xi$ is the order at the discrete valuation determined by $\mathfrak{p}$, and γ is a fixed number, $0 < \gamma < 1$. Thus

$$H(\xi) = H_W(\xi) = (1/\gamma)^{d(\xi)}$$

where $d(\xi) = \deg(\xi)_\infty$ is the degree of the divisor of poles of ξ. More generally, if $(\xi_0, \ldots, \xi_n)$ is a point in $\mathbf{P}^n$ over K, then

$$H(P) = (1/\gamma)^{\deg \sup_i (\xi_i)_\infty}.$$

Thus in function fields, it is convenient to take the log to the base γ.

For each $\mathfrak{p}$ we suppose our absolute value extended to $\overline{K}$ in some fixed way. *Then the statement we gave above of Roth's theorem holds in the present situation.* This is seen as follows. First, one reduces the situation to the case where K is of dimension 1 over k, by taking generic hyperplane sections. Let L_u be a generic hyperplane over k, and $(w_1, \ldots, w_n)$ an affine generic point of W over k. Let $t = u_1 w_1 + \ldots + u_n w_n$ and let $k' = k(u_1, \ldots, u_n, t)$ (cf. [4], Ch. VIII, § 6). The generic hyperplane section $W' = W \cdot L_u$ is defined over k', assuming that $\dim W \geq 2$. If $\mathfrak{p}$ is a prime rational divisor of W over k, then $\mathfrak{p}' = \mathfrak{p} \cdot L_u$ is a prime rational divisor of W' over k'. If $\xi \in K$ is a function on W, and ξ' the induced function on W', then $(\xi')_\infty = (\xi)_\infty \cdot L_u$. Thus the height remains invariant by going over to generic hyperplane sections :

$$H_W(P) = H_{W'}(P)$$

if P is a point in $\mathbf{P}^n$ rational over K. (Geometrically speaking, the point P gives rise to a rational map of W into $\mathbf{P}^n$ and the induced rational map of W' into $\mathbf{P}^n$.)

The absolute value $|\;|_\mathfrak{p}$ described previously extends to the field $\overline{K}(u_1, \ldots, u_n)$ in such a way that it is trivial on $k(u_1, \ldots, u_n, t)$, and corresponds to the prime divisor $\mathfrak{p}'$. Consequently, we see that if we prove Roth's theorem for the field $K' = K(u_1, \ldots, u_n)$ viewed as function field over the constant field k', relative to the model W' (which is projective and non-singular in codimension 1) then it will follow for (K, k, W). This brings us to the function fields in one variable.

As for those, a prime $\mathfrak{p}$ is then a conjugate set of points over k. One sees immediately that we may go over to the algebraic closure of k, and then, in terms of orders, to prove the theorem in the following form :

Let K be a function field of one variable over an algebraically closed constant field k of characteristic 0. Let S be a finite set of primes of K over k. Let α be algebraic over K. Then the degrees $\deg(\beta)_\infty$ of elements β in K satisfying the inequality

$$\sum_{\mathfrak{p} \in S} \mathrm{ord}_\mathfrak{p}(\alpha - \beta) \geq \varkappa \deg(\beta)_\infty \qquad\qquad (\varkappa > 2)$$

are bounded.

Actually, one gets a finiteness statement, because of the following remark : Let β_1, β_2 be solutions of the above inequality such that $d(\beta_1) = d(\beta_2) = d$ is > 0. Then $\beta_1 = \beta_2$. Indeed, we get

$$\mathrm{ord}_\mathfrak{p}(\beta_1 - \beta_2) \geq \varkappa d.$$

But $\deg(\beta_1 - \beta_2)_\infty \leq 2d$. If $\beta_1 \neq \beta_2$, then $\beta_1 - \beta_2$ has more zeros than poles, which is impossible.

We observe that $d(\beta) = 0$ if and only if β is constant, i.e. lies in k. (In terms of heights, this means $H(\beta) = 1$.) Thus finally, we can state Roth's theorem in dimension 1 in the following form, say for one prime p :

Let α be algebraic over K. There is only a finite number of elements $\beta \in K$ which are not constant (i.e. not in k) such that

$$\mathrm{ord}_p(\alpha - \beta) \geqq \varkappa d(\beta)$$

if $\varkappa > 2$.

The proof of Roth's theorem for function fields in one variable is essentially the same as Roth's own proof. One must use the Riemann-Roch theorem precisely in the place where Roth does his counting to get his crucial polynomial. By the way, in number fields at this point, it is best to use the known estimates giving the number of algebraic integers in given parallelotopes (as in Artin-Whaples Theorem 4 [1]). For an exposition, cf. mimeographed notes to appear in the near future.

§ 2. A geometric formulation of Roth's theorem.

In this section we give a formulation of Roth's theorem which is adapted to the use we wish to make of it afterwards. We let K be a *global field* : This means a number field, or a function field over a constant field k which we assume of characteristic zero for this section. In the function field case, heights are taken with respect to a model as described in § 1.

THEOREM 1. *Let W be a complete non-singular curve defined over K. Let z, y be two non-constant functions in $K(W)$, and let r be the largest of the orders of the zeros of z. Assume that y has no zero or pole among the zeros of z, and that y gives an injective mapping of this set of zeros into $\overline{K}$. Let $\varkappa$ be a number > 2, and $c > 0$. Then the points $Q \in W_K$ such that*

$$|z(Q)|_{\mathfrak{p}} \leqq \frac{c}{H(y(Q))^{\varkappa r}}$$

have bounded height H_y.

PROOF. Without loss of generality, we may assume that $K(z, y) = K(W)$. If necessary, we may consider the complete non-singular curve which is a model of $K(z, y)$ instead of W. Our assumptions will still be valid for this curve. We may also assume that the values $|y(Q)|_{\mathfrak{p}}$ are bounded. Indeed, if there is an infinite sequence of points Q whose height $H_y(Q) = H(y(Q))$ tends to infinity, and satisfying the above inequality, but with $|y(Q)|_{\mathfrak{p}}$ unbounded, then we may consider $1/y$ instead of y, together with an infinite subsequence of such points Q. Let Φ be the set of zeros of z. Since y has no pole in Φ, it is integral over the local ring $\mathfrak{o}$ of the point $z = 0$ in the function field $K(z)$. Let $F(Y)$ be its irreducible equation over $\mathfrak{o}$. Then

$$F(Y) \equiv G(Y) \qquad (\mathrm{mod} \ z)$$

where $G(Y)$ is a polynomial with coefficients in K, leading coefficient 1, and mod z means modulo the maximal ideal of $\mathfrak{o}$ generated by z.

By hypothesis, y induces an injection $Q \to y(Q)$ of Φ into $\overline{K}$. The multiplicity of a root of $G(Y)$ is thus $\leq$ the multiplicity of a point on W in the inverse image of $z = 0$, this being the multiplicity of a zero of z. (One can see this formally for instance as follows : Let $(y^{(1)})$ be an affine generic point of W over K all of whose coordinates are integral over $\mathfrak{o}$. If $(y^{(1)}, \ldots, y^{(M)})$ is a complete set of conjugates of $(y^{(1)})$ over $K(z)$, then the cycle on W which is the inverse image of $z = 0$ consists of a specialization $(\bar{y}^{(1)}, \ldots, \bar{y}^{(M)})$ of $(y^{(1)}, \ldots, y^{(M)})$ over $z \to 0$. The conjugates of y correspond to the conjugates $(y^{(1)}, \ldots, y^{(M)})$, and one can then use [4], Theorem 2 of Chapter I, § 4, applied to the polynomial $F(Y)$.)

We can write

$$F(Y) = G(Y) + zA(z, Y)$$

where $A(z, Y)$ is a polynomial in Y with coefficients in $\mathfrak{o}$. Since $A(o, Y)$ is defined, so is $A(z(Q), Y)$ for small values of $|z(Q)|_{\mathfrak{p}}$, which is all that we need to consider. Thus the values

$$|A(z(Q), y(Q))|_{\mathfrak{p}}$$

remain bounded since we could assume that $|y(Q)|_{\mathfrak{p}}$ remains bounded. Since $F(y) = 0$, we get an estimate for $G(y(Q))$, namely

$$|G(y(Q))|_{\mathfrak{p}} \leq c_1 |z(Q)|_{\mathfrak{p}}$$

$$\leq \frac{c_2}{H(y(Q))^{\kappa r}}$$

which puts us precisely in the situation described in § 1, and concludes the proof.

§ 3. Behaviour of heights under projection.

We use the same notation as in [7]. For this section, K is a global field. Property 1 F of [7] is still clearly valid without the restriction dim $K = 1$, and so are the other Properties 2, 3, 4 and Theorem 3, where dim $K = 1$ is not assumed.

Let V be a variety defined over K. For each morphism $\varphi : V \to \mathbf{P}^n$ (everywhere defined rational map) defined over K, we have height functions on V_K and $V_{\overline{K}}$, namely

$$H_\varphi(P) = H(\varphi(P)) \quad \text{and} \quad h_\varphi(P) = h(\varphi(P)).$$

We do not repeat here the discussion establishing the correspondence between maps of V into $\mathbf{P}^n$ and linear systems on V, but to fix the notation, if V is complete, normal, if X is a divisor on V rational over K, and $\mathscr{L}(X)$ is its complete linear system, and if we assume that $\mathscr{L}(X)$ is without fixed point, then we denote by H_X (or h_X) the height associated with any map into projective space arising from this linear system. These are well defined up to equivalence.

Foremost among the properties of heights is Property 4 (due to Weil) that the heights associated with two linear systems without fixed points whose divisors are linearly equivalent to each other (i.e. having the same complete linear system) are equivalent. This property will be used constantly in what follows.

The local behaviour at a point on a variety is represented by local parameters, and it is frequently more convenient to deal with these than with the coordinates in a projective embedding. In the following discussion (which depends only on Property 4 of heights), we make a generic projection and compare the heights arising from the embedding and its projection. We adjust the discussion to the immediate application we have in mind, and thus restrict ourselves to the case of curves.

Let W be a projective non-singular curve defined over K. The height h (or H) is taken with respect to this embedding. Let $(y_0, \ldots, y_n)$ with $y_0 = 1$ be functions in $K(W)$ determining our given embedding. Let Φ be a finite set of points of W in $\overline{K}$. Then there is some polynomial equation such that if $(a_0, \ldots, a_n, b_0, \ldots, b_n)$ are elements in k not satisfying this equation, then the function

$$y = \frac{a_0 y_0 + \ldots + a_n y_n}{b_0 y_0 + \ldots + b_n y_n}$$

has the following properties :

The function y is not constant, and has no zero or pole among the points of Φ.

If h_y is the height determined by the mapping of W into $\mathbf{P}^1$ arising from the function y, then

$$h_y \sim h,$$

and thus $H_y \sim H$ (as functions on W_K).

The mapping

$$Q \rightarrow y(Q)$$

gives an injection of Φ into $\overline{K}$.

These properties are easily proved. Indeed, let Q be a point of W in $\overline{K}$. If t is a function of order 1 at Q, then each y_i has an expansion as a power series in t, say $y_i = \xi_i t^{e_i} + \ldots$ with an integer e_i, which may be negative. We see that there is a polynomial G_Q (linear) such that for any set of elements $(a_0, \ldots, a_n)$ in k for which $G_Q(a) \neq 0$, then $a_0 y_0 + \ldots + a_n y_n$ has order e at Q, where $e = \inf e_i$. Taking Q from a finite set Φ, we then take the product of the G_Q for $Q \in \Phi$, and achieve the same thing for all $Q \in \Phi$.

If $\mathfrak{a}_0$ denotes the sup of the polar divisors of our given y_i, then we see that almost all linear combinations $a_0 y_0 + \ldots + a_n y_n$ have precisely $\mathfrak{a}_0$ as polar divisor. Furthermore, applying the above remarks to zeros instead of poles, and taking into account that the linear system determined by $(1, y_1, \ldots, y_n)$ is without fixed point, we see that we can make a sufficiently general choice of a_i and b_i such that the function y has no zero or pole in Φ, and its divisor

$$(y) = (y)_0 - (y)_\infty$$

is such that $(y)_\infty$ lies in the above linear system. According to Property 4 of heights, it follows that h_y is equivalent to h.

To insure that the map $Q \to y(Q)$ is injective, we select among the y_i (for each Q) that function having the highest order pole at Q and denote it by y_Q. All quotients y_i/y_Q are defined at Q, and we have

$$y(Q) = \frac{a_0 w_0(Q) + \ldots + a_n w_n(Q)}{b_0 w_0(Q) + \ldots + b_n w_n(Q)}$$

where $w_i = y_i/y_Q$. (Strictly speaking, each w_i should carry Q also as an index.) We can choose the b_i so that the denominator does not vanish, and the condition that $y(Q) \neq y(Q')$ when $Q \neq Q'$ are two distinct points of Φ is immediately seen to be implied by the non-vanishing of a polynomial in the a_i and b_i. This concludes the proof of our three statements.

REMARK. Given a subfield E of K(W) containing K, and such that K(W) is finite separable algebraic over E, then it is clear that in addition to the above conditions, we can also require y to be a generator of K(W) over E.

Putting the results of § 2 together with the technique of generic projections, we get a more useful version of Theorem 1 :

THEOREM 1'. *Let* W *be a projective non-singular curve defined over a global field of characteristic* 0. *Let* z *be a non-constant function in* K(W), *and let* r *be the largest of the orders of the zeros of* z. *Let* $\varkappa$ *be a number* > 2 *and* $c > 0$. *Then the points* $Q \in W_K$ *such that*

$$|z(Q)|_\mathfrak{p} \leqq \frac{c}{H(Q)^{\varkappa r}}$$

have bounded height.

PROOF. We let Φ be the set of zeros and poles of W, and apply Theorem 1, taking into account the properties of H_y and its relation to H, the height taken relative to the given embedding of W in a projective space.

§ 4. Another property of heights.

Let K be a global field. As pointed out already, linear equivalence of linear systems gives rise to equivalent height functions. We shall now prove that algebraic equivalence gives rise to quasi-equivalent height functions. We need a lemma from pure algebraic geometry.

LEMMA. *Let* V *be a complete non-singular variety. Let* X *be a divisor on* V *such that some multiple* eX *is ample* (e *an integer* > 0). *Then there exists an integer* $e' > 0$ *such that for any divisor* Z *on* V *algebraically equivalent to* 0, *the divisor* $Z + e'$X *is ample.*

PROOF. Let $\hat{A}$ be the Picard variety of V. We can always find a Poincaré divisor D on $V \times \hat{A}$ which is positive : If V is an abelian variety, this is Theorem 10 of Chapter IV,

329

§ 4, [5], and otherwise, making a generic translation on D, the pull-back method of Weil gives a positive Poincaré divisor on $V \times \hat{A}$ (*ibid.*, Theorem 1 of Chapter VI, § 1).

By hypothesis, $Z \sim {}'D(a) - {}'D(o)$ for some point $a \in \hat{A}$. The intersections are defined after making a generic translation on D. It is well known that there exists an integer $e_1 > o$ such that $-{}'D(o) + e_1 X$ is ample (see for instance [9], Lemma 1). Furthermore, the divisors ${}'D(a)$ as a ranges over $\hat{A}$ are all algebraically equivalent to each other, are positive divisors, and have the same projective degree. Hence there exists an integer $e_2 > o$ such that ${}'D(a) + e_2 X$ is ample, again by [9], Lemma 1. From this our lemma is immediate.

PROPERTY 5. *Let* V *be a complete non-singular variety defined over* K. *Let* X, Y $> o$ *be two positive divisors on* V, *rational over* K. *Assume that a positive multiple of each is ample, and that the linear systems* $\mathscr{L}(X)$ *and* $\mathscr{L}(Y)$ *are without fixed points. If* X *and* Y *are algebraically equivalent, then* h_X *and* h_Y *are quasi-equivalent (and so are* H_X *and* H_Y).

PROOF. Using property 4, we shall reduce our assertion to a statement concerning linear equivalence classes of divisors on V. By the lemma, there exists an integer $e > o$ such that for all $n > o$ we have

$$n(X - Y) + eX \sim Z_n$$

where Z_n is a positive divisor on V, and $\mathscr{L}(Z_n)$ is without fixed points. Since $n(X - Y) + eX$ is rational over K, one may take Z_n rational over K. We get $nX + eX \sim nY + Z_n$, and taking heights,

$$h_X^{n+e} \sim h_Y^n h_{Z_n}.$$

Since $h_{Z_n}(P) \geqq 1$ for all P, taking an n-th root, we see that given $\varepsilon > o$, there exists a number $c > o$ such that

$$c h_X(P)^{1+\varepsilon} \geqq h_Y(P)$$

if we take n sufficiently large. The other inequality is obtained in a similar way, or by symmetry.

When V is a curve, the statement is due to Siegel [14] whose proof we essentially imitate here, except that of course Siegel uses the Riemann-Roch theorem where we have used the Picard variety (see for instance [15], p. 435). In the case of curves, algebraic equivalence is determined by the degree of the divisor, and hence if $\deg(X) = d$ and $\deg(Y) = d'$, then h_X is quasi equivalent to $h_Y^{d/d'}$.

In particular, we note that if a set of points on the curve V has bounded height in some projective embedding, then it has bounded height in every projective embedding : The notion of a set of bounded height is independent of the embedding.

§ 5. **Inequalities from the theory of heights.**

We come now to the proof of the diophantine theorem proper.

Let C *be a non-singular curve, of genus* $\geqq 1$, *imbedded in some projective space over the global field* K. *The height* H *as a function on* C_K *is determined by this embedding.*

Let x be a function on C, *defined over* K *and not constant.* Then (I, x) determines a mapping of C into $\mathbf{P}^1$, and the corresponding linear system is of degree

$$r = [\mathrm{K}(\mathrm{C}) : \mathrm{K}(x)],$$

a divisor in it being, for instance, the divisor of poles of x. We assume that $\mathrm{K}(\mathrm{C})$ over $\mathrm{K}(x)$ is separable. (At the very end, and only then, do we need characteristic o, to contradict Roth's theorem.)

Let S *be a finite set of primes of* K, *containing the archimedean primes, and let* R *be a subring of* K *all of whose elements are* p-*integral for* p *not in* S. *Let* $\mathfrak{R}$ *be the set of points of* C *rational over* K, *and such that* $x(\mathrm{P})$ *lies in* R. *We wish to prove that the height of the points in* $\mathfrak{R}$ *is bounded.* We assume the contrary, derive a list of inequalities which eventually contradict Roth's theorem. We let $\mathfrak{R}_1$ be a subsequence of $\mathfrak{R}$ such that the height of the points in $\mathfrak{R}_1$ tends to infinity.

By assumption, we have $|\xi|_{\mathfrak{p}} \leq \mathrm{I}$ for $\mathfrak{p} \notin \mathrm{S}$ and $\xi \in \mathrm{R}$. Hence for all $\mathrm{P} \in \mathfrak{R}_1$, we get

$$\mathrm{H}(x(\mathrm{P})) = \prod_{\mathfrak{p} \in \mathrm{S}} \sup(\mathrm{I}, |x(\mathrm{P})|_{\mathfrak{p}}^{\mathrm{N}_{\mathfrak{p}}})$$

where $\mathrm{N}_{\mathfrak{p}}$ is the local degree in number fields, and I in function fields. Let $\mathrm{N} = [\mathrm{K} : \mathbf{Q}]$ in number fields, and I in function fields. Let s be the number of primes in S. Rewriting our product in terms of the absolute values, we see that we have at most $\mathrm{N}s$ terms in it, of type

$$\sup(\mathrm{I}, |x(\mathrm{P})|_{\mathfrak{p}}).$$

Consequently, for each $\mathrm{P} \in \mathfrak{R}_1$, there exists one $\mathfrak{p}$ in S such that $|x(\mathrm{P})|_{\mathfrak{p}} \geq \mathrm{H}(x(\mathrm{P}))^{1/\mathrm{N}s}$. Hence there exists an infinite subset $\mathfrak{R}_2$ of $\mathfrak{R}_1$ such that for some $\mathfrak{p}$ in S and all points P in $\mathfrak{R}_2$ we have

$$\mathrm{H}(x(\mathrm{P}))^{1/\mathrm{N}s} \leq |x(\mathrm{P})|_{\mathfrak{p}}.$$

In view of Property 5, we can compare $\mathrm{H}(x(\mathrm{P}))$ and $\mathrm{H}(\mathrm{P})$. If d is the degree of C in its given projective embedding, and φ the mapping into $\mathbf{P}^1$ given by the function x, we conclude that there is a number $c_3 > 0$ depending on ε such that for P in C_{K} we have

$$\mathrm{H}(\mathrm{P})^{r/d - \varepsilon} \leq c_3 \mathrm{H}(x(\mathrm{P})).$$

Combining this with the previous inequality, we see that there is a number $\rho > 0$ such that for some $\mathfrak{p} \in \mathrm{S}$ and all $\mathrm{P} \in \mathfrak{R}_2$ we have for suitable $c_4 > 0$:

$$\mathrm{H}(\mathrm{P})^{\rho} \leq c_4 |x(\mathrm{P})|_{\mathfrak{p}}.$$

This inequality will be improved by going over to a covering of C, derived from the weak Mordell-Weil theorem. Furthermore, the arguments will prove the following improvement of Theorem I', for curves of genus $\geq \mathrm{I}$.

331

THEOREM 2. *Let* K *be a global field of characteristic* 0. *Let* ρ, c *be numbers* > 0. *Let* C *be a curve of genus* ≥ 1 *defined over* K, *and* x *a non-constant function in* $K(C)$. *Let* $\mathfrak{p}$ *be a prime of* K. *Then the height of points* P *in* C_K *such that*

$$|x(P)|_{\mathfrak{p}} \leq \frac{c}{H(P)^{\rho}}$$

is bounded.

(To apply the theorem, use the function $1/x$ instead of x.)

§ 6. Inequalities from the Mordell-Weil theorem.

We assume that C is embedded in its Jacobian J over K, and take J embedded in projective space. We take the induced embedding on C. For any point $P \in J_K$ we let $H(P)$ be the height determined by our embedding, which remains fixed throughout.

In view of the definition of the height, and of $|\ |_{\mathfrak{p}}$, we may, without loss of generality, in the function field case, assume that the constant field is algebraically closed. This insures that for an integer $m > 1$ we have J_K/mJ_K finite.

PROPOSITION 1. *Let* m *be an integer* > 0, *unequal to the characteristic of* K. *Let* $\mathfrak{S}$ *be an infinite set of rational points of* C *in* K. *Then there exists an unramified covering* $\omega : W \to C$ *defined over* K, *an infinite set of rational points* $\mathfrak{S}'$ *of* W *in* K *such that* ω *induces an injection of* $\mathfrak{S}'$ *into* $\mathfrak{S}$, *and a projective embedding of* W *over* K *such that* $H \circ \omega$ *is quasi-equivalent to* H^{m^2}. (*Of course, in* $H \circ \omega$ *the* H *refers to the height on* C, *while in* H^{m^2} *it refers to the height on* W.)

PROOF. Let $a_1, \ldots, a_l$ be representatives of cosets of J_K/mJ_K. Infinitely many $P \in S$ lie in the same coset, and so there exists one point, say a_1, and infinitely many points Q in J_K such that $mQ + a_1$ lies in $\mathfrak{S}$. We let $\mathfrak{S}'$ be this infinite set of points Q. The covering $\omega : J \to J$ given by $\omega u = mu + a_1$ is unramified, and its restriction W to C is non-degenerate, i.e. is an irreducible covering of the same degree [6]. The inverse image of a point in C lies in W. Thus $\mathfrak{S}'$ is actually a subset of W_K. Restricting one's attention to a subset of $\mathfrak{S}'$ guarantees that ω induces an injection on this subset.

To prove the relation concerning the heights, we may work on the Jacobian itself since we take W in the projective embedding induced by that of J. If X is a hyperplane section of J, then

$$\omega^{-1}(X) = (m\delta)^{-1}(X_{-a_1}) \equiv m^2 X$$

where $\equiv$ is the equivalence of the square, known to be the same as algebraic equivalence. But $h_{\omega^{-1}(X)} \sim h_X \circ \omega$ by Property 3 of heights (which is trivial) and $h_{\omega^{-1}(X)}$ behaves essentially like $h_{m^2 X} \sim h_X^{m^2}$ by Property 5. Our proposition is now clear, since for rational points in K, these equivalences are valid for H.

We note that $x(P) = x(\omega Q)$. Let z be the function on W such that $z(Q) = x(\omega Q)$. Let $\varkappa$ be a number > 2 and let m be large enough such that $m^2 \rho > \varkappa r$. Then for a suitable constant c_5 our inequality becomes

$$|z(Q)|_{\mathfrak{p}} \leq \frac{c_5}{H(Q)^{\varkappa r}}$$

which is precisely the case treated in Theorem $1'$. The fact that W is unramified over C guarantees us that the orders of the zeros of z are bounded by r. This concludes the proof of the original statement and of Theorem 2.

It is striking that the use we made of the Jacobian is formally analogous to the one in class field theory [6]. In that case, Artin's reciprocity law was reduced to a formal computation in the isogeny $u \to u^{(q)} - u$ of the Jacobian. In the present case, the heart of the proof is reduced to a formal computation of heights in the isogeny $u \to mu + a$.

§ 7. Extensions of finite type.

We shall extend the Siegel finiteness statement to rings of finite type over **Z**, and its analogue in function fields, including the relative case. We reduce our theorem to the case of dimension 1 (number fields, or function fields of one variable). We begin by giving a criterion which allows us to lower the field of definition of a curve.

Let K *be a function field over the constant field* K_0, which need not be algebraically closed, but which we assume to be *of characteristic* 0. If its dimension is > 1, we select a projective normal model, relative to which we take our heights.

Let C be a complete non-singular curve defined over K. If $\Re$ is a subset of C_K, and the height of the points in $\Re$ is bounded in some projective embedding of C, then it is bounded in any other, as mentioned above. The next proposition deals with such sets. For the K/K_0-trace, cf. [5], Chapter VIII.

PROPOSITION 2. *Let* K, K_0 *be as above, and* C *a complete non-singular curve of genus* ≥ 1, *defined over* K. *Let* $\Re$ *be an infinite subset of* C_K *of bounded height. Regard* C *as embedded in its Jacobian* J *over* K, *and let* (B, τ) *be a* K/K_0-trace *of* J. *Then* τ *is an isomorphism. The points of* $\Re$ *lie in a finite number of cosets of* τB_{K_0}, *and if infinitely many of them lie in one coset, so are of type* $a + \tau b$ *where* a *is some point of* J_K *and* b *ranges over an infinite subset of* B_{K_0}, *then* $C_0 = \tau^{-1}(C_{-a})$ *is defined over* K_0, *and* τ *induces an isomorphism of* C_0 *onto* C_{-a}.

PROOF. We may assume J, B embedded in projective space. Using the auxiliary model of K over K_0 as in [7], we see from Theorem 3 and Proposition 2 of [7] that the points of $\Re$ lie in a finite number of cosets of τB_{K_0}. Say infinitely many lie in the coset $a + \tau B_{K_0}$. We know that τ establishes an isomorphism between B and $\tau(B)$ by Cor. 2 of Th. 9, Ch. VIII, [5]. Since infinitely many points of τB_{K_0} lie in C_{-a}, it follows that their K-closure in τB or in J is precisely C_{-a}. Put $C_0 = \tau^{-1}(C_{-a})$. Then C_0 is a curve contained in B, and τ induces an isomorphism τ_0 of C_0 onto C_{-a}. Furthermore, C_0 contains infinitely many points b of B rational over K_0. It is then a trivial matter to conclude that C_0 is defined over K_0 because these infinitely many points are both K_0 and K-dense in C_0. Since τB contains a translation of C, it follows that $\tau B = J$ is the Jacobian, i.e. τ is an isomorphism.

REMARK. If we do not assume characteristic 0 in Proposition 2, then τ is merely bijective, and C_0 may be defined over a purely inseparable extension of K_0.

333

COROLLARY. *Let* K, K_0 *be as above. Let* C *be a complete non-singular curve of genus* ≥ 2 *defined over* K, *and let* $\mathfrak{R}$ *be an infinite subset of* C_K *consisting of points of bounded height. Then there exists a curve* C_0 *defined over* K_0 *and a birational transformation* $T : C_0 \to C$ *defined over* K *such that all but a finite number of points of* $\mathfrak{R}$ *are images under* T *of rational points of* C_0 *in* K_0.

PROOF. Since the genus is ≥ 2, the curve cannot be equal to any translation of itself in its Jacobian. Hence there can only be one coset having infinitely many points of C and we apply the proposition.

Combining our corollary with the results obtained in the previous section, we get the relative formulation of Siegel's theorem.

THEOREM 3. *Let* K *be a function field over a constant field* k *of characteristic* 0. *Let* R *be a subring of* K *of finite type over* k. *Let* C *be a complete non-singular curve of genus* ≥ 1 *defined over* K, *and* φ *a non-constant function on* C *also defined over* K. *Let* $\mathfrak{R}$ *be the subset of* C_K *consisting of those points* P *such that* $\varphi(P) \in R$. *If* $\mathfrak{R}$ *is infinite, then there exists a curve* C_0 *defined over* k, *and a birational transformation* $T : C_0 \to C$ *defined over* K. *If the genus is* ≥ 2, *then all but a finite number of points of* $\mathfrak{R}$ *are images under* T *of points of* C_{0k}. *If the genus is* 1, *then the points of* $\mathfrak{R}$ *lie in a finite number of cosets of* $T(C_{0k})$.

To deal with the absolutely algebraic case, we need a specialization argument.

THEOREM 4. *Let* K *be a field of finite type over* $\mathbf{Q}$, *and* R *a subring of* K *of finite type over* $\mathbf{Z}$. *Let* C *be a non-singular curve of genus* ≥ 1 *defined over* K, *and let* φ *be a function in* $K(C)$ *which is not constant. Let* $\mathfrak{R}$ *be the subset of* C_K *consisting of points* P *such that* $\varphi(P) \in R$. *Then* $\mathfrak{R}$ *is finite.*

PROOF. We may assume C projective non-singular. Suppose $\mathfrak{R}$ infinite. Let k be the algebraic closure of $\mathbf{Q}$ in K. Then by Theorem 3, C is birationally equivalent over K to a curve C_0 defined over k. If the genus of C is 1, we restrict our attention to an infinite subset of points of $\mathfrak{R}$ which lie in the same coset of $T(C_{0k})$. Then, without loss of generality, we may assume $C = C_0$, and that we have infinitely many points of C in K such that $\varphi(P) \in R$, where φ is a function on C defined over K. We shall now prove our theorem by induction on the dimension of K over $\mathbf{Q}$.

Let F be a subfield of K containing k, and such that the dimension of K over F is 1. There exists a discrete valuation ring $\mathfrak{o}$ of K containing F and R whose residue class field $E = \mathfrak{o}/\mathfrak{m}$ is finite over F, and such that the reduction φ' of φ mod $\mathfrak{m}$ is a non-constant function $\varphi' : C \to \mathbf{P}^1$ (of the same degree as φ). For any point Q of C_K, we get a specialized point Q' in C_E, and $\varphi'(Q') = \varphi(Q)'$, using the compatibility of intersections and reductions, i.e. formally, using the graphs :

$$[\Gamma_\varphi \cdot (Q \times \mathbf{P}^1)]' = \Gamma'_\varphi \cdot (Q' \times \mathbf{P}^1)$$

the left hand side being $Q \times \varphi(Q)$ and the right hand side being $Q' \times \varphi'(Q')$. This yields infinitely many points Q' of C_E such that $\varphi'(Q')$ lies in the ring R', image of R in

334

the homomorphism $\mathfrak{o} \to \mathfrak{o}/\mathfrak{m}$.   Since E is of finite type over $\mathbf{Q}$, of dimension one less than that of K, and since R' is still of finite type over $\mathbf{Z}$, this concludes the proof.

REMARK.   The theorem could also be proved by using Néron's theorem which implies that say for an affine curve $f(X, Y) = 0$ of genus ≥ 1 with coefficients in R, there exists a homomorphism $R \to R'$ of R into a ring R' contained in a number field such that if $f'(X, Y) = 0$ is the specialized curve, then its genus is also ≥ 1 and the homomorphism $R \to R'$ induces an injection of the points of f in R into those of f' in R'. (See [11], Th. 6.)

Columbia University, New York.

BIBLIOGRAPHY

[1] E. Artin and G. Whaples, Axiomatic characterization of fields by the product formula, *Bull. Am. Math. Soc.*, vol. 51, n° 7 (1945), pp. 469-492.

[2] C. Chabauty, Sur les équations diophantiennes liées aux unités d'un corps de nombres algébriques fini, thèse, *Annali di Math.*, 17 (1938), pp. 127-168.

[3] — Sur les points rationnels des variétés algébriques dont l'irrégularité est supérieure à la dimension, *Comptes rendus Académie des Sciences*, Paris, 212 (1941), pp. 1022-1024.

[4] S. Lang, Introduction to algebraic geometry, *Interscience*, New York, 1959.

[5] — Abelian varieties, *Interscience*, New York, 1959.

[6] — Unramified class field theory over function fields in several variables, *Annals of Math.*, vol. 64, n° 2 (1956), pp. 285-325.

[7] S. Lang and A. Néron, Rational points of abelian varieties in function fields, *Am. J. of Math.*, vol. 81, n° 1 (1959), pp. 95-118.

[8] K. Mahler, Über die rationalen Punkte auf Kurven vom Geschlecht Eins, *J. Reine angew. Math.*, Bd. 170 (1934), pp. 168-178.

[9] T. Matsusaka, On algebraic families of positive divisors..., *J. Math. Soc. Japan*, vol. 5, n° 2 (1953), pp. 118-136.

[10] L. J. Mordell, On the rational solutions of the indeterminate equation of the third and fourth degrees, *Proc. of the Cambridge Philos. Soc.*, 21 (1922).

[11] A. Néron, Problèmes arithmétiques et géométriques rattachés à la notion de rang d'une courbe algébrique dans un corps, *Bull. Soc. Math. France*, 80 (1952), pp. 101-166.

[12] D. Ridout, The p-adic generalization of the Thue-Siegel-Roth theorem, *Mathematika*, 5 (1958), pp. 40-48.

[13] K. F. Roth, Rational approximations to algebraic numbers, *Mathematika*, 2 (1955), pp. 1-20.

[14] C. L. Siegel, Über einige Anwendungen Diophantischer Approximationen, *Abh. Preussischen Akademie der Wissenschaften*, Phys. Math. Klasse (1929), pp. 41-69.

[15] A. Weil, Arithmetic on algebraic varieties, *Annals of Math.*, vol. 53, n° 3 (1951), pp. 412-444.

[16] — L'arithmétique sur les courbes algébriques, *Acta Mathematica*, 52 (1928), pp. 281-315.

Reçu le 20 mars 1960.

271

1960. — Imprimerie des Presses Universitaires de France. — Vendôme (France)
ÉDIT. N° 26 039   IMPRIMÉ EN FRANCE   IMP. N° 16 336

272

SOME THEOREMS AND CONJECTURES IN DIOPHANTINE EQUATIONS

BY

SERGE LANG

Reprinted from the
BULLETIN OF THE AMERICAN MATHEMATICAL SOCIETY
July 1960, Vol. 66, No. 4
Pp. 240-249

SOME THEOREMS AND CONJECTURES IN DIOPHANTINE EQUATIONS

BY SERGE LANG

The theory of diophantine equations may be regarded as the natural continuation of algebraic geometry proper: Having once obtained a general theory of algebraic equations in several variables over essentially arbitrary ground fields (or rings), one tries to get statements depending on the special arithmetic structure of the coefficient domain. By definition, this becomes diophantine analysis. We shall list a few of the theorems and conjectures which arise in this direction.

Let k be a field, and $f(X_1, \cdots, X_n)$ a polynomial, also written $f(X)$, with coefficients in k. The equation $f = 0$ defines an algebraic set, i.e. the set of all n-tuples $(x_1, \cdots, x_n)$ in some algebraically closed field containing k, such that $f(x) = 0$. Such a point (x) in n-space is said to be a zero of f. It is said to be a rational point in k if all x_i lie in k. If f is a form (i.e. a homogeneous polynomial) then one views f as defining an algebraic set in projective space, and one considers nontrivial zeros, that is zeros such that not all x_i are 0. A nontrivial zero then defines a point in projective space, which is rational over k if again the coordinates can be chosen in k.

More generally, one considers systems of equations, or varieties (meaning an absolutely irreducible algebraic set). If V is a variety defined over a field k, then a point in it is rational over k if it has a set of coordinates in k.

The basic coefficient domain is that of the rational numbers Q or the integers Z. It is but a step from this to a finite extension k of Q (called a number field) or the ring of integers I_k of k instead of Z. We have primes $\mathfrak{p}$ associated with such fields: They are the absolute values which either induce the ordinary absolute value on Q (called archimedean primes) or the p-adic absolute value, defined by a prime number p:

$$\left| p^r m/n \right|_\mathfrak{p} = 1/p^r$$

if $m, n \in Z$, $mn \neq 0$, and $p \nmid mn$. The latter are called finite primes. One can then form the completion $k_\mathfrak{p}$ under the prime $\mathfrak{p}$, which is called a $\mathfrak{p}$-adic field, and is the field of real or complex numbers if $\mathfrak{p}$ is

---

An address delivered before the February Meeting of the Society in New York on February 27, 1960 by invitation of the Committee to Select Hour Speakers for Eastern Sectional Meetings; received by the editors March 16, 1960.

240

archimedean. (It is also becoming standard to consider field and ring extensions of finite type.)

Because of the topology and the completeness, a $\mathfrak{p}$-adic field gives rise to simpler diophantine problems than a number field, and one tries to reduce certain classes of diophantine problems to $\mathfrak{p}$-adic ones. Let us observe right away that a variety (say affine) defined over a number field k has a rational point in all but a finite number of $\mathfrak{p}$-adic fields $k_\mathfrak{p}$. (Remember V is absolutely irreducible.) This is easily seen, for instance as follows: By cutting V with sufficiently general hyperplane sections, one reduces the question to the case where V is a curve. One then reduces mod $\mathfrak{p}$. By the Riemann hypothesis in function fields (i.e. Weil's theorem [25]) the curve has a simple point mod $\mathfrak{p}$ for all but a finite number of primes, and this simple point can be refined to a $\mathfrak{p}$-adic point, say by Hensel's lemma.

1. **Hasse's theorem.** Let us begin with essentially the simplest type of variety, that defined by a quadratic equation. The main result here is Hasse's theorem (for an exposition, see for instance [26]). *Let k be a number field, f a quadratic form with coefficients in k. Then f has a nontrivial zero in k if and only if f has a nontrivial zero in each $k_\mathfrak{p}$.*

This is supplemented by a useful $\mathfrak{p}$-adic criterion: If $\mathfrak{p}$ is a finite prime, then *every quadratic form in 5 variables over the $\mathfrak{p}$-adic field $k_\mathfrak{p}$ has a nontrivial zero in $k_\mathfrak{p}$.*

These two theorems are typical examples of the following general principles: To get a global theorem from local ones, and to get solutions if the number of variables is large. As a corollary, we see that every quadratic form in 5 variables over a number field k, which is indefinite for every real embedding of the number field, has a nontrivial zero in k.

One may try to embed the above statements in theories concerning either forms of higher degree than 2, or concerning principal homogeneous spaces. Let us discuss the first.

2. **Quasi-algebraic closure.** There is a theorem of Peck [17] that *a form over a number field k of degree d, in n variables has a nontrivial zero in k if n is sufficiently large compared to d, and if k is totally imaginary (i.e. has no embedding into the reals).*

The condition that k be totally imaginary is essential: A sum of squares in a real field never has a nontrivial zero, and in fact, Artin gives an example of a form which is indefinite, has a $\mathfrak{p}$-adic zero for all $\mathfrak{p}$, has arbitrarily many variables, and still no zero in k, namely

$$(X_1^2 + \cdots + X_n^2)^2 - 2(Y_1^2 + \cdots + Y_m^2)^2 = 0$$

over the rational numbers (substituting sums of squares in $U^2 - 2T^2$).

Artin's substitution method is also used to reduce a system of simultaneous equations to one equation (especially with reference to quasi-algebraic closure, see below). One also knows how to reduce a system over a finite extension E of k to a system of equations in k (linearizing by means of a basis). For both of these, cf. [6; 7].

The above reductions are " multiplicative." This is important, because one may ask whether by restricting one's attention to forms of odd degree one does not recover the desired conclusion even when the field is real. Taking function fields over the reals as ground fields, I gave precise criteria under which such forms have nontrivial zeros [7]. The analogue of Peck's theorem for forms of odd degree and any number field was proved by Birch [1].

The Birch-Peck theorem holds for a large number of variables. Except for quadratic forms, one has no precise bound in number fields, but Artin has at least made conjectures concerning it. He defines a field K to be *quasi-algebraically closed* (QAC) if every form with coefficients in K, of degree d, in n variables, with $n > d$, has a nontrivial zero in K. A field which is quasi-algebraically closed does not admit division algebras of finite degree above it. Tsen proved that a function field in one variable over an algebraically closed constant field has no such division algebras. Analysing Tsen's proof, Artin was led to make the above remark, to define quasi-algebraic closure, to realize that Tsen's proof actually showed that such a function field was quasi-algebraically closed, and in view of Wedderburn's theorem, to conjecture that *finite fields are* QAC. This was proved by Chevalley [5]. Furthermore, it is known from class field theory that the field Ω obtained by adjoining all roots of unity to the rationals admits no finite division algebra above it. This and the analogy with function fields (a function field in one variable over a finite field to which one adjoins all roots of unity becomes a function field over an algebraically closed constant field) led him to *conjecture that Ω is* QAC. Thus, for instance, every form of degree d in n variables over the rationals with $n > d$ would have a nontrivial zero in some cyclotomic field.

How about number fields proper? In this case, Artin suggested that probably the condition $n > d$ has to be replaced by $n > d^2$ (always provided the field is totally imaginary). This is true in the analogous case of function fields over finite fields. At any rate, Artin conjectured that *a cubic form in* 10 *variables over the rationals $\mathbf{Q}$ has a nontrivial* *zero in* Q. Artin also made the analogous local conjectures. The one concerning the roots of unity is proved in my thesis [6], and the case of cubic forms is settled by Lewis [12]. The $\mathfrak{p}$-adic case proper remains open, in spite of the fact that the analogue for power series in one variable over a finite field is easily taken care of [6] with $n > d^2$. By the way, in each case the condition $n > d$ or $n > d^2$ is easily seen to be best possible. None of the global conjectures has yet been proved.

One can consider function fields over number fields or $\mathfrak{p}$-adic fields as ground fields themselves, and extend to those and to power series fields the same type of result. The condition $n > d$ (or $n > d^2$) has to be replaced by $n > d^i$ where i goes up with the number of variables. Although one can settle the function field case [6], the case of power series in several variables also remains open.

Finally, to go back to number fields, it seems to me reasonable to expect that a form with $n > d$ at least has a nontrivial zero in all but a finite number of $\mathfrak{p}$-adic fields. As pointed out previously, there is a problem here only if the form is not absolutely irreducible.

3. **Principal homogeneous spaces.** Let us return to quadratic forms. Let f, g be two quadratic forms over a number field k, in the same number of variables. They are equivalent over k if there exists a matrix T with coefficients in k such that $Tf = g$. It follows immediately from Hasse's theorem that if f, g are equivalent over every $k_\mathfrak{p}$ then they are equivalent over k (see Witt [26]).

Observe that the set of transformations T such that $Tf = g$ is a principal homogeneous space over the orthogonal group of f, which operates simply transitively on this set. More generally, let G be a group variety defined over a field k. A variety V is said to be a principal homogeneous space of G over k if V is defined over k, and we are given over k an everywhere defined rational map of $G \times V$ into V such that for every point $v \in V$, the map $x \to xv$ of G into V establishes an isomorphism of G onto V (for the structure of algebraic variety). Cf. Weil [22], who was the first to call attention to principal homogeneous spaces in relation to diophantine analysis.

This is precisely the situation we have with our quadratic forms f, g and they are equivalent over k if and only if the principal homogeneous space has a rational point. This aspect of Hasse's theorem may therefore be formulated by saying that *a principal homogeneous space over the orthogonal group has a rational point in k if and only if it has a rational point in every* $k_\mathfrak{p}$. Serre has suggested that this may remain true for any semi-simple group G, not only the orthogonal group. [*He later found a counterexample*].

The conclusion that the existence of a $\mathfrak{p}$-adic point for all $\mathfrak{p}$ implies the existence of a rational point in the number field k does not hold when one considers other types of group varieties, for instance an abelian variety, or for concreteness an elliptic curve. Selmer [18] has given examples of elliptic curves over the rationals, namely $3X^3+4Y^3 +5Z^3=0$, which have a point in every $\mathcal{Q}_p$ but not a rational point in $\mathcal{Q}$. Every such curve can be regarded as a principal homogeneous space over its Jacobian. (See also Cassels [2].)

Here again, before dealing with the global theory, one studies the local one, over a $\mathfrak{p}$-adic field. Over the reals, this is the way one can interpret a paper of Witt [27]. Over $\mathfrak{p}$-adic fields, Shafarevic [19] and Tate [21] have considered the question, and obtained a classification theorem in the case of elliptic curves.

In many cases, the principal homogeneous spaces and the existence of birational correspondences between curves had been studied by Chatelet [4], who pointed out their connection with cohomology. Let G be a group variety and V a principal homogeneous space defined over a field k. Let K be a Galois extension of k, with Galois group $\mathfrak{g}=\mathfrak{g}_{K/k}$, in which V has a rational point v_0. For each $\sigma\in\mathfrak{g}$, the point σv_0 lies in V, and hence there exists a unique element x_σ in G, rational over K, such that $x_\sigma\sigma v_0=v_0$. One verifies that (x_σ) is a 1-cocycle of $\mathfrak{g}$ in G_K, i.e. that $x_\sigma\sigma x_\tau=x_{\sigma\tau}$. Defining coboundaries in the obvious way, one obtains a cohomology set $H^1(\mathfrak{g}, G_K)$. One sees immediately that V has a rational point in k if and only if its associated cohomology class in $H^1(\mathfrak{g}, G_K)$ is trivial. Going to the injective limit to the separable algebraic closure of k, one is led to study the set $H^1(k, G)$, limit of the $H^1(\mathfrak{g}_{K/k}, G_K)$, and whose elements are in bijective correspondence with the isomorphism classes of principal homogeneous spaces of G over k. (Cf. [11]). If G is commutative, then $H^1(k, G)$ is of course a group, called the first cohomology group, and one can define also the higher dimensional ones.

When the ground field is a $\mathfrak{p}$-adic field ($\mathfrak{p}$ finite) Tate [21] has obtained a duality theorem: *Let A be an abelian variety defined over $k_\mathfrak{p}$. Then $H^1(k_\mathfrak{p}, A)$ is dual to the compact group of rational points in $k_\mathfrak{p}$ of the Picard variety of A.* He has also obtained a complete analysis of the cohomology involved for a coefficient module which arises from the points of a commutative group variety which is of multiplicative type (i.e. becomes a product of multiplicative groups over the algebraic closure), and for abelian varieties, both for the limit cohomology and in finite layers. The former, in finite layers, are dual to the modules arising in the Nakayama-Tate theorem of class field theory [15]. For *an abelian variety A over $k_\mathfrak{p}$, he shows that $H^r(k_\mathfrak{p}, A)=0$ if*

$r > 1$, *and if K is finite Galois over $k_{\mathfrak{p}}$, then $H^r(\mathfrak{g}_{K/k_{\mathfrak{p}}}, A_K)$ is dual to $H^{1-r}(\mathfrak{g}_{K/k_{\mathfrak{p}}}, \hat{A}_K)$ where $\hat{A}$ is the Picard variety,* and H now denotes Tate's cohomology functor with $-\infty < r < \infty$.

One may consider more generally the limit cohomology (say in dimension 1) with arbitrary coefficients: Let F be a group on which the Galois group Γ_k of the algebraic closure of k over k acts continuously (regarding Γ_k as compact with Krull topology, and F as discrete). (Cf. Bourbaki seminar, 1959, Exposé on Tate's work.) One can build $H^1(\Gamma_k, F)$ as a limit set (group if F is commutative) just as with the connected group varieties. If k is a $\mathfrak{p}$-adic field, and F is commutative and finitely generated (over $\mathbf{Z}$) then Tate has shown that this $H^1(\Gamma_k, F)$ is finite. If F is a finite group (not necessarily commutative) it is actually easy to prove the finiteness statement directly, using the fact that a $\mathfrak{p}$-adic field has only a finite number of extensions of given degree. If F is finitely presented (i.e. given by a finite number of generators and relations) the answer is not known.

Going over to the global case, one sees that $H^1(k, A)$ is a large group. Shafarevic has given examples of elliptic curves A over the rationals $\mathbf{Q}$ such that $H^1(k, A)$ has elements of arbitrarily high period [19]. Over a suitable number field, this can be done more easily [11]. On the other hand, Shafarevic and Tate have been led to conjecture that *if A is an abelian variety defined over a number field k, then the subgroup of $H^1(k, A)$ consisting of those elements which split over every $k_{\mathfrak{p}}$ is finite.* Stated in geometric terms, this means that the set of isomorphism classes of principal homogeneous spaces over A defined over k, which have a rational point in every $\mathfrak{p}$-adic field is finite.

Tate has proved the analogous statment for the cohomology arising from groups of multiplicative type, or from coefficient modules which are finitely generated (over $\mathbf{Z}$). As in the local case, if F is a finite group (not necessarily commutative), one can give a direct proof of the analogous fact, without using class field theory, and with all but a finite number of $\mathfrak{p}$, instead of all $\mathfrak{p}$. (One uses that in a Galois extension of degree > 1, infinitely many primes do not split completely.)

As Chatelet perceived [3; 4], the noncommutative cohomology also arises when one asks for conditions under which two varieties become isomorphic to each other. For instance, let V, W be two projective nonsingular varieties defined over k. Let $T: V \to W$ be an isomorphism defined over the Galois extension K of k, with group $\mathfrak{g}$. Then $T^{-1}T^\sigma$ is a cocycle of $\mathfrak{g}$ with coefficients in the group of automorphisms of V defined over K, and V is isomorphic to W over k if and only if this cocycle splits (cf. Weil [24]). This situation is similar to that which arose from Hasse's theorem. Chatelet saw that from class field theory, one can deduce the theorem that *if a variety over a number field becomes isomorphic to projective space over every $\mathfrak{p}$-adic field $k_\mathfrak{p}$, then it is isomorphic to projective space over k.* The cocycle one gets is in the projective group, connected with the multiplicative group (handled in class field theory) through the exact sequence with the full linear group, whose Galois cohomology in dimension 1 is trivial.

The group of automorphisms of a variety (projective nonsingular) is still somewhat of a mystery in general, although one knows that it is a group extension of an algebraic group by a discrete group (Matsusaka). One may ask whether this discrete group (which in general is not commutative) is finitely presented. For curves of genus ≥ 2, it is classical that it is finite, and hence *if V is a curve of genus ≥ 2 defined over a number field k, then the set of curves (up to k-isomorphism) which are defined over k, and are isomorphic to V over every $k_\mathfrak{p}$, is a finite set.*

For all the above groups, one can look at the Galois cohomology for varieties defined over number fields, and ask in each case whether that part of the first cohomology set which splits at all $\mathfrak{p}$ is finite. Supposing for instance that the conjecture of Shafarevic-Tate is true for abelian varieties, can it be extended to all algebraic groups defined over a number field? Can the preceding theorem be extended to all varieties (projective nonsingular), beginning with elliptic curves, or analogously, can it be extended to forms of arbitrary degree, considering only the group of all linear automorphisms?

I would like to conclude this discussion of principal homogeneous spaces by pointing out that *over a finite field, every homogeneous space of a group variety has a rational point* [8].

4. Curves. We have seen in Hasse's theorem, and the theory of quasi-algebraic closure, that equations with many variables have a tendency to have solutions. In the opposite direction, equations with few variables have a tendency not to have any, or at any rate rather few. In this connection, one has again some theorems and some conjectures.

Foremost among the theorems is the following one of Siegel's [20]: *A curve $f(X, Y) = 0$ defined over a number field has only a finite number of integral points (i.e. points (x, y) whose coordinates are integers of that field) if its genus is ≥ 1.* For curves of genus 1 over the rationals, Mahler has extended this to points having only a finite number of prime numbers in their denominators [13], and actually

I can extend the theorem to a curve of genus ≥ 1 defined over a field of finite type over $\mathbf{Q}$, with points having their coordinates in a subring of finite type over $\mathbf{Z}$ [9].

Mordell [14] has conjectured that actually *the curve will have only a finite number of rational points (in a number field) if its genus is at least* 2.

A curve of genus 1 is an elliptic curve, and if it has a rational point its rational points form a group (coming from the addition formula for elliptic functions). If this group contains one element of infinite order, then it can not be finite, of course. For curves of genus 1 over the rationals $\mathbf{Q}$, Mordell proved that *this group is finitely generated* [14]. This was extended by Weil [23] to the group of rational points of an abelian variety over a number field, and this Mordell-Weil theorem has applications to geometric problems of algebraic geometry (Néron [16], see also [10]).

A curve of genus ≥ 2 can always be embedded in its Jacobian J, an abelian variety, over a field in which it has a rational point. The group of rational points J_k being finitely generated if k is a number field, one sees Mordell's conjecture in the following light: The intersection of this finitely generated group with the curve should be finite, the curve being of lower dimension than J (if its genus is ≥ 2) and thus rather thinly distributed in J. One may even ask whether it might not be true that Mordell's conjecture could be extended to any subvariety of an abelian variety, which does not contain the translation of an abelian subvariety.

Thus, for curves of genus ≥ 2 and abelian varieties, the conjecture (resp. theorem) asserts that there are as few rational points as is compatible with the obvious structure of the variety under consideration. The Fermat curve $X^n + Y^n = 1$ has genus ≥ 2 if $n \geq 4$, and thus falls under Mordell's conjecture.

In this connection, I would conjecture that Siegel's finiteness statement concerning integral points should in fact be true for affine subsets of abelian varieties, or at least an affine subset which is the complement of a hyperplane section in some projective embedding. The theorem for curves should then be obtainable by pull-back from the Jacobian.

Finally, there is a remarkable conjecture of Siegel, at the end of [20], which I quote: "Die Untersuchungen der vorangehenden Paragraphen geben die Möglichkeit, eine Schranke für die Anzahl der Lösungen der diophantischen Gleichung $f(X, Y) = 0$ als Funktion der Koeffizienten von f explicit aufzustellen, falls diese Gleichung nur endlich viele Lösungen besitzt. Man kann nun vermuten, dass sich sogar eine Schranke finden lässt, die nur von der Anzahl der Koeffizienten abhängt; doch dürfte dies recht schwer zu beweisen sein." Siegel then goes on to give evidence for this conjecture, by treating "allerdings sehr spezielle Resultate."

BIBLIOGRAPHY

1. B. J. Birch, *Homogeneous forms of odd degree in a large number of variables*, Mathematika vol. 4 (1957) pp. 102–105.

2. J. W. Cassels, *Arithmetic on curves of genus* 1, J. Reine Angew. Math. vol. 202 (1959) pp. 52–99.

3. F. Chatelet, *Méthode Galoisienne et courbes de genre* 1, Ann. Univ. Lyon, vol. 9 (1946) pp. 40–49.

4. ———, *Variations sur un thème de Poincaré*, Ann. Sci. Ecole Norm. Sup. vol. 59 (1944) pp. 249–300.

5. C. Chevalley, *Démonstration d'une hypothèse de M. Artin*, Abh. Math. Sem. Univ. Hamburg vol. 11 (1935) pp. 73–75.

6. S. Lang, *On quasi algebraic closure*, Ann. of Math. vol. 55 (1952) pp. 373–390.

7. ———, *The theory of real places*, Ann. of Math. vol. 57 (1953) pp. 380–391.

8. ———, *Algebraic groups over finite fields*, Amer. J. Math. vol. 78 no. 3 (1956) pp. 555–563.

9. ———, *Integral points on curves*, to appear.

10. S. Lang and A. Néron, *Rational points of abelian varieties in function fields*, Amer. J. Math. vol. 81, no. 1 (1959) pp. 95–118.

11. S. Lang and J. Tate, *Principal homogeneous spaces over Abelian varieties*, Amer. J. Math. vol. 80, no. 3 (1958) pp. 659–684.

12. D. J. Lewis, *Cubic homogeneous polynomials over p-adic number fields*, Ann. of Math. vol. 56, no. 3 (1952) pp. 473–478.

13. K. Mahler, *Über die rationalen Punkte auf Kurven vom Geschlecht Eins*, J. Reine Angew. Math. vol. 170 (1934) pp. 168–178.

14. L. J. Mordell, *On the rational solutions of the indeterminate equation of the third and fourth degrees*, Proc. Cambridge Philos. Soc. vol. 21 (1922) pp. 179–192.

15. T. Nakayama, *Cohomology of class field theory and tensor product modules*, Ann. of Math. vol. 65 (1957) pp. 255–267.

16. A. Néron, *Problèmes arithmétiques et géometriques rattachés à la notion de rang d'une courbe algébrique dans un corps*, Bull. Soc. Math. France vol. 80 (1952) pp. 101–166.

17. L. J. Peck, *Diophantine equations in algebraic number fields*, Amer. J. Math. vol. 71 (1949) pp. 387–402.

18. E. Selmer, *The diophantine equation $ax^3 + by^3 + cz^3 = 0$*, Acta Math. vol. 85 (1951) pp. 203–362.

19. I. Shafarevic, *Birational equivalence of elliptic curves* and *Exponents of elliptic curves*, Doklady Akad. Nauk SSSR vol. 114 (1957) pp. 267–270 and pp. 714–716.

20. C. L. Siegel, *Über einige Anwendungen Diophantischer Approximationen*, Abh. Preuss. Akad. Wiss. Phys. Math. Kl. (1929) pp. 41–69.

21. J. Tate, *Galois cohomology of abelian varieties over p-adic fields*, to appear. Cf. Bourbaki seminar, 1957.

22. A. Weil, *Algebraic groups and homogeneous spaces*, Amer. J. Math. vol. 77, no. 3 (1955) pp. 493–512.

23. ——, *L'arithmétique sur les courbes algébriques*, Acta Math. vol. 52 (1928) pp. 281–315.

24. ——, *The field of definition of a variety*, Amer. J. Math. vol. 78, no. 3 (1956) pp. 509–524.

25. ——, *Sur les courbes algébriques et les variétés qui s'en déduisent*, Paris, Hermann, 1948.

26. E. Witt, *Theorie der quadratischen Formen in beliebigen Körper*, J. Reine Angrew. Math. vol. 176 (1937) pp. 31–44.

27. ——, *Zerlegung reeller algebraische Funktionen in Quadrate*, J. Reine Angew. Math. vol. 171 (1934) pp. 4–11.

COLUMBIA UNIVERSITY

ON A THEOREM OF MAHLER

Serge Lang

In trying to extend some theorems of diophantine approximation to number fields or function fields, one meets certain difficulties associated with the presence of units. By using a very simple trick, one overcomes these difficulties. In this paper I limit myself to one example.

Let K be a field of characteristic zero with a family of absolute values $|\ |_{\mathfrak{p}}$ indexed by a set $\{\mathfrak{p}\}$, and let K^* be the multiplicative group of K. We assume that for $\beta \varepsilon K^*$, we have $|\beta|_{\mathfrak{p}} = 1$ for all but a finite number of $\mathfrak{p}$. We suppose that there are integers $N_{\mathfrak{p}} \geqslant 1$, such that, if we set $\|\beta\|_{\mathfrak{p}} = |\beta|_{\mathfrak{p}}^{N_{\mathfrak{p}}}$, then the product formula holds:

$$\prod_{\mathfrak{p}} \|\beta\|_{\mathfrak{p}} = 1, \quad \beta \varepsilon K^*.$$

We can then define the *height* of an element of K, namely:

$$H(\beta) = \prod_{\mathfrak{p}} \sup\,(1, \|\beta\|_{\mathfrak{p}}).$$

We shall use two obvious properties of the height. First, if $\beta_1 \varepsilon K^*$ is fixed, then on K^* the functions $H(\beta_1 \beta)$ and $H(\beta)$ are equivalent, *i.e.* each is less than a constant multiple of the other. Second, we have $H(\beta^n) = H(\beta)^n$ for an integer $n \geqslant 1$.

We shall now assume that Roth's theorem holds, and state it as follows. Let $|\ | = |\ |_{\mathfrak{p}}$ be any one of our absolute values, and assume it is extended in some way to the algebraic closure of K. Let α be algebraic over K. Let c be a number > 0, and $\chi > 2$. Then the elements $\beta \varepsilon K$, such that

$$|\alpha - \beta| \leqslant \frac{c}{H(\beta)^{\chi}},$$

have bounded height.

Let us consider the following statement which generalizes a well-known theorem of Mahler [1].

Let α be algebraic over K. Let Γ be a finitely generated subgroup of K^*. Let χ and c be numbers > 0. Then the elements $\beta \varepsilon \Gamma$ such that

$$|\alpha - \beta| \leqslant \frac{c}{H(\beta)^{\chi}}$$

have bounded height.

The proof is done by a direct reduction to Roth's theorem. Let n be an integer such that $n\chi > 2$. If our elements β do not have bounded height, we can find a sequence of them whose height tends to infinity, and an infinite number of them will lie in the same coset of Γ/Γ^n, say the coset of β_1. This gives us solutions ξ in K of the inequality

$$|\alpha - \beta_1\,\xi^n| \leqslant \frac{c}{H(\beta_1\,\xi^n)^\chi}$$

with unbounded heights, and β_1 fixed. Let $\lambda^n = \alpha/\beta_1$. In view of the two properties mentioned above, we get solutions of the inequality

$$|\lambda^n - \xi^n| \leqslant \frac{c'}{H(\xi)^{n\chi}},$$

for a suitable constant c'. We have

$$|\lambda^n - \xi^n| = \Pi\,|\zeta\lambda - \xi|$$

the product being taken over all n-th roots of unity ζ. If one of these factors becomes small, *i.e.* ξ approximates $\zeta\lambda$ for some fixed ζ, then the other terms stay bounded away from 0, and we thus get solutions with unbounded height for the inequality

$$|\zeta\lambda - \xi| \leqslant \frac{c_1}{H(\xi)^{n\chi}},$$

with a suitable constant c_1. This contradicts Roth's theorem.

For a proof of Roth's theorem in number fields and function fields see my forthcoming book "Arithmetic Geometry".

Reference.

1. K. Mahler, " Ein Analogon zu einem Schneiderschen Satz ", *Proc. Kon. Akad. Wetensch. Amsterdam*, 39 (1936), 633–640.

Harvard University,
    Cambridge, Mass.

(Received on the 12th of July, 1960.)

Éléments de géométrie algébrique. Par A. Grothendieck, rédigés avec la collaboration de J. Dieudonné. Publications de l'Institut des Hautes Études Scientifiques No. 4, Paris, 1960. 228 pp. 27 NF.

The present work, of which Chapters 0 and I are now appearing together, is one of the major landmarks in the development of algebraic geometry. It plans to cover eventually everything that is known in algebraic geometry over arbitrary ground rings, and of course a lot more besides. A tentative list of its chapters is as follows:

Chapter
- I. Le langage des schémas.
- II. Étude globale élémentaire de quelques classes de morphismes.
- III. Cohomologie des faisceaux algébriques cohérents. Applications.
- IV. Étude locale des morphismes.
- V. Procédés élémentaires de construction de schémas.
- VI. Technique de descente. Méthode générale de construction de schémas.
- VII. Schémas de groupes, espaces fibrés principaux.
- VIII. Étude differentielle des espaces fibrés.
- IX. Le groupe fondamental.
- X. Résidus et dualité.
- XI. Théories d'intersection, classes de Chern, théorème de Riemann-Roch.
- XII. Schémas abeliens et schémas de Picard.
- XIII. Cohomologie de Weil.

The list is subject to modifications, especially in so far as later chapters are concerned, partly because much of the research needed to complete these chapters remains to be done.

To give the prospective reader some idea of the size of the work, suffice it to say that Chapter I is 134 pages long, that subsequent chapters are expected to be at least as long (probably around 150 pages each), that all chapters are regarded as being open (i.e., subject to additions such as are deemed necessary in the course of the writing), and that Chapters 0 and I together weigh 1 and 3/4 pounds in their present form.

In order to get a more specific idea of what is to come, one should consult first Grothendieck's address to the International Congress at

Edinburgh, 1958, and also the whole series of talks at Bourbaki seminars given in the past two years (available at the Institut Henri Poincaré, 11 Rue Pierre Curie, Paris) in which he has given a sketch of the proofs of important results to appear in later chapters. These talks will provide the necessary motivation to the whole work. They are written concisely, directly, and excitingly. Such motivation could not be given in the actual text, which is written very lucidly, is perfectly organized, and very precise. Thanks are due here to Dieudonné, without whose collaboration the labor involved in writing and publishing the work would have been insurmountable.

Before we go into a closer description of the contents of Chapters 0 and I, it is necessary to say a few words explaining why the present treatise differs radically in its point of view from previous ones.

1. Most of algebraic geometry up to now has been concerned with varieties, say over arbitrary fields. It includes some results on algebraic families of varieties, but such results are few in number, and it has become increasingly clear in recent years that one was facing serious difficulties in dealing with such algebraic systems. For example, the geometer is able to attach to a fixed variety other geometric objects, say a Picard variety. It is then a problem to show that if one has an algebraic system of varieties, the Picard varieties can be associated in such a way that they move along with the varieties, following the same parameter variety, even when special members of the family are degenerate. The tools available at present to deal with such a problem are recognized to be deficient (although of course in special cases, interesting results have been obtained, especially for non-degenerate fibers).

2. In applications to number theory, it has been realized for some time that the reduction mod $\mathfrak{p}$ of a variety defined over a number field was completely analogous to the situation of algebraic systems, a fiber being such a reduction. Although it was possible here again to give an ad hoc definition and results having useful applications to interesting special problems, the theory was technically disagreeable to apply, to say the least.

In order to deal efficiently with the above two points, it was necessary to incorporate from the start into the foundations the notion of a variety defined over a ring, not necessarily Noetherian, and having nilpotent elements (say to reduce mod $\mathfrak{p}^n$, or to describe degenerate fibers in a system). This meant that a variety could not be regarded any more as a model of a "function field," and thus that it should be defined starting with a local description supplemented by a method for gluing local pieces together (sheaves being the natural tool here).

3. The classical tools available were impotent to deal with the problem of defining the homology and homotopy functors to which one is accustomed in topology, and having similar properties. The necessity of having the homology functor, say, was made clear by Weil, who pointed out that if one has it, then the structure of the zeta function for non-singular projective varieties defined over finite fields follows immediately from the Lefschetz fixed point formula. In order to have this, a minimum requirement is that the homology groups H_n associated with a variety V be modules, or vector spaces having characteristic 0 (no matter what the characteristic of the field of definition of V is!).

4. The study and classification of non-abelian coverings of varieties, and in particular the determination of the fundamental group, was completely outside the range of available methods, except for varieties defined over the complex numbers where one could use transcendental methods.

The above list could be expanded, but it gives a good idea why a new approach to algebraic geometry was needed.

Let us now give a closer look at the contents of Chapters 0 and 1.

Chapter 0 is intended to include results of commutative algebra needed for the geometric applications. They are more or less well known, but it is difficult to give references for them. The reader should skip this chapter until he meets a place where he needs it. He should start reading Chapter I immediately. For this, he needs to know only what a ring is (commutativity and unit element are always assumed), and the definition of a ring of fractions, which runs as follows. Let A be a ring, S a subset of A closed under multiplication and containing 1. One considers equivalence classes of pairs (a, s) with $a \in A$ and $s \in S$ such that $(a, s) \sim (a', s')$ if there exists $s_1 \in S$ such that $s_1(s'a - sa') = 0$. The equivalence class of (a, s) is denoted by a/s, and these form a ring in the obvious way. This ring is denoted by $S^{-1}A$, and is 0 if S contains nilpotent elements. The most important case is that where S is the complement of a prime ideal $\mathfrak{p}$, so that $S^{-1}A = A_\mathfrak{p}$ is the local ring at $\mathfrak{p}$.

We recall that a ringed space is a pair (X, O_X) consisting of a topological space X and a sheaf of rings O_X. Ringed spaces form a category: A morphism $(X, O_X) \rightarrow (Y, O_Y)$ is a pair consisting of a continuous map $\phi: X \rightarrow Y$ and a contravariant map $\psi: O_Y \rightarrow O_X$ compatible with f. If we denote by O_x the fiber of O_X above a point $x \in X$, then ψ induces a homomorphism $\psi_x: O_{\phi(x)} \rightarrow O_x$. The ringed space (X, O_X) is called a local ringed space if all the rings O_x are local rings. If (X, O_X) and (Y, O_Y) are local ringed spaces, a morphism (ϕ, ψ) above is called local if the inverse image of the maximal ideal of O_x by ψ_x is the maximal ideal of $O_{\phi(x)}$. The local ringed spaces and the local morphisms then form a category. It is a subcategory of this one which is of interest to the algebraic geometer.

Namely, given a ring A, its spectrum $X = \operatorname{spec}(A)$ is the topological space (T_0 but not T_1) whose points are the prime ideals of A with Zariski topology (the set of primes containing a given ideal is closed). One views X as a ringed space, the sheaf being that of the local rings $A_{\mathfrak{p}}$. It is thus a local ringed space, called an affine scheme. A prescheme is a local ringed space (X, O_X) such that every point admits an open neighborhood U such that $(U, O_X | U)$ is isomorphic to an affine scheme. The preschemes form a category, the morphisms being the local morphisms.

To simplify the notation, one sometimes omits the structure sheaf O_X and the map ψ, just writing for instance $\phi: X \to Y$ to indicate a morphism in the category of preschemes.

Let Γ be the functor "section". For each open subset U of X, ΓU is the ring of sections of O_X over U. Given a morphism $\phi: X \to Y$, we have a homomorphism $\Gamma(\phi): \Gamma Y \to \Gamma X$. The converse is true for affine schemes, and in fact affine schemes Y are characterized among preschemes by the fact that for each prescheme X the map $\phi \to \Gamma(\phi)$ of $\operatorname{Mor}(X, Y)$ into $\operatorname{Hom}(\Gamma Y, \Gamma X)$ is an isomorphism. (One could actually let X range over local ringed spaces.) Furthermore, if $Y = \operatorname{spec}(A)$, then ΓY is naturally isomorphic to A.

The other main result of Chapter I is then given: It is the proof that products exist in the category of preschemes. Let us recall some terminology in abstract categories. Let C be a category, and S an object in C. We denote by C_S the category of objects over S, i.e. pairs (X, f) where X is in C and f is a morphism $f: X \to S$ in C, called the structural morphism. Given two objects $f: X \to S$ and $g: Y \to S$ in C_S, a morphism ϕ in C_S is a morphism $\phi: X \to Y$ in C which is such that the diagram

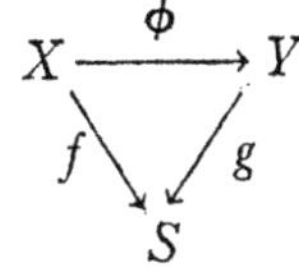

is commutative.

In the category of preschemes, the object S plays the role of a ground object (ground field, ground ring, ground anything you want vastly generalized, parameter object, etc.).

A product of two objects (X, f) and (Y, g) over S consists of an object (written $X \times_S Y$) and two morphisms

$$\phi: X \times_S Y \to X$$

$$\psi: X \times_S Y \to Y$$

making the following diagram commutative and satisfying the obvious universal mapping property for such pairs of maps:

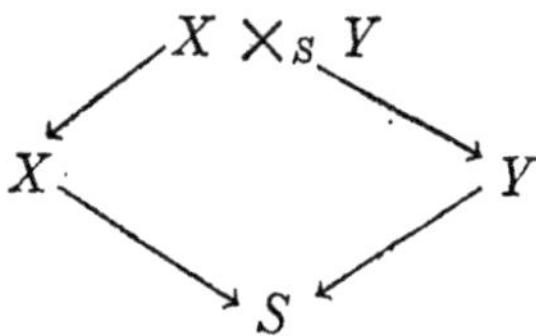

It is uniquely determined, up to a unique isomorphism.

If A, B, R are three rings, and A, B are algebras over R, then the product of the two affine schemes $\mathrm{spec}(A)$ and $\mathrm{spec}(B)$ over $\mathrm{spec}(R)$ is $\mathrm{spec}(A \otimes_R B)$, the morphisms involved being the obvious ones. This is practically immediate from the definitions, and the existence proof in the general case is carried out by gluing local pieces together.

One can consider the product non-symmetrically. Viewing S as ground object, let $S' = Y$ be viewed as an extension of it. Then $X \times_S S'$ (sometimes written $X^{S'}$) may be viewed as an object over S', called the pull back of X by the morphism $g: S' \to S$. This pull back involves as a special case the extension of ground field or ring, and also reduction mod p, or the process of taking a fiber. For instance, if $S = \mathrm{spec}(Z)$ (Z the integers), then for each prime p, we have a morphism

$$\mathrm{spec}(Z/pZ) \to \mathrm{spec}(Z)$$

and thus for each prescheme X over Z, we get its fiber over Z/pZ, namely $X \times_Z \mathrm{spec}(Z/pZ)$.

Having constructed products, one gets a diagonal morphism

$$X \to X \times X$$

(the product without subscript being always over $\mathrm{spec}(Z)$). One says that X is a scheme if this morphism is closed (obvious definition).

Most of the rest of Chapter I is devoted to defining certain classes of morphisms in the category of preschemes (immersions, closed immersions, local immersions, morphisms of finite type, proper morphisms, separated morphisms, etc. and in subsequent chapters affine morphisms, projective morphisms, flat morphisms, unramified morphisms, simple morphisms, ad lib.) and of proving standard properties concerning the composition and products of such special classes of morphisms. Namely, given a category C, let us say that a subclass C' of morphisms of C is distinguished if it has the following properties:

(i)   If f, g are in C' and can be composed, so is fg.

(ii)   If $f: X \to S$ is in C' and $g: Y \to S$ is in C, then the pull back of f by g is in C'.

(iii)   If both f and g are in C', so is $f \times_S g$.

(iv)   If f and g can be composed, g is in C' and gf is in C', then f is in C'.

The general rule is that all particular types of morphisms defined in Chapter I (and subsequently) will form a distinguished subclass, except possibly under certain conditions of finiteness and separation. There is no point in going into the specific details here. We wish merely to indicate the way the system works.

Chapter I concludes with an extended discussion of quasi-coherent sheaves, and formal schemes, those arising essentially from completions of topological rings, and playing an important role in local analytic (algebraic) questions. They are not used until Chapter III, which will include Zariski's theory of holomorphic functions and the connectedness theorem, and the reader may skip that part until he needs it.

One more notion appears in Chapter I, worthy of notice for the implications it has concerning the point of view of the work. Again it is best to describe it in an abstract category C. Let A be a fixed object in C and let X vary in C. Then

$$F_A: X \to \mathrm{Mor}(X, A)$$

is a (contravariant) functor from C into the category of sets, denoted Ens. We may also denote $\mathrm{Mor}(X, A)$ by $A(X)$ and in our category of preschemes, we think of it as giving the set of points of A in X. (To justify this, think of A as an affine variety V over a field k, and let Γ be its finitely generated algebra of functions over k. Let K range over fields containing k. Then points of V in K are in bijective correspondence with homomorphism of Γ into K, i.e. morphisms of $\mathrm{spec}(K)$ into $\mathrm{spec}(\Gamma)$. Here, $\mathrm{spec}(K)$ consists of one point, and the local ring above it is just K itself.)

Given a functor $F: C \to \mathrm{Ens}$ of C into the category of sets, Grothendieck calls F representable if it is isomorphic to a functor of type F_A. (The functors of one category into another form themselves a category, the morphisms being the obvious ones.) It is then immediate that the object A is uniquely determined, up to a unique isomorphism.

Observe that the definition of products has been made in accordance with the representation functor, i.e. to satisfy the formula

$$(X \times_S Y)(T) \approx X(T) \times_{S(T)} Y(T)$$

for all objects T, the fiber product on the right being the usual one in the theory of sets (pairs of points projecting on the same point in $S(T)$).

This notion of representable functor allows one to transport to any category standard notions like group, ring, etc. For instance, an object G is called a group object if one is given two morphisms $G \times G \to G$ (composition) and $G \to G$ (inverse) such that the representation functor into the category of sets defines a group structure on the set $G(X)$ for each X. (We have assumed finite products exist, but a rephrasing would do away with this.)

It is one of the most basic ones of mathematics. To give an example from topology: On the category of CW complexes, the functor H_π^n is representable by $K(\pi, n)$. Or on the category of reasonable topological spaces, the functor K (classes of vector bundles) is also representable by the classifying space.

In algebraic geometry, Grothendieck reformulates certain classical problems in terms of the representation of functors, for instance the problem of constructing Picard schemes. Given X over S, the Picard functor consists in associating to each T over S the divisor classes of X which are rational over T. (This can of course be made precise.) The Picard scheme, if it exists, represents this functor. Grothendieck has recently obtained a fairly general condition on functors in the category of schemes under which he can prove that a functor is representable. This point of view marks a complete discontinuity with those preceding it and in a certain sense, is the first essentially new approach having entered algebraic geometry since the Italian school.

A theorem is not true any more because one can draw a picture, it is true because it is functorial.

To conclude this review, I must make a remark intended to emphasize a point which might otherwise lead to misunderstanding. Some may ask: If Algebraic Geometry really consists of (at least) 13 Chapters, 2,000 pages, all of commutative algebra, then why not just give up?

The answer is obvious. On the one hand, to deal with special topics which may be of particular interest only portions of the whole work are necessary, and shortcuts can be taken to arrive faster to specific goals. Thus one may expect a period of coexistence between Weil's

Foundations and *Elements*. Only history will tell if one buries the other. Projective methods, which have for some geometers a particular attraction of their own, and which are of primary importance in some aspects of geometry, for instance the theory of heights, are of necessity relegated to the background in the local viewpoint of *Elements*, but again may be taken as starting point given a prejudicial approach to certain questions.

But even more important, theorems and conjectures still get discovered and tested on special examples, for instance elliptic curves or cubic forms over the rational numbers. And to handle these, the mathematician needs no great machinery, just elbow grease and imagination to uncover their secrets. Thus as in the past, there is enough stuff lying around to fit everyone's taste. Those whose taste allows them to swallow the *Elements*, however, will be richly rewarded.

S. Lang

A TRANSCENDENCE MEASURE FOR E-FUNCTIONS

Serge Lang

Siegel defined an E-function to be a function which admits a power series expansion

$$E(z) = \sum_{n=0}^{\infty} \beta_n \frac{z^n}{n!}$$

with complex coefficients β_n, belonging to some algebraic number field K (of finite degree over the rationals $\mathbf{Q}$), satisfying the following conditions:

(i) Given $\epsilon > 0$, the maximum of the absolute values of the conjugates of β_n, satisfies

$$\overline{|\beta_n|} = O(n^{n\epsilon}), \quad n \to \infty.$$

(ii) There exists a sequence of integers d_n in $\mathbf{Z}$, $d_n > 0$, which are also $O(n^{n\epsilon})$, and such that $d_n \beta_k$ is an algebraic integer for $k \leqslant n$.

We shall consider E-functions $f_1, \ldots, f_s$, satisfying a linear differential equation

$$X' = Q^* X,$$

where $X = (X_1, \ldots, X_s)$ is a vector, and $Q^* = (Q_{ij}^*)$, $i, j = 1, \ldots, s$, is a square matrix of rational functions in $K(z)$. We shall prove the following result.

THEOREM. *Let $f_1, \ldots, f_s$ be E-functions, algebraically independent over $K(z)$, and satisfying a linear differential equation as above. Let α be an algebraic number, which is distinct from 0, and from the poles of the rational functions Q_{ij}^*. Let $g(X_1, \ldots, X_s)$ be a polynomial with coefficients in $\mathbf{Z}$, of degree d and height $G \neq 0$. Then*

$$\left| g\big(f_1(\alpha), \ldots, f_s(\alpha)\big) \right| \geqslant aG^{-bd^s},$$

where a is a number > 0, depending on the f_j, s, Q^*, α, d, and b depends on the degree $N = [K(\alpha): \mathbf{Q}]$ and on s.

We recall that the height of g is the maximum of the absolute value of its coefficients, which may be assumed relatively prime.

The theorem extends to E-functions Siegel's result on the Bessel function [3], and the classical transcendence measure for e^α. The proof follows Siegel's own proof, with the following modifications. On the one hand, we must of course use Shidlovsky's result, which allows us to replace the complicated normality condition of Siegel by the algebraic independence of $f_1, \ldots, f_s$. This will be recalled as Lemma 4 below. On the other hand, we combine the method of Siegel's original paper [3] together with the exposition in his book [4]. We assume that the reader is familiar with §§1, 2, 3, 5 and 6 of [4], Chapter II, and we keep the notation of that chapter.

In the proof we meet various constants which will not be computed explicitly, and will be numbered consecutively, $a_1, a_2, \ldots$.

Siegel's Lemma 4 will be replaced by Shidlovsky's result, which can be stated as follows.

LEMMA 4. *Let $E_1, \ldots, E_m$ be E-functions, linearly independent over $K(z)$, and satisfying a linear differential equation $Y' = QY$, where Q is a matrix of rational functions in $K(z)$. Let $P_1, \ldots, P_m$ be polynomials in $K[z]$ and let*

$$R_1 = P_1 E_1 + \ldots + P_m E_m.$$

Let T be a polynomial denominator for the rational functions Q_{ij}, and define inductively

$$R_k = T R'_{k-1} = P_{k1} E_1 + \ldots + P_{km} E_m.$$

Let r be the rank of the matrix (P_{kj}), and suppose $r < m$. Then

$$\operatorname{ord} R_1 \leqslant r(\max \deg P_j) + a_1,$$

where a_1 is a positive number depending only on $E_1, \ldots, E_m$, Q (and not on the P_j).

(Shidlovsky's constant a_1 is the only one in the proof of the theorem which cannot be easily estimated.)

We shall now describe the entire proof, giving complete arguments only in those parts differing from Siegel's book [4].

Given E-functions $E_1, \ldots, E_m$, we can find polynomials $P_1, \ldots, P_m$ with integral coefficients in K, of degree $\leqslant 2n - 1$, such that the maximum of the absolute values of their conjugates is $O(n^{(2+\epsilon)n})$ for $n \to \infty$, and such that the power series

$$R = P_1 E_1 + \ldots + P_m E_m$$

has a zero of order $\geqslant (2m - 1)n$ (Siegel's Lemma 2). The ν-th coefficient of R has then the order of magnitude

$$\nu^{\epsilon\nu} O(n^{2n}) \quad \text{for} \quad \nu \to \infty.$$

If we form inductively the sequence

$$R_{k+1} = T R'_k,$$

then it is a simple matter to make the following estimates:

$$|R_k(\alpha)| = O(n^{(3+\epsilon)n - (2m-2)n}),$$

$$|P_{kj}(\alpha)| = O(n^{(3+\epsilon)n}), \quad j = 1, \ldots, m,$$

with $k \leqslant n + a_2$ for some constant a_2.

Furthermore, using Lemma 4, we conclude at once that the rank of the matrix (P_{kj}) must be m as soon as n is larger than a_1.

We now start with our functions $f_1, \ldots, f_s$ assumed algebraically independent over $K(z)$. Given a polynomial $g(X_1, \ldots, X_s)$ of degree d and height $G \neq 0$, we select an integer $\tau > 0$, such that

$$1 > 4N\left(1 - \left\{\frac{\tau N - 1}{\tau N}\right\}^{\sigma}\right), \quad 1 \leqslant \sigma \leqslant s,$$

where $N = [K(\alpha) : \mathbf{Q}]$. We let

$$m = \binom{s + \tau N d}{s}, \quad v = \binom{s + \tau N d - d}{s},$$

and let $w = m - v$. Then there are exactly m monomials

$$f_1^{\nu_1} \ldots f_s^{\nu_s}$$

of degree $\leqslant \tau N d$, and these monomials are E-functions, which we denote by $E_1, \ldots, E_m$. The differential equation $X' = Q^* X$ gives rise to the differential equation $Y' = QY$, the rational functions Q_{ij} being linear combinations of the Q_{ij}^* with integer coefficients.

There are exactly v polynomials of type

$$f_1^{\nu_1} \ldots f_s^{\nu_s} g(f_1, \ldots, f_s),$$

with $\nu_1 + \ldots + \nu_s \leqslant \tau N d - d$. We denote these functions by

$$\psi_1, \ldots, \psi_v.$$

We have inductively

$$R_k = P_{k1} E_1 + \ldots + P_{km} E_m$$

and Siegel's Lemma 5 states that the matrix $\left(P_{kj}(\alpha)\right)$, with $j = 1, \ldots, m$ and $k \leqslant n + a_2$, has rank m for some constant a_2. We can write

$$\begin{aligned}
\psi_1 &= \lambda_{11} E_1 + \ldots + \lambda_{1m} E_m, \\
&\cdots \\
\psi_v &= \lambda_{v1} E_1 + \ldots + \lambda_{vm} E_m,
\end{aligned} \tag{1}$$

with coefficients λ_{ij}, which are also coefficients of the original polynomial g. Their absolute values are therefore bounded by G. We can find then $m - v = w$ functions among the R_k $(k \leqslant n + a_2)$, say

$$\begin{aligned}
\phi_1 &= P_{k_1 1}(\alpha) E_1 + \ldots + P_{k_1 m}(\alpha) E_m, \\
&\cdots \\
\phi_w &= P_{k_w 1}(\alpha) E_1 + \ldots + P_{k_w m}(\alpha) E_m,
\end{aligned} \tag{2}$$

such that the determinant Δ of the matrix consisting of the (λ) and the $P_{k_i j}(\alpha)$ is not zero. We may assume without loss of generality that $E_1 = 1$ (*i.e.* it is the monomial of degree 0). Then

$$\Delta = A_1 \psi_1(\alpha) + \ldots + A_v \psi_v(\alpha) + B_1 \phi_1(\alpha) + \ldots + B_w \phi_w(\alpha), \tag{3}$$

where the coefficients (A) and (B) are obvious subdeterminants of Δ.

Each P_{kj} has degree $\leqslant 2n-1+(k-1)q$ (where q depends on Q). We may view Δ as a polynomial in α, and as such, its degree is therefore at most $(q+2)n+a_3$ for suitable integer $a_3 > 0$ (for our range $k \leqslant n+a_2$). If c is an integer $\neq 0$ in $\mathbf{Z}$, such that $c\alpha$ is an algebraic integer, then $c^{(q+2)n+a_3}\Delta$ is an algebraic integer, and the absolute value of its norm is therefore $\geqslant 1$. Thus, letting $\mathbf{N}$ be the norm from $\mathbf{K}(\alpha)$ to $\mathbf{Q}$, we have

$$1 \leqslant a_4{}^n \, \mathbf{N}(\Delta).$$

We shall give an upper bound for the expression on the right, by using the preceding estimates. One conjugate of Δ can be estimated by using (1), (2) and (3). In expressions involving for instance $O(n^{(3+\epsilon)n})$ we take $\epsilon = 1$. Then we have

$$\left| A_1\psi_1(\alpha)+\ldots+A_v\psi_v(\alpha) \right| \leqslant a_5 \left| g\left(f(\alpha) \right) \right| G^{v-1} n^{4nw},$$

and

$$\left| B_1\phi_1(\alpha)+\ldots+B_w\phi_w(\alpha) \right| \leqslant a_6 \, G^v \, n^{4nw} \, n^{-(2m-2)\,n}.$$

The other conjugates of Δ are bounded by

$$a_7 \, G^v \, n^{4nw}.$$

Hence finally, we get

$$1 \leqslant a_8 \, G^{Nv} \, n^{4nwN} \left[\frac{\left| g\left(f(\alpha) \right) \right|}{G} + \frac{1}{n^{n(2m-2)}} \right]. \tag{4}$$

We take n to be the smallest integer $> a_1$, and such that

$$n^n > 2a_8 \, G^N.$$

Recall that $w = m-v$. *We contend that*

$$2m-2-4N(m-v) > v.$$

It will certainly suffice to prove that $m > 4N(m-v)$, in view of the equality

$$2m-2-4N(m-v)-v = m-2-(4N-1)(m-v).$$

We have

$$s! \, m = (\tau N d+1)\ldots(\tau N d+s) = (\tau N)^s \, d^s + \xi_{s-1} (\tau N)^{s-1} d^{s-1} + \ldots,$$

$$s! \, v = (\tau N-1)^s \, d^s + \xi_{s-1}(\tau N-1)^{s-1} d^{s-1}+\ldots,$$

where $\xi_{s-1}, \ldots, \xi_0$ are integers depending only on s. We estimate each

$$(\tau N)^\sigma - (\tau N-1)^\sigma$$

using the definition of τ. Our contention follows at once.

If we now substitute $2a_8 \, G^N$ for n^n we get an upper bound of $1/2$ for the second term on the right of (4). Transposing this $1/2$ to the left of (4), yields

$$\left| g\left(f(\alpha) \right) \right| \geqslant (G^{Nv} \, n^{4nwN})^{-1},$$

under our assumption that $n^n > 2a_8 G^N$.   Since n is chosen to be smallest, satisfying this inequality, we have

$$(n-1)^{n-1} \leqslant 2a_8 G^N.$$

From this one sees at once that n^n is of the order of magnitude of $G^N \log (G)$, or G^{N+1} say.   Thus we obtain an inequality

$$\left| g\big(f(\alpha)\big) \right| \geqslant \frac{a_9}{G^{Nv} \, G^{(N+1)Nw}} \cdot$$

Both v and w are obviously of type bd^s with a suitable constant b and our theorem is proved.

It is, in fact, easy to see that w has degree $s-1$ in both N and d, and consequently that our constant b is of type $cN^{s+1}d^s$, where c is a constant depending only on s.

References.

1. K. Mahler, *On a theorem of Shidlovsky*, Mimeographed notes (Amsterdam, 1959).
2. A. V. Shidlovsky, " On a criterion of algebraic independence . . . " (in Russian) *Izvestia Akademia Nauk USSR*, 23 (1959), 35–66.
3. C. L. Siegel, " Über einige anwendungen diophantischer approximationen ", *Abhandlungen der Preussischen Akademie der Wissenschaften* (1929), 1–41.
4. ———, " Transcendental numbers ", *Annals of Mathematics Studies No. 16* (Princeton, 1949).

Columbia University,
    New York, U.S.A.

(Received on the 30th of April, 1962.)

Topology, Vol. 1, pp. 313–318. Pergamon Press, 1962. Printed in Great Britain.

TRANSCENDENTAL POINTS ON GROUP VARIETIES

SERGE LANG

(*Received* 11 *August* 1962)

LET V be a variety (algebraic and irreducible), defined over a field K. If P is a point of V with affine co-ordinates $(x_1, \dots , x_n)$ then we denote by $K(P)$ the field $K(x_1, \dots , x_n)$. We say that P is rational over K if $K(P) = K$, algebraic over K if $K(P)$ is algebraic over K, and transcendental over K if $K(P)$ is not algebraic over K.

It is a classical result that if α is algebraic over the rational numbers $\mathbf{Q}$ and $\alpha \neq 0$, then e^α is transcendental. Cartier conjectured that the analogous statement should be true for an arbitrary group variety, and the purpose of this paper is to prove this conjecture.

THEOREM 1. *Let G be a group variety defined over the field of all algebraic numbers* $\mathbf{K}$. *Let $\mathfrak{g}$ be its Lie algebra at the origin, with its structure of $\mathbf{K}$-vector space. Let α be a non-zero element of $\mathfrak{g}$ which is rational over $\mathbf{K}$, and let* $\exp = \exp_G$ *be the exponential function on G. Assume that $\exp(t\alpha)$ is not an algebraic function of t. Then $\exp(\alpha)$ is transcendental over* $\mathbf{K}$.

Suppose that G is a linear group variety, thereby admitting a global matrix representation over $\mathbf{K}$. A tangent vector at the origin is simply a matrix M, and $\exp(M)$ is given by the series

$$\sum M^\nu/\nu!.$$

If B is an invertible matrix, then

$$\exp(B^{-1}MB) = B^{-1} \exp(M)B.$$

Assume that B is rational over $\mathbf{K}$. We observe that $\exp(M)$ is rational over $\mathbf{K}$ if and only if $\exp(B^{-1}MB)$ is rational over $\mathbf{K}$. Any matrix M rational over $\mathbf{K}$ can be conjugated into a matrix consisting of blocks

$$\begin{pmatrix} \alpha & & & & \\ & \alpha & & & N \\ & & \ddots & & \\ & 0 & & \alpha & \\ & & & & \alpha \end{pmatrix}$$

where α is an algebraic number, i.e. blocks of type $\alpha I + N$ where N is nilpotent. Therefore $\exp(\alpha I + N)$ is equal to $e^\alpha I$ times a rational matrix, and is transcendental if and only if $\alpha \neq 0$. Consequently, in the linear case, Theorem 1 is equivalent with the classical result concerning the ordinary exponential function.

313

In general, G contains a maximal linear subgroup L such that $G/L = A$ is an abelian variety. Let $\pi : G \to A$ be the canonical projection. Then π is an algebraic mapping. Let π_* be the induced homomorphism on the Lie algebras. If $\pi_* \alpha = 0$, then α is contained in the Lie algebra of L, and the theorem reduces to the linear case. If $\pi_* \alpha \neq 0$, then it suffices to prove that

$$\pi \circ \exp_G(\alpha) = \exp_A \circ \pi_*(\alpha)$$

is transcendental, and our theorem is reduced to the case where G is an abelian variety. We note that in that case, the exponential map is always a transcendental function. In the case of elliptic curves, for instance, it is given by the $\wp$-function.

In the case of an abelian variety A, we get a global representation by means of theta functions (cf. Baily [1]). In other words, there exists a holomorphic homomorphism

$$\mathbf{C}^d \overset{\Theta}{\to} A$$

inducing a holomorphic isomorphism of the torus ($\mathbf{C}^d$ modulo the period lattice) onto A. If $(z_1, \ldots, z_d)$ are the co-ordinates of $\mathbf{C}^d$, then the partial derivatives $\partial/\partial z_i$ are a basis of the Lie algebra of invariant derivations on A, over the complex numbers. Making a suitable change of co-ordinates in $\mathbf{C}^d$, we may assume that this basis is defined over $\mathbf{K}$, i.e. if f is a meromorphic function on A defined over $\mathbf{K}$, then $\partial f/\partial z_i$ is defined over $\mathbf{K}$ for all i. A tangent vector is then represented by co-ordinates

$$\alpha = (\alpha_1, \ldots, \alpha_d)$$

and is rational over $\mathbf{K}$ if and only if all α_i are elements of $\mathbf{K}$. The exponential function is represented by Θ, i.e.

$$\exp(t\alpha) = \Theta(t\alpha).$$

Thus Theorem 1 can be formulated as follows.

THEOREM 2. *Let A^d be an abelian variety of dimension d, defined over the field of algebraic numbers $\mathbf{K}$. Let*

$$\Theta : \mathbf{C}^d \to A$$

be the homomorphism given by the theta functions, inducing an isomorphism of the complex torus onto A. Assume that the derivations $\partial/\partial z_i$ ($i = 1, \ldots, d$) are defined over $\mathbf{K}$. If $\alpha \in \mathbf{C}^d$ is a complex vector $\neq 0$ such that all α_i lie in $\mathbf{K}$, then $\Theta(\alpha)$ is transcendental over $\mathbf{K}$. In particular, the periods are transcendental.

We shall now make several remarks concerning Theorems 1 and 2.

(i) Given any complex analytic isomorphism Ψ from the complex torus onto A, the map $\Psi\Theta^{-1}$ is a complex analytic automorphism of A. Since the law of composition on A is assumed to be defined over $\mathbf{K}$, $\Psi\Theta^{-1}$ must be algebraic and defined over $\mathbf{K}$ (Chow's theorem), so that Theorem 2 applies to any such map Ψ instead of Θ, provided always that the $\mathbf{K}$-rationality of the Lie algebra is preserved by Ψ. We use Θ because the series expansion for the theta functions show that they are entire functions of order 2, a growth condition which is used in the proof. In the case of elliptic curves, one usually takes the map given by $(\wp, \wp')$, and the result is due to Schneider. Furthermore, using Theorem 1, we also get the analogue of Theorem 2 for a commutative group variety which is a quotient of $\mathbf{C}^d$. In other words: Let $\phi : \mathbf{C}^d \to A$ be a complex analytic homomorphism onto a commutative group variety A, having a discrete subgroup as kernel, and preserving the $\mathbf{K}$-rationality of the Lie algebra. Let $\alpha \in \mathbf{C}^d$ be a complex vector $\neq 0$, which is algebraic, and such that the curve $\phi(t\alpha)$ is not an algebraic function of t. Then $\phi(\alpha)$ is transcendental.

(ii) We shall prove Theorem 2 following the classical method of Gelfond and Schneider (cf. Siegel [5], Schneider [3] and [4]), properly formulated. The exponential function satisfies a differential equation, and Theorem 3 below contains the heart of the proof, using only the differential equation and not the group law. It contains [4] (but not conversely). It is however insufficient to cover also Siegel's result on the Bessel function [5]. In order to get both, one would need a statement such as the following.

Let V be a projective or affine non-singular variety defined over a number field K. Let $f : \mathbf{C} \to V$ be a holomorphic map, which is not an algebraic function, and such that $f(0)$ is a rational point of V over K. Assume that f is the integral curve of a (time-dependent) algebraic vector field on V, defined over K, and having no singular point. If α is algebraic then $f(\alpha)$ is transcendental.

It is probably also necessary to assume that f can be represented by quotients of holomorphic functions of finite order, or some other such restriction.

(iii) Finally, I would also conjecture that if

$$\phi : \mathfrak{H}_n \to V$$

is the Baily mapping giving the transcendental-algebraic representation of the moduli variety over $\mathbf{Q}$ [1], then $\phi(\alpha)$ is transcendental whenever α is algebraic, provided that the abelian variety associated with the moduli $\phi(\alpha)$ has a trivial ring of endomorphisms. In dimension 1, this would amount to the result of Schneider that $j(\tau)$ is transcendental whenever τ is algebraic, and not a quadratic imaginary irrationality.

Proof. We assume that $\exp(\alpha)$ is rational over $\mathbf{K}$ and derive a contradiction.

In Theorems 1 and 2, all algebraic objects are defined over a finitely generated extension of $\mathbf{Q}$, (for instance the algebraic group law on A). Hence without loss of generality we may replace $\mathbf{K}$, by a number field K of finite degree over $\mathbf{Q}$. In particular, $\exp(\alpha)$ is rational over K, and so is $\exp(t\alpha)$ for infinitely many algebraic points t in K, e.g. $t = 1, 2, \dots$. All we need is a finite number of such points, say m, with

$$m > 20[K : \mathbf{Q}].$$

In Theorem 2, let

$$\Theta = (\theta_0, \dots, \theta_N)$$

be the projective co-ordinates of our map. We can find a suitable linear combination of $\theta_0, \dots, \theta_N$ with coefficients in K which does not vanish at any of our m points. After a projective change of co-ordinates over K, we may assume that it is θ_0. We let

$$f_i = \theta_i/\theta_0 \qquad i = 1, \dots, N.$$

Then each f_i is an abelian function, holomorphic at our m points, and $(f_1, \dots, f_N)$ are the co-ordinates of an affine open subset of A. Since each $\partial f_i/\partial z_j$ is holomorphic whenever

$f_1, \ldots, f_N$ are holomorphic, such a partial derivative, which is an abelian function defined over K, must lie in the ring $K[f_1, \ldots, f_N]$.

We let $\bar{\theta}_i$ and $\bar{f}_i$ be the induced functions of one variable t, obtained by taking $\Theta(t\alpha)$. Then the ring

$$K[\bar{f}_1, \ldots, \bar{f}_N]$$

is stable under the derivation $D = d/dt$.

The exponential map $\Theta(t\alpha)$ cannot be an algebraic function of t. Hence the transcendence degree of the field

$$K(t, \bar{f}_1, \ldots, \bar{f}_N)$$

over K is ≥ 2. The next theorem will therefore conclude the proof.

THEOREM 3. *Let K be a number field of finite degree over $\mathbf{Q}$. Let $K[g_1, \ldots, g_N]$ be a finitely generated ring of meromorphic functions on $\mathbf{C}$, of finite order ρ (i.e. quotients of entire functions of finite order ρ). Assume that the field $K(g_1, \ldots, g_N)$ has transcendence degree ≥ 2 over K, and that the differential operator $D = d/dt$ maps the ring into itself. Let $w_1, \ldots, w_m$ be distinct complex numbers not lying among the poles of the g_i such that*

$$g_i(w_\nu) \in K$$

for $i = 1, \ldots, N$ and $\nu = 1, \ldots, m$. Then $m \leq 10\rho \cdot [K : \mathbf{Q}]$.

In the course of the proof, we shall make estimates, and it is convenient to introduce some terminology.

By the *size* of a set of elements of K we shall mean the maximum of the absolute values of all conjugates of these elements. By the size of a polynomial with coefficients in K, we shall mean the size of its set of coefficients. A *denominator* for a set of elements of K will be any positive rational integer whose product with every element of the set is an algebraic integer. We define in a similar way a denominator for a polynomial with coefficients in K.

The next lemma gives us local estimates. The proof will be concluded afterwards by comparing these with global estimates.

LEMMA 1. *Let K be a number field of finite degree over $\mathbf{Q}$. Let $g_1, \ldots, g_N$ be functions, holomorphic on a neighborhood of a point $w \in \mathbf{C}$, and assume that $D = d/dt$ maps the ring $K[g_1, \ldots, g_N]$ into itself. Assume that $g_i(w)$ lies in K for all i. Then there exists a number C_1 such that, for any polynomial*

$$g = P(g_1, \ldots, g_N)$$

with coefficients in K, of degree $\leq r$ in each variable g_i, and size $\leq C_2$, we have for all positive integers k,

$$\text{size}\,(D^k g(w)) \leq r^k k! C_2 C_1^{k+r}.$$

Furthermore, if C_2 is also a denominator for P, we can select C_1 such that $C_2 C_1^{k+r}$ is a denominator for $D^k g(w)$.

Proof. We introduce variables $T_1, \ldots, T_N$ so that $P = P(T_1, \ldots, T_N)$. We can also write

$$Dg_i = P_i(g_1, \ldots, g_N)$$

with some polynomial P_i having coefficients in K. We observe that P is dominated by

$$C_2(1 + T_1 + \ldots + T_N)^r$$

which has positive coefficients. We define formally

$$DT_i = P_i(T_1, \ldots, T_N)$$

and observe that there exist an integer h and a constant C_3 independent of g such that each P_i is dominated by $C_3(1 + T_1 + \ldots + T_N)^h$. The k-th derivative $D^k P$ is dominated by the k-th derivative of $C_2(1 + T_1 + \ldots + T_N)^r$, and a simple inductive argument gives us the desired bound for the size of $D^k g(w)$. The second assertion of the lemma is proved by a trivial induction which we leave to the reader. We observe that the lemma could also be formulated completely algebraically.

We return to the proof of Theorem 3. Say that g_1, g_2 are algebraically independent over K, and call them f, g. Let r be a positive integer divisible by $2m$. We shall let r tend to infinity later. Let

$$F = \sum a_{ij} f^i g^j \qquad 0 \leqq i, j < r$$

have coefficients a_{ij} in K. Let $n = r^2/2m$. We can select the a_{ij} not all equal to zero, and such that

$$D^k F(w_v) = 0$$

for $0 \leqq k < n$ and $v = 1, \ldots, m$. Indeed, we have to solve a system of mn linear equations in $r^2 = 2mn$ unknowns. Note that $mn/(2mn - mn) = 1$. Using Lemma 2 on page 37 of Siegel [5] and Lemma 1, we can in fact take the a_{ij} to be algebraic integers, whose size is bounded by

$$0(r^n n! C_1^{n+r}) \leqq 0(n^{3n})$$

for $n \to \infty$. (The exponent 3 is of course very wasteful.)

Since f, g are algebraically independent over K, our function F is not identically zero. We let s be the smallest integer such that all derivatives of F up to order $s - 1$ vanish at all points $w_1, \ldots, w_m$, but such that $D^s F$ does not vanish at one of the w_v, say w_1. Then $s \geqq n$. We let

$$\gamma = D^s F(w_1) \neq 0.$$

Then γ is an element of K, and by Lemma 1, it has a denominator which is bounded by $0(s^{6s})$ for $s \to \infty$. Let c be this denominator. The norm of $c\gamma$ from K to $\mathbf{Q}$ is then a non-zero rational integer. Each conjugate is bounded by $0(s^{9s})$. Consequently, we get

$$(1) \qquad\qquad 1 \leqq \left| N_Q^K(c\gamma) \right| \leqq 0(s^{9s[K:\mathbf{Q}]}) |\gamma|$$

where $|\gamma|$ is the fixed absolute value of γ, which will now be estimated very well by a global argument.

Let θ be an entire function of order ρ such that θf and θg are entire and $\theta(w_1) \neq 0$. Then $\theta^{2r} F$ is entire. We consider the entire function

$$H(t) = \frac{\theta(t)^{2r} F(t)}{\prod_{v=1}^{m} (t - w_v)^s}.$$

Then $H(w_1)$ differs from $D^s F(w_1)$ by obvious factors, bounded by $C_4^s s!$. By the maximum modulus principle. its absolute value is bounded by the maximum of H on a large circle of radius R. If we take R large, then $t - w_v$ has approximately the same absolute value as R, and consequently, on the circle of radius R, $H(t)$ is bounded in absolute value by an expression of type

$$\frac{C_5^{2rR^\rho}}{R^{ms}} \cdot$$

We select $R = (ms)^{1/2\rho}$. We then get the estimate

$$|\gamma| \leq \frac{s^s C_6^s}{s^{ms/2\rho}}$$

which, combined with inequality (1) gives us the desired bound on m. This concludes the proof.

Instead of d/dt we could also have taken $D = \lambda \cdot d/dt$ with some complex number λ, always assuming that the ring of functions is stable under D. This yields the Gelfond–Schneider theorem (which is not a special case of Theorem 1). Indeed, let α, β be algebraic, $\alpha \neq 0$, 1 and β irrational. Take $D = (\log \alpha)^{-1} \, d/dt$ and $g_1 = e^{t \log \alpha}$, $g_2 = e^{\beta t \log \alpha}$. Since β is irrational, these functions are algebraically independent, and hence $e^{\beta \log \alpha} = \alpha^\beta$ is transcendental. Similarly, to prove the transcendence of e^α for algebraic $\alpha \neq 0$, we take the two functions t and $e^{\alpha t}$.

Finally, we could assume that the algebraic differential equation of Theorem 3 depends algebraically on time t, or say for simplicity, that Dg_i lies in the ring $K(t)[g_1, \ldots , g_N]$. The values w_v must then be assumed distinct from the poles of the rational functions of t occurring as coefficients of the differential equation. The proof of Theorem 3 extends trivially to any of the above-mentioned generalizations.

REFERENCES

1. W. Baily: On the theory of theta functions, the moduli of abelian varieties, and the moduli of curves *Ann. Math., Princeton* **75** (1962), 342–381.
2. A. O. Gelfond: *Transcendental and algebraic numbers* (Translated by Leo F. Boron), Dover, New York, 1960.
3. T. Schneider: Transcendenzuntersuchungen periodischer Funktionen, *J. reine angew. Math.* **172** (1934), 65–74. (Note: The proofs of Schneider's subsequent paper on abelian integrals are invalid because of an improper use of Cauchy's formula in several variables.)
4. T. Schneider, *Einfuhrung in die transcendenten Zahlen* (Chapter II, Satz 13), Springer, Berlin, 1957.
5. C. L. Siegel: Transcendental numbers, *Ann. Math., Princeton*, Studies 16, (1949).

Columbia University,
New York.

Fonctions Implicites et Plongements Riemanniens

par Serge Lang
(d'après J. Nash [2] et J. Moser [1])

1. Métriques induites

Soit X une variété différentiable compacte. Un morphisme $u \colon X \to V$ de X dans une autre variété est un plongement si et seulement si elle est injective, et son application tangente est une injection en chaque point. Supposons que ceci soit le cas, et que V soit munie d'une métrique (i.e. d'un champ de tenseurs bilinéaires symétriques) g. Alors u induit une métrique sur X, notée $u^*(g)$. Si X est aussi munie d'une métrique g_1, on dira que u est un plongement métrique si $u^*(g) = g_1$. Si les métriques sont définies positives ($g > 0$ et $g_1 > 0$), on dit qu'elles sont riemanniennes, et que u est un plongement riemannien.

Les métriques sur X forment un espace vectoriel $\mathrm{Met}(X)$. Si l'on considère X comme une variété de classe C^p ($0 \le p < \infty$) cet espace vectoriel est même un espace de Banach, car on peut prendre le sup des dérivées d'ordre $\le p - 1$ (par rapport à un nombre fini de cartes locales recouvrant X) pour définir la norme. Ceci étant, les métriques riemanniennes forment alors un ouvert $\mathrm{Ri}(X)$ dans ce Banach, et par conséquent, on se doit de considérer $\mathrm{Ri}(X)$ comme variété (de dimension infinie), dont l'espace tangent n'est autre que $\mathrm{Met}(X)$.

Notons $M^p(X, V)$ l'ensemble des morphismes (= applications différentiables) de classe C^p de X dans V. Si $V = \mathbf{R}^n$, alors $M^p(X, \mathbf{R}^n)$ est un espace vectoriel, et un espace de Banach pour la p-norme des dérivées $\le p < \infty$. Le sous-ensemble des plongements noté $\mathrm{Em}^p(X, \mathbf{R}^n)$ est ouvert dans $M^p(X, \mathbf{R}^n)$. Si on considère $\mathbf{R}^n$ comme muni de sa métrique ordinaire ρ, alors on a une application

$$f \colon M^p(X, \mathbf{R}^n) \to \mathrm{Met}^{p-1}(X)$$

qui, à tout morphisme u, associe $u^*(\rho) = f(u)$. En outre, f applique l'ouvert des plongements dans l'ouvert des métriques riemanniennes,

$$f \colon \mathrm{Em}^p(X, \mathbf{R}^n) \to \mathrm{Ri}^{p-1}(X).$$

Le théorème de Nash, selon lequel toute variété riemannienne admet un plongement riemannien dans un espace euclidien, peut s'exprimer alors en disant que f est surjective (pour un choix convenable de n en fonction de d).

Pour démontrer le théorème, GARSIA m'a fait remarquer qu'on peut se borner au cas où X est un tore. En effet, toute variété compacte se plonge dans un espace euclidien, et son image est contenue dans une boule ouverte bornée qu'on peut plonger dans une carte locale sur le tore. On obtient alors X comme sous-variété fermée du tore, et la métrique riemannienne donnée s'étend à une métrique riemannienne sur le tore, d'abord localement, puis globalement par une partition de l'unité. Il suffit alors de plonger le tore métriquement dans un $\mathbf{R}^n$ pour résoudre le problème pour X.

Pour des raisons techniques, il est fréquemment plus facile de traiter le tore qu'une variété quelconque. Par exemple, dans ce cas, soit $u \colon T^d \to \mathbf{R}^n$ un plongement. La métrique $f(u) = g$ sur T induite par ρ s'écrit

$$g_{ij}(x) = \left\langle \frac{\partial u}{\partial x_i}, \frac{\partial u}{\partial x_j} \right\rangle \qquad i, j = 1, \ldots, d$$

si $(x_1, \ldots, x_d)$ sont les coordonnées (globales!) de T et si u est représenté comme fonction périodique sur $\mathbf{R}^d$.

L'application f admet alors une dérivée

$$f' \colon M^p(T, \mathbf{R}^n) \to \mathrm{Met}^{p-1}(T)$$

qui s'écrit

$$(1) \qquad f'(u)v = \left\langle \frac{\partial u}{\partial x_i}, \frac{\partial v}{\partial x_j} \right\rangle + \left\langle \frac{\partial u}{\partial x_j}, \frac{\partial v}{\partial x_i} \right\rangle$$

par rapport à ces coordonnées.

L'espace des plongements n'est d'ailleurs pas le plus agréable à regarder, et c'est vraiment aux plongements non-dégénérés qu'on en a. On dit qu'un plongement $u \colon X^d \to \mathbf{R}^n$ est *non-dégénéré* si, en chaque point, il existe des coordonnées locales $(x_1, \ldots, x_d)$ telles que la matrice

$$(2) \qquad \begin{bmatrix} \dfrac{\partial u}{\partial x_i} \\[2mm] \dfrac{\partial^2 u}{\partial x_i \partial x_j} \end{bmatrix}$$

soit carrés et de déterminant non nul en chaque point. Cela implique, en particulier, que $n = d + d(d+1)/2$, et que la matrice ait n colonnes et n lignes. On voit immédiatement que cette condition est indépendante du choix de coordonnées locales (un changement de coordonnées ne fait qu'introduire des relations linéaires, donc ne peut que réduire le rang de la matrice, d'où égalité de rang par symétrie).

Notre application f induit alors une application (encore notée f)

$$f \colon ND^p(X, \mathbf{R}^n) \to \mathrm{Ri}^{p-1}(X).$$

Il est évident que les plongements non-dégénérés forment aussi un ouvert dans $M^p(X, \mathbf{R}^n)$. On cherche alors à décrire la structure de l'application f. NASH démontre que c'est une application ouverte, et c'est cette partie de la démonstration qui représente la contribution la plus substantielle de son article. C'est la seule que nous considérons ici.

MOSER [1] a repris la question, et suivant la méthode de NASH, l'exprime sous la forme d'un théorème de fonctions implicites que nous étudierons plus loin en détail. Il est naturel de se demander si f est localement une projection. Cette question n'est pas résolue.

Nous n'entrerons pas dans la seconde partie de la démonstration de NASH. Elle consiste à montrer, d'abord, qu'il existe des plongements non-dégénérés, et, ensuite, que toute métrique riemannienne s'approche par une métrique induite par un tel plongement. A ceci, il faut encore ajouter une astuce pour tenir compte de certaines uniformités en ce qui concerne les sup de dérivées supérieures. Notre but ici est uniquement d'exposer la relation entre le théorème des fonctions implicites et le problème du plongement.

2. Fonctions implicites et métriques

Soit $p \geq 3$, et notons E_p l'espace de Banach $M^p(T, \mathbf{R}^n)$. Soit F l'espace de Banach $\mathrm{Met}^{p-1}(T)$. On a l'application

$$f: E_p \to F$$

et sa dérivée f'. On notera que cette derivée admet une extension linéaire à E_{p-2}, donnée par la formule (1),

$$E_{p-2} \supset E_p \xrightarrow{f} F.$$

Soit $\overset{\circ}{u}$ un plongement non-dégénéré de T dans $\mathbf{R}^n$. Nous allons définir pour chaque u suffisamment voisin de $\overset{\circ}{u}$ (pour la p-topologie) un opérateur linéaire continu

$$L(u): F \to E_{p-2}$$

tel que $f'(u) \circ L(u) = \mathrm{id}$, c'est-à-dire un inverse à droite pour $f'(u)$. A chaque métrique h il existe une application $v: T \to \mathbf{R}^n$ et une seule qui satisfasse au système de conditions linéaires:

$$(3) \qquad -2\left\langle \frac{\partial^2 u}{\partial x_i \partial x_j}, v \right\rangle = h_{ij},$$

$$(4) \qquad \left\langle \frac{\partial u}{\partial x_i}, v \right\rangle = 0.$$

La condition (4) exprime le fait que v est perpendiculaire à l'espace tangent au tore dans $\mathbf{R}^n$. La condition (3) s'obtient en prenant la dérivée de (4) et en la combinant avec la relation

$$(5) \qquad f'(u)v = h.$$

Cela montre que l'on peut écrire

$$L(u)h = Q(u_x, u_{xx})h$$

avec une matrice Q de fonctions rationnelles en les u_x, u_{xx} dont le dénominateur n'est autre que le déterminant de (2). Par conséquent, v est de classe C^{p-2} et Q est continue, en tant qu'application

$$B_a(\overset{\circ}{u}, E_p) \to E_{p-2}$$

définie sur une petite boule de rayon a autour de $\overset{\circ}{u}$ dans E_p.

Pour regagner les deux dérivées que nous avons perdues, il faudra des opérateurs de régularisation qu'on décrira au prochain paragraphe. Le dernier paragraphe donnera alors le théorème de fonctions implicites qui, moyennant des hypothèses de dérivabilité supplémentaires, nous permettra de démontrer le résultat suivant.

THÉORÈME 1. *Soit $p \geq 3$. Soit $\overset{\circ}{u}$ un plongement non-dégénéré de T dans $\mathbf{R}^n$ de classe $p + 44$, et posons $\overset{\circ}{g} = f(\overset{\circ}{u})$. Soit $A > 0$ et $\varepsilon > 0$. Il existe $\delta > 0$ tel que, si g est de classe $p + 44 - 2 = p + 42$, si $\|g\|_{p+42} < A$, et $\|g - \overset{\circ}{g}\|_{p-1} < \delta$, alors $g = f(u)$ avec un u de classe C^p, tel que*

$$\|u - \overset{\circ}{u}\|_p < \varepsilon.$$

En fait, ce théroème ne devrait être qu'un lemme, et on devrait pouvoir démontrer dans le cas C^∞ que l'application

$$f: ND^\infty(T, \mathbf{R}^n) \to \mathrm{Met}^\infty(T)$$

est ouverte, et même possède une section locale en chaque point $\overset{\circ}{u}$. Je m'excuse de ne pas avoir été capable de le faire. En tous cas, on obtient le u désiré comme limite d'une suite

$$u_{n+1} = u_n - T_{\xi_{n+1}} L(u_n)[f(u_n) - g]$$

avec des opérateurs T_ξ définis au prochain paragraphe.

3. Opérateurs de régularisation

THÉORÈME 2. *Soient $K_0 \supset K_1 \supset K_2 \supset \cdots$ les espaces de Banach des fonctions de classe $C^0, C^1, C^2, \ldots$ sur $\mathbf{R}^d$, à supports compacts. Pour tout nombre $\xi > 1$, il existe une application linéaire*

$$T_\xi: K_0 \to K_\infty = \bigcap K_p$$

satisfaisant aux conditions suivantes:

$$\|T_\xi v\|_{r+s} \;\leq\; c(r,s)\xi^s \|v\|_r$$

$$\|(I - T_\xi)v\|_r \;\leq\; c(r,s)\xi^{-s}\|v\|_{r+s}$$

$r, s \geq 0$, $v \in K_{r+s}$. *On note $c(r,s)$ une constante ne dépendant que de r et s, indépendante de ξ, v.*

Démonstration. Désignons par $x, y, \ldots$ des points de $\mathbf{R}^d$, $x = (x_1, \ldots, x_d)$. Posons $|x| = \sup |x_i|$. Soit φ_0 une fonction C^∞ sur $\mathbf{R}$ telle que $0 \leq \varphi_0 \leq 1$, et

$$\varphi_0(y) = \begin{cases} 1 & \text{si } y < 1/2 \\ 0 & \text{si } y \geq 1. \end{cases}$$

Définissons $\varphi(x) = \varphi_0(x_1) \cdots \varphi_0(x_d)$ et soit $\psi = \hat{\varphi}$ sa transformée de Fourier. Alors ψ est analytique, on a:

(i) Pour chaque $\nu, \mu \geq 0$ la fonction

$$|x|^\nu |\psi^{(\mu)}(x)|$$

est bornée par $|\varphi^{(\nu+\mu)}|$ (à une constante près).

(ii) Si on pose $k = (k_1, \ldots, k_d)$ et $x^k = x_1^{k_1} \cdots x_d^{k_d}$ alors

$$\int x^k \psi(x)\, dx = \begin{cases} 1 & \text{si } k = 0 \\ 0 & \text{si } k \neq 0. \end{cases}$$

La première assertion se montre facilement en dérivant ψ, et la seconde en dérivant la transformée de Fourier de ψ et en employant la formule d'inversion, plus le fait que φ est constante en un voisinage de l'origine.

On définit T_ξ par la convolution

$$(T_\xi u)(x) = \xi^d \int \psi(\xi(x - y))u(y)\, dy.$$

La première inégalité est évidente du fait que les opérateurs de dérivation commutent à la convolution. Quant à la seconde, relative à $I - T_\xi$, on se ramène aussitôt au cas où $r = 0$, puis on évalue la convolution en évaluant

$$u(x) - \int \psi(z)u\left(x - \frac{z}{\xi}\right) dz = \int \psi(z)\left[u(x) - u\left(x - \frac{z}{\xi}\right)\right] dz$$

en développant $u(x) - u(x - \frac{z}{\xi})$ suivant sa série de Taylor, il reste:

$$\left| u(x) - u\left(x - \frac{z}{\xi}\right) - \sum_{|k|<s} a_k(x)z^k \right| \leq \frac{|z|^s}{|\xi|^s} C \|u\|_s.$$

On trouve l'estimation voulue en se servant de (ii).

On démontre l'existence de tels opérateurs de régularisation sur une variété compacte en la plongeant dans un espace euclidien. Dans la cas particulier du tore, on pourrait aussi prendre la convolution sur le tore.

4. Théorème abstrait de fonctions implicites

Supposons donnés quatre espaces de Banach

$$E_- \supset E_+ \supset E_\flat \supset E_\sharp$$

qui, dans les applications, correspondraient à des espaces fonctionnels,

$$E_{p-\sigma} \supset E_p \supset E_{p+\beta} \supset E_{p+\beta+\lambda}.$$

(Dans la discussion du §2, on aurait $\sigma = \lambda = 2$ et $\beta = 42$.) On suppose donnée une application linéaire pour chaque entier positif ξ,

$$T_\xi \colon E_0 \to E_\sharp$$

et on n'aura besoin que d'une partie des hypothèses du théorème 2, à savoir qu'il existe des nombres α, β, $\sigma > 0$ et $C > 2$ tels que T_ξ soit un opérateur borné, les bornes entre les espaces variés étant données par:

$$|T_\xi| \leq C\xi^\alpha \qquad \text{pour } E_- \to E_\sharp$$
$$|I - T_\xi| \leq C\xi^{-\beta} \qquad \text{pour } E_\flat \to E_+$$
$$|T_\xi| \leq C\xi^\sigma \qquad \text{pour } E_- \to E_+.$$

On supposera qu'il existe un nombre s, $1 < s < 2$, tel que, si l'on pose $\tau = (s\sigma + 1)/(2 - s)$, alors

$$s\beta - \alpha \geq s^2(\sigma + \tau) + 1.$$

Dans l'application du §2, on prend $s = 3/2$ et $\beta = 42$.

THÉORÈME 3. *Les notations étant comme ci-dessus, prenons un nombre $0 < a < 1/2$ et $u_0 \in E_+$. Soit $f \colon B_{2a}(u_0, E_+) \to F$ une application de classe C^2 dans un espace de Banach F, dont les dérivées f' et f'' sont bornées par C dans la boule B_{2a}. Supposons que, pour tout u dans la boule $B_a(u_0, E_+)$,*

(i) *Il existe une application linéaire $L(u) \colon F \to E_-$, aussi bornée par C, et une extension linéaire de $f'(u)$ à l'image de $L(u)$ telles que*

$$f'(u) \circ L(u) = \text{identité}.$$

(ii) *Si $u \in B_a(u_0, E_+)$ et $u - u_0 \in E_\sharp$, alors $L(u)f(u)$ est dans $E_\flat$ et*

$$|L(u)f(u)|_\flat \leq C \sup(1, |u - u_0|_\sharp).$$

Alors il existe $\varepsilon > 0$ dépendant de α, β, σ, s, C tel que, si $|f(u_0)| < \varepsilon$, il existe $u \in B_a(u_0, E_+)$ tel que $f(u) = 0$.

Remarques. Dans les applications, on applique le théorème à l'application $f_g(u) = f(u) - g$ pour g voisin de $g_0 = f(u_0)$. La condition (ii) dans le cas particulier du §2 se démontre facilement au moyen de formules d'interpolation pour les dérivées, de sorte que notre théorème 3 implique immédiatement le théorème 1.

Nous prenons ξ_0 suffisamment grand, et nous posons $\xi_n = \xi_0^{s^n}$. Nous supposons en outre que $|f(u_0)| \leq a/C^3 \cdot \xi_0^{-s(\sigma+\tau)}$. Nous montrerons alors que la suite

$$u_{n+1} = u_n - T_{\xi_{n+1}} L(u_n) f(u_n)$$

converge dans $B_a(u_0, E_+)$ vers un élément w tel que $f(w) = 0$. On démontrera récursivement que

$$(1n) \qquad\qquad |f(u_n)| \;\leq\; a/C^3 \cdot \xi_{n+1}^{-(\sigma+\tau)}$$

$$(2n) \qquad\qquad |u_{n+1} - u_n|_+ \;\leq\; a\xi_{n+1}^{-\tau}$$

$$(3n) \qquad\qquad |u_n - u_0|_\sharp \;\leq\; a\xi_n^{\alpha}.$$

D'après $(2n)$, on voit que la suite est une suite de Cauchy, que tous les u_n sont dans la boule $B_a(u_0, E_+)$, ξ_0 étant grand. D'après $(1n)$ on voit que la suite u_n converge vers un zéro de f. On n'a mis $(3n)$ que pour faciliter la récurrence.

Pour $n = 0$, la condition $(1n)$ est une hypothèse, $(3n)$ est évidente, et pour $(2n)$, on estime $|u_1 - u_0|_+$ en employant successivement la borne de T_ξ de E_- à E_+, la borne absolue de $L(u_0)$, et l'hypothèse sur la borne de $|f(u_0)|$. Pour démontrer les conditions $n + 1$, on a:

$1(n + 1)$. Par le fait que f est deux fois différentiable,

$$|f(u_{n+1}) - f(u_n) - f'(u_n)(u_{n+1} - u_n)| \leq C|u_{n+1} - u_n|_+^2.$$

L'expression à gauche est égale à

$$|f(u_{n+1}) - f(u_n) - f'(u_n)T_{\xi_{n+1}}L(u_n)f(u_n)|.$$

Écrivons $T = T - I + I$. Alors le terme en $f(u_n)$ disparaît. On obtient ce qu'on veut en employant la borne de $I - T$, la borne absolue de f' et la condition (ii), ainsi que l'hypothèse que $s\beta - \alpha \geq s^2(\sigma + \tau) + 1$.

$2(n + 1)$. C'est comme pour $n = 0$, à savoir: on se sert successivement de la borne pour $T_{\xi_{n+2}}$, de la borne uniforme pour $L(u_{n+1})$, et de l'hypothèse de récurrence pour $f(u_{n+1})$.

$3(n + 1)$. Par récurrence, on trouve

$$|u_{n+1} - u.|_\sharp \leq |u_{n+1} - u_n|_\sharp + |u_n - u_0|_\sharp.$$

Pour le premier terme, on estime successivement la borne de $T_{\xi_{n+1}}$ de E_- à $E_\sharp$, la borne uniforme de $L(u_n)$ et finalement la borne de $f(u_n)$ d'après l'hypothèse de récurrence. Pour le second terme, on se sert de l'hypothèse de récurrence. Une estimation grossière donne ce qu'on veut.

Bibliographie

[1] Moser, J. A new technique for the construction of solutions of nonlinear differential equations. *Proc. Nat. Acad. Sc.* U.S.A., t. 47, 1961, pp. 1824–1831.

[2] Nash, J. The embedding problem for Riemannian manifolds. *Annals of Math.*, Series 2, t. 63, 1956, pp. 20–63.

DIOPHANTINE APPROXIMATIONS ON TORUSES.

By Serge Lang.

The main purpose of this paper is to make precise certain analogies between diophantine approximations on the multiplicative group and elliptic curves, and to show how these are related with approximations on the additive group by means of the exponential mapping. We shall state two conjectures. To relate their exponential and logarithmic formulations, it is best to prove first some results relating the exponential and logarithmic heights.

1. Heights on the multiplicative group. Let K be a number field, and let M_K be the set of absolute values on K extending the standard, normalized absolute values on the rational numbers $\boldsymbol{Q}$. (cf. [3], Chapter III.) If $|\ |_v$ is in M_K, and α is an element of K, we write

$$v(\alpha) = \log \| \alpha \|_v$$

where $\| \alpha \|_v = | \alpha |_v^{N_v}$. Instead of saying that $|\ |_v$ is in M_K, we shall also say that v is in M_K. By the product formula, we have

$$\sum_v v(\alpha) = 0$$

for any $\alpha \in K^*$ ($=$ multiplicative group of K). If $(\alpha_0, \cdots, \alpha_n)$ is a point in projective n-space with coordinates in K, we define its *height* to be

$$\sum_v \sup_i (0, v(\alpha_i)).$$

(This is the logarithm of the height as used frequently in the past literature.) If P is the point represented by $(\alpha_0, \cdots, \alpha_n)$, we shall write $h(P)$ for its height (again taking the log of the notation in [3]). If α is an element of K, we let $h(\alpha)$ be the height of the point $(1, \alpha)$ in projective 1-space.

We shall use the standard Vinogradov notation for orders of magnitude. If ϕ, ψ are two real functions on a set, whose values are ≥ 0, we write $\phi \ll \psi$ if there exists a constant $c > 0$ such that $\phi \leq c\psi$. We write $\phi \gg \ll \psi$ if $\phi \ll \psi$ and $\psi \ll \phi$.

Let Γ be a finitely generated subgroup of the multiplicative group K^* of K. There exists a finite set $S \supset S_\infty$ of absolute values in M_K, such that Γ

Received November 1, 1963.

is contained in the S-units K_S, i.e. those elements of K^* having absolute
value equal to 1 outside S. Let $s+1$ be the cardinality of S, and map

$$L: K_S \to \mathbf{R}^{s+1}$$

by the log, such that

$$L(\beta) = (v_1(\beta), \cdots, v_{s+1}(\beta)),$$

if v_j are the elements of S. The kernel of L consists precisely of the roots
of unity in K. The image of L is a lattice of dimension s in the hyperplane
H defined by the equation

$$\sum_{j=1}^{s+1} x_j = 0.$$

We now define a function $\mathbf{F}$ on $\mathbf{R}^{s+1}$ by the condition

$$\mathbf{F}(X) = \sum_{j=1}^{s+1} \sup(0, x_j).$$

Then for any $t \geq 0$ we have $\mathbf{F}(tX) = t\mathbf{F}(X)$, and for any two vectors X, Y
we have $\mathbf{F}(X + Y) \leq \mathbf{F}(X) + \mathbf{F}(Y)$. We view $\mathbf{F}$ as restricted to the hyperplane H. Then for X in H, the condition $\mathbf{F}(X) = 0$ implies $X = 0$. Finally,
for X in H, we have $\mathbf{F}(-X) = \mathbf{F}(X)$. Hence $\mathbf{F}$ is a norm function on H,
and in particular, induces a norm function on any subspace of H.

The image of Γ under L is a discrete subgroup of $\mathbf{R}^{s+1}$. Let W be the
vector space generated by this image. Then $L(\Gamma)$ is a lattice of maximal
rank, say m, in the vector space W. Let $B_1, \cdots, B_m$ be a basis of $L(\Gamma)$.
For any vector X in W, we can write

$$X = y_1 B_1 + \cdots + y_m B_m$$

We define

$$\mathbf{G}(X) = \sup |y_i|.$$

Then $\mathbf{G}$ is a norm on W. Since two norms on a finite dimensional vector
space are always of the same order of magnitude, we conclude:

PROPOSITION 1. *We have* $\mathbf{F} \gg\ll \mathbf{G}$ *on the space* W.

Let $\beta_1, \cdots, \beta_m$ be free generators of Γ modulo its torsion group. For
each β in Γ we can write

$$\beta = \beta_1^{\nu_1} \cdots \beta_m^{\nu_m} \tau$$

with suitable integers $\nu_1, \cdots, \nu_s$ and a root of unity τ. Define

$$\nu = \nu(\beta) = \sup |\nu_i|.$$

From Proposition 1, we obtain immediately the

COROLLARY. *As functions on* Γ, *we have* $v \gg\!\!\ll h$.

We shall now give an asymptotic formula for the number of units of bounded height. This amounts to counting lattice points in certain regions. It seems to me that for such problems, the criterion given by Davenport in [1] is rather hard to apply, and that it is better to proceed as follows. One proves that the boundary of the region can be parametrized by a finite number of unit cubes whose dimension is smaller than that of the region. For homogeneous problems, one can then use the following result.

PROPOSTION 2. *Let D be a bounded set in $\mathbf{R}^m$ and assume that its boundary is contained in a finite union of pieces, each of which is the image of the unit $(m-1)$-cube under a mapping satisfying a Lipschitz condition. Let Λ be a lattice in $\mathbf{R}^m$ of dimension m. Then for $t \to \infty$, the number of lattice points in tD is equal to*

$$\frac{\operatorname{Vol} D}{\operatorname{Vol} F}\, t^m + O\left(t^{m-1}\right)$$

where Vol *means volume in $\mathbf{R}^m$, and F is a fundamental domain for the lattice.*

Proof. We recall that the boundary is taken in the sense of point set topology. It is the set of points having arbitrary small neighborhoods intersecting both the set D and its complement.

Let $\lambda_1, \cdots, \lambda_m$ be a basis for Λ over $\mathbf{Z}$, and let F be the canonical fundamental domain consisting of all vectors

$$t_1\lambda_1 + \cdots + t_m\lambda_m, \qquad\qquad 0 \leqq t_j < 1.$$

If λ is an element of the lattice, and if the translation F_λ of F intersects tD but is not contained in tD, then F_λ intersects the boundary of tD. To see this, let a be a point in $F_\lambda \cap tD$ and let b be a point in F_λ but not in tD. Since F_λ is convex, the line segment between a and b is contained in F_λ. There exists at least one point on this line segment which intersects the boundary of tD. Indeed, if a is an interior point of tD, we consider the set of points on the segment lying in the interior of tD. It is non-empty and open. If no point of this segment lies in the closure of tD but not in the interior of tD, then our set is also closed, contradiction, since the segment is connected, and b lies outside tD.

Let:

$$n(t) = \text{number of translates } F_\lambda \text{ contained in } tD,$$
$$n'(t) = \text{number of translates } F_\lambda \text{ intersecting } tD,$$
$$b(t) = \text{number of translates } F_\lambda \text{ intersecting } \partial(tD).$$

Then $0 \leqq n'(t) - n(t) \leqq b(t)$. Each translate F_λ of F contains precisely one lattice point, namely λ. If we can prove that

$$b(t) = O(t^{m-1}),$$

then the assertion of the proposition follows at once from the inequality

$$\mathrm{Vol}(F)n(t) \leqq \mathrm{Vol}(tD) = t^m \,\mathrm{Vol}(D) \leqq \mathrm{Vol}(F)n'(t).$$

We now prove that $b(t) \ll t^{m-1}$. Let $g: I^{m-1} \to \boldsymbol{R}^m$ be a mapping of the $(m-1)$-cube I^{m-1} satisfying a Lipschitz condition, and parametrizing one piece of the boundary ∂D. Since $\partial(tD) = t\partial(D)$, the map tg parametrizes the corresponding piece of $\partial(tD)$. It suffices to prove our assertion for this piece. Cut up the sides of I^{m-1} into intervals of equal length $1/[t]$. This determines a partition of I^{m-1} into $\ll t^{m-1}$ small cubes, each such small cube having a diameter $\leqq c_1/t$ for some constant c_1. The image of such a small cube under g has diameter $\leqq c_2/t$ (where c_2 is equal to c_1 times a Lipschitz constant for g). Hence the image of such a small cube under tg has diameter $\leqq c_2$. Hence such an image meets only a bounded number of translates F_λ of our fundamental domain. Since we have at most $c_3 t^{m-1}$ small cubes, we have proved our assertion.

Note. It would be a useful criterion if one knew that a compact subset of a real algebraic set in $\boldsymbol{R}^m$ of algebraic dimension $< m$ satisfied the hypothesis of Proposition 2. Because of the singularities, it does not seem easy to prove, and is one source of hand-waving in the literature. *Added in proof:* I understand from Hironaka that he can prove this statement.

We shall apply Proposition 2 to count units in number fields.

Let S be a finite subset of M_K containing the archimedean absolute values. Let $s+1$ be the cardinality of S, and let K_S be the S-units, i. e. the element of K^* whose absolute value outside S is equal to 1. Then K_S modulo a finite group of roots of unity is a free abelian group on s generators.

PROPOSITION 3. *Let Γ be a finitely generated subgroup of K_S. For each $t > 0$, let $n_\Gamma(t)$ be the number of elements of Γ whose height is $\leqq t$. Then there is a constant $c > 0$ such that*

$$n_\Gamma(t) = ct^m + O(t^{m-1})$$

where m is the rank of Γ.

Proof. As before, we map K_S into Euclidean $(s+1)$-space $\boldsymbol{R}^{s+1}$ by the log mapping:

$$L: K_S \to \boldsymbol{R}^{s+1}$$

such that

$$L(\beta) = (v_1(\beta), \cdots, v_{s+1}(\beta)),$$

if v_j are the elements of S. The kernel of L consists precisely of the roots of unity in K. The image of L is a lattice of dimension s in the hyperplane H defined by the equation

$$\sum_{j=1}^{s+1} x_j = 0.$$

Let $D(t)$ be the region consisting of points $(x_1, \cdots, x_{s+1})$ contained in the hyperplane, and satisfying the relation

$$\sum_{j=1}^{s+1} \sup(0, x_j) \leqq t.$$

Then $D(t) = tD(1)$, and up to a root of unity, the elements of K_S of height $\leqq t$ can be identified with the lattice points contained in $D(t)$. The set of vectors X in the hyperplane such that no x_j is equal to 0 and

$$\sum_{j=1}^{s+1} \sup(0, x_j) < 1$$

is open in H. Hence the boundary of $D(1)$ consists of the vectors X in $D(1)$ satisfying

$$\sum_{j=1}^{s+1} \sup(0, x_j) = 1,$$

except for pieces where some $x_j = 0$, which can be dealt with by induction. Let T denote a subset of S (identified with the integers $1, \cdots, s+1$) which is neither empty nor equal to S. The points $(x_1, \cdots, x_{s+1})$ satisfying

$$x_\nu \geqq 0, \ \sum_{\nu \in T} x_\nu = 1 \quad \text{and} \quad x_\mu \leqq 0, \ \sum_{\mu \notin T} x_\mu = -1$$

are contained in the boundary, and the boundary is the union of such sets of points as T ranges over subsets of S as above, together with pieces with some $x_j = 0$ which can be dealt with inductively.

Let W be the subspace generated by the elements of Γ over the real numbers. Then Γ is a discrete subgroup of W, of rank m, i.e. Γ is a lattice of maximal rank in W. We also observe that W is a subspace of H. Let $D_W(t) = D(t) \cap W$. The interior (open) of $D(t)$ intersected with W is open in W. Hence the boundary of $D_W(t)$ is contained in the intersection of the boundary of $D(t)$ and W. Since $D_W(t) = tD_W(1)$, we are led to considering the finite union of pieces described by a set T as above, intersected with W, and we must show that each one of these can be parametrized by an $(m-1)$-cube. If we can do this, we can then apply Proposition 1 to conclude our proof.

Without loss of generality, say that T consists of the first p coordinates for some integer p. Let

$$B = (1, \cdots, 1, 0, \cdots, 0) \quad \text{and} \quad B' = (0, \cdots, 0, 1, \cdots, 1)$$

be the vectors having coordinates equal to 1 in the first p places (resp. the last $s+1-p$). Let $A_0 = B + B'$. Then H is the hyperplane perpendicular to A_0. We can find vectors $A_1, \cdots, A_{s-m}$ in H such that $A_0, A_1, \cdots, A_{s-m}$ are mutually perpendicular, and such that W consists of all vectors of H perpendicular to $A_1, \cdots, A_{s-m}$. Let E be the set of vectors X such that

$$X \cdot B = 1 \quad \text{and} \quad X \cdot B' = -1.$$

Then $D(1)$ is contained in E, and is obtained from E by imposing additional inequalities, namely $x_\nu \geqq 0$ for $\nu \in T$ and $x_\mu \leqq 0$ for $\mu \notin T$. The intersection $E \cap W$ is the set of vectors X such that:

$$X \cdot B = 1, \ X \cdot B' = -1, \ X \cdot A_1 = 0, \cdots, X \cdot A_{s-m} = 0.$$

We must now consider two cases.

Case 1. The vectors B, B', $A_1, \cdots, A_{s-m}$ are linearly independent. Then E is the translation of an $(m-1)$-dimensional space, and applying the additional inequalities to E yields an $(m-1)$-dimensional closed, bounded intersection of planes and half-planes which can be parametrized by an $(m-1)$-cube. (We leave it to the reader to select some parametrization.)

Case 2. The vectors B, B', $A_1, \cdots, A_{s-m}$ are linearly dependent. Since $A_0, \cdots, A_{s-m}$ are mutually perpendicular, and $A_0 = B + B'$, there exists numbers b, $a_1, \cdots, a_{s-m}$ such that

$$B' = bB + a_1 A_1 + \cdots + a_{s-m} A_{s-m}.$$

Let X be in $E \cap W$. Taking the scalar product with X, we obtain

$$X \cdot B' = -1 = b.$$

Hence $B + B' = A_0$ is a linear combination of $A_1, \cdots, A_{s-m}$, which is impossible. We conclude that in Case 2, the intersection $E \cap W$ is empty, and the intersection of the boundary of $D(1)$ with W corresponding to the piece relative to T is empty.

In both cases, we see that $\partial D(1) \cap W$ admits a parametrization by an $(m-1)$-cube. Thus Proposition 1 is applicable and Proposition 2 is proved.

(*Note.* The analogous counting theorem for elements of K of bounded height is considerably harder to prove, and has been done by S. Schanuel in [4].)

2. Approximation on the multiplicative group. Let G be the multiplicative group. Let Γ be a finitely generated subgroup of G_K, where K is a number field. Let ϕ be a non-constant rational function on G, defined over K, and let r be the maximum of the multiplicities of its zeros on G (so distinct from 0 or ∞). Let m be the rank of Γ.

CONJECTURE 1. *The height of points P in Γ which are bounded away from 0 and ∞ and satisfy the inequality*

$$(1) \qquad |\phi(P)| \ll \frac{1}{h(P)^{rm+\epsilon}} \qquad (\epsilon > 0)$$

is bounded.

We shall transform inequality (1). The function field $K(G)$ is generated by a single function, and ϕ can be viewed as a rational function. A point P is represented by an element of K, say β, and if $\phi(\beta)$ is small in absolute value, then there is some root $\alpha \neq 0$ of the rational function ϕ such that β is close to α. If β is close to α, then its distance from any other zero or pole of ϕ is bounded away from 0 (approximately by the distance of α itself from another zero or pole). The multiplicity of α in a factorization of ϕ is at most r. The worst possible case is that in which this multiplicity is r. In that case, $|\phi(\beta)|$ is approximately equal to $|\alpha - \beta|^r$ (up to a constant factor). Hence our inequality amounts to

$$|\alpha - \beta| \ll \frac{1}{h(\beta)^{m+\epsilon}}.$$

If one replaces $h(\beta)^{m+\epsilon}$ by $e^{h(\beta)\epsilon}$ in the right hand side of our inequality, then the statement is actually a known theorem of Mahler (cf. for instance [3], Corollary of Theorem 1, Chapter VI, § 1).

The above approximation can be transferred to the additive group. Let $\beta_1, \cdots, \beta_m$ be free generators of Γ modulo its torsion group (i.e. the group of roots of unity). Then we can write

319

$$\beta = \beta_1{}^{\nu_1} \cdots \beta_m{}^{\nu_m}\tau$$

for some torsion element τ. Furthermore, $|\alpha - \beta|$ is of the same order of magnitude as $|\log \alpha - \log \beta|$. Since τ ranges over a finite set, in dealing with solutions of the inequality (1), we may assume that we have always the same τ. Let $u_0 = \log(\alpha\tau)$. Taking the corollary of Proposition 2 into account, we may rewrite our inequality in the form

$$\left| u_0 - \nu_1 \log \beta_1 - \cdots - \nu_m \log \beta_m + \nu_{m+1} 2\pi i \right| \ll \frac{1}{\nu^{m+\epsilon}},$$

where the log is one fixed value of the logarithm.

We have therefore transferred our diophantine approximation on the multiplicative group over a number field into an inhomogeneous approximation on the additive group. The period $2\pi i$ of the exponential function contributes one term to the sum on the left, and gives rise to $m + 1$ free choices of the coefficients $\nu_1, \cdots, \nu_{m+1}$. A standard application of Dirichlet's principle shows that m cannot be replaced by a smaller exponent on the right hand side. In fact, given real numbers $\xi_1, \cdots, \xi_m$ and an integer $\nu > 0$, there exist integers $\nu_1, \cdots, \nu_m$ not all 0, such that

$$\left| \nu_1 \xi_1 + \cdots + \nu_m \xi_m \right| \ll \frac{1}{\nu^{m-1}},$$

and $|\nu_j| \leqq \nu$. (This has content of course only when $\xi_1, \cdots, \xi_m$ are linearly independent over the rationals.) Hence the exponent rm in the conjecture cannot be improved upon.

Khintchine's theorem and suitable generalizations by Schmidt ([5] Theorem 2) imply the following: For almost all sets $\xi_1, \cdots, \xi_m, \eta$ of real numbers the solutions of the inequality

$$\left| \eta + \nu_1 \xi_1 + \cdots + \nu_m \xi_m \right| \ll \frac{1}{\nu^{m-1+\epsilon}}$$

(where $\nu = \max |\nu_j|$) are finite in number. Thus our conjecture essentially asserts that the logarithms of algebraic numbers behave like "almost all numbers". It may be viewed as expressing a measure of linear independence for logarithms of algebraic numbers. One conjectures of course that $\log \beta_1, \cdots, \log \beta_m$ are in fact algebraically independent, and one could make an analogous conjecture for their measure of algebraic independence.

Actually, in view of the fact that the period of the exponential is pure imaginary, namely $2\pi i$, the approximation in its additive form separates into two simultaneous approximations, concerning the real part and the pure imaginary. If $u_0, \log\beta_1, \cdots, \log\beta_m$ are all real, then to get a small number on the left hand side of our inequality, we must have $v_{m+1} = 0$. In that extreme case, the exponent m will be wasteful, and should be replaced by $m - 1$.

Gelfond [2] has obtained transcendence measures for quotients of logarithms of algebraic numbers which improve the Mahler result in that case, for instance of the type

$$\mid v_1 \log \alpha + v_2 \log \beta \mid \gg \frac{1}{v^{(\log v)^{2+\epsilon}}} .$$

for algebraic α, β multiplicatively independent. This is still far short of what is to be expected, but Gelfond's clever method might be a step in the right direction. In any case, it is a first indication that the Mahler result is a priori too coarse.

I have had some preliminary computations comparing the absolute values of expressions of type

$$\mid v_1 \log 2 + v_2 \log 3 + v_3 \log 5 \mid$$

with $1/v^{2+\epsilon}$ (letting $v = \max \mid v_j \mid$), and $\epsilon = 1/4$. They were carried out only for $v \leq 50$. They seem to indicate a tendency for the solutions to thin out, but one should carry them much further to get a significant computational support for the conjecture that there is only a finite number.

Actually, the computations suggest one other possibility. For each v, let $\psi(v)$ be the number of solutions of the inequality

$$\mid v_1 \log \beta_1 + \cdots + v_n \log \beta_n \mid \leq \frac{1}{v^{n-1}}, \qquad\qquad \mid v_i \mid \leq v$$

where $\beta_1, \cdots, \beta_n$ are, say, real algebraic numbers. Then in the above case, except for $v = 1$, the values of $\psi(v)$ were 1, 2, or 3. Thus there is a vague possibility that $\psi(v)$ might be bounded. It would also be interesting to push the computations quite far, to see how $\psi(v)$ behaves asymptotically. Especially for this function, v ranging up to 50 cannot be regarded as really significant.

3. Heights on elliptic curves. Aside from the additive and multiplicative groups, the only other group variety of dimension 1 is an elliptic curve. We shall be interested to the analogues of Proposition 3 and Conjecture 1 for such curves. In this section we deal with the analogue of Proposition 3, and I shall reproduce results of Tate (unpublished at the moment, as usual). It will in fact be just as easy to work on abelian varieties.

Two functions on a set will be called *equivalent* if their difference is bounded in absolute value.

THEOREM 1. (Tate) *Let A be an abelian variety in projective space, defined over the number field K. Then there exists a unique positive definite quadratic form q and a linear form l on A_K such that h is equivalent to $q + l$, as functions on A_K.*

Proof. We consider the function f on $A \times A$ given by $f(P, Q) = h(P + Q) - h(P) - h(Q)$. We contend that as a function of each variable, f is essentially linear, i.e. for all P_1, P_2, Q the expression

$$| f(P_1 + P_2, Q) - f(P_1, Q) - f(P_2 Q) |$$

is bounded. This expression is equal to

$$| h(P_1 + P_2 + Q) - h(P_1 + P_2) - h(P_1 + Q)$$
$$- h(P_2 + Q) + h(P_1) + h(P_2) + h(Q) |.$$

Let $\phi_{123}: A \times A \times A \to A$ be the sum. Let ϕ_{12} be the composite map

$$A \times A \times A \xrightarrow{\ \pi_{12}\ } A \times A \xrightarrow{\ s_2\ } A$$

where π_{12} is the projection on the first two factors and s_2 is the sum, and define ϕ_{13}, ϕ_{23} similarly. Let ϕ_i $(i = 1, 2, 3)$ be the projection on the i-th factor. Let H be a hyperplane section giving the projective embedding of A and the height. Then by the theorem of the square,

$$\phi_{123}^{-1}(H) - \phi_{12}^{-1}(H) - \phi_{13}^{-1}(H) - \phi_{23}^{-1}(H)$$
$$+ \phi_1^{-1}(H) + \phi_2^{-1}(H) + \phi_3^{-1}(H)$$

is linearly equivalent to 0. It is a divisor on $A \times A \times A$, and the height associated with this divisor is therefore equivalent to 0, by one of the main properties of heights (cf. [3], Property 3, Chapter IV, § 2). This height is precisely the function expressing the linearity of f in the $_1$rst variable.

We now define

$$b(P, Q) = \lim_{n \to \infty} \frac{1}{2^{2n}} f(2^n P, 2^n Q).$$

It is a trivial exercise to verify that $b(P, Q)$ is actually bilinear, and is equivalent to $f(P, Q)$.

We let $q(P) = \frac{1}{2} b(P, P)$. It is a trivial exercise to verify that $h - q$ is equivalent to a unique linear functional l. Thus it follows that h is equivalent to $q + l$. Since the height is always ≥ 0 and there is only a finite number of points with bounded height, it follows that q is positive definite. The uniqueness statements are trivially proved.

COROLLARY 1. (Tate) *If the hyperplane section H of A giving the projective embedding satisfies $H = H^{-}$ (transform of H under the automorphism $P \rightsquigarrow -P$ of A), then the linear function l is equal to 0.*

Proof. Easy exercise, using the theorem of the square.

COROLLARY 2. *Let Γ be a finitely generated subgroup of A_K. Let $P_1, \cdots, P_m$ be free generators of Γ modulo the torsion group, and for each P in Γ, write*

$$P = v_1 P_1 + \cdots + v_m P_m + Q$$

where Q is an element of the torsion group. Let $v = v(P) = \max |v_j|$. As function on Γ, we have $h \gg \ll v^2$.

Proof. Obvious.

COROLLARY 3 (Néron). *Let Γ be as in Corollary 2. The number of points P in Γ such that $q(P) \leq t$ is equal to*

$$ct^{m/2} + O(t^{(m-1)/2}) \qquad\qquad \textit{for } t \to \infty$$

and some constant $c > 0$.

Proof. In the case of a positive definite quadratic form q on Γ, we can view q as defined on the tensor product of Γ with the real numbers, and the condition of parametrizability stated in Proposition 2 is satisfied in the present situation, because the unit sphere relative to the quadratic form is a compact submanifold of dimension $m - 1$ in m-space.

4. Approximation on elliptic curves. Let A be an abelian variety of dimension 1 defined over the number field K. We can represent A (or rather its complex points) as a quotient of the complex 1-space, say by means of a $\wp$-function associated with A, with g_2, g_3 in K. If u is a complex number, we denote by $\exp(u)$ the point whose projective coordinates are $(1, \wp(u), \wp'(u))$. Conversely, if P is a point on A, we let $\log P$ be some fixed u such that $\exp(u) = P$ (we choose u in a fixed, bounded fundamental domain).

Let ϕ be a non-constant rational function on A, defined over the number field K and let r be the maximum of the multiplicities of its zeros. Le Γ be a finitely generated subgroup of A_K, of rank m.

5

CONJECTURE 2. *The height of points P in Γ which satisfy the inequality*

$$|\phi(P)| \ll \frac{1}{h(P)^{\frac{1}{2}r(m+1)+\epsilon}}$$

is bounded.

If one replaces the right hand side of the inequality by $e^{h(P)\epsilon}$, then it is known that there is only a finite number of P satisfying the corresponding inequality. This is essentially the formalization of one step in the Siegel argument that an elliptic curve has only a finite number of integral points, taking Roth's Theorem into account (cf. Theorem 1 of [3], Chapter VII, § 1).

The rational function ϕ may be viewed as a doubly periodic elliptic function, so that we have $\phi(P) = \Phi(u)$, for u complex. Since the complex points of A form a compact set, any infinite sequence of points P whose height tends to infinity and satisfies our inequality must have a point of accumulation P which will be a zero of ϕ. The worst case from our point of view is when Φ has a zero of order r at P_0. In that case, Φ has a power series expansion near $u_0 = \log P_0$ given by

$$\Phi(u) = c_r(u - u_0)^r + \cdots$$

for some constant $c_r \neq 0$.

Let $P_1, \cdots, P_m$ be free generators of A_K modulo the torsion group. Let $u_j = \log P_j$. Let ω_1, ω_2 be two fundamental periods. Let

$$P = v_1 P_1 + \cdots + v_m P_m + Q$$

where Q is in the torsion group. Without loss of generality, we may assume that we always have the same point Q, for P satisfying our inequality. Then P is close to P_0, and there exist integers v_{m+1}, v_{m+2} such that

$$v_1 u_1 + \cdots + v_m u_m + v_{m+1}\omega_1 + v_{m+2}\omega_2$$

is close to $u_0 = \log(P_0 - Q)$. Hence our inequality amounts to

$$|-u_0 + v_1 u_1 + \cdots + v_m u_m + v_{m+1}\omega_1 + v_{m+2}\omega_2| \ll \frac{1}{h(P)^{\frac{1}{2}(m+1)+\epsilon}}.$$

This is an inhomogeneous approximation problem near u_0. By Tate's theorem, h and v^2 are of the same order of magnitude (letting v as usual be $\max|v_j|$). Hence our inequality may be rewritten as

$$|-u_0 + v_1 u_1 + \cdots + v_m u_m + v_{m+1}\omega_1 + v_{m+2}\omega_2| \ll \frac{1}{v^{m+1+\epsilon}}.$$

Thus our conjecture asserts that from the point of view of diophantine approximations, the logs of algebraic points on our elliptice curve behave again like " almost all numbers ".

Actually, the two periods ω_1 and ω_2 are linearly independent over the reals. This would seem to indicate that the exponent $m + 1$ should be replaced by m, just as for the multiplicative group. This also raises the following question. Given a non-zero period ω, are there infinitely many real values of t such that $\wp(t\omega)$ are algebraic and linearly independent over the integers? The extent to which m can be further lowered depends on the existence of such values of t.

REFERENCES.

[1] H. Davenport, " On a principle of Lipschitz ", *Journal of the London Mathematical Society*, vol. 26 (1951), pp. 179-183.

[2] A. O. Gelfond, *Transcendental and algebraic numbers*, translated by Leo F Boron, Dover, 1960.

[3] S. Lang, *Diophantine Geometry*, Interscience, New York, 1962.

[4] S. Schanuel, " Heights in number fields ", *to appear*.

[5] W. Schmidt, "A metrical theorem in diophantine approximation ", *Canadian Journal of Mathematics*, vol. 22 (1960), pp. 619-631.

Bull. Soc. math. France,
93, 1965, p. 177 à 192.

REPORT ON DIOPHANTINE APPROXIMATIONS (*);

BY

Serge LANG.

The theory of transcendental numbers and diophantine approximations has only few results, most of which now appear isolated. It is difficult, at the present stage of development, to extract from the literature more than what seems a random collection of statements, and this causes a vicious circle : On the one hand, technical difficulties make it difficult to enter the subject, since some definite ultimate goal seems to be lacking. On the other hand, because there are few results, there is not too much evidence to make sweeping conjectures, which would enhance the attractiveness of the subject.

With these limitations in mind, I have nevertheless attempted to break the vicious circle by imagining what would be an optimal situation, and perhaps recklessly to give a coherent account of what the theory might turn out to be. I especially hope thereby to interest algebraic geometers in the theory.

1. Measure theoretic results.

Let α be a real number. We denote by $\|\alpha\|$ its distance from the origin on $\mathbf{R}/\mathbf{Z}$ (reals mod 1), i. e. its distance from the closest integer. If q is an integer, then

$$\|q\alpha\| = |q\alpha - p|$$

for some integer p, uniquely determined if this absolute value is small enough.

We begin by quoting an old result of Dirichlet.

Let α be a real number and N a positive integer. There exists an integer q, $0 < q \leq N$ such that $\|q\alpha\| < 1/q$.

(*) This work was partially supported by the National Science Foundation under Grant NSF GP-1904.

To prove this, cut up the interval $[0, 1]$ into N equal segments of length $1/N$, and consider the $N + 1$ numbers

$$0\alpha, \quad 1\alpha, \quad 2\alpha, \quad \ldots, \quad N\alpha.$$

Two of them must lie in the same segment (mod 1), say $r\alpha$ and $s\alpha$ with $r < s$. We put $q = s - r$ to get

$$\| q\alpha \| < \frac{1}{N} \leq \frac{1}{q}.$$

We note that the inequality $|q\alpha - p| < 1/q$ is equivalent with

$$\left| \alpha - \frac{p}{q} \right| < \frac{1}{q^2}.$$

We are interested in estimates of $\| q\alpha \|$ determined somewhat more generally as follows. Let ψ be a positive function of a real variable, monotone decreasing to 0. A theorem of KHINČIN asserts :

Assume that

$$\sum_{q=1}^{\infty} \psi(q)$$

converges. Then for almost all $\alpha \in \mathbf{R}$ (i. e. outside a set of measure 0), there is only a finite number of solutions to the inequality

$$\| q\alpha \| < \psi(q).$$

Here again, the proof is quite simple. Given $\varepsilon > 0$, select q_0 such that

$$\sum_{q \geq q_0}^{\infty} \psi(q) < \varepsilon.$$

We may restrict our attention to those numbers α lying in the interval $[0, 1]$. Consider those for which the inequality has infinitely many solutions. For each $q \geq q_0$, consider the intervals of radius $\psi(q)/q$, surrounding the rational numbers

$$\frac{0}{q}, \quad \frac{1}{q}, \quad \ldots, \quad \frac{q-1}{q}.$$

Every one of our α will lie in one of these intervals because for such α we have

$$\left| \alpha - \frac{p}{q} \right| < \frac{\psi(q)}{q}.$$

The measure of the union of these intervals is bounded by the sum

$$\sum_{q \geq q_0} q \, \frac{2\psi(q)}{q} < 2\varepsilon$$

as was to be shown.

Suppose thirdly that the sum $\sum \psi(q)$ diverges. Then we have the following recent theorem of Schmidt [9] :

For almost all α the number of solutions $\lambda(B)$ for the inequality $\|q\alpha\| < \psi(q)$ with $0 < q \leq B$ is given asymptotically by

$$\lambda(B) \sim c_1 \sum_{q=1}^{B} \psi(q).$$

for some constant $c_1 > 0$. (One can in fact take $c_1 = 2$ for almost all numbers.)

The proof is too long to be given here.

The main object of the theory of diophantine approximations is to determine a wide class of numbers, which may be called classical numbers, which will behave like almost all numbers. We shall now proceed to discuss such a class.

2. Classical numbers.

The classical numbers are essentially those which appear as values of classical functions or mappings, (e. g. exponential, automorphic, zeta, spherical, solutions of classical differential equations, etc.) with algebraic arguments, and similarly for their inverse functions, under suitable normalizations. The above mentioned functions are to be taken in an extended sense, e. g. Γ-functions and Γ'/Γ are to be viewed as functions of zeta type. Abelian functions are to be viewed as functions of exponential type. One must also deal with iterations of these functions, say to deal with numbers like e^e and α^β.

In a classical situation, one meets an open subset U of some complex space and a map $f : U \to V$ of U into an algebraic variety, defined over a number field. We shall give examples below, with suitable normalizations, and recall as we go along certain classical transcendence results.

EXAMPLES :

(i) Let $f : \mathbf{C} \to \mathbf{C}^*$ be given by $f(t) = e^t$. Here $\mathbf{C}^*$ is the multiplicative group, and a classical theorem of Hermite-Lindemann asserts that if α is algebraic $\neq 0$ then e^α is transcendental. The inverse function of f is the ordinary log.

(ii) Let $f: \mathbf{C}^n \to A_\mathbf{C}$ be the universal covering map of an abelian variety A defined over a number field. (One may realize f explicitly by theta functions.) We normalize f so that the origin in $\mathbf{C}^n$ maps on the origin on A, and so that the derivative $f'(o)$ at the origin is algebraic. It is known that if P is algebraic $\neq o$ in $\mathbf{C}^n$, then $f(P)$ is a transcendental point on A [3]. The inverse map of f is given by abelian integrals.

(iii) Let V be a non-singular curve of genus $\geqq 2$ defined over a number field, and let U be a disc, centered at the origin in $\mathbf{C}$. Let $f: U \to V_\mathbf{C}$ be the universal covering map, normalized so that the origin goes to an algebraic point, and again $f'(o)$ is algebraic. One conjectures that if P is an algebraic point of the disc, $\neq o$, then $f(P)$ is transcendental. As a side question (already related to moduli), one can ask if the radius of the disc is transcendental, and whether the radius can be defined as some real function on a suitable moduli space.

(iv) Let $f: S^n \to V$ be the moduli mapping from the Siegel upper half space to the variety of moduli. The fact that f represents the moduli normalizes it automatically. In dimension 1, one represents f explicitly by the modular function j. We recall that if τ is algebraic and the abelian variety associated with $j(\tau)$ has no complex multiplication, then $f(\tau) = j(\tau)$ is transcendental (SCHNEIDER).

(v) Let J_λ be the Bessel function with an algebraic parameter λ, and take $f = (J_\lambda, J_\lambda')$, so that $f: \mathbf{C} \to \mathbf{C}^2$ is the integral curve of a differential equation, normalized so that the initial conditions are algebraic and the coefficients of the equation are rational functions with algebraic numbers as coefficients. We recall that if λ is such that J_λ, J_λ' are algebraically independent (as functions over the field of rational functions), then $J_\lambda(x)$ and $J_\lambda'(x)$ are algebraically independent for any algebraic $x \neq o$ (SIEGEL). It is natural here to view $\mathbf{C}^2$ as an algebraic variety with an algebraic differential equation over it.

Some of the above examples satisfy a differential equation and some do not, but all would be accepted as " classical ".

We shall now show how, given a classical mapping, we can generate a field of numbers with it.

Let f be a mapping as above and let Ω be the smallest field generated from the rational numbers by performing the following operations inductively, and iterating them :

Taking algebraic closure.

Adjoining values of f and its inverse function with the argument in the field obtained inductively after a finite number of steps.

We are concerned with the numbers of Ω from the point of view of diophantine approximations. We note that Ω is denumerable as would be any field generated in a manner similar to the one we have described. We now expect (real) irrational numbers in Ω to behave like almost all numbers, with respect both to the second and third results recalled in paragraph 1. For this, one must make certain obvious restrictions on the function ψ, and we shall discuss these in the next section.

Here we make further remarks concerning known results and evidence in the direction of our expectations.

Liouville numbers do not behave like almost all numbers, and there are non-countably many of these.

A set of measure 1 on the unit circle (normalized to have length 1) is such that its sum with itself is the whole circle. For a moment, let us define a number to be *ordinary* if there is only a finite number of solutions to the standard inequality

$$\| q\alpha \| < \frac{1}{q^{1+\varepsilon}}$$

for every $\varepsilon < 0$. Then from the above remark, we see that ordinary numbers cannot form a field. Hence the fact that we deal with a field is essentially dependent on the manner of generating our numbers. We note that it is not known whether, if all elements of a field are ordinary, then the elements of its algebraic closure are ordinary. For the rational numbers, this is Roth's theorem. It is likely, however, that to prove such a result in general, one has to make a stronger assumption on the field elements, i. e. one has to load the hypothesis.

Since we wish to iterate our functions (or mappings), there should be some kind of inductive results, which, assuming that certain approximation properties are satisfied by all numbers of a certain field, similar properties are satisfied by those obtained in one of the two ways described above, starting with the given field. In particular, applying one of our two operations to the numbers of our given field, the conjecture asserts in particular that we cannot obtain a Liouville number as a value. To obtain a inductive result, it will be necessary to deal with a loaded inductive assumption, involving what is commonly known as a measure of lineari ndependence. We shall discuss this below. As special cases, one could consider the fields $\mathbf{Q}(e)$ or $\mathbf{Q}(\pi)$.

We note that it is hard to predict if one must make some restriction in the generation of a field Ω with respect to maxing mappings of various " types ", for instance, applying an exponential function to the value of a zeta function, or considering a number like $e + \zeta(3)$. For definiteness, one may therefore take a special case, letting Ω be the field generated by values of e^t and $\log t$, and their iteration, starting with the algebraic numbers. In some sense, this should also be the simplest case to treat.

3. The convergent case.

Let us now consider more closely a number α in our field Ω.

Generally speaking, it is a problem to determine a class of functions ψ (convex, monotone decreasing to o with convergent sum) such that the inequality

$$o < \| q\alpha \| < \psi(q)$$

has only a finite number of solutions. Very few results are known. We shall make a list of them.

To begin with, we have Roth's theorem, taking α algebraic and $\psi(q) = 1/q^{1+\varepsilon}$ for any $\varepsilon > o$. Taking such a ψ is in a sense the coarsest way of making the series converge.

It is an old theorem of Popken that e satisfies a similar result, and even a stronger one, namely there exists an absolute constant c' such that for all sufficiently large integers q,

$$\| qe \| > q^{-1-\frac{c'}{\log\log q}}.$$

In fact, Popken's theorem asserts that if $q_1, \ldots, q_m$ are integers and $q = \max |q_i|$ is sufficiently large, then

$$\| q_1 e + \ldots + q_m e^m \| > q^{-m-\frac{c'}{\log\log q}}$$

(c' then depends on m).

This result was improved by MAHLER. *I have analysed Mahler's proof* (reproduced in SCHNEIDER [11]), *and observed that the proof applies to the numbers e^α with α rational $\neq o$.* When α is irrational algebraic, one obtains a result depending on the degree of α, and this leads one to hope that a mixture of the Thue-Siegel-Roth techniques with those techniques used in the theory of transcendental numbers might lead one to stronger results. In other words, one must consider an approximating sequence, and argue combinatorially on this sequence.

In this connection, SIEGEL himself observes that for the *Bessel function, the inequality*

$$\| q_1 J_0(\alpha) + q_2 J_0'(\alpha) \| < \frac{1}{q^{2+\varepsilon}}$$

has only a finite number of solutions whenever α is rational, $\neq o$, and similarly for a polynomial in $J_0(\alpha)$, $J_0'(\alpha)$ [12]. Here again, the problem is open for algebraic α, or when one deals with J_λ, and λ is algebraic irrational. SIEGEL also mentions the analogous result for e, and linear combinations of powers of e. *We note in passing that the finiteness property for the inequality*

$$\| q J_0(\alpha) \| < \frac{1}{q^{1+\varepsilon}}$$

has not yet been proved. Siegel's method does not give it as it stands.

For numbers of type α^β (with α, β algebraic, $\alpha \neq 0$, 1 and β irrational), or for values of p-functions with algebraic g_2, g_3, or their inverse functions with algebraic arguments, one has only a much weaker result, of the following type : If ξ is a number of the kind just mentioned, then GELFOND and FELDMAN [2] have shown that

$$\| q\xi \| > \frac{c_1}{q^{(\log q)^2+\varepsilon}}.$$

It seems clear to me, however, that all the above results point to the same direction, i. e. the finiteness statement as for almost all numbers with the function $\psi(q) = 1/q^{1+\varepsilon}$.

For more general functions ψ, the situation is more complicated. First, let ξ be a number with the following property :

$(\bigstar)$ *There exists a constant $c > 0$ such that for all integers $q > 0$ we have*

$$\| q\xi \| > \frac{c}{q}.$$

Then for any monotone decreasing ψ such that $\sum \psi(q)$ converges, the inequality

$$\| q\xi \| < \psi(q)$$

has only a finite number of solutions.

Proof. — Suppose that ψ is a monotone decreasing function such that $\psi(q) > 1/q$ for infinitely many q, say $q_1 < q_2 < \ldots$. Then we contend that $\sum \psi(q)$ diverges. Indeed, let us make ψ smaller by supposing that

$$\psi(q) = \frac{1}{q_n}$$

for $q_{n-1} < q \leq q_n$. Then

$$\sum \psi(q) \geq \frac{1}{q_1} + (q_2 - q_1)\frac{1}{q_2} + (q_3 - q_2)\frac{1}{q_3} + \ldots.$$

Take $n = n_1$ large. The first n terms of this series have a lower bound given by

$$(q_2 - q_1)\frac{1}{q_n} + \ldots + (q_n - q_{n-1})\frac{1}{q_n} = \frac{q_n - q_1}{q_n}.$$

Thus for n large, we get a contribution $> 1/2$ to our sum. We repeat this procedure with a number n_2 which will give a contribution greater than

$$\frac{q_{n_2} - q_{n_1}}{q_{n_2}} > 1/2$$

to our sum, and so on with n_3, ... In this manner, we see that the sum diverges.

S. SCHANUEL *has pointed out to me that the converse is also true, i. e. if α is a number such that for every smooth convex monotone decreasing function ψ with convergent sum the inequality*

$$\| q\alpha \| < \psi(q)$$

has only a finite number of solutions, then α satisfies $(\star)$.

To prove this, SCHANUEL argues as follows. Suppose α does not satisfy $(\star)$. Then we can find a sequence of integers q_i, with $1 < q_1 < q_2 < \ldots$ such that $\| q_i \alpha \| < 1/2^i q_i$. Let

$$\psi(t) = \sum \frac{e}{2^i q_i} e^{-\frac{t}{q_i}}.$$

Then $\psi(q_j) > 1/2^j q_j$, and the sum (or integral) for ψ converges. This achieves what we wanted. (Also, Schanuel's function is as good as possible from the point of view of convexity.)

It is a problem to determine specific numbers which have, and have not, property $(\star)$. It is trivial to prove that quadratic numbers have this property, and hence behave maximally well in the convergent case. It is unknown whether any other algebraic numbers (irrational) have the property.

One is thus faced with a problem in two directions concerning the finiteness statement : For which numbers does it apply, and for what class of functions ψ.

We note that the set of numbers satisfying $(\star)$ has measure 0, and so the description of functions ψ for which the finiteness statement holds appears as subtle. If the general philosophy that classical numbers behave like almost all numbers holds in the present instance, then one would expect only the real quadratic numbers among them to satisfy $(\star)$. Admittedly, the range in which one tries to guess the answer here is delicate.

It is already clear from the ideas of the Thue-Siegel-Roth proof that in a certain sense, the difficulty of extending the proof to the usual transcendental numbers does not lie so much in the transcendentality as in an intrinsic weakness of the structure of the proofs, even for algebraic numbers. That such a weakness exists is clear, since the proof is unable to decide whether algebraic numbers satisfy property $(\star)$, or if they satisfy the finiteness property with respect to a function like $\psi(q) = 1/q(\log q)^{1+\varepsilon}$.

It is therefore necessary to start investigations of the theory from the beginning, and to have essentially completely new ideas which would exhibit the finiteness by means of a more canonical combinatorial treatment of the approximating sequence.

4. Asymptotic approximations.

Suppose that ψ is smooth, convex, strictly decreasing to o (for sufficiently large numbers), and that its integral to infinity diverges. Modulo an additional restriction on ψ which will be discussed below, one expects that *for all classical numbers α (real and irrational), the number $\lambda(B)$ of solutions of the inequality*

$$\| q\alpha \| < \psi(q), \qquad o < q \leq B$$

is given asymptotically by

$$\lambda(B) \sim c_1 \int_1^B \psi(t)\, dt,$$

with some constant c_1 (possibly depending on α, ψ).

That some restriction is needed on ψ is clear from the possibility of property $(\star)$. We are thus led to introduce the function

$$\omega(t) = t\,\psi(t).$$

If $(\star)$ is satisfied, one must then assume that ω is not decreasing to o. Furthermore, it is entirely reasonable to require that ω is not oscillating, and is itself convex. It is then naturel to split the theory into two cases, according as ω is constant, or strictly increasing to infinity (for all sufficiently large t). When we take ω constant, we must assume that there actually exist infinitely many solutions for the inequality

$$\| q\alpha \| < \frac{\omega(t)}{t},$$

so that for definiteness, one may take $\omega(t) = c \geq 1$. There are various ways of preventing ω from oscillating.

Actually, it is not clear if there is an *a priori* characterization of those functions ψ for which everything works out as expected. However, one expects all reasonable functions (built up out of exponentials, logs, essentially the elementary functions in a finite number of steps) to be acceptable, provided there is no oscillation. The point is that any idea for a proof will carry with it an explicit error term which will determine automatically the range of validity of the asymptotic estimate.

This is in fact precisely the situation which occurs in [8] where the expected theorem is proved for quadratic irrationalities (real), with a definite error term which shows the estimate to be good when ω is a power of the log, or iterated logs (i. e. ω can grow quite slowly). However, the estimate is not valid for all ω.

Even for quadratic irrationalities, the problem is open when $\omega(t) = t^{\delta}$ with $0 < \delta < 1$. When $\omega(t) = ct$ with $0 < c < 1$, then the situation is the classical one of equidistribution, and the answer is known. There exist several papers dealing with it by HECKE, BEHNKE, HARDY-LITTLEWOOD, etc. (cf. *Abhandl. math. Sem. Hamburg. Univ.*, t. 1, 1922 et t. 2, 1923).

Some machine computations for a few classical numbers (e, π, $e + \pi$, $\log 2$, $\log 3$, γ) tend to support the present conjecture [1]. In any case, the classical transcendental numbers seem to be no different from the present point of view than the algebraic numbers. It should also be pointed out that no paper had considered the asymptotic problem for specific numbers previous to [8].

5. Generalizations.

Let $\alpha_1, \ldots, \alpha_m$ be (real) numbers in our field Ω. Then one may study the inequality

$$\| q_1 \alpha_1 + \ldots + q_m \alpha_m \| < \frac{1}{q^m}$$

or

$$\| q_1 \alpha_1 + \ldots + q_m \alpha_m \| < \psi(q),$$

with a suitable function ψ, subject to similar restrictions as before. Here we put $q = \max |q_i|$, and the exponent m generalizes the exponent 1 considered previously. Similarly, the convergence condition now applies to the sum of $\psi(q)$ taken for all m-tuples $(q_1, \ldots, q_m)$, with $\max |q_i| \leq B$.

If the sum converges, then one expects a finite number of solutions for the inequality

$$0 < \| q_1 \alpha_1 + \ldots + q_m \alpha_m \| < \psi(q).$$

If the sum diverges, and if $1, \alpha_1, \ldots, \alpha_m$ are linearly independent over the rationals, or equivalently, linearly independent mod $\mathbf{Z}$ on the circle, then one expects the usual asymptotic estimate for the number of solutions of the inequality

$$\| q_1 \alpha_1 + \ldots + q_m \alpha_m \| < \psi(q), \qquad 0 < q \leq B.$$

(Cf. SCHMIDT's paper again for the corresponding theorem holding almost everywhere.)

In the present context we therefore see the theory of transcendental numbers as determining which classical numbers are linearly independent or alge-

braically independent, and the theory of diophantine approximations then gives quantitative results concerning such numbers. Quantitative results are known as measures of linear independence or measures of transcendence. If α is a classical transcendental number, then one may put

$$\alpha_i = \alpha^i,$$

and a measure of linear independence for $\alpha, \ldots, \alpha^m$ becomes a measure of transcendence for α.

One can also work on the torus. Let L be a lattice in $\mathbf{R}^n$, having a basis whose elements consist of vectors with coordinates in our field Ω. For any vector X in $\mathbf{R}^n$, define $\|X\|$ to be its distance from the origin on the torus $\mathbf{R}^n/L$. One then considers the inequality

$$0 < \|q_1 X_1 + \ldots + q_m X_m\| < \psi_{m,n}(q)$$

with a suitable function $\psi_{m,n}$. This is a problem in simultaneous approximations of vectors.

As stated above, the problem is on $\mathbf{R}^n$. However, it has applications to elliptic curves and abelian varieties [6]. For instance, if A is an abelian variety defined over a number field K, and $f : \mathbf{C}^n \to A_{\mathbf{C}}$ is as in Example (ii) of paragraph 2 the representation of $A_{\mathbf{C}}$ as a quotient of the universal covering space, then we may identify $\mathbf{C}^n$ with $\mathbf{R}^{2n}$. If $P = f(X)$, we also write $X = \log P$. Taking $P_1, \ldots, P_m$ points of A_K (i. e. algebraic points) linearly independent over $\mathbf{Z}$, we see that the approximation question concerning

$$\|q_1 P_1 + \ldots + q_m P_m\| < \psi_{m,n}(q)$$

becomes equivalent with an approximation as described above, with vectors X_i having transcendental coordinates. A similar situation exists with respect to the multiplicative group [6], and one obtains in this way generalizations of statements of SIEGEL and MAHLER in diophantine geometry, concerning integral points (*cf.* for instance [5], Corollary of Theorem 1, Chapter VI, § 1 and Theorem 1 of Chapter VII, § 1). For instance, the theorem in diophantine approximations needed to prove Siegel's finiteness of integral points on curves of genus ≥ 1 over number fields, is the following :

Let V be a complete non-singular curve of genus ≥ 1 defined over a number field K. Let g be a non-constant rational function in $K(V)$. Let I_K be the ring of algebraic integers of K. Let $c > 0$. Then the set of points P in V_K which are not poles of g and such that

$$|g(P)| \leq \frac{1}{H(P)^c}$$

is finite.

For elliptic curves, the inhomogeneous analogues of conjectures expressed in this paper would imply the finiteness only under the assumption that

$$|g(P)| \leq \frac{1}{(\log H(P))^{\frac{rm}{2}+\varepsilon}}$$

where r is the maximum of the multiplicities of the zeros of g, m is the rank of A_K, and H is the height in a fixed projective embedding.

I cannot resist mentioning other applications to diophantine geometry. In [5], I have proved the following statement : Let Γ_0 be a finitely generated multiplicative group of complex numbers, say. Let $f(X, Y) = 0$ be the equation of a curve (irreducible) and assume that there exist infinitely many points (x, y) such that $x, y \in \Gamma_0$ and $f(x, y) = 0$. Then f has a " multiplicative " structure, i. e. there exist integers n, $m \neq 0$ and non-zero constants a, b such that we have identically $f(at^n, bt^m) = 0$. It follows then easily that f consists of at most two monomials.

As a special case, it follows that *if g is a rational function with complex coefficients, and if there exist infinitely many elements $x \in \Gamma_0$ such that $g(x) \in \Gamma_0$ then g is of type aX^n for some integer n and some constant a.*

The proof uses the ideas of Siegel's theorem, combined with an additional combinatorial argument on coverings. Thus in effect, the proof depends on the above-mentioned result in diophantine approximations.

Let Γ be the multiplicative group of complex numbers z such that some integral power z^m lies in Γ_0 (some $m \neq 0$). *I would conjecture that the same results as above hold when Γ_0 is replaced by Γ.* As a special case, one has the following very elementary statement :

Let g be a rational function with complex coefficients, and assume that there exist infinitely many roots of unity ζ such that $g(\zeta)$ is a root of unity. Then g is of type $g(X) = aX^n$ for some integer n, and some root of unity a.

A proof for this last statement was shown to me by Ihara, Serre and Tate. We can reformulate and generalize the above statements as follows :

Let A be a group variety in characteristic 0 which is either an abelian variety or a product of multiplicative groups, or a group extension of an abelian variety by such a product. Let Γ_0 be a finitely generated subgroup, and let Γ be the subgroup of points $x \in A$ such that there exists an integer $n \neq 0$ such that $nx \in \Gamma_0$. Let V be an irreducible algebraic curve in A, and assume that the intersection of V with Γ is infinite. Then V is the translation of a group subvariety.

The above formulation implies the Mordell conjecture, as pointed out in [5]. It also implies a conjecture of Mumford, who, a few years ago, asked me the following question : If a curve embedded in its Jacobian contains infinitely many points of finite order, is the curve of genus 1 ? (The question was also raised independently by MANIN in his work on Picard-Fuchs equations. MANIN pointed out that although it takes infinitely many algebraic equations to define the points of finite order on an abelian variety, it takes only a finite number of differential equations.) The question concerning our polynomial $f(X, Y)$ and roots of unity is the analogous question for the multiplicative group. As an example, one always has the straight line $X + Y = 1$. The theorem proved in [5] shows that this line contains only a finite number of points whose coordinates lie in a finitely generated multiplicative group.

Other generalizations are possible. One can consider approximations $|q\alpha - p|$ where q, p lie in a number field K, and similarly for

$$q_0 \alpha_0 + \ldots + q_m \alpha_m.$$

One lets q be the height of the point $(q_0, \ldots, q_m)$ in projective space. When dealing with points on abelian varieties, one could let the q_i range over the ring of endomorphisms.

Finally, one can ask for approximations questions on group varieties and homogeneous spaces or transformation spaces, defined over number fields. For instance let G be a group variety and V a homogeneous space defined over the number field K. Let x_0, y_0 be points of V with coordinates in our field Ω. One can then ask for those points $g \in G$, rational over K, such that dist $(gx_0, y_0) < \psi(g)$, where dist is the distance in a suitably normalized metric on V, and ψ is a function of the height of g. When G operates on itself by translation, and is commutative, we are led to considering an inequality

$$\text{dist}(x, x_0) = \| x_0 - x \| < \psi(x)$$

which looks formally like the inequality that is usually written down in the simplest case of approximation on the circle. This gives rise to homogeneous or inhomogeneous approximation problems in globalized setting, on linear groups or abelian varieties.

6. Relation with transcendental numbers.

We shall conclude this report by pointing out a more technical connection between the theory of diophantine approximations, and the theory of transcendental numbers, due to GELFOND, whose result is as follows.

THEOREM. — *Let σ be a strictly monotone increasing real function tending to infinity, and assume that there is a number $a_0 > 1$ such that $\sigma(N + 1) < a_0 \sigma(N)$ for all integers $N > N_0$. Let w be a complex* *number. Assume that for each integer $N > N_0$ there exists a non-zero polynomial F_N with integer coefficients such that*

$$| F_N(w) | < e^{-C \sigma^2 (N)}$$

where $C = 50\, a_0^2$, and

$$\max(\deg F_N, \log | F_N |) \leq \sigma(N),$$

Then w is algebraic.

As usual, $| F_N |$ denotes the maximum of the absolute value of the coefficients. Since in his book [2], GELFOND gives the proof of a weaker result, it is worth while to summarize roughly the argument here. First one proves that if F is a polynomial in one variable with integer coefficients, relatively prime, and σ, C are numbers > 0 such that

$$| F(w) | < e^{-\sigma^2 C}$$

and $\deg F < \sigma$, $\log | F | < \sigma$, then there exists an irreducible factor P of F (with integer coefficients) such that

$$| P(w) | < e^{-\frac{\sigma^2 C}{2 s}}$$

with some integer $s \leq \sigma$, and $\max(\deg P, \log | P |) \leq \sigma/s$.

The proof proceeds first by factoring F into relatively prime polynomials which are powers of irreducible polynomials, and using the estimate given by the resultant of the factors expressed as a determinant. One sees that each factor must have a small absolute value, and then one takes some s-th root with $s \leq \sigma$.

To prove the theorem, one can then assume the coefficients of each F_N to be relatively prime, and $F_N(w) \neq 0$. Given a sufficiently large integer q (say $q > q_0$) we can find an irreducible factor P_q of F_q such that

$$| P_q(w) | < e^{-\frac{\sigma^2 (q) C}{2 s}}$$

with some integer $s \leq \sigma(q)$, and

$$\max(\deg P, \log | P_q |) \leq \frac{1}{s} \sigma(q).$$

Let x_q be the number such that

$$\sigma(x_q) = \max(\deg P_q, \log | P_q |).$$

Then x_q goes to infinity with q, and trivially

$$| P_q(w) | < e^{-\frac{\sigma^2 (x_q) G}{2 s}}$$

Now find an integer N such that

$$\sigma(N-1) < \frac{\sigma(x_q)}{4\,a_0} \leq \sigma(N).$$

Then $\sigma(N) \leq a_0\sigma(N-1) \leq \sigma(x_q)/4$. Take $F = F_N$ so that

$$|F(w)| < e^{-\sigma^2(N)\,C}$$

and $\max(\deg F, \log|F|) \leq \sigma(N)$. Then P_q and F are relatively prime. Otherwise P_q divides F, whence

$$\max(\deg P_q, \log|P^q|) < \sigma(x_q)$$

which is impossible.

The resultant R of P_q and F is not zero and is an integer. Using an easy estimate arising from the expression of the resultant as a determinant, we find

$$|R| \leq [\,|P_q(w)| + |F(w)|\,]\,e^{\varepsilon\,\sigma^2(x_q)}$$

for q large. In the estimate for $F(w)$, we can replace $\sigma(N)$ by $\sigma(x_q)/4\,a_0$. This makes the resultant less than 1, contradiction.

GELFOND, in his book, proves the result only with some function instead of the constant C. It is also easy to see that the theorem applies when one deals with a rational function or an algebraic function instead of a polynomial. This is useful when one deals with a function field other than the rational field.

We note that $\max(\deg F, \log|F|)$ is essentially a height function on polynomials, which measures the speed with which both the degree and the coefficients tend to infinity. The exponent σ^2 is the " correct " one in the optic of results holding almost everywhere. GELFOND applies his theorem to prove that certain numbers are algebraically independent. Indeed, under certain circumstances connected with values of exponential functions, one knows that a number w is transcendental, and one wants to prove that it is algebraically independent of another number y. Assuming the contrary, one can construct a sequence F_N as in the theorem to lead to a contradiction, thereby giving the algebraic independence of w and y.

I have tried to use Gelfond's method to prove that e, π are algebraically independent, but an application of known ideas in the theory leads to a sequence F_N satisfying an inequality weaker than the needed one (i. e. the exponent of σ is not quite 2.) Still, Gelfond's theorem gives a good approach to these questions.

One can conjecture a generalization, using a sequence of polynomials F_N in several variables, and an n-tuple of complex numbers $(w_1, \ldots, w_n)$. In that case, the conclusion should be that $w_1, \ldots, w_n$ are algebraically *dependent, provided that in Gelfond's inequality we take the exponent σ^{n+1} instead of σ^2, and the constant C depends only on σ and n (not on the chosen numbers).*

If one tries to apply the theorem for one variable inductively, one obtains some result, but with an exponent for σ which is much too large to be of interest.

The possible generalization of Gelfond's theorem to several variables gives an interesting direction for the problem of diophantine approximations, when the degree of the polynomial varies, together with its coefficients. In the discussions of preceding sections, we kept the degree constant. The theory of transcendental numbers shows that the more general behaviour also has to be considered.

BIBLIOGRAPHY.

[1] ADAMS (W.) and LANG (S.). — *J. für reine und angewandte Math.* (to appear).

[2] GELFOND (A. O.). — *Transcendental and algebraic numbers.* — New York, Dover Publications, 1960 (Dover Books on advanced Mathematics). [See the bibliography at the end for further references to GELFOND and FELDMAN's papers.]

[3] LANG (S.). — Transcendental points on group varieties, *Topology*, Oxford, t. 1, 1962, p. 313-318.

[4] LANG (S.). — Algebraic values of meromorphic functions, *Topology*, Oxford (to appear).

[5] LANG (S.). — *Diophantine geometry.* — New York, Interscience Publishers, 1962 (Interscience Tracts in pure and applied Mathematics, 11).

[6] LANG (S.). — Diophantine approximations on toruses, *Amer. J. of Math.*, t. 86, 1964, p. 521-533.

[7] LANG (S.). — *Transzendenten Zahlen.* Lecture notes by W. Scharlau. — Bonn, 1964.

[8] LANG (S.). — Asymptotic approximations to quadratic irrationalities, *Amer. J. of Math.*, 1965 (to appear).

[9] SCHMIDT (W. M.). — A metrical theorem in diophantine approximation, *Canadian J. of Math.*, t. 12, 1960, p. 619-631.

[10] SCHMIDT (W. M.). — Metrical theorems on fractional parts of sequences, *Trans. Amer. math. Soc.*, t. 110, 1964, p. 493-518.

[11] SCHNEIDER (T.). — *Einführung in die transzendenten Zahlen.* — Berlin, Springer-Verlag, 1957 (Die Grundlehren der mathematischen Wissenschaften, 81). [For references to Mahler, Siegel, Popken, cf. bibliography at the end of Schneider's book.]

[12] SIEGEL (C. L.). — Über einige Anwendungen diophantischer Approximationen, *Abhandl. Preuss. Akad. Wiss.*, 1929, n° 1, 70 pages. [cf. especially p. 28.]

(Manuscrit reçu le 7 janvier 1965.)

Serge LANG,
Professor,
Department of Mathematics,
Columbia University,
New York 27, N. Y. (États-Unis).

341

Lang, Serge
1965
Annali di Matematica pura ed applicata
(IV), Vol. LXX, pp. 229-234

Division points on curves.

by Serge Lang (a New York)

In memory of Guido Castelnuovo, in the recurrence of the first centenary of his birth.

Summary. - *A curve contained in a product of multiplicative groups passing through the origin, and containing infinitely many torsion points is a subgroup. The analogous statement on abelian varieties is discussed and reduced to an analogue of the irreducibility of the cyclotomic equation.*

Let A be an abelian variety, or a product of multiplicative groups (or a group extension of these), defined over the complex numbers, say. Let Γ_0 be a finitely generated subgroup of A, and let V be a curve (subvariety of dimension 1) in A, passing through the origin, say. When A is an abelian variety, the Mordell conjecture has been expressed as asserting that *if the intersection of V with Γ_0 is infinite, then V is a group subvariety* (cf. [2], [3]). When A is a product of multiplicative groups, then I proved the analogous statement in [2], [3].

A few years ago, Mumford asked me the following question:

If a curve in its Jacobian contains infinitely many points of finite period, is the curve of genus 1? The same question arose in Manin's investigations of the Picard-Fuchs equations [4]. At the time I did not see how to make a conjecture which would include all the above statements, but now it seems to me that one can formulate such a conjecture as follows: *Let Γ be the division group of Γ_0, i.e. the group of points P on A such that there exists some integer $n \geq 1$ (depending on P) such that nP lies in Γ_0. If the intersection of V with Γ is infinite, then V is a group subvariety.* In other words, from a diophantine point of view, when one considered previously a finitely generated group, one may as well consider its division group. Taking Γ_0 to be the unit element of A yields the special case when Γ consists of all points of finite period on A.

When A is a product of multiplicative groups, then the conjecture admits a particularly elementary formulation:

Let Γ_0 be a finitely generated multiplicative group of complex numbers, and let Γ be the group of complex numbers z such that z^n lies in Γ_0 for some n (depending on z).

Let $f(X, Y) = 0$ be a curve in the plane (absolutely irreducible) passing through the multiplicative origin, i.e. $f(1, 1) = 0$. If there exist infinitely many elements $x, y \in \Gamma$ such that $f(x, y) = 0$, then the curve is actually a group, whence f is a polynomial of type $aX^m + bY^n = 0$ or $X^n Y^m + c = 0$.

As a special case, one then has the statement: *Let $g(X)$ be a rational function, and assume that there exist infinitely many elements $x \in \Gamma$ such that $g(x) \in \Gamma$. Then $g(X) = cX^n$ for some constant c and some integer n.*

When we take elements in Γ_0 instead of Γ, then of course this is a special case of the theorem proved in [2]. For instance, the line

$$aX + bY = c$$

with $abc \neq 0$ has only a finite number of points with coordinates in Γ_0.

IHARA, SERRE, and TATE have shown to me how to prove the conjecture in the case Γ consists of roots of unity, and I shall reproduce here IHARA's and TATE's proofs. (SERRE's proof is similar to TATE's.) I shall then describe a geometric interpretation for these proofs, and indicate how they lead to certain «irreducibility» criteria on abelian varieties, related to recent work of SERRE concerning the GALOIS group of division points of elliptic curves over number fields [5].

§ 1. IHARA'S PROOF. – Let us begin with IHARA's proof, given only for a rational function. Let g be a rational function, and assume that there exist infinitely many roots of unity ζ such that $g(\zeta)$ is a root of unity. Then g has coefficients in some cyclotomic field. (Proof: Let F be the field obtained by adjoining al roots of unity to the rationals, let K be the field obtained from F by adjoining the coefficients of g. If $K \neq F$, there exists an isomorphism σ of K over F such that $g^\sigma \neq g$, and then g^σ, g take on the same values at infinitely many roots of unity, which is impossible). We may assume that this field of coefficients of g is finite over Q, generated by a primitive $m-th$ root of unity, say $k = Q(\zeta_m)$. After a multiplicative translation, we may also assume that $g(1) = 1$.

Let $\{\zeta_n\}$ be a sequence of roots of unity such that $g(\zeta_n)$ is a root of unity, with $n = n_1, n_2, \ldots, n \to \infty$, and each ζ_n is a primitive $n-th$ root of unity. For each n, there exists an automorphism σ_n of $Q(\zeta_n)$ over Q such that

$$\sigma_n \zeta_n = e^{2\pi i/n}.$$

Extend σ_n to F. Then

$$(\sigma_n g)(\sigma_n \zeta_n) = \sigma_n(g(\zeta_n))$$

is contained in the field $Q(\zeta_m, \zeta_n)$, and hence is a $2mn - th$ root of unity (not necessarily primitive). For infinitely many n, the restriction of σ_n to k

induces the same isomorphism on k. Taking a subsequence of $\{n\}$, and dealing with a conjugate of g if necessary, we can assume without loss of generality that for infinitely many n there exists an integer d_n such that

$$g(e^{2\pi i/n}) = e^{2\pi i d_n/2mn} .$$

The roots of unity $e^{2\pi i/n}$ approach 1 as n tends to infinite. Without loss of generality, we may assume that $2\pi i d_n/2mn$ approaches 0. We have (by the mean value theorem):

$$|e^{2\pi i d_n/2mn} - 1| = |g(e^{2\pi i/n}) - g(1)| \leqq C |e^{2\pi i/n} - 1|$$

for some constant $C > 0$, whence

$$|d_n/2mn| \leqq C' |1/n|$$

for some other constant $C' > 0$. This implies that the numbers d_n are bounded, and hence taking a subsequence of n if necessary, that they are all equal to the same number. But then one concludes at once that there exist infinitely many roots of unity $\zeta(= e^{2\pi i/n})$ and a fixed integer D such that

$$g(\zeta)^{2m} = \zeta^D,$$

whence $g(X)^{2m} = X^D$ identically, as was to be shown.

§ 2. Tate's proof. – Let $f(X, Y)$ be an irreducible polynomial in $C[X, Y]$, and assume that there exist infinitely many pairs of roots of unity $\zeta = (\zeta', \zeta'')$ such that $f(\zeta) = 0$. Then the coefficients of f lie in some field $Q(\zeta_m) = k$, generated over Q by a primitive $m - th$ root of unity. (As before, one sees this by considering conjugates of f over the field F obtained by adjoining all roots of unity to the rationals).

Let n be the period of ζ (i.e. the least common multiple of the periods of ζ', ζ''). Let d be a positive integer prime to n. There exists an automorphism σ of $Q(\zeta)$ such that $\sigma\zeta = \zeta^d$. If in addition $d \equiv 1 \pmod m$, then σ can be extended to an automorphism of $k(\zeta)$ inducing the identity on k. Then $f(\zeta^d) = 0$, so that ζ is a zero of $f(X, Y)$ and also of $f(X^d, Y^d)$. But

$$[k(\zeta) : k] \geqq \varphi(n)/m.$$

Applying any automorphism τ of $k(\zeta)$ over k, we find that $\tau\zeta$ is also a common zero of these two polynomials, which have therefore at least $\varphi(n)/m$ zeros in common. However, by Bezout's theorem, these polynomials have at most $(\deg f)^2 d$ common zeros, unless $f(X, Y)$ divides $f(X^d, Y^d)$. As soon as n is

large enough, we can use the prime number theorem giving the existence of
of primes in arithmetic progressions to find a prime number d satisfying the
above conditions, such that d is much smaller than $\varphi(n)$. Hence we conclude
that $f(X, Y)$ divides $f(X^d, Y^d)$, and it is then an exercise to show that
$f(X, Y) = 0$ defines a subgroup variety of $C^* \times C^*$. This concludes the proof,
which also shows that n is bounded in terms of deg f and m.

Note that we can avoid the congruence condition $d \equiv 1 \pmod{m}$ by using
the following variation of Tate's argument. Let $r = [k : Q]$. We extend σ to
an automorphism of $k(\zeta)$. Then σ^r induces the identity on k, and

$$\sigma^r \zeta = \zeta^{d^r}.$$

Then $f(\zeta^{d^r}) = 0$, so that ζ is a zero of $f(X, Y)$ and also $f(X^{d^r}, Y^{d^r})$. We can
then argue as before, using only the lemma:

Lemma. – *Given an integer s, there exists an integer n_0 such that for all*
$n > n_0$, *there exists a prime number p not dividing n, such that $p^s \leq n$.*

Proof. – The worst case occurs when n is a product of distinct primes,
in which case the assertion is an immediate consequence of the fact $\pi(N)$ is
of the order of magnitude of $N/\log N$.

§ 3. Abelian varieties. – We shall now see how this variation can be
formulated on abelian varieties.

Let A be an abelian variety, defined over the complex numbers, and let
V be a curve (subvariety of dimension 1) in A. We assume that V contains
infinitely many torsion points. After a translation, we may assume that V
passes through the origin. Let k be a field of definition for A and V, finitely
generated over the rationals. We shall reduce the proof that V is of genus
1 to a statement analogous to the irreducibility of the cyclotomic equation.

Let m be an integer ≥ 1. Let $\lambda_m : A \longrightarrow A$ be multiplication by m. As a
cycle, $\lambda_m(V) = \mu \cdot V^{(m)}$, where $V^{(m)}$ consists of all points mx, with $x \in V$.
Then $\mu \cdot V^{(m)}$ is algebraically equivalent to $m^2 \cdot V$. If $V \neq V^{(m)}$, then $V \cap V^{(m)}$
has at most $m^2(\deg V)^2$ points, by a routine generalization of Bezout's theorem.
(Cf. for instance [2], Lemma 4, Chapter III, § 3. We can view V and $V^{(m)}$ as
divisors on their sum in A).

If $V = V^{(m)}$, then λ_m gives an unramified covering of V over itself, of
degree m^2, and hence V is of genus 1, so is an abelian subvariety.

Let Γ be the group of torsion points of A. We reduce the proof of the
conjecture in this case to the following hypothesis:

(*) *Let A be an abelian variety defined over k. There exists an integer*
$c \geq 1$ *with the following property. Let x be a point of period n on A. Let G_n*

be the multiplicative group of integers prime to n, mod n. Let G be the sub-group of G_n consisting of those integers d such that dx is conjugate to x over k. Then

$$(G_n : G) \leq c.$$

To apply (*), suppose that there exist points x_n of period n, $n \to \infty$, lying on V. Let d be a positive integer prime to n. By (*), there exists an automorphism σ of $k(x)$ over k such that $\sigma x = d^r x$, where r is a positive integer bounded by c. Then

$$\sigma x = d^r x \in V \cap V^{(d^r)}.$$

Furthermore, if τ is in the group of automorphisms of $k(x_n)$ over k, then

$$\tau d^r x \in V \cap V^{(d^r)}.$$

If $V \neq V^{(d^r)}$, we obtain the inequalities, using (*):

$$\frac{\varphi(n)}{c} \leq \text{Number of points on } V \cap V^{(d^r)} \leq d^{2r}(\deg V)^2.$$

We note that $\varphi(n) \geq n^{1/2}$ for sufficiently large n. By the lemma, taking d to be a sufficiently small prime number not dividing n, we get a contradiction as soon as n is sufficiently large, as desired.

At present, very little is known concerning (*). Recent work of SERRE has been concerned with the size of the GALOIS groups of period points of elliptic curves over number fields, and SERRE has ben able to prove «finite index» property for most elliptic curves, when n is a prime power [5]. SERRE also tells me that property (*) is true when elliptic curve has complex multi-plication. Furthemore, if the elliptic curve has a transcendental j-invariant over the rationals, then the truth of (*) follows from results of IGUSA [1]. Nothing seems to be known in more general cases.

§ 4. APPENDIX. - We conclude by a remark concerning the exercise about $f(X^d, Y^d)$ made previously. Let G be a commutative group variety (in characteristic 0), and let V be a curve on G, passing through the origin. Assume that there exists an integer $d > 1$ such that for all $x \in V$, the point dx also lies in V. (We write the group law additively). Then V is a subgroup of G. Indeed, let p be a prime number dividing d. If k_0 is a field of defini-tion for G and V, finitely generated over the rationals, we can embed k_0 in a finite extension of the p-adic field Q_p, and let k denote the completion of the algebraic closure of Q_p. If x is a point of V_k sufficiently close to the origin, then dx lies in V_k, and the points $d^n x$ approach 0 as n tends to infinity (for the p-adic topology on V_k). Taking the inverse image by the exponential map of a sufficiently small neighborhood U of 0 in V_k, we find on the tangent space at the origin that $\exp^{-1}(U)$ has an infinite intersection with a straight line, having 0 as point of accumulation. This implies that $\exp^{-1}(U)$ contains a small (infinite) subgroup, and hence that U contains a small subgroup. Since V is a curve, this small subgroup is ZARISKI dense in V, and hence V is a group, as was to be shown.

BIBLIOGRAPHY

[1] J. IGUSA, *Fiber systems of elliptic curves*, American Journal of Mathematics 81 (1959), pp. 453-476.

[2] S. LANG, *Diophantine Geometry*, Intersience, 1962.

[3] — —, *Integral points on curves*, Publication l'I.H.E.S., No. 6 (1960) pp. 27-43.

[4] J. MANIN, *Rational points of algebraic curves over function fields* (in Russian), Izvestia Akademia Nauk 27 (1963), pp 1395-1440.

[5] J. P. SERRE, *Groupes de Lie l-adiques attachés aux courbes elliptiques*, Colloque de Clermond-Ferrand (1964) pp. 197-212.

Topology Vol. 3, pp. 183–191 Pergamon Press, 1965. Printed in Great Britain

ALGEBRAIC VALUES OF MEROMORPHIC FUNCTIONS

SERGE LANG

(*Received 6 September* 1963)

WE SHALL extend the main theorem of [1] to functions of several variables, showing that under certain circumstances, they cannot take too many algebraic values on certain sets of complex numbers.

We recall that an entire function of n variables $g(U) = g(u_1, \ldots, u_n)$ is said to be of order $\leq \rho$ if there exists a constant c such that $|g(U)| \leq c^{R^\rho}$ on the domain $|u_i| = R$ for all sufficiently large R. (We omit the usual ε which is irrelevant for what follows.) A meromorphic function will be said to be of order $\leq \rho$ if it can be written as a quotient of two entire functions of order $\leq \rho$. If P is point in $\mathbf{C}^n$ and f a meromorphic function of order $\leq \rho$ we say that f is defined at P if we can write $f = g/h$ where g, h are entire of order $\leq \rho$ and $h(P) \neq 0$.

THEOREM 1. *Let K be a number field, $g_1, \ldots, g_M$ meromorphic functions of order $\leq \rho$. Assume that the ring $K[g_1, \ldots, g_M]$ is mapped into itself by the partial derivatives $D_1, \ldots, D_n$, and that its transcendence degree over K is $\geq n + 1$. Let $\mathfrak{S}$ be a finite set of points in $\mathbf{C}^n$ at which all the g_μ are defined, and such that $g_\mu(P) \in K$ for all μ and all $P \in \mathfrak{S}$. Suppose in addition that after some change of co-ordinate system, the set $\mathfrak{S}$ becomes a product*

$$S_1 \times \ldots \times S_n$$

where each S_i is a set of m distinct complex numbers. Then

$$m \leq b\rho[K : \mathbf{Q}]$$

with a constant b depending only on n, and easily estimated.

The proof of Theorem 1 will make use of the ideas of Schneider [2] who has shown me how to justify his application of Cauchy's formula in several variables (cf. Lemma 2 below). Thus the arguments of Schneider's paper [2] are justified (contrary to what I thought in [1]).

The hypothesis concerning the set of points where the functions take on algebraic values applies to a lattice, and to periodic functions (i.e. abelian functions) with respect to this lattice.

As a corollary of Theorem 1, we shall obtain a statement concerning moduli. Let A be an abelian variety defined over the number field K, and let

$$\Theta : \mathbf{C}^n \longrightarrow A$$

be a representation of the abelian variety A as a quotient of complex n-space. We assume throughout that the partial derivatives are defined over K. Let $\Omega_1, \ldots, \Omega_{2n}$ be fundamental periods, let Ω be the $n \times 2n$ period matrix, viewing the Ω_i as column vectors, and let $\mathbf{P}$ be a principal matrix such that Ω and $\mathbf{P}$ satisfy the Riemann relations. After making an *integral* change of co-ordinates, we can assume that $\mathbf{P}$ has the usual canonical form

$$\mathbf{P} = \begin{pmatrix} 0 & \mathbf{D} \\ -\mathbf{D} & 0 \end{pmatrix}.$$

Then $\Omega = (W_1, W_2)$, each W_1, W_2 being an $n \times n$ matrix, and W_1 is invertible.

The matrix $W_1^{-1} W_2$ is the moduli point which is usually associated with the abelian variety in the Siegel upper half space H_n. However, Theorem 2 below is concerned with $W_2 W_1^{-1}$. In his paper [2], Schneider makes a remark concerning the applicability of his theorem to the transcendence of the moduli point (p. 113, top), but does not specify what he takes as moduli point. Presumably, he had Theorem 2 in mind.

THEOREM 2. *Let A be as above, and let $\Theta : \mathbf{C}^n \longrightarrow A$ be as above. Let Ω be a period matrix, normalized so that the principal matrix has the usual canonical form. Let $T = W_2 W_1^{-1}$. If T is algebraic, then T (viewed as a linear transformation) maps the period lattice tensored with $\mathbf{Q}$ into itself.*

(When $n = 1$, then $T = \tau$, $W_1 = \omega_1$, and $W_2 = \omega_2$, in the usual notation.)

To prove Theorem 2, let $\Phi(U) = \Theta(TU)$ and assume that T is algebraic. After extending K if necessary, we may assume all components of T lie in K. We view U as a vertical vector for this proof. Then $\Omega_1, \ldots, \Omega_n$ are periods for Φ and Θ, and generate a lattice such that for any point P in this lattice, $\Phi(P)$ and $\Theta(P)$ are points of A rational over K. A change of basis in $\mathbf{C}^n$ transform lattice points into a set of points as in Theorem 1, namely

$$(j_1, \ldots, j_n) \qquad\qquad 0 \leqq j_i \leqq m.$$

Taking m sufficiently large, we dehomogenize the projective maps Φ and Θ to obtain two rings of functions:

$$K[f_1, \ldots, f_N] \qquad \text{and} \qquad K[g_1, \ldots, g_M]$$

giving corresponding maps into affine space and defined at the above points. Applying Theorem 1, we conclude that the product map

$$(\Theta, \Phi) : \mathbf{C}^n \longrightarrow A \times A$$

has in fact dimension n, i.e. that Φ is algebraic over Θ. For a large integer r it follows that $\Phi(rU)$ is rational over $\Theta(U)$ and hence that *all* periods of $\Theta(U)$ are also periods of $\Phi(rU)$. Hence T maps the period lattice tensored with $\mathbf{Q}$ into itself, as was to be shown.

When the abelian variety (defined over the number field) has only trivial endomorphisms, it would be interesting to determine the transcendence degree of the point T (in the Siegel upper half space).

In Theorem 1, there may be other types of sets $\mathfrak{S}$ on which functions $g_1, \ldots, g_M$ as in the theorem cannot take on values in K when $\mathfrak{S}$ is too large. Some condition must be put on $\mathfrak{S}$, however. For instance, the functions $z_1, z_2, e^{z_1 - z_2}$ take on algebraic values whenever z_1, z_2 are algebraic and $z_1 = z_2$. Even though in 'natural cases' one expects transcendence degrees to be at least n, some sort of additional condition must be assumed. For instance, already with functions of one variable, we have algebraically independent functions

$$t, e^t, e^{t^2}, e^{t^3}, \ldots$$

whose values have transcendence degree 1 whenever t is an integer. Finally, if $Q(t)$ is a polynomial with integer coefficients, then $e^{Q(t)}$ takes on algebraic values at the finite number of algebraic zeros of Q.

It is also possible to give a more significant example of degeneracy when dealing with abelian varieties. Let $C_1, \ldots, C_m$ be elements of $\mathbf{C}^n$ which are linearly independent over $\mathbf{Q}$. Assume that the ring of endomorphisms of A is trivial. In particular, A is simple. Let

$$\Phi_m : \mathbf{C}^{mn} \longrightarrow A^m$$

be the mapping given by $\Phi_m(t) = (\Theta(tC_1), \ldots, \Theta(tC_m))$. Thus $\Phi_m(t)$ is the exponential map passing through the vector $(C_1, \ldots, C_m)$. We contend that this map has algebraic dimension mn, i.e. that it is Zariski-dense in A^m. We prove this by induction on m. For $m = 1$, it follows from the simplicity of A. Assume it for m. If the dimension of the map with $m + 1$ factors is not $n(m \not\equiv 1)$, then $\Phi_{m+1}(t)$ is contained in an abelian subvariety of A^{m+1} whose projection on the first m factors is A^m, and which is algebraic over A^m. There exists an integer $r \neq 0$ such that

$$\Theta(rtC_{m+1})$$

is rational over $\Phi_m(t)$, i.e.

$$(\Theta(tC_1), \ldots, \Theta(tC_m), \Theta(rtC_{m+1}))$$

lies in an abelian subvariety B of A^{m+1} which is isomorphic to A^m under projection, and whose projection on the last factor is A. Hence there exists a homomorphism of A^m onto A, and since A has only trivial endomorphisms, there exist integers $r_1, \ldots, r_m$ such that

$$\Theta(trC_{m+1}) = r_1\Theta(tC_1) + \ldots + r_m\Theta(tC_m).$$

This implies that

$$rC_{m+1} = r_1 C_1 + \ldots + r_m C_m,$$

a contradiction.

We apply the above to the $2n$ periods $\Omega_1, \ldots, \Omega_{2n}$. The dimension of the map Φ_{2n} is then $2n^2$, but the period $(\Omega_1, \ldots, \Omega_{2n})$ is mapped onto 0 by Φ_{2n}. According to the Riemann relations, the transcendence degree of the period matrix cannot be $2n^2$.

One may raise the question whether such degeneracy can occur in the case of a simple abelian variety, possibly under the additional assumption that the ring of endomorphisms is trivial. In other words, suppose that A is a simple abelian variety. Let α be a non-zero algebraic tangent vector (i.e. an element of $\mathbf{C}^n$ which is algebraic). Then one may ask whether the transcendence degree of the point

$$(\alpha, \Theta(\alpha))$$

in $\mathbf{C}^n \times A$ is $\geq n$.

It would be desirable to have a general conjecture also for products. For instance if A has dimension 1, and our mapping is represented by the Weierstrass function with algebraic g_2, g_3, let ω_1, ω_2 be two fundamental periods. If A has no complex multiplication, then it is reasonable to expect that ω_1 and ω_2 are algebraically independent. One would see this by looking at the representation of $A \times A$ as a quotient of $\mathbf{C}^2$.

Conceivably, the only reasonable way of formulating the most general conjecture is to take the moduli variety into account, i.e. given a ring of endomorphisms, to consider directly the holomorphic mapping from the product $H_n \times \mathbf{C}^n$, of the Siegel upper half space and $\mathbf{C}^n$, onto the algebraic 'fiber space' having a natural projection on the variety of moduli, and whose fibers are Kummer varieties. It is not easy to understand the literature on the subject, especially as it relates the algebraic description of this fiber space, and its trancendental description.

We shall now give the proof of Theorem 1. We need some lemmas, to make estimates.

Let $\{\alpha\}$ be a set of elements in K. A *denominator* for this set is a positive integer a such that $a\alpha$ is an algebraic integer for all α. By the *size* of the set, we shall mean the maximum of the absolute values of all conjugates of all elements of the set. If Q is a polynomial with coefficients in K, its size is the size of the set of coefficients.

LEMMA 1. *Let K be a number field and $g_1, \ldots, g_M$ functions of n complex variables, holomorphic at a point A, and such that the ring $K[g_1, \ldots, g_M]$ is mapped into itself by the partial derivatives $D_1, \ldots, D_n$. Let I_K be the ring of algebraic integers, and Q a polynomial in M variables of total degree $\leqq d$, with coefficients in I_K. Let D be a differential operator*

$$D = D_1^{k_1} \ldots D_n^{k_n}$$

of order $k = k_1 + \ldots + k_n$. Then

$$\text{size } D(Q(g_1, \ldots, g_M))(A) \leqq k!\, d^k c_1^{k+d} \text{ size } (Q),$$

and there exists a denominator for $D(Q(g_1, \ldots, g_M))(A)$ bounded by c_2^{d+k}. (Here, c_1, c_2 are constants depending only on K, $g_1, \ldots, g_M$.)

Proof. The proof uses the usual estimating techniques. Let Q, Q' be two polynomials in variables $X_1 \ldots, X$ and assume Q' has *real coefficients* $\geqq 0$. Write:

$$Q = \sum \beta_{(i)} X_1^{i_1} \ldots X_M^{i_M}$$

$$Q' = \sum \beta'_{(i)} X_1^{i_1} \ldots X_M^{i_M}.$$

We say that Q' *dominates* Q if

$$|\beta_{(i)}| \leqq \beta'_{(i)} \qquad \text{all } (i),$$

and we write $Q \leqq Q'$. Then one verifies trivially that

$$Q_1 \leqq Q_1' \quad \text{and} \quad Q_2 \leqq Q_2' \quad \text{implies} \quad Q_1 + Q_2 \leqq Q_1' + Q_2' \quad \text{and} \quad Q_1 Q_2 \leqq Q_1' Q_2'.$$

Furthermore, for any differential operator D as above, we have

$$DQ_1 \leqq DQ_1'.$$

If Q has degree d, then

$$Q \leqq (1 + X_1 + \ldots + X_M)^d \text{ size } (Q).$$

From these remarks and a simple induction, the proof is easily carried out.

Next we discuss some applications of Cauchy's formula. First, we consider the case of one variable z. Let $E(z)$ be an entire function of one variable. Let S be a set of m distinct complex numbers. Let R be a positive number such that $|y| < R$ for all $y \in S$. For each $y \in S$, suppose given an integer $\lambda(y) \geq 0$. Let

$$Q(z) = \prod_{y \in S} (z - y)^{\lambda(y)},$$

and for each $y \in S$, let

$$Q_y^*(z) = \frac{Q(z)}{(z - y)^{\lambda(y)}}.$$

Then Q_y^* is defined and not zero at y. Let

$$\Delta_y = (d/dz)^{\lambda(y) - 1}$$

if $\lambda(y) \geq 1$, and 0 otherwise. Then

$$\int_{|z| = R} \frac{E(z)}{Q(z)} \, dz = \frac{1}{2\pi i} \sum_{y \in S} \frac{1}{(\lambda(y) - 1)!} \, \Delta_y(E/Q_y^*)(y).$$

If we deal with a function of several variables, then we can form a repeated integral over a similar quotient, and obtain an analogous formula. We shall need a special case, as in Schneider [2].

LEMMA 2. *Let $E(z_1, \ldots, z_n)$ be an entire function. Let each one of $S_1, \ldots, S_n$ be a set of m complex numbers. Let $\Delta_i = \partial/\partial z_i$. Let σ be an integer > 0. Assume that for all differential operators Δ of order $\leq \sigma$ and all points $Y \in S_1 \times \ldots \times S_n$ we have*

$$\Delta E(Y) = 0.$$

Let $(\lambda) = (\lambda_1, \ldots, \lambda_n)$ be a vector of integers ≥ 0 such that $\lambda_1 + \ldots + \lambda_n = \sigma + 1$, and assume that

$$\Delta^{(\lambda)} E(A) \neq 0$$

for some A in $S_1 \times \ldots \times S_n$. Let $S_i' = S_i - \{a_i\}$. Let

$$Q(Z) = (z_1 - a_1)^{\lambda_1 + 1} \ldots (z_n - a_n)^{\lambda_n + 1} \prod_i \prod_{y_i \in S'_i} (z_i - y_i)^{\lambda_i}.$$

Let $R > \|Y\|$ for all Y. Then

$$\int \ldots \int \frac{E(Z)}{Q(Z)} \, dZ = \frac{1}{\lambda!} \frac{1}{(2\pi i)^n} \frac{\Delta^{(\lambda)} E(A)}{Q_A^*(A)}$$

where the repeated integral is taken over $|z_i| = R$, and

$$Q_A^*(Z) = \frac{Q(Z)}{(z_1 - a_1)^{\lambda_1 + 1} \ldots (z_n - a_n)^{\lambda_n + 1}}.$$

Proof. Taking the repeated integral and using the remarks we made above for one variable, we obtain a sum over vectors Y involving differential operators. However, when evaluated at vectors Y, they will give only a zero contribution because of our assumption that $\Delta E(Y) = 0$, except for the term belonging to A, which is the only term for which a differential operator of order $\sigma + 1$ will occur. Up to powers of $2\pi i$ and factorials, this term is of type

$$\Delta^{(\lambda)}(E/Q_A^*)\big|_{Z=A}.$$

Our lemma now follows from the next remark.

LEMMA 3. *Let E_1, E_2 be two functions holomorphic at a point A. Assume that $\Delta E_1(A) = 0$ for all differential operators Δ of order $\leq \sigma$, and let $\Delta^{(\lambda)}$ be an operator of order $\sigma + 1$. Then*

$$\Delta^{(\lambda)}(E_1 E_2)(A) = \Delta^{(\lambda)} E_1(A) \cdot E_2(A).$$

Proof. Using repeatedly the rule for the derivative of a product, with respect to $\Delta_1, \ldots, \Delta_n$, we find that all terms will be 0 except the term which applies all the differential operations to E_1 and not to E_2. This yields what we wanted.

We now carry out the proper part of the proof of Theorem 1, and we shall follow the same pattern of proof as in [1], taking into account the additional parameter n. Constants depending only on n will be denoted by $b_1, b_2, \ldots$. Constants depending on K, $g_1, \ldots, g_M$ and $\mathfrak{S}$ will be denoted by $c_1, c_2, \ldots$.

For any positive integer r, we have

$$b_1 r^n \leq \binom{r+n}{n} \leq b_2 r^n.$$

We let b_3 be a positive number such that $b_2 b_3^n = 1/2$.

Let $g_1, \ldots, g_M$ be our given functions in n variables $U = (u_1, \ldots, u_n)$. We let $D_i = \partial/\partial u_i$. Say $g_1, \ldots, g_{n+1}$ are algebraically independent over K. We let q be a large integer which will tend to infinity later. We contend that there exists a function (not identically zero)

$$G = \sum \alpha_{(i)} g_1^{i_1} \cdots g_{N+1}^{i_{N+1}}$$

with coefficients $\alpha_{(i)}$ which are algebraic integers in K, with

$$0 \leq i_1, \ldots, i_{n+1} \leq q,$$

such that $DG(P) = 0$ for all differential operators D of order at most

$$r = \left[\frac{b_3 q^{\frac{n+1}{n}}}{m}\right]$$

and all $P \in \mathfrak{S}$, and finally, the size of the $\alpha_{(i)}$ is bounded by

$$c_4^r r^{b_5 r}.$$

We view the coefficients $\alpha_{(i)}$ as unknowns. The conditions that $DG(P) = 0$ amount to linear equations in the $\alpha_{(i)}$ with coefficients in K, namely

$$D(g_1^{i_1} \cdots g_{n+1}^{i_{n+1}})(P).$$

Trivially, we have
$$\text{Number of unknowns} \geq q^{n+1}.$$

The number of differential operators of order $\leq r$ is given by the binomial coefficient.

Hence

$$\text{Number of equations} = \binom{r+n}{n}m^n \leq b_2 r^n m^n \leq \tfrac{1}{2}q^{n+1}.$$

Finally, using Lemma 1, we see that the size and denominator of the coefficients of our linear equations are bounded by

$$c_3^r r^{b_4 r}.$$

By a standard lemma (cf. [3, Chapter II, Lemma 2]), we can solve our equations non trivially in the desired manner. Since the functions $g_1, \ldots, g_{n+1}$ are algebraically independent over K, it follows that G is not identically zero.

Let σ be the smallest positive integer such that

$$DG(P) = 0$$

for all differential operators D of order $\leq \sigma$ and all $P \in \mathfrak{S}$. Then $\sigma \geq r$, and there exists an n-tuple $(s) = (s_1, \ldots, s_n)$ such that $s_1 + \ldots + s_n = \sigma + 1$, and

$$D^{(s)}G(\bar{A}) \neq 0$$

for some $\bar{A} \in \mathfrak{S}$. We let $\gamma = D^{(s)}G(\bar{A})$. Then $\gamma \in K$, $\gamma \neq 0$, and by Lemma 1, *the size and denominator of γ is bounded by*

$$c_5^\sigma \sigma^{b_6 \sigma}.$$

We shall now estimate one conjugate of γ very well using Lemma 2.

Let B be the $n \times n$ matrix giving the change of coordinates mentioned in the theorem, so that $U = ZB$, $Z = (z_1, \ldots, z_n)$, both U, Z are row vectors. Then $g_i(U) = f_i(Z)$ and

$$G(U) = G(ZB) = F(Z).$$

We let $\Delta_i = \partial/\partial z_i$. Then there are linear combinations $L_1, \ldots, L_n$ of $\Delta_1, \ldots, \Delta_n$ with complex coefficients such that

$$D_i G(U) = L_i F(Z),$$

and hence

$$D^{(s)}G(U) = L^{(s)}F(Z).$$

We can write

$$L^{(s)} = L_1^{s_1} \ldots L_n^{s_n} = \sum_{(\lambda)} \xi_{(\lambda)} \Delta_1^{\lambda_1} \ldots \Delta_n^{\lambda_n}$$

with complex coefficients $\xi_{(\lambda)}$, and $\lambda_1 + \ldots + \lambda_n = \sigma + 1$.

We let A be the point in $S_1 \times \ldots \times S_n$ corresponding to $\bar{A}$ under the linear transformation. We then have

$$\gamma = D^{(s)}G(\bar{A}) = L^{(s)}F(A) = \sum \xi_{(\lambda)}\Delta_1^{\lambda_1} \ldots \Delta_n^{\lambda_n}F(A).$$

We let Y range over $S_1 \times \ldots \times S_n$ (the transform of $\mathfrak{S}$ under our linear transformation). Since no point Y lies in the poles of the functions f_i, we can find an entire function θ of order $\leq \rho$ such that $\theta(A) \neq 0$, and θf_i is entire of order $\leq \rho$. We let $H(Z) = \theta(Z)^{(n+1)q}$. Then

$$E(Z) = F(Z)H(Z)$$

is an entire function, and $H(A) \neq 0$. Furthermore,

$$\Delta E(Y) = 0$$

for all differential operators Δ of order $\leq \sigma$, and by Lemma 3,

$$\Delta^{(\lambda)}E(Y) = \Delta^{(\lambda)}F(Y)H(Y).$$

By Lemma 2, we obtain

$$|\Delta^{(\lambda)}F(A)| \leq \lambda!(2\pi)^n \frac{|Q_A^*(A)|}{|H(A)|}\left|\int \cdots \int \frac{E(Z)}{Q(Z)}\,dZ\right|$$

taking the repeated integral on the circles of radius

$$R = \sigma^{1/(n+1)\rho}.$$

We can estimate the integral by taking the maximum of the absolute value of $E(Z)/Q(Z)$ on the circles of radius R. We note that $1/Q(Z)$ gives essentially a contribution bounded by

$$2^{m\sigma} \cdot \frac{1}{R^{m\sigma}},$$

as soon as σ is large enough with respect to the points Y. One can also estimate $E(Z)$, and one finds a bound of type $c_6^{q\,R^\rho}$. We note that $q \leq (m\sigma/b_3)^{n/(n+1)}$, because $r \leq \sigma$. Thus $E(Z)$, on the circles of radius R, is bounded by c_7^σ.

We therefore find:

$$|\Delta^{(\lambda)}F(A)| \leqq \sigma^\sigma \frac{c_8^\sigma}{R^{m\sigma}}.$$

The $\xi_{(\lambda)}$ are easily estimated, and the number of terms in the sum expressing γ in terms of derivatives of F at A is also easily estimated.   We note that $|H(A)|$ is the nq power of a constant. The terms $\lambda!$ and $|Q_A^*(A)|$ are trivially estimated. We obtain finally our very good estimate for γ, namely

$$|\gamma| \leqq \frac{c_9^\sigma}{R^{m\sigma}}.$$

We also know that the size and denominator of γ is bounded by $c_5^\sigma \sigma^{b_6\sigma}$. We multiply γ by a positive integer denominator, and take the norm, i.e. the product of all conjugates. If $[K:\mathbf{Q}]$ is the degree of K over $\mathbf{Q}$, then for one conjugate we use our very good estimate, and for all other $[K:\mathbf{Q}]-1$ conjugates, we use the estimate for the size. Using the fact that the norm is a rational integer whose absolute value is $\geqq 1$, and replacing $[K:\mathbf{Q}]-1$ by $[K:\mathbf{Q}]$ we find:

$$1 \leqq \frac{c_{10}^\sigma \sigma^{b_7[K:\mathbf{Q}]\sigma}}{R^{m\sigma}}.$$

We multiply by $R^{m\sigma}$, use our definition of R, and take logs. We obtain:

$$\frac{m\sigma}{(n+1)\rho}\,(\log \sigma - [K:\mathbf{Q}]\log c_{10}) \leqq b_7[K:\mathbf{Q}]\sigma \log \sigma.$$

We divide both sides of this inequality by $\sigma \log \sigma$ and let σ tend to infinity. We find

$$m \leqq b_8 \rho [K : \mathbf{Q}],$$

as was to be proved.

REFERENCES

1. S. LANG: Transcendental points on group varieties, *Topology* **2** (1962), 313–318.
2. T. SCHNEIDER: Zur Theorie der Abelschen Funktionen und Integrale, *J. reine angew. Math.* (1941), 110–128.
3. C. L. SIEGEL: Transcendental numbers, *Ann. Math. Studies*, Princeton, 1949.

Columbia University,
New York 27, N.Y., U.S.A.

Am. J. Math., LXXXVII No. 2, 1965

ASYMPTOTIC APPROXIMATIONS TO QUADRATIC IRRATIONALITIES, I.

By Serge Lang.

A recent theorem of Schmidt [2] states that for almost all (real) numbers β, the number of solutions of

$$| q\beta - p | < \frac{1}{q}$$

with integers p, q and $0 < q \leq B$ is asymptotic to $c_1 \log B$ $(B \to \infty)$ with some number $c_1 > 0$.

One could ask whether a similar estimate is not true for algebraic numbers, and also for the numbers entering in the theory of transcendental numbers (essentially those numbers in the field generated over the rationals by values of the classical functions suitably normalized, taking algebraic closure). Similarly, one can ask for extensions to other group varieties (beside the additive group). For instance, let A be an elliptic curve defined over a number field, and P a rational point. Take a suitably normalized complex analytic isomorphism of A with a complex torus (corresponding to algebraic g_2, g_3 if A is parametrized by a $\wp$-function), and let $\| \ \|$ denote the distance from the origin. One can ask for an estimate of the q such that $\| qP \| < 1/q$, and similarly for linear combinations of several points. On abelian varieties, lifting to complex n-space gives a problem concerning linear combinations of vectors.

No result of any kind except for Schmidt's theorem seems to be known regarding such asymptotic estimates. Certain machine computations for a few of the classical numbers $(e, \pi, \log 2, \cdots)$ have a tendency to support an affirmative answer [1]. It would seem quite difficult to prove for algebraic numbers, let alone transcendental ones. Remarkably enough, it has never been noticed that the result is true for quadratic numbers. The literature mentions mostly what seems to me a freak behaviour, namely that if β is quadratic and c is sufficiently small positive number, then $| q\beta - p | < c/q$ has only a finite number of solutions, and one gets the false impression that quadratic numbers somehow misbehave. Nevertheless:

THEOREM. *Let β be a quadratic real irrational number. Let c be a*

Received August 31, 1964.

481

15

number ≥ 1. *For any integer* $B > 0$, *let* $\lambda(B)$ *be the number of integers* q *such that* $|q| \leq B$ *and*

$$0 < q\beta - p < \frac{c}{|q|}$$

for some integer p. *There exist numbers* $c_1 > 0$ *and* $c_2 > 0$ *such that for all positive integers* B *we have*

$$|\lambda(B) - c_1 \log B| \leq c_2.$$

In other words, $\lambda(B) = c_1 \log B + O(1)$.

A similar assertion holds if we replace $q\beta - p$ by $p - q\beta$ in our inequality, and hence we also get the asymptotic estimate for the number of solutions of the inequality

$$|q\beta - p| < \frac{c}{q}$$

with $0 < q \leq B$.

The proof will be carried out by a straightforward brute force argument. We note that the number c_1 depends on c and β, even though in Schmidt's theorem, one could take c_1 to be the same for all numbers outside a set of measure 0.

We shall now prove the theorem.

Let D be an integer > 1, square free, and let

$$\alpha = \begin{cases} \sqrt{D} & \text{if } D \equiv 2, 3 \pmod 4 \\ \dfrac{1 + \sqrt{D}}{2} & \text{if } D \equiv 1 \pmod 4. \end{cases}$$

Then $1, \alpha$ form a basis of the algebraic integers of $\mathbf{Q}(\alpha)$ over $\mathbf{Z}$. We let $\beta = a\alpha + b$, with rational a, b and $a \neq 0$. We let d be a positive integer such that da and db are both integers. Then $d\beta$ is an algebraic integer. We let $\bar{\beta}$ be the conjugate of β over $\mathbf{Q}$.

We denote by N the norm from $\mathbf{Q}(\alpha)$ to $\mathbf{Q}$.

We shall use the phrase "sufficiently large (resp. small)" to mean "greater (resp. smaller) than a constant depending only on α, β, c".

LEMMA 1. *There exists an integer* $k > 0$ *having the following property. An integer* q *(with* $|q|$ *sufficiently large) is such that*

$$(*) \qquad\qquad 0 < q\beta - p < \frac{c}{|q|}$$

for some p, if and only if there exists p such that $q\beta - p$ is positive, sufficiently small, and

$$(**) \qquad |N(q\beta - p)| \leqq \frac{k}{d^2}.$$

Proof. We have to distinguish cases, depending on whether $cd^2(\beta - \bar\beta)$ is or is not an integer. (It could be an integer only if c is equal to $t\sqrt{D}$ for some rational number t.)

Suppose that $cd^2(\bar\beta - \beta)$ is not an integer. Then we take

$$k = [cd^2(\bar\beta - \beta)].$$

First assume (*). Then

$$|N(qd\beta - dp)| < \frac{cd^2}{|q|} |q\bar\beta - p| = cd^2 |\bar\beta - \frac{p}{q}|.$$

If p/q is close to β then $\bar\beta - p/q$ is close to $\bar\beta - \beta$. Since the norm of an algebraic integer is an integer, we conclude that

$$|N(qd\beta - dp)| \leqq k,$$

thereby proving (**).

Secondly, assume (**), and also that $q\beta - p$ is positive sufficiently small. Then

$$0 < q\beta - p \leqq \frac{k}{d^2 |q\bar\beta - p|}$$

$$\leqq \frac{c}{|q|} \frac{k}{cd^2 |\bar\beta - \frac{p}{q}|}.$$

The quotient $k/cd^2 |\bar\beta - \beta|$ is a fixed number < 1. For $|q|$ sufficiently large, it is clear that the right hand side of our inequality is $< c/|q|$, thereby proving (*).

Suppose that $cd^2(\bar\beta - \beta)$ is an integer. We must distinguish two sub-cases, depending on whether $\beta < \bar\beta$ or $\bar\beta < \beta$.

If $\beta < \bar\beta$, then for $|q|$ sufficiently large we have

$$\frac{p}{q} < \beta < \bar\beta.$$

We take k as before, and the first part of the argument runs as before. To conclude the argument in the second part, we now use the fact that

$$\bar\beta - \frac{p}{q} > \bar\beta - \beta.$$

If $\bar{\beta} < \beta$, then for $|q|$ sufficiently large, we have

$$\bar{\beta} < \frac{p}{q} < \beta.$$

This time, we take $k = cd^2 \,|\, \bar{\beta} - \beta \,| - 1$. In the first part of the argument, we have

$$\left| \bar{\beta} - \frac{p}{q} \right| < |\beta - \beta|$$

and the desired conclusion follows. The second part of the argument is carried out as before, thereby proving the lemma.

In view of the lemma, we are reduced to counting the number of integers q such that there exists p for which $q\beta - p$ is positive, sufficiently small, and

$$|N(qd\beta - dp)| \leqq k.$$

Let m be an integer, $1 \leqq m \leqq k$. We shall prove that our asymptotic estimate holds for the number of solutions of

$$|N(qd\beta - dp)| = m$$

with $q\beta - p$ positive, sufficiently small, provided that there exists at least one solution. Adding up these estimates for $m = 1, \cdots, k$ and using the fact that our original inequality actually has infinitely many solutions, we obviously obtain a proof of our theorem.

Our final step is to reduce our problem to counting certain units. Let $\xi,\ \xi'$ be two algebraic integers in $\boldsymbol{Q}(\alpha)$. We say that they are *equivalent* if there exists a positive unit u such that $u\xi = \xi'$. If $\xi,\ \xi'$ are equivalent then

$$|N(\xi)| = |N(\xi')|.$$

Furthermore, there is only a finite number of equivalence classes of algebraic integers in $\boldsymbol{Q}(\alpha)$ having a given norm. To prove our theorem, it will suffice to prove that the number of algebraic integers ξ satisfying the following conditions has the desired asymptotic estimate.

(1) ξ lies in a given equivalence class.

(2) ξ is positive.

(3)   There exist integers q, p such that $\xi = qd\beta - dp$.

(4) ξ is sufficiently small.

LEMMA 2. *Let $q_0,\ p_0$ be integers, $q_0 \neq 0$, and let $\xi_0 = q_0 d\beta - dp_0$. The* *set of units u such that $u\xi_0$ can be written in the form $qd\beta - dp$ with integers q, p, is a group.*

Proof. Let $a' = da$ and $b' = db$. We write a unit u as $u = x\alpha + y$, with integers x, y. We shall prove that the condition stated in the lemma is equivalent with a congruence condition on x.

We have

$$\xi_0 = q_0 a'\alpha + q_0 b' - dp_0.$$

Then

$$u\xi_0 = (x(q_0 b' - dp_0) + yq_0 a') \cdot \alpha + xq_0 a' D + y(q_0 b' - dp_0),$$

and we must find a necessary and sufficient condition that this expression is of type

$$qa'\alpha + qb' - dp,$$

with integers q, p. This amounts to the pair of conditions:

$$x(q_0 b' - dp_0) + yq_0 a' = qa'$$

$$xq_0 a' D + y(q_0 b' - dp_0) = qb' - dp.$$

The first one simply means that a' divides the left hand side. Let w be the g.c.d. of a' and $(q_0 b' - dp_0)$. Write $a' = wa_0$. Then the first condition is equivalent with $a_0 \mid x$ (provided $q_0 b' - dp_0 \neq 0$, a case we leave to the reader). We shall write $x = a_0 x^*$.

The first condition being satisfied, our second condition yields another divisibility condition, namely that d divides

$$x^*\left(a_0 q_0 a' D - b'\frac{q_0 b' - dp_0}{w}\right).$$

Let t be the g.c.d. of d and the expression in parentheses which we have just obtained. Write $d = td_0$. Then our last condition amounts to

$$d_0 \mid x^*.$$

Hence finally, our two conditions are equivalent with the divisibility

$$a_0 d_0 \mid x.$$

Since $u^{-1} = \pm\,\bar{u}$, it now follows at once that the units satisfying our divisibility condition form a group, as contended.

Let q_0, p_0 be integers, $q_0 \neq 0$, and let

$$\xi_0 = q_0 d\beta - dp_0$$

be in a given equivalence class. Assume $\xi_0 > 0$. The set of units u such that $u\xi_0$ satisfies the first three conditions (1), (2), (3) is the subgroup S of positive units in Lemma 2. Furthermore, $u\xi_0$ is sufficiently small if and only if u is sufficiently small. Hence we are reduced to counting the number of units u sufficiently small in S such that, when we write

$$u\xi_0 = qd\beta - dp$$

with integers q, p, then $|q| \leqq B$.

The group S either consists of one element, or is infinite cyclic. Assume that S is infinite. One sees at once that there exist two constants $k_1 > 0$, $k_2 > 0$ having the following property. Given any u in S, and writing $u\xi_0$ as above, we have

$$k_1 \max(|u|, |\bar{u}|) \leqq |q| \leqq k_2 \max(|u|, |\bar{u}|).$$

LEMMA 3. *There is a number $k_3 > 0$ such that the number of units u in S satisfying*

$$\max(|u|, |\bar{u}|) \leqq B$$

is equal to $k_3 \log B + O(1)$.

Proof. As usual, map a unit u in S on the vector

$$(\log|u|, \log|\bar{u}|).$$

Then S gets embedded on an infinite cyclic discrete subgroup of the straight line in the plane define by

$$\log|u| + \log|\bar{u}| = 0.$$

Our assertion is then obvious.

Let k_4 be a number > 0. We note that the number of units u in S such that

$$\max(|u|, |\bar{u}|) \leqq k_4 B$$

and the number of units u in S such that

$$\max(|u|, |\bar{u}|) \leqq B$$

differ by a bounded term. In view of the lemma and the remarks preceding it, we see that our theorem is completely proved.

We note that the number c was chosen > 1 only for definiteness. Any

$c > 0$ for which the given inequality has infinitely many solutions would do just as well. Lemma 1 and the fact that the norm of an algebraic integer must be an integer show precisely how small we can take c and still get infinitely many solutions.

COLUMBIA UNIVERSITY.

REFERENCES.

[1] W. Adams and S. Lang, " Some computations in diophantine approximations ", *to appear.*
[2] W. Schmidt, "A metrical theorem in diophantine approximation ", *Canadian Journal of Mathematics* (1959), pp. 619-631.

Am. J. Math., LXXXVII No. 2, 1965

ASYMPTOTIC APPROXIMATIONS TO QUADRATIC IRRATIONALITIES, II.

By Serge Lang.

Let β be a (real) quadratic irrationality, and c a number $\geqq 1$. In the preceding paper, we determined asymptotically the number of integral solutions of the inequality

$$|q\beta - p| < \frac{c}{|q|}$$

for $|q| \leqq B$ and $B \to \infty$. We shall now consider the more refined inequality

$$|q\beta - p| < \psi(q)$$

where ψ is a suitable function. Again one expects that in general the number is asymptotic to

$$\Psi(B) = c_1 \int_1^B \psi(t)\, dt,$$

with some constant $c_1 > 0$, but we shall have to make a growth assumption on ψ in order to obtain this result.

When $\psi(x) = \rho$ with $0 < \rho < 1$, the problem is one of equidistribution for the numbers $q\beta$ on the circle, and has been considered before, notably by Hecke [1], who introduced the corresponding Dirichlet generating series, proved that it is meromorphic, and obtained a rather good error term. Hecke in fact uses a method which allows him to deal with any real number β such that for every $\epsilon > 0$ we have

$$|q\beta - p| > \frac{1}{q^{1+\epsilon}}$$

for all but a finite number of q, and hence in view of Roth's theorem, Hecke's theorem applies to algebraic numbers. (See the last theorem in Hecke's paper.)

Although Hecke's method works therefore rather well in his case, it cannot work for the counting problem with a function ψ, unless one knows something about the analytic behaviour of the functions which would play in the present questions a role similar to the L-functions in the theory of prime numbers.

Received October 8, 1964.

488

The method used in the present paper does not apply to Hecke's case as it stands, but works in other cases, for instance when

$$\psi(x) = \frac{(\log x)^m}{x}$$

with any $m > 0$. Thus our result is in a certain sense complementary to Hecke's.

Generally speaking, the study of the asymptotic distribution of field elements (as distinguished from ideals), would seem to deserve more attention than it has up to now.

1. Statement of the theorem. Let ψ be a real function of a real variable t, defined for $t > 0$, monotone decreasing to 0. We let $\omega(t) = t\psi(t)$, so $\psi(t) = \omega(t)/t$. We shall assume that ω is of class C^1, positive, strictly monotone increasing to infinity, and say that $\omega(t) = O(t^{\frac{1}{3}})$. (Our result will not apply to an ω growing too fast.)

As a matter of notation, we let f be the inverse function of ω. It is convenient to extend the domain of definition of ψ and ω to negative t by letting $\psi(t) = \psi(|t|)$, and similarly for ω.

We shall prove a theorem concerning norms:

THEOREM 1. *Let β be a real quadratic irrationality. Let β_1, β_2 be a basis of $\mathbf{Q}(\beta)$ over $\mathbf{Q}$, with β_1, β_2 algebraic integers. Let $\lambda(B)$ be the number of integral pairs (q_1, q_2) such that*

$$|q_1\beta_1 + q_2\beta_2| \leq 1, \qquad\qquad |q_1| \leq B$$

and

$$|N(q_1\beta_1 + q_2\beta_2)| \leq \omega(q_1).$$

Then there exists a constant $c_1 > 0$ such that for $B \to \infty$ we have

$$\lambda(B) = c_1 \int_1^B \psi(t)\,dt + O(\omega(B) + (\log B)\omega(B)^{\frac{1}{2}}).$$

When ω does not grow too fast, the integral is usually asymptotic to a constant times $\omega(B)\log B$, so that the error term is of a lower order of magnitude. For example, if $\omega(t) = (\log t)^\rho$ with $0 < \rho < 1$, we get the asymptotic estimate

$$c_1 \frac{(\log B)^{\rho+1}}{\rho+1} + O((\log B)^\rho + (\log B)^{(\rho/2)+1}).$$

Similarly when $\omega(t) = \log\log t$, or further iterated logs. However, if $\omega(t) = t^\delta$ with $0 < \delta \leq 1$, our result does not assert anything because the error term is of the same order of magnitude as the main term.

We shall now show how Theorem 1 implies a theorem concerning approximations to β. Given $0 < c < 1$, there is only a finite number of $\xi = q\beta - p$ such that $c \leq |\xi| \leq 1$, and

$$|\bar{\xi}| \leq \omega(q)/c.$$

Indeed, we note that $\max(|q|, |p|)$ and $\max(|\xi|, |\bar{\xi}|)$ are of the same order of magnitude. Hence as $\max(|q|, |p|) \to \infty$, it follows that $|\bar{\xi}|$ grows faster than $\omega(q)$.

Suppose that (q, p) is a pair of integers such that

$$(1) \qquad |q\beta - p| < \frac{\omega(q)}{q}.$$

Let d be an integer > 0 such that $d\beta$ is an algebraic integer. Then

$$(2) \qquad \begin{aligned} |N(qd\beta - dp)| &< \frac{\omega(q)}{|q|} d^2 |q\bar{\beta} - p| \\ &< \omega(q) d^2 \left(|\bar{\beta} - \beta| + \frac{\omega(q)}{q^2}\right). \end{aligned}$$

Let $\omega_1(t) = \omega(t) d^2(|\bar{\beta} - \beta| + \omega(t)/t^2)$. Then ω_1 is usually also strictly increasing for sufficiently large t. A simple sufficient condition, for instance, is that $\omega'(t) > 1/t^2$, as one sees at once by taking the derivative.

The set of solutions of (1) is contained in the set of solutions of (2). If we apply Theorem 1, taking $\beta_1 = d\beta$ and $\beta_2 = d$, using the function ω_1 instead of ω, we find that the number of solutions of (2), up to the given error term, is the same as the number of solutions of

$$(3) \qquad |N(qd\beta - dp)| \leq \omega(q) d^2 |\bar{\beta} - \beta|.$$

Conversely, let $\omega_2(t) = \omega(t) d^2(|\bar{\beta} - \beta| - 2\omega(t)/t^2)$. Let (q, p) be a pair of integers such that

$$(4) \qquad |N(qd\beta - dp)| \leq \omega_2(q),$$

and such that $|q\beta - p| \leq 1$, $|q|$ is sufficiently large. Then in fact, $|q\beta - p|$ is small, and a simple computation shows that $|q\beta - p|$ satisfies (1). The function $\omega_2(t)$ is strictly increasing (no extra condition is needed this time). Thus we can apply Theorem 1 again, and find that the number of solutions of (4), up to the given error term, is the same as the number of solutions of (3). Since the desired solutions of (1) are squeezed in between, we see that Theorem 1 implies:

THEOREM 2. *Let β be a real quadratic irrationality. Let ω, ψ be as* *before, assume in addition that $\omega'(t) > 1/t^2$ for all t sufficiently large. There exists a constant $c_1 > 0$ such that the number of integral solutions (q, p) for the inequality*

$$|q\beta - p| < \psi(q)$$

with $|q| \leqq B$ is equal to

$$c_1 \int_1^B \psi(t)\,dt + O(\omega(B) + (\log B)\omega(B)^{\frac{1}{2}}).$$

2. Proof of the theorem. We prove Theorem 1. Let us introduce some notation. Let L be the module generated over the integers $\mathbf{Z}$ by β_1, β_2. Denote by $\lambda_L(B)$ the number of elements of L satisfying the conditions stated in Theorem 2.

We embed K in $\mathbf{R}^2$ as usual. If $\xi \in K$, we map ξ on the vector $(\sigma\xi, \bar\sigma\xi)$ where σ, $\bar\sigma$ are the conjugate embeddings of K in $\mathbf{R}$. We shall identify ξ with $\sigma\xi$, and thus write this vector $(\xi, \bar\xi)$. Under this mapping, we see that L is embedded on a lattice or rank 2 in $\mathbf{R}^2$.

Let $\mathfrak{o}_L$ be the subring of K consisting of all elements γ such that $\gamma L \subset L$. Then $\mathfrak{o}_L$ is a subring of the ring of algebraic integers I_K, of rank 2 over $\mathbf{Z}$. It is easily seen that the group of units U_L of $\mathfrak{o}_L$ is a subgroup of finite index in the group of units U_K of I_K.

By an L-ideal we shall mean a submodule $\mathfrak{a} \neq 0$ of L such that $\mathfrak{o}_L\mathfrak{a} = \mathfrak{a}$. If $\mathfrak{a}$ is a principal L-ideal, and ξ, ξ' are two generators of $\mathfrak{a}$ over $\mathfrak{o}_L$, then there exists a unit $u \in U_L$ such that $\xi' = u\xi$. We denote $\mathfrak{o}_L\xi$ by (ξ). If $\mathfrak{a} = (\xi)$, we define $N\mathfrak{a} = |N(\xi)|$.

LEMMA 1. *The number of principal L-ideals $\mathfrak{a}$ such that $N\mathfrak{a} \leqq B$ is equal to $c_L B + O(B^{\frac{1}{2}})$, with some constant $c_L > 0$.*

Proof. The argument is entirely similar to the classical one when $L = I_K$. A careful treatment (which simplifies considerably in the special case under consideration) for the classical case, without hand waving, will be found in Schanuel [3]. It consists in estimating the number of lattice points in a homogeneously expanding region, and we shall omit it here.

If $\xi \in L$ and $\xi = q_1\beta_1 + q_2\beta_2$, then we let $q(\xi) = |q_1|$.

LEMMA 2. *There exist constants c_2, $c_3 > 0$, depending only on L, having the following property. For any principal L-ideal $\mathfrak{a}$, the number of $\beta \in L$ such that $(\xi) = \mathfrak{a}$, $|\xi| \leqq 1$, $q(\xi) \leqq B$, and $|N(\xi)| \leqq \omega(q(\xi))$, differs from*

$$c_2(\log B - \log f(N\mathfrak{a}))$$

by a term bounded by c_3.

Proof. Let T be the set of $\xi \in L$ such that $(\xi) = \mathfrak{a}$. We map T into $\mathbf{R}^2$ by the usual log mapping,

$$l: \xi \rightarrow (\log |\xi|, \log |\bar{\xi}|).$$

If $\xi_0 \in T$, then $l(T) = l(\xi_0) + l(U_L)$, and $l(U_L)$ is a discrete subgroup of the hyperplane (line) $x + y = 0$ in $\mathbf{R}^2$. Thus $l(T)$ is the translation by $l(\xi_0)$ of this subgroup on the line $x + y = \log N\mathfrak{a}$. If we let $l(\xi) = (x, y)$, then our conditions on ξ can be expressed by saying that

$$x \leqq 0, \qquad \log q(\xi) \leqq \log B,$$

and

$$f(N\mathfrak{a}) \leqq q(\xi) \leqq B.$$

There are two constants $k_1, k_2 > 0$ such that for all $\xi \in L$, $\xi \neq 0$ with $|\xi| \leqq 1$ we have

$$k_1 |\bar{\xi}| \leqq q(\xi) \leqq k_2 |\bar{\xi}|.$$

On the straight line $x + y = \log N\mathfrak{a}$, a point (x, y) of $l(T)$ is such that $y = \log |\bar{\xi}|$. The number of such points with $x < 0$ and

$$\log f(N\mathfrak{a}) + \log k_1 \leqq y \leqq \log B + \log k_2$$

is equal to

$$c_2 (\log B - \log f(N\mathfrak{a})) + O(1).$$

This proves our lemma.

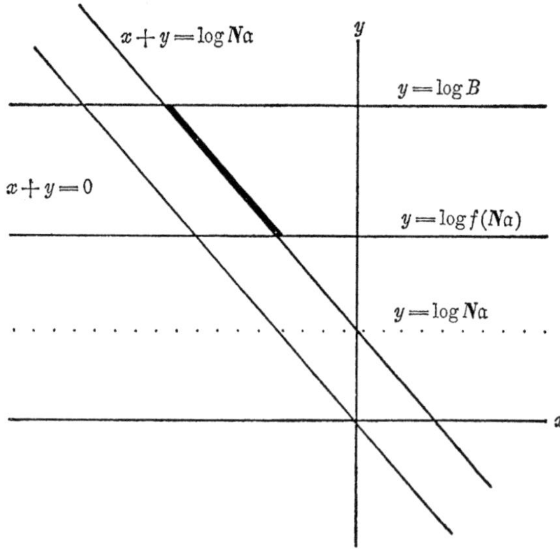

The above picture is a diagram of the preceding proof. One determines the lattice points on the line $x + y = \log N\mathfrak{a}$ between $y = \log f(N\mathfrak{a})$ and $y = \log B$.

In view of Lemmas 1 and 2, our number $\lambda_L(B)$ is given by

$$c_2^{-1}\lambda_L(B) = \sum_{N\mathfrak{a} \leq \omega(B)} (\log B - \log f(N\mathfrak{a})) + O(\omega(B)).$$

We shall conclude the proof by summing by parts to get the desired asymptotic expression.

For each positive integer ν, let $A(\nu)$ be the number of principal L-ideals $\mathfrak{a}$ such that $N\mathfrak{a} \leq \nu$. Let us denote by $S(B)$ the sum on the right hand side of our expression for $c_2^{-1}\lambda_L(B)$. Then

$$S(B) = \sum_{\nu=1}^{[\omega(B)]} \log\left(\frac{B}{f(\nu)}\right) (A(\nu) - A(\nu-1))$$

$$= (\log B - \log f([\omega(B)])) A([\omega(B)])$$

$$+ \sum_{\nu=1}^{[\omega(B)]-1} (\log f(\nu+1) - \log f(\nu)) A(\nu).$$

We now write $A(\nu) = c_L\nu + O(\nu^{\frac{1}{2}})$. Then, up to a constant factor c_L, the main term of our expression becomes

$$(\log B - \log f([\omega(B)])) [\omega(B)] + \sum_{\nu=1}^{[\omega(B)]-1} (\log f(\nu+1) - \log f(\nu))\nu.$$

Unwinding this sum by parts again, we obtain

$$(\log B) [\omega(B)] - \sum_{\nu=1}^{[\omega(B)]} \log f(\nu).$$

Since f and hence $\log f$ are strictly increasing, we have

$$\sum_{\nu=1}^{[\omega(B)]} \log f(\nu) = \int_1^{\omega(B)} \log f(t)\,dt + O(\log B)$$

$$= (\log B)\omega(B) - \int_1^{\omega(B)} \frac{t}{f(t)}\, f'(t)\,dt + O(\log B),$$

using integration by parts. Hence the main term in our expression is equal to

$$\int_1^{\omega(B)} \frac{t}{f(t)}\, f'(t)\,dt + O(\log B).$$

Changing variables, putting $t = \omega(\tau)$, we find that this main term is the desired one, i. e.

$$\int_{f(1)}^{B} \frac{\omega(t)}{t}\, dt + O(\log B).$$

There remains to estimate the error term which we had previously, namely

$$(\log B - \log f([\omega(B)]))[\omega(B)]^{\frac{1}{2}}$$

$$+ \sum_{\nu=1}^{[\omega(B)]-1} (\log f(\nu+1) - \log f(\nu))\nu^{\frac{1}{2}}.$$

Unwinding this sum as in the preceding case, we get

$$(\log B)[\omega(B)]^{\frac{1}{2}} + \sum_{\nu=1}^{[\omega(B)]} \log f(\nu)\,(\nu^{\frac{1}{2}} - (\nu-1)^{\frac{1}{2}}).$$

Each term in the sum is positive. We replace $f(\nu)$ by $f(\omega(B)) = B$, and see that our error is bounded by a constant times $(\log B)\omega(B)^{\frac{1}{2}}$, as was to be shown.

3. Remarks. One sees at once from the various constructions of the proof that $c_1 = c_2 c_L$, and hence that c_1 depends only on L, i. e. β (or β_1, β_2), and not on the auxiliary functions ψ or ω. Thus for functions ψ such that the error term is small compared to the main term, this main term depends linearly on ψ.

In the preceding paper, we consider the number of $\xi \in L$ satisfying the inequality $|N(\xi)| \leq k$ with some constant k, equal to a positive integer. We can apply the method of proof for Theorem 1, using the units of $\mathfrak{o}_L$ (this being a slight variation of the method used in [2]). One then sees that the number of solutions of this inequality with $q(\xi) \leq B$, $|\xi| \leq 1$, is equal to

$$c_2 A(k) \log B + O(1),$$

where c_2 is a constant determined by the length of a fundamental domain for the units, and $A(k)$ is the number of L-ideals $\mathfrak{a}$ such that $N\mathfrak{a} \leq k$.

From this one also sees that our restrictions on the function ω are necessary. Indeed, ω cannot tend to 0 in view of the existence of the constant $c > 0$ such that $|q\beta - p| > c/|q|$. Furthermore, ω cannot oscillate, because using Lemma 1 of the preceding paper, if we let ω oscillate very slightly near a critical value of the constant k then it follows at once that there is no asymptotic value for the number of solutions of our inequality (the constant $A(k)$ jumps by discrete amounts). Thus in dealing with quadratic irrationalities, one must assume that ω is increasing. If ω is bounded, this amounts to the constant case. We are thus led to assuming that ω is strictly increasing to infinity.

Of course the problem arises to determine whether a similar phenomenon occurs for algebraic numbers of degree greater than 2, or for the classical transcendental numbers. Generally speaking, I expect the classical numbers to behave like almost all numbers. For instance, when

$$\psi(q) = 1/q^{1+\epsilon}$$

as in Roth's theorem, it is known that there is only a finite number of solutions for the inequality $|\, qe - p\,| < \psi(q)$ (Mahler-Popken). Using the same method as Mahler, I can show that if α is rational and $\neq 0$, then e^α also satisfies this property with the above function ψ, or even a slightly better one, as in Mahler's proof (cf. Schneider [4]). For arbitrary ψ, the answer is not known, even in the case of algebraic numbers.

Finally, it should be mentioned that the brute force and essentially simple-minded methods of the present paper have of course no chance of succeeding when dealing with the approximation problem to algebraic numbers of higher degree (and even less to transcendental ones). I see no hints anywhere (in Roth's proof or elsewhere) for a method which would succeed in these more general cases. The combinatorial structure of all known proofs of measures of irrationality or transcendence never exhibits the appearance of the integral

$$\int_1^B \psi(t)\, dt$$

or the sum $\Sigma\, \psi(q)$. This seems to me to be the main reason why the technique of the Thue-Siegel-Schneider-Roth proof, involving a polynomial in several variables, fails to apply for numbers other than algebraic ones. For instance, instead of making certain coefficients equal to 0 in the proof, when dealing with values of say e^t at algebraic α, one could require only that such values are very small, by imposing a zero of high order on the functions $F^{(i)}(e^t, \cdots, e^t)$. However, the linear equations involved in this question have too many variables to allow one to solve them in the frame of the present structure of the proof.

COLUMBIA UNIVERSITY.

———————

REFERENCES.

[1] E. Hecke, "Über analytische Funktionen und die Verteilung von Zahlen mod eins," *Abhandlungen aus dem Mathematischen Seminar der Universität Hamburg,* Bd. 1 (1921), pp. 54-76.

[2] S. Lang, "Asymptotic approximation to quadratic irrationalities I," *American Journal of Mathematics,* vol. 87 (1965), pp. 481-487.

[3] S. Schanuel, "Heights in number fields," *to appear.*

[4] T. Schneider, *Einführung in die tranzcendente Zahlen,* Springer Verlag (1957), p. 88, and p. 91, where the argument with rational a works. When a is algebraic, the method gives only a weaker theorem, involving the degree of a.

Journal für die reine und angewandte Mathematik

Herausgegeben von **Helmut Hasse** und **Hans Rohrbach**

Verlag Walter de Gruyter & Co., Berlin 30

Sonderabdruck aus Band 220 Heft 3/4, 1965. Seite 163 bis 173

Some computations in diophantine approximations[1])

By *W. Adams* at Berkeley and *S. Lang* at New York

Let $w_1, \ldots, w_m$ be (real) numbers, linearly independent over the rationals. Let B be an integer > 0 and c a number > 0. We define $\lambda(B, c)$ to be the number of solutions of the inequality

$$| q_1 w_1 + \cdots + q_m w_m | \leq \frac{c}{\max | q_i |^{m-1}}$$

with integers q_i $(i = 1, \ldots, m)$ such that $| q_i | \leq B$. The theory of diophantine approximations is concerned among other things with the study of this function $\lambda(B, c)$. For special numbers, nothing seems to be known about it except for quadratic irrationalities [1]. On the other hand, there is a theorem of Schmidt [2] stating that for almost all m-tuples, there is a number c_1 such that $\lambda(B, c)$ is asymptotic to $c_1 \log B$.

We have carried out computations giving this function for large values of B (up to 10^4 or 10^6) and a few classical numbers like e, π, $\log 2$, γ, $e + \pi$, $\dfrac{\sqrt{5} - 1}{2}$. These computations in each case have a tendency to confirm the expected behaviour, and we thought it would be useful to have them available in the literature. Most of them are for the usual case $| q w + p |$ (i. e. $m = 2$).

Of course, one knows the asymptotic theorem for $\dfrac{\sqrt{5} - 1}{2}$ [1], but we have included it for comparison with the other numbers, about which nothing is known.

One could ask for the behaviour of another function $\tau(B, c)$, equal to the number of solutions of the inequality

$$| q_1 w_1 + \cdots + q_m w_m | \leq \frac{c}{B}$$

with integers q_i such that $| q_i | \leq B$. This is a more delicate function, and we merely observe that in all cases which we have computed below, its values are small, depending on c, ranging between 2 and 8, except in the case of π, where there is a well known disturbance corresponding to one unusually good approximation with $q = -113$ and $p = 355$.

The tables are easy to read. On the top of each page, we indicate the value for $w_1, w_2, \ldots$ and c. In case we deal with $q_1 w_1 + q_2$, we write it as $q w + p$.

---

[1]) We are much indebted to Columbia University for use of the machine, IBM 794, Project No. UR7HB01.

The columns from left to right give:

The values for q_1, q_2, ... (or q, p).

In the third column, the maximum number N such that the absolute value of the sum is $\leq c/N^{m-1}$.

The absolute value of the sum itself. The symbols $E\text{–}01$ at the end of a number mean that this number should be multiplied by 10^{-1}. Similarly, $E\text{–}02$ means multiplication by 10^{-2}, etc.

Finally, in the last three columns, we give certain values for B, $\lambda(B, c)$ and $\lambda(B, c)/\log B$. Since c is fixed at the top of the page, we write $\lambda(B)$ instead of $\lambda(B,c)$. Values of B were selected, for which a new solution for the inequality occurs.

The quotient $\lambda(B)/\log B$ is the one which should be more or less constant. Its variations seem to be small, and also seem to follow some wave pattern. It is hard to tell whether there is anything significant to this. We have computed it by slide rule, and rounded it off to one decimal.

If one plots the graph of $\lambda(B)$ against B, one finds that it is a step function, which fits a curve $c_1 \log B$ rather well.

For simplicity of programming, a few uninteresting solutions of the inequality in some cases have been omitted from the first four columns. These always occur at the beginning, with values of q_1, q_2, ... equal to 0, 1, 2, or 3. However, their existence was taken into account when computing the function $\lambda(B)$.

We observe finally that we usually took a number c which is such that given an integer $n > 0$ there always exists a solution for an inequality (say)

$$| qw + p | < \frac{c}{n}$$

with $| q | \leq n$. This allows one to make a check on the computations to see that the required solution has been found by the machine.

$$w = e, \quad c = 1 + e$$

q	p	N	$\lvert qw + p \rvert$	B	$\lambda(B)$	$\dfrac{\lambda(B)}{\log B}$
−1	2	5	7.1828182E−01	10	9	3.9
−1	3	13	2.8171817E−01	20	11	3.7
−2	5	8	4.3656365E−01	30	12	3.5
−2	6	6	5.6343634E−01	40	13	3.5
−3	8	24	1.5484548E−01	50	14	3.6
−4	11	29	1.2687268E−01	70	15	3.5
−7	19	132	2.7972799E−02	90	16	3.6
−11	30	37	9.8899887E−02	110	17	3.6
−14	38	66	5.5945598E−02	200	18	3.4
−18	49	52	7.0927087E−02	300	19	3.3
−25	68	86	4.2954288E−02	390	20	3.4
−32	87	248	1.4981489E−02	580	21	3.3
−39	106	286	1.2991310E−02	1080	22	3.2
−71	193	1868	1.9901794E−03	1270	23	3.2
−110	299	337	1.1001130E−02	1460	24	3.3
−142	386	934	3.9803587E−03	2730	25	3.2
−213	579	622	5.9705381E−03	4180	26	3.1
−394	1071	1222	3.0404128E−03	5450	27	3.1
−465	1264	3540	1.0502334E−03	8170	28	3.1
−536	1457	3955	9.3994594E−04	20510	29	2.9
−1001	2721	33714	1.1028750E−04	23230	30	3.0
−1537	4178	4481	8.2965844E−04	25950	31	3.1
−2002	5442	16857	2.2057501E−04	49180	32	3.0
−3003	8163	11238	3.3086251E−04	75120	33	3.0
−7543	20504	22141	1.6793342E−04	98350	34	3.0
−8544	23225	64502	5.7645921E−05	147520	35	3.0
−9545	25946	70633	5.2641582E−05	468490	36	2.8
−18089	49171	743012	5.0043345E−06	517660	37	2.8
−27634	75117	78054	4.7637251E−05	566830	38	2.9
−36178	98342	371506	1.0008669E−05	1084490	39	2.8
−54267	147513	247670	1.5013007E−05	1651310	40	2.8
−172346	468485	489086	7.6025026E−06	2168970	41	2.8
−190435	517656	1431122	2.5981571E−06			
−208524	566827	1545336	2.4061301E−06			
−398959	1084483	19357453	1.9208528E−07			
−607483	1651310	1679450	2.2139866E−06			
−797918	2168966	9678726	3.8417056E−07			

$$w = \pi, \quad c = 1 + \pi$$

q	p	N	$\lvert qw + p \rvert$	B	$\lambda(B)$	$\dfrac{\lambda(B)}{\log B}$
−1	3	29	1.4159265E−01	10	10	4.3
−1	4	4	8.5840733E−01	20	11	3.7
−2	6	14	2.8318531E−01	30	13	3.8
−3	9	9	4.2477795E−01	50	14	3.6
−6	19	27	1.5044408E−01	70	15	3.5
−7	22	467	8.8514247E−03	90	16	3.6
−8	25	31	1.3274123E−01	340	17	3.0
−14	44	233	1.7702850E−02	360	18	3.1
−21	66	155	2.6554275E−02	380	19	3.2

q	p	N	$\|qw + p\|$	B	$\lambda(B)$	$\dfrac{\lambda(B)}{\log B}$
−28	88	116	3.5405699E−02	710	20	3.1
−106	333	469	8.8212804E−03	1070	21	3.1
−113	355	137391	3.0144353E−05	1420	22	3.0
−120	377	466	8.8815690E−03	1780	23	3.1
−226	710	68695	6.0288706E−05	2130	24	3.1
−339	1065	45797	9.0433060E−05	2490	25	3.2
−452	1420	34347	1.2057741E−04	2840	26	3.3
−565	1775	27478	1.5072176E−04	3200	27	3.3
−678	2130	22898	1.8086612E−04	3550	28	3.4
−791	2485	19627	2.1101047E−04	3910	29	3.5
−904	2840	17173	2.4115483E−04	4260	30	3.6
−1017	3195	15265	2.7129917E−04	4620	31	3.7
−1130	3550	13739	3.0144353E−04	4970	32	3.8
−1243	3905	12490	3.3158788E−04	5330	33	3.9
−1356	4260	11449	3.6173224E−04	5680	34	3.9
−1469	4615	10568	3.9187659E−04	6040	35	4.0
−1582	4970	9813	4.2202094E−04	6390	36	4.1
−1695	5325	9159	4.5216529E−04	6750	37	4.2
−1808	5680	8586	4.8230965E−04	104000	38	3.3
−1921	6035	8081	5.1245400E−04	104400	39	3.4
−2034	6390	7632	5.4259835E−04	208400	40	3.3
−2147	6745	7231	5.7274271E−04	312700	41	3.2
−33102	103993	216504	1.9129322E−05	521100	42	3.2
−33215	104348	375994	1.1015029E−05	625400	43	3.2
−66317	208341	510407	8.1142933E−06	833800	44	3.2
−99532	312689	1427780	2.9007206E−06	1146500	45	3.2
−165849	521030	794391	5.2135438E−06	1980200	46	3.2
−199064	625378	713890	5.8014411E−06	2292900	47	3.2
−265381	833719	1790708	2.3128232E−06	3126600	48	3.2
−364913	1146408	7044754	5.8789737E−07			
−630294	1980127	2400864	1.7250422E−06			
−729826	2292816	3522377	1.1757948E−06			
−995207	3126535	3642097	1.1371449E−06			

$$w = \gamma, \; c = 1 + \gamma$$

q	p	N	$\|qw + p\|$	B	$\lambda(B)$	$\dfrac{\lambda(B)}{\log B}$
−1	2	10	1.5443133E−01	10	9	3.9
−2	3	5	2.6835300E−01	20	12	4.0
−2	4	5	3.0886266E−01	30	13	3.8
−3	5	13	1.1392167E−01	50	14	3.6
−4	7	38	4.0509654E−02	60	15	3.7
−7	12	21	7.3412020E−02	80	16	3.7
−8	14	19	8.1019307E−02	100	17	3.7
−11	19	47	3.2902366E−02	130	18	3.7
−15	26	207	7.6072874E−03	150	19	3.8
−26	45	62	2.5295079E−02	250	20	3.6
−30	52	103	1.5214575E−02	280	21	3.7
−41	71	89	1.7687792E−02	400	22	3.7
−56	97	156	1.0080504E−02	520	23	3.7
−71	123	637	2.4732171E−03	790	24	3.6

q	p	N	$\|qw \div p\|$	B	$\lambda(B)$	$\dfrac{\lambda(B)}{\log B}$
−86	149	307	5.1340703E−03	1190	25	3.5
−142	246	318	4.9464341E−03	1580	26	3.5
−157	272	592	2.6608532E−03	4870	27	3.2
−228	395	8405	1.8763610E−04	5260	28	3.3
−299	518	690	2.2855810E−03	5660	29	3.3
−456	790	4202	3.7527221E−04	10520	30	3.2
−684	1185	2801	5.6290831E−04	10920	31	3.3
−912	1580	2101	7.5054441E−04	16170	32	3.3
−2807	4863	7117	2.2158384E−04			
−3035	5258	46460	3.3947739E−05			
−3263	5653	10262	1.5368836E−04			
−6070	10516	23230	6.7895478E−05			
−6298	10911	13171	1.1974063E−04			
−9333	16169	18383	8.5792886E−05			

$$w = e + \pi, c = 1 + w$$

q	p	N	$\|qw + p\|$	B	$\lambda(B)$	$\dfrac{\lambda(B)}{\log B}$
−1	5	7	8.5987448E−01	10	7	3.0
−1	6	48	1.4012552E−01	20	8	2.7
−2	12	24	2.8025103E−01	40	9	2.4
−6	35	43	1.5924689E−01	50	11	2.8
−7	41	358	1.9121374E−02	90	12	2.7
−8	47	56	1.2100414E−01	260	13	2.3
−14	82	179	3.8242748E−02	300	14	2.5
−43	252	270	2.5397272E−02	340	15	2.6
−50	293	1093	6.2758975E−03	630	16	2.5
−57	334	534	1.2845477E−02	920	17	2.5
−107	627	1044	6.5695792E−03	1840	18	2.4
−157	920	23358	2.9368167E−04	2760	19	2.4
−314	1840	11679	5.8736333E−04	3680	20	2.4
−471	2760	7786	8.8104500E−04	4600	21	2.5
−628	3680	5839	1.1747267E−03	19613	22	2.2
−785	4600	4671	1.4684083E−03	20533	23	2.3
−3347	19613	63176	1.0858254E−04	40146	24	2.3
−3504	20533	37060	1.8509913E−04			
−6851	40146	89652	7.6516590E−05			

$$w = \log\log 2 + \pi,\ c = 1 + w$$

q	p	N	$\|qw + p\|$	B	$\lambda(B)$	$\dfrac{\lambda(B)}{\log B}$
−1	2	4	7.7507973E−01	10	9	3.9
−1	3	16	2.2492026E−01	20	11	3.7
−2	5	6	5.5015946E−01	30	12	3.5
−2	6	8	4.4984053E−01	40	13	3.5
−3	8	11	3.2523920E−01	50	14	3.6
−4	11	37	1.0031893E−01	70	15	3.5

22*

q	p	N	$\mid qw \div p \mid$	B	$\lambda(B)$	$\dfrac{\lambda(B)}{\log B}$
-5	14	30	1.2460133E-01	90	16	3.6
-9	25	155	2.4282403E-02	120	17	3.6
-13	36	49	7.6036528E-02	140	18	3.6
-18	50	77	4.8564805E-02	230	19	3.5
-22	61	72	5.1754125F-02	340	20	3.4
-31	86	137	2.7471723E-02	700	21	3.2
-40	111	1183	3.1893203E-03	810	22	3.3
-49	136	178	2.1093082E-02	920	23	3.4
-80	222	591	6.3786406E-03	1720	24	3.2
-120	333	394	9.5679609E-03	2630	25	3.2
-249	691	733	5.1464809E-03	4350	26	3.1
-289	802	1928	1.9571606E-03	6980	27	3.1
-329	913	3063	1.2321597E-03	11320	28	3.0
-618	1715	5206	7.2500097E-04	15660	29	3.0
-947	2628	7443	5.0715869E-04	22630	30	3.0
-1565	4343	17329	2.1784228E-04	26980	31	3.1
-2512	6971	13048	2.8931642E-04			
-4077	11314	52817	7.1474140E-05			
-5642	15657	25791	1.4636814E-04			
-8154	22628	26408	1.4294828E-04			
-9719	26971	50405	7.4893998E-05			

$$w = \log\log 31, \quad c = 1 \div w$$

q	p	N	$\mid qw \div p \mid$	B	$\lambda(B)$	$\dfrac{\lambda(B)}{\log B}$
-1	2	2	7.6627796E-01	10	12	5.2
-2	2	4	4.6744407E-01	20	14	4.7
-2	3	4	5.3255592E-01	30	15	4.4
-3	3	3	7.0116611E-01	40	16	4.3
-3	4	7	2.9883389E-01	60	17	4.2
-4	5	34	6.5111854E-02	80	18	4.1
-5	6	13	1.6861018E-01	100	19	4.1
-8	10	17	1.3022371E-01	140	20	4.1
-9	11	21	1.0349833E-01	190	21	4.0
-13	16	58	3.8386472E-02	230	22	4.0
-17	21	83	2.6725382E-02	330	23	4.0
-30	37	191	1.1661090E-02	420	24	4.0
-47	58	148	1.5064292E-02	650	25	3.9
-60	74	95	2.3322180E-02	740	26	3.9
-77	95	656	3.4032016E-03	1070	27	3.9
-107	132	270	8.2578886E-03	1480	28	3.8
-154	190	328	6.8064032E-03	1800	29	3.9
-184	227	460	4.8546870E-03	2540	30	3.8
-261	322	1538	1.4514855E-03	3280	31	3.8
-338	417	1144	1.9517161E-03	4340	32	3.8
-522	644	769	2.9029709E-03	5880	33	3.9
-599	739	4465	5.0023061E-04	7620	34	3.8
-860	1061	2348	9.5125487E-04	10160	35	3.8
-1198	1478	2232	1.0004612E-03			
-1459	1800	4952	4.5102425E-04			

q	p	N	$\mid qw + p \mid$	B	$\lambda(B)$	$\dfrac{\lambda(B)}{\log B}$
−2058	2539	45394	4.9206364E−05			
−2657	3278	4065	5.4943697E−04			
−3517	4339	5559	4.0181788E−04			
−4116	5078	22697	9.8412728E−05			
−6174	7617	15131	1.4761909E−04			
−8232	10156	11348	1.9682546E−04			

$$w = \frac{\sqrt{5}-1}{2},\ c = 1,\ 0 < qw + p$$

q	p	N	$\mid qw + p \mid$	B	$\lambda(B)$	$\dfrac{\lambda(B)}{\log B}$
−1	0	1	6.1803398E−01	10	3	1.3
−2	1	4	2.3606797E−01	20	4	1.3
−5	3	11	9.0169942E−02	40	5	1.3
−13	8	29	3.4441853E−02	90	6	1.3
−34	21	76	1.3155617E−02	240	7	1.3
−89	55	199	5.0249987E−03	610	8	1.2
−233	144	521	1.9193787E−03	1600	9	1.2
−610	377	1364	7.3313743E−04	4200	10	1.2
−1597	987	3571	2.8003358E−04			
−4181	2584	9349	1.0696331E−04			

$$w_1 = \log 3,\ w_2 = \log 2,\ c = \log 3 + \log 2$$

q	p	N	$\mid qw_1 + pw_2 \mid$	B	$\lambda(B)$	$\dfrac{\lambda(B)}{\log B}$
−1	2	6	2.8768207E−01	1	6	8.6
−2	3	15	1.1778303E−01	5	8	5.0
−3	5	10	1.6989903E−01	10	10	4.4
−4	6	7	2.3556607E−01	20	13	4.4
−5	8	34	5.2116001E−02	30	14	4.1
−7	11	27	6.5667034E−02	40	15	4.1
−10	16	17	1.0423200E−01	50	16	4.1
−12	19	132	1.3551033E−02	70	17	4.0
−17	27	46	3.8564967E−02	90	18	4.1
−24	38	66	2.7102067E−02	110	19	4.0
−29	46	71	2.5013934E−02	150	20	4.0
−41	65	156	1.1462901E−02	170	21	4.1
−53	84	858	2.0881324E−03	240	22	4.0
−65	103	114	1.5639166E−02	260	23	4.1
−94	149	191	9.3747685E−03	320	24	4.2
−106	168	429	4.1762647E−03	410	25	4.2
−147	233	245	7.2866362E−03	490	26	4.2
−159	252	286	6.2643971E−03	570	27	4.3
−200	317	344	5.1985038E−03	1060	28	4.0
−253	401	576	3.1103715E−03	1540	29	4.0
−306	485	1752	1.0222391E−03	2110	30	3.9
−359	569	1680	1.0658932E−03	3170	31	3.9
−665	1054	41044	4.3654110E−05	4220	32	3.8

q	p	N	$\lvert qw_1 + pw_2 \rvert$	B	$\lambda(B)$	$\dfrac{B(\lambda)}{\log B}$
−971	1539	1830	9.7858500E−04	5270	33	3.9
−1330	2108	20522	8.7308221E−05	6330	34	3.9
−1995	3162	13681	1.3096233E−04	23680	35	3.5
−2660	4216	10261	1.7461644E−04	24730	36	3.6
−3325	5270	8208	2.1827055E−04	25790	37	3.7
−3990	6324	6840	2.6192466E−04	50510	38	3.5
−14936	23673	28970	6.1848688E−05	75240	39	3.5
−15601	24727	98477	1.8194580E−05	101020	40	3.5
−16266	25781	70376	2.5459532E−05	125750	41	3.5
−31867	50508	246630	7.2649564E−06	176260	42	3.5
−47468	75235	163936	1.0929623E−05	302000	43	3.5
−63734	101016	123315	1.4529913E−05	427740	44	3.4
−79335	125743	488928	3.6646670E−06	478250	45	3.5
−111202	176251	497672	3.6002748E−06	603990	46	3.5
−190537	301994	27819424	6.4406776E−08	905990	47	3.4
−269872	427737	480483	3.7290738E−06	1207976	48	3.4
−301739	478245	506736	3.5358826E−06	1509970	49	3.4
−381074	603988	13909712	1.2881355E−07			
−571611	905982	9277333	1.9313302E−07			
−762148	1207976	6954856	2.5762711E−07			
−952685	1509970	5566400	3.2188836E−07			

$$w_1 = \log 3, \quad w_2 = \log 5, \quad w_3 = \log 7, \quad c = \frac{w_1 + w_2 + w_3}{3}$$

q_1	q_2	q_3	N	$\lvert \text{sum} \rvert$	B	$\lambda(B)$	$\dfrac{\lambda(B)}{\log B}$
−2	0	1	2	2.5131443E−01	10	13	5.7
−2	−1	2	4	8.5157807E−02	20	16	5.3
−1	−3	3	4	8.9195578E−02	30	18	5.3
−8	−3	7	19	4.1589966E−03	40	20	5.4
−3	−4	5	19	4.0377704E−03	50	23	5.9
−11	−7	12	113	1.2122627E−04	60	25	6.1
−6	−8	10	13	8.0755408E−03	70	26	6.1
−757	−8	434	1171	1.1312253E−06	90	27	6.0
−1	−9	8	9	1.6272308E−02	150	28	5.6
−19	−10	19	19	4.2802230E−03	180	29	5.6
−14	−11	17	19	3.9165441E−03	220	30	5.6
−22	−14	24	79	2.4245255E−04	290	31	5.5
−33	−21	36	65	3.6367882E−04	360	32	5.4
−44	−28	48	56	4.8490510E−04	430	33	5.4
−10	−50	47	80	2.4150278E−04	700	34	5.2
−21	−57	59	113	1.2027651E−04	730	35	5.3
−32	−64	71	1278	9.4976534E−07	760	36	5.4
−43	−71	83	112	1.2217604E−04	790	37	5.5
−789	−72	505	2923	1.8146011E−07	830	38	5.7
−64	−128	142	903	1.8995307E−06	860	39	5.8
−821	−136	576	1420	7.6830519E−07			
−96	−192	213	737	2.8492960E−06			
−853	−200	647	950	1.7180705E−06			
−128	−256	284	639	3.7990613E−06			
−160	−320	355	571	4.7488267E−06			
−192	−384	426	521	5.6985919E−06			

q_1	q_2	q_3	N	$|\text{sum}|$	B	$\lambda(B)$	$\dfrac{\lambda(B)}{\log B}$
3	−2	0	4	7.6961040E−02			
34	−22	−1	46	7.2640789E−04			
5	−1	−2	13	8.1967671E−03			
693	−120	−292	715	3.0307561E−06			
725	−56	−363	863	2.0809908E−06			
−1	2	−1	2	1.7435338E−01			
−1	3	−2	3	1.6211885E−01			
−1	43	−35	65	3.6272906E−04			
−2	5	−3	11	1.2234538E−02			
−7	6	−1	8	2.0431305E−02			
−12	36	−23	56	4.8395534E−04			
−23	29	−11	50	6.0518160E−04			
−174	149	−25	189	4.3001952E−05			

$$w_1 = \log 3,\ w_2 = \log 5,\ w_3 = \log 7,\ c = w_1 + w_2 + w_3$$

q_1	q_2	q_3	N	$|\text{sum}|$	B	$\lambda(B)$	$\dfrac{\lambda(B)}{\log B}$
−2	0	1	4	2.5131443E−01	10	32	13.9
−7	0	4	7	9.3354574E−02	20	41	13.7
−2	−1	2	7	8.5157807E−02	30	46	13.5
−4	−1	3	5	1.6615662E−01	40	53	14.4
−0	−2	2	2	6.7294447E−01	50	57	14.6
−1	−2	2	3	4.2566781E−01	60	62	15.1
−2	−2	3	3	4.2163004E−01	70	63	14.8
−4	−2	4	5	1.7031562E−01	80	65	14.8
−6	−2	5	7	8.0998811E−02	90	66	14.7
−13	−2	9	19	1.2355764E−02	100	67	14.6
−1	−3	3	7	8.9195578E−02	110	69	14.7
−8	−3	7	33	4.1589966E−03	120	70	14.6
−1	−4	4	4	2.4727665E−01	130	71	14.6
−3	−4	5	33	4.0377704E−03	150	72	14.4
−5	−5	7	7	8.1120037E−02	160	73	14.4
−0	−6	5	7	7.2923270E−02	170	74	14.4
−16	−6	14	23	8.3179933E−03	180	75	14.4
−11	−7	12	195	1.2122627E−04	190	76	14.5
−6	−8	10	24	8.0755408E−03	210	78	14.6
−757	−8	434	2028	1.1312253E−06	220	81	15.0
−1	−9	8	16	1.6272308E−02	240	82	15.0
−19	−10	19	32	4.2802230E−03	270	83	15.0
−14	−11	17	34	3.9165441E−03	280	84	14.9
−9	−12	15	19	1.2113311E−02	290	85	15.0
−4	−13	13	15	2.0310078E−02	310	86	15.0
−22	−14	24	138	2.4245255E−04	340	87	14.9
−17	−15	22	24	7.9543145E−03	360	90	15.3
−30	−17	31	32	4.4014493E−03	390	91	15.3
−25	−18	29	35	3.7953179E−03	450	92	15.2
−33	−21	36	113	3.6367882E−04	510	93	14.9
−44	−28	48	97	4.8490510E−04	520	94	15.0
−55	−35	60	87	6.0613137E−04	540	95	15.1
−66	−42	72	79	7.2735764E−04	570	97	15.3
−270	−43	188	340	4.0152656E−05	600	98	15.3

q_1	q_2	q_3	N	$\mid \text{sum} \mid$	B	$\lambda(B)$	$\dfrac{\lambda(B)}{\log B}$
−10	−50	47	138	2.4150278E−04	630	99	15.4
−21	−57	59	196	1.2027651E−04	640	101	15.6
−32	−64	71	2213	9.4976534E−07	670	102	15.7
−43	−71	83	195	1.2217604E−04	700	103	15.7
−789	−72	505	4236	1.8146011E−07	730	104	15.8
−54	−78	95	138	2.4340231E−04	760	105	15.8
−65	−85	107	112	3.6462859E−04	790	106	15.9
−31	−107	106	113	3.6177929E−04	830	107	15.9
−302	−107	259	344	3.9202891E−05	860	108	16.0
−42	−114	118	139	2.4055302E−04	890	109	16.0
−53	−121	130	197	1.1932674E−04	920	110	16.2
−64	−128	142	1565	1.8995307E−06	950	111	16.2
−75	−135	154	194	1.2312580E−04			
−821	−136	576	2461	7.6830519E−07			
−334	−171	330	348	3.8253126E−05			
−96	−192	213	1278	2.8492960E−06			
−853	−200	647	1645	1.7180705E−06			
−128	−256	284	1106	3.7990613E−06			
−885	−264	718	1320	2.6678358E−06			
−160	−320	355	989	4.7488267E−06			
−917	−328	789	1134	3.6176010E−06			
−192	−384	426	903	5.6985919E−06			
−949	−392	860	1009	4.5673665E−06			
−224	−448	497	836	6.6483572E−06			
−256	−512	568	782	7.5981227E−06			
−288	−576	639	737	8.5478879E−06			
2	−1	0	2	5.8778666E−01			
2	−2	0	2	1.0216512E−00			
3	−1	−1	4	2.5951119E−01			
12	−7	−1	12	2.8628072E−02			
34	−22	−1	80	7.2640789E−04			
5	−1	−2	23	8.1967671E−03			
8	−3	−2	8	6.8764273E−02			
10	−2	−4	16	1.6393534E−02			
37	−18	−6	37	3.3113625E−03			
565	−376	−8	825	6.8298175E−06			
45	−15	13	74	8.4763416E−04			
56	−8	−25	69	9.6886044E−04			
206	−85	−46	332	4.2052187E−05			
217	−78	−58	242	7.9174088E−05			
597	−312	−79	889	5.8800520E−06			
238	−21	−117	336	4.1102422E−05			
629	−248	−150	971	4.9302868E−06			
661	−184	−221	1081	3.9805214E−06			
693	−120	−292	1239	3.0307561E−06			
725	−56	−363	1495	2.0809908E−06			
−1	2	−1	5	1.7435338E−01			
−1	3	−2	5	1.6211885E−01			
−1	43	−35	113	3.6272906E−04			
−2	2	−1	2	9.2425889E−01			
−2	5	−3	19	1.2234538E−02			
−3	3	−1	3	4.1343327E−01			
−4	4	−1	6	9.7392346E−02			
−4	10	−6	13	2.4469075E−02			
−5	7	−3	8	6.4726502E−02			

q_1	q_2	q_3	N	sum	B	$\lambda(B)$	$\dfrac{\lambda(B)}{\log B}$
−7	6	−1	15	2.0431305E−02			
−9	11	−4	11	3.2665842E−02			
−12	36	−23	98	4.8395534E−04			
−15	32	−18	36	3.5538151E−03			
−23	29	−11	87	6.0518160E−04			
−26	25	−6	36	3.4325888E−03			
−31	26	−4	31	4.7641783E−03			
−35	65	−34	65	1.0891369E−03			
−46	58	−22	62	1.2103632E−03			
−57	51	−10	59	1.3315895E−03			
−110	277	−167	321	4.4901483E−05			
−142	213	−96	325	4.3951718E−05			
−153	206	−84	245	7.7274557E−05			
−163	156	−37	168	1.6422823E−04			
−174	149	−25	328	4.3001952E−05			
−185	142	−13	243	7.8224322E−05			
−327	355	−109	368	3.4272604E−05			
−359	291	−38	363	3.5222370E−05			
−437	632	−276	661	1.0628878E−05			
−469	568	−205	693	9.6791134E−06			
−501	504	−134	730	8.7293481E−06			
−533	440	−63	773	7.7795827E−06			

Bibliography

[1] *S. Lang*, Asymptotic approximations, to appear.

[2] *W. Schmidt*, A metrical theorem in diophantine approximation, Canadian Journal of Mathematics **11** (1959). 619—631.

Columbia University

Eingegangen 21. September 1964

Topology Vol. 5, pp. 363–370. Pergamon Press, 1966. Printed in Great Britain

ALGEBRAIC VALUES OF MEROMORPHIC FUNCTIONS—II

SERGE LANG

(*Received* 14 *December* 1965)

WE CONTINUE THE STUDY of the possible distributions of numbers at which certain meromorphic functions take on algebraic values. In particular, a 1-parameter subgroup of a linear group or an abelian variety which contains sufficiently many algebraic points must itself be an algebraic subgroup.

§1. PRELIMINARIES

Let K be a number field (finite extension of $\mathbf{Q}$), and let $\alpha \in K$. By a denominator for α, we mean a positive integer d such that $d\alpha$ is an algebraic integer. If B is real >0, we shall say size $(\alpha) \leqq B$ if there exists a denominator d for α such that $\log d \leqq B$, and if

$$\log \max_{\sigma} |\sigma\alpha| \leqq B$$

for all embeddings σ of K into $\mathbf{C}$ (i.e. all conjugates of α have logs of absolute value bounded by B). For any $\alpha \neq 0$ in K, we have

$$-2[K : \mathbf{Q}] \text{ size}(\alpha) \leqq \log |\sigma\alpha|.$$

If r is a positive integer, then size $(\alpha^r) \leqq r$ size(α).

Next we recall that an entire function (of a complex variable) is said to be of order $\leqq \rho$ if there is a constant $C > 0$ such that

$$|F|_R = \max_{|t| = R} |F(t)| \leqq C^{R^\rho}$$

for all R sufficiently large. (We omit the usual ε which is irrelevant for what follows.) If F is such a function, then the number of zeros of F in a circle of radius R is $O(R^\rho)$.

A meromorphic function f is said to be of order $\leqq \rho$ if it can be expressed as a quotient of entire functions of order $\leqq \rho$. If S is a set of complex numbers, we say that such a meromorphic function f is *defined* on S if we can write $f = g/h$ where g, h are entire of order $\leqq \rho$ and no point of S is a zero of h. For simplicity, we shall always assume that ρ is an integer $\geqq 1$.

We took the log in our definition of the size because the estimates which occur in subsequent proofs always occur in the exponent if we don't take the log. Furthermore, these estimates will be independent of constant factors. Thus we use the following notation. In dealing with two positive functions λ_1, λ_2 defined on some set, we write $\lambda_1 \ll \lambda_2$ if there exists a constant $C > 0$ such that $\lambda_1 \leqq C\lambda_2$. This is the usual Vinogradov notation. We write $\lambda_1 \gg \ll \lambda_2$ to mean the obvious thing.

Let S be a set of complex numbers, expressed as a union $S = \bigcup_{n=1}^{n} S_n$ such that $S_n \geqq S_{n+1}$ for all n. One says that the subsets $\{S_n\}$ form a filtration of S. We shall assume throughout when dealing with such filtrations that there is a constant $C > 0$ such that for all n and all $z \in S_n$, we have $|z| \leqq Cn$.

Let f be a meromorphic function defined on S, with values in a number field K. We shall say that f is of *arithmetic order* $\leqq \rho$ on S (or more accurately, with respect to the filtration $\{S_n\}$) if there is a constant $C \geqq 1$ such that the following conditions are satisfied.

AO 1.   For all n and $z \in S_n$, we have size $f(z) \leqq Cn^\rho$.

AO 2.   There is an entire function h of order $\leqq \rho$, such that hf is entire, h has no zero in S, and for all n, $xz \in S_n$,

$$\log |1/h(z)| \leqq Cn^\rho.$$

Using our notation $\ll$, we can write our conditions in the form

$$\text{size } f(z) \ll n^\rho \qquad \text{and} \qquad \log |1/h(z)| \ll n^\rho,$$

for $z \in S_n$, and $n \to \infty$.

§2.   VALUES OF MEROMORPHIC FUNCTIONS

The following theorem is similar to theorems of Schneider (cf. Schneider's book *Einführung in die Transzendenten Zahlen*, p. 49).

THEOREM 1.   *Let f, g be meromorphic functions of order $\leqq \rho$. Let $S = \bigcup S_n$ be as above, and assume that f, g are defined on S, with values in the number field K, and of arithmetic order $\leqq \rho$ on S. Let λ be a number > 2. If* card $(S_n) \gg \ll n^{\lambda\rho}$ *for $n \to \infty$, then f, g are algebraically dependent over K.*

Proof.   Let $C_1, C_2 > 0$ be such that

$$C_1 n^{\lambda\rho} \leqq \text{card}(S_n) \leqq C_2 n^{\lambda\rho}$$

for all n. Without loss of generality, we may assume (for convenience) that C_2 is an integer, and is a square. We shall deal with a large integer n, taken to be a square, and which will tend to infinity later.

Let $r = 2C_2^{1/2} n^{\lambda\rho/2} \ll n^{\lambda\rho/2}$.

We can find algebraic integers a_{ij} not all zero in K, such that the function

$$F = \sum_{i,j=1}^{r} a_{ij} f^i g^j$$

has a zero at every point $z \in S_n$. This amounts to solving linear equations in r^2 unknowns, and we have

$$\text{number of equations} \leqq C_2 n^{\lambda\rho} \leqq r^2/4.$$

The coefficients of these equations are the values

$$f(z)^i g(z)^j$$

with $z \in S_n$. By hypothesis, we have

$$\text{size of coefficients} \ll n^\rho r.$$

Hence by a standard, easy lemma of Siegel ([8], cf. also [4]) we can find the a_{ij} such that

$$\text{size}(a_{ij}) \leqq n^\rho r$$

for all n sufficiently large.

If f, g are algebraically independent over K, then F is not identically zero. Let s be the smallest integer such that $f(z) = 0$ for all $z \in S_s$ but $F(w) \neq 0$ for some $w \in S_{s+1}$. (Such an s exists because in a circle of radius s we know that F has $O(s^\rho)$ zeros for $s \to \infty$) Then $s \geqq n$, and trivial estimates yield

$$\text{size } F(w)^{\cdot} \ll s^\rho r.$$

We note that $s^\rho r \ll s^{\lambda \rho}$.

We now estimate $|F(w)|$. Let h be the entire function satisfying condition $AO\ 2$ with respect to both f and g. (If h_1 satisfies this condition with respect to f, and h_2 satisfies it with respect to g, then $h = h_1 h_2$ will serve our purposes.) Then $h^{2r}F$ is an entire function having zeros at all elements of S_s. We use the expression

$$F(w) = \left. \frac{h(t)^{2r}F(t)}{h^{2r}(w)\Pi(t-z)} \Pi(w-z) \right|_{t=w}$$

where the products are taken for z in S_s. We apply the maximum principle, and bound the function on the right on a circle of radius $R = s^{\lambda/2}$. We estimate three things:

First, $h^{2r}F$ using the fact that h, hf, hg are entire of order $\leqq \rho$, and using the expression

$$h^{2r}F = \sum_{i,j=1}^{r} a_{ij}h^{2r-i-j}(hf)^i(hg)^j.$$

We take the log of the absolute value, and find

$$\sup_{|t|=R} \log |h^{2r}(t)F(t)| \ll s^{\lambda \rho/2} r \ll s^{\lambda \rho}.$$

Second, we have the upper bound (by $AO\ 2$),

$$\log \frac{1}{|h^{2r}(w)|} \ll s^\rho r \ll s^{\lambda \rho}.$$

Third, and most important, we have

$$\log \sup_{|t|=R} \prod_{z \in S_s} \frac{|w-z|}{|t-z|} \ll s^{\lambda \rho} - \frac{\lambda-2}{2} s^{\lambda \rho} \log s.$$

Combining these three estimates, we get the upper bound

$$\log |F(w)| \ll s^{\lambda \rho} - \frac{\lambda-2}{2} s^{\lambda \rho} \log s$$

(because $3s^{\lambda\rho} \ll s^{\lambda\rho}$). For s sufficiently large, this contradicts the fact that

$$-\text{size } F(w) \ll \log |F(w)|,$$

thereby proving Theorem 1.

Remark. The theorem holds also if the condition card $(S_n) \gg$ $\ll n^{\lambda\rho}$ is replaced by card $(S_n) \gg n^{\lambda\rho}$, since from sets S_n satisfying this second condition, we can at once obtain sets S'_n satisfying the first by throwing away some of the points, proceeding inductively, to insure that $S'_n \subset S_n$.

COROLLARY 1. *Let β_1, β_2 be complex numbers linearly independent over the rationals, and let z_1, z_2, z_3 be complex numbers, also linearly independent over the rationals. Then not all numbers*

$$e^{\beta_i z_v} \qquad (i = 1,2 \text{ and } v = 1,2,3)$$

are algebraic.

Proof. Let $f(t) = e^{\beta_1 t}$ and $g(t) = e^{\beta_2 t}$. Let S_n be the set of linear combinations

$$k_1 z_1 + k_2 z_2 + k_3 z_3$$

with $1 \le k_v \le n$. The theorem then applies.

COROLLARY 2. *Let β be irrational. Then there are at most two multiplicatively independent algebraic numbers $\alpha \ne 0$ such that α^β is algebraic.*

Proof. Let $\beta_1 = 1$, $\beta_2 = \beta$, $z_v = \log \alpha_v$.

It was pointed out to me by E. G. Straus that the corollaries above have been known to Siegel for some time (cf. L. ALAOGLU and P. ERDÖS, On highly composite and similar numbers, *Trans. Am. Math. Soc.* **56** (1944), 445) but they do not appear explicitly otherwise in the literature. I was led to them by a recent question of Serre: If y is real, and x^y is algebraic for all positive rational numbers x, then y is rational. This application arose in Serre's determination of certain characters of ideal classes, and will be discussed by Serre elsewhere. It also arose in the above mentioned paper.

§3. GENERAL EXPONENTIAL FUNCTIONS: THE LINEAR GROUP

Theorem 1 can be applied to give information concerning the general exponential map on group varieties. In this section, we deal with the linear group.

First, a remark on 1-parameter subgroups. Let G be a group variety, and let $\varphi : \mathbf{C} \to G_{\mathbf{C}}$ be a 1-parameter subgroup. This means that φ is a complex-analytic homomorphism of $\mathbf{C}$ into G. Then of course, complex analytically, φ is an analytic curve in G. However, it may well be that algebraically, φ has dimension > 1. Indeed, if say G is embedded in some projective space, and $(\varphi_0, \ldots, \varphi_N)$ are the projective coordinates of φ, then by the algebraic dimension of φ we shall mean the number of algebraically independent coordinate functions φ_i/φ_0 $(i = 1, \ldots, N)$. This is the same as the dimension of the smallest group subvariety of G containing $\varphi(\mathbf{C})$, i.e. containing the 1-parameter subgroup.

THEOREM 2. *Let G be a linear group variety defined over the field of all algebraic numbers. Let $\varphi : \mathbf{C} \to G_{\mathbf{C}}$ be a 1-parameter subgroup of G, of algebraic dimension ≥ 2. Let Γ be a subgroup of $\mathbf{C}$ having at least three linearly independent elements over $\mathbf{Q}$. Then $\varphi(\Gamma)$ cannot be contained in the group of algebraic points of G.*

Proof. We can represent G globally as a group of matrices, so that for some matrix M,

$$\varphi(t) = \exp(tM) = \sum t^{\mu}M^{\mu}/\mu!$$

Let z_{ν} $(\nu = 1, 2, 3)$ be complex numbers linearly independent over $\mathbf{Q}$, such that $\varphi(z_{\nu})$ is algebraic, say with components in a number field K. Suppose that M is an $m \times m$ matrix. Then we can write

$$\varphi(t) = (\varphi_{ij}(t)), \qquad 1 \leq i,j \leq m,$$

with entire functions φ_{ij} which are of order ≤ 1. If φ has algebraic dimension ≥ 2, then at least two among the functions φ_{ij} are algebraically independent over K, say f, g.

Let S_n be the set of linear combinations

$$k_1 z_1 + k_2 z_2 + k_3 z_3$$

with $1 \leq k_{\nu} \leq n$. Condition AO 1 is easily verified, because

$$\varphi(k_1 z_1 + k_2 z_2 + k_3 z_3) = \varphi(z_1)^{k_1}\varphi(z_2)^{k_2}\varphi(z_3)^{k_3},$$

and it is easy to make the necessary estimates on the size of a product of matrices to see that AO 1 is true. Condition AO 2 is always true for entire functions (take $h = 1$), and consequently, we can apply Theorem 1 to obtain the desired contradiction.

COROLLARY. *Let G be a linear group variety defined over a number field K. Let Γ be a commutative subgroup of G_K, containing at least three elements independent over $\mathbf{Z}$. If Γ is contained in a 1-parameter subgroup of G, then this 1-parameter subgroup is a group subvariety (i.e. algebraic) of dimension 1.*

§4. ABELIAN VARIETIES

The estimates needed to verify that the coordinate functions of a 1-parameter subgroup of a linear group satisfy the hypotheses of Theorem 1 turned out to be quite easy to make. In order to do the same thing for abelian varieties, we must exert more effort, both in the analytic and arithmetic directions.

Let A be an abelian variety defined over the field of algebraic numbers. We can represent the complex analytic group $A_{\mathbf{C}}$ as a quotient of complex d-space (if $d = \dim A$), say by the map

$$\Theta : \mathbf{C}^d \to A_{\mathbf{C}},$$

which can be given explicitly in terms of theta functions. If $\alpha \in \mathbf{C}^n$, and $\alpha \neq 0$, then the curve $\varphi : \mathbf{C} \to A_{\mathbf{C}}$ such that $\varphi(t) = \Theta(t\alpha)$ is a 1-parameter subgroup of A, and all 1-parameter subgroups of A can be so described.

THEOREM 3. *Let A be an abelian variety defined over the field of algebraic numbers. Let $\varphi : \mathbf{C} \to A_{\mathbf{C}}$ be a 1-parameter subgroup of A. Let Γ be a subgroup of $\mathbf{C}$ having at least seven linearly independent elements over $\mathbf{Q}$. If φ has algebraic dimension ≥ 2, then $\varphi(\Gamma)$ cannot be contained in the group of algebraic points of A.*

Proof. Let $\Theta = (\theta_0, \ldots, \theta_N)$ be the representation of Θ by theta functions, giving an embedding of $A_{\mathbf{C}}$ into projective N-space. We induce the functions $\theta_0, \ldots, \theta_N$ on the analytic curve φ, and obtain entire functions $(f_0, \ldots, f_N)$ in one variable, of order ≤ 2, which realize our 1-parameter map φ, that is $\varphi(t) = (f_0(t), \ldots, f_N(t))$. Without loss of generality, we may assume that the divisor of zeros of θ_0 does not pass through the origin, so that in particular, f_0 is not identically zero. Let D be a small disc of radius $\delta > 0$ around the origin in $\mathbf{C}$, so that no zero of f_0 lies in D. Then $|f_0(t)|$ is bounded away from O on D, and hence $\log |1/f(t)|$ is bounded from above on D.

By hypothesis, at least two of the functions among $f_1/f_0, \ldots, f_N/f_0$ are algebraically independent over the algebraic numbers, say f and g.

We may assume that A is defined over a number field K, and that Γ is generated by seven elements z_ν $(\nu = 1, \ldots, 7)$ linearly independent over $\mathbf{Q}$, such that $\varphi(z_\nu)$ is contained in A_k for $\nu = 1, \ldots, 7$. Let S_n be the set of linear combinations

$$k \cdot z = k_1 z_1 + \cdots + k_7 z_7$$

with $-n \leq k_\nu \leq n$, and such that $k \cdot z$ lies in D. By routine techniques, one verifies that card $(S_n) \gg n^5$ or $\gg n^6$, depending on whether at least two of the elements z_ν are linearly independent over $\mathbf{R}$ or not. (For instance, suppose that z_6, z_7 are linearly independent over $\mathbf{R}$, and let Λ be the lattice generated by them. We consider $k_1 z_1 + \cdots + k_5 z_5$ on $\mathbf{C}/\Lambda = \mathbf{R}^2/\Lambda$. We cut up a fundamental domain into approximately $1/\delta$ small square of sides approximately equal to $\delta^{1/2}$, and use the Dirichlet box principle on the $\gg n^5$ elements consisting of linear combinations $k \cdot z$ with $|k| \leq n/C$ for some large constant C, depending on the z_ν. Subtracting one element from all the others crowded in a small box, we obtain essentially δn^5 elements in D. By being more careful and using translations by z_6, z_7, using points $k \cdot z$ lying outside small circles around the zeros of the entire function f_0, and using the minimum modulus principle for entire functions, one can probably shrink 7 to 5, but this is a secondary matter here.) For definiteness, let us assume that we are in the case

$$\text{card } (S_n) \gg \ll n^5.$$

Let $S = \bigcup S_n$. We already know that our two functions f, g satisfy condition AO 2 on S since in fact, they have the bounded denominator $h = f_0$ on D. We shall prove in the next lemma that they satisfy condition AO 1, and thus are of arithmetic order ≤ 2 on S. Applying Theorem 1 then gives the contradiction which proves Theorem 3.

LEMMA 1. *Let A be an abelian variety defined over a number field K. Let $\varphi : \mathbf{C} \to A_{\mathbf{C}}$ be a 1-parameter subgroup, represented by projective coordinates $(f_0, \ldots, f_N)$. Say f_0 is not identically zero. Let $z_1, \ldots, z_m \in \mathbf{C}$ be linearly independent over $\mathbf{Q}$, such that $\varphi(z_\nu) \in A_k$ for $\nu = 1, \ldots, m$. Let S_n^* be the set of linear combinations*

$$k_1 z_1 + \cdots + k_m z_m, \qquad 1 \leq k \leq n$$

and let Z_n be the subsets of S_n consisting of the zeros of f_0. Let S_n be the complement of Z_n in

S_n^*. Let $S = \bigcup S_n$. Let $f = f_i/f_0$ for some $i = 1, \ldots, N$. Then f satisfies condition $AO\ 1$ on S, with $\rho = 2$.

PROOF. We shall use the quadratic form of Néron–Tate [6], [7]. Let h be the (logarithmic) height, defined on the group of rational points A_K of A in K. Then $h = q + l + O(1)$, for some positive definite quadratic form q, and some linear form l. If we write $P_v = \varphi(z_v)$, and

$$P = k_1 P_1 + \cdots + k_m P_m = \varphi(k_1 z_1 + \cdots + k_m z_m),$$

with $1 \leq k_v \leq n$, then we see that

$$(1) \qquad\qquad h(P) \ll n^2 \quad \text{for} \quad n \to \infty.$$

What we need is to bound the size of a single coordinate function $f(z)$ for $z \in S_n$ in terms of the height. This is a trivial technical matter. Indeed, the height of a point $P = (\xi_0, \ldots, \xi_N)$ in projective space over K is defined by

$$h(P) = \sum_v \sup_j v(\xi_j)$$

where the sum is taken over the set of normalized absolute values on K, and $v(\xi_j) = \log \|\xi_j\|_v$ (notation as in [5], Chapter III, except that we take the logarithm). From this definition, it is clear that if say $\xi_0 \neq 0$, and $\alpha = \xi_i/\xi_0$ is the i-th affine coordinate of our point, then

$$(2) \qquad\qquad h(\alpha) = h(\xi_0, \xi_i) \leq h(P).$$

Let $d(\alpha)$ be the leading coefficient of the irreducible polynomial satisfied by α over the *integers* $\mathbf{Z}$, with relatively prime coefficients. Then it is well known that $d(\alpha) \ll h(\alpha)$, for $\alpha \in K$. (See for instance [5], Proposition 4 of Chapter III, §2. The implied constant depends on the degree of α over $\mathbf{Q}$.) Furthermore $d(\alpha)$ is a denominator for α, that is $d(\alpha)\alpha$ is integral over $\mathbf{Z}$. From this and the estimate for the height, we conclude that

$$(3) \qquad\qquad \operatorname{size}(\alpha) \ll h(\alpha)$$

for $\alpha \in K$. Putting (1), (2), (3) together, we find that f satisfies condition $AO\ 1$ on S with $\rho = 2$, as was to be shown.

As we said before, Lemma 1 completes the proof of Theorem 3.

COROLLARY. *Let A be an abelian variety defined over a number field K. Let Γ be a finitely generated subgroup of A_K of rank ≥ 7. If Γ is contained in a 1-parameter subgroup of A, then this 1-parameter subgroup is an abelian subvariety of dimension 1 (an elliptic curve) in A.*

Remark 1. Theorems 1, 2, 3 extend to the p-adic case. In that case, they are purely, local theorems, on some disc around the origin.

Remark 2. In the study of abelian varieties, certain theorems (like the Mordell–Weil theorem, and Siegel's theorem concerning integral points on curves) were originally proved using theta functions. However, these theorems dealt only with the algebraic-arithmetic aspects of the situation, and when an algebraic theory of abelian varieties was developed (by Weil), it became clear that one could give expositions for the proofs entirely within the algebraic context. (Néron did it for the Mordell–Weil theorem in his thesis, and

I did it for the Siegel theorem. cf. [5].) In other words, the theta functions were used only as a convenient tool to write down the group law on abelian varieties, and this tool became superfluous when one saw how to formulate the group law purely algebraically.

Besides purely algebraic theorems on abelian varieties, there are other types, for instance those concerned with the complex analytic structure (as in Weil's book on Kähler manifolds). In the present context, we deal still with a third aspect of abelian varieties, namely the direct study of the properties of their transcendental parametrizations, with an arithmetic point of view, in which this parametrization occupies a central position.

REFERENCES

1.   A. O. GELFOND: *Algebraic and Transcendental Numbers*, Dover, New York, 1960.
2.   S. LANG: Transcendental points on group varieties, *Topology* **2** (1962), 313–318.
3.   S. LANG: Algebraic values of meromorphic functions, *Topology* **3** (1965), 183–191.
4.   S. LANG: Algebra, appendix.
5.   S. LANG: *Diophantine Geometry*, Interscience, New York, 1962.
6.   S. LANG: Diophantine approximations on toruses, *Am. J. Math.* **86** (1964), 521–533.
7.   A. NÉRON: Quasi-fonctions et hauteurs sur les variétés abéliennes, *Ann. Math.* September (1965), 349–331.
8.   C. L. SIEGEL: Transcendental numbers, *Ann. Math. Studies* 16, p. 37, Princeton, New Jersey, 1949.
9.   E. C. TITCHMARSH: *The Theory of Functions*, Chapter VIII. Oxford, 1932.
10.   A. WEIL: *Variétés Kahlériennes*, Actualités Sci. Ind. 1267, Hermann, Paris, 1958.

Columbia University

Partially supported by NSF grant GP 3426

Reprinted from the Proceedings of the National Academy of Sciences
Vol. 55, No. 1, pp. 31–34.   January, 1966.

ASYMPTOTIC DIOPHANTINE APPROXIMATIONS*

By Serge Lang

COLUMBIA UNIVERSITY

Communicated by S. Eilenberg, November 5, 1965

1. *Statement of the Theorem.*—Let ω, ψ be positive functions of a real variable, such that $\psi(t) = \omega(t)/t$.   We assume that ψ is decreasing ≤ 1, and ω is increasing, not necessarily strictly.   Let α be a real irrational number.   We let $\lambda = \lambda_{\alpha,\psi}$ be the function such that $\lambda(B)$ is the number of solutions in integers q, p for the inequalities

$$0 \leq q\alpha - p < \psi(q) \qquad \text{and} \qquad 1 \leq q < B, \tag{1}$$

for $B \to \infty$.   We sometimes abbreviate $q\alpha - p$ by $R(q\alpha)$.   Let

$$\Psi(B) = \int_1^B \psi(t)dt.$$

In reference 3, I determined λ asymptotically when α is a quadratic irrationality and ω is constant, or strictly increasing satisfying some other growth condition.

Schmidt[7] has shown how to remove the additional condition on ω, and he obtains an error term bounded by a constant times

$$\int_1^B \frac{\omega(t)^{1/2}}{t} \, dt$$

which is always $o(\Psi(B))$ when ω is increasing to infinity. On the other hand, Adams[1] has obtained λ asymptotically when $\alpha = e$ with $\omega = 1$, and shown that in this case, $\lambda(B)$ is asymptotic to a constant times $[(\log B)/(\log \log B)]^{1/2}$. This shows that, contrary to what happens for almost all numbers,[6] there may be an exceptional behavior for *constant* ω, and special numbers α. We shall prove, however, that as soon as ω is increasing sufficiently fast, then one recovers the expected asymptotic behavior of λ.

Let g be a positive function, assumed to be increasing. We shall say that α is of *type g* if for all sufficiently large integers B, there exists a solution in relatively prime integers q, p of the inequalities

$$0 < |q\alpha - p| < 1/q \quad \text{and} \quad B/g(B) \leqq q < B.$$

THEOREM. *Let α be an irrational real number of type g. Assume that ω is increasing to infinity, and that $\omega(t)^{1/2} g(t)/t$ is decreasing for all t sufficiently large. Then*

$$\lambda(B) = \Psi(B) + O\left(\int_1^B \frac{\omega(t)^{1/2}g(t)}{t} \, dt\right).$$

Remark: If h is a function decreasing to 0, then the integral from 1 to B of $\omega(t)h(t)/t$ is always $o(\Psi(B))$, so that the error term given in the theorem gives the asymptotic result $\lambda \sim \Psi$ when $g(t) \leqq \omega(t)^{1/2}h(t)$. Examples of this will now be given.

2. *Applications.*—To apply the theorem, one makes use of the type of α, which is known in a number of cases.

Example 1: Assume that g is constant. This is equivalent to saying that there exists $c > 0$ such that $|q\alpha - p| > c/q$ for all sufficiently large q. Then we can take g constant, and ω to be any increasing function to infinity. This special case was also obtained by Cassels.

Example 2: Assume that α satisfies the finiteness condition for the inequality $0 < |q\alpha - p| < 1/q^{1+\epsilon}$ for every $\epsilon > 0$. Then we can take $g(t) = t^\epsilon$ and the error term is $o(\Psi)$ if $\omega(t) \geqq t^\eta$ for some $\eta > 0$. This applies to algebraic numbers by the Thue-Siegel-Roth theorem.

Example 3: More generally, let $\{p_n/q_n\}$ be the sequence of principal convergents to α, and let f be some increasing function such that for all n sufficiently large,

$$\frac{1}{q_n f(q_n)} \leqq |q_n\alpha - p_n| \leqq \frac{1}{q_{n+1}}.$$

Then $q_{n+1} \leqq q_n f(q_n)$. Given B we find n such that $q_n \leqq B \leqq q_{n+1}$. We can take $g = f$ by using the relation

$$B/f(B) \leqq B/f(q_n) \leqq q_n \leqq B.$$

Conversely, under a growth condition for g, it can be shown that $|q\alpha - p| > c/qg(q)$ for q, p relatively prime and some constant c.

As a concrete example, let $\alpha = e$. From Adams' paper, one sees that g can be taken $g(t) = c(\log t)/(\log \log t)$ for some constant c, and $\omega \leq g/h$. In view of Adams' result, this is a best possible lower bound on ω.

Example 4: Suppose, for instance, that $f(q_n) = (\log q_n)^{1+\epsilon}$. Then we can take $g(t) = (\log t)^{1+\epsilon}$. By Khintchine's convergence theorem, this holds for almost all numbers.

We see that the theorem essentially reduces the problem of asymptotic approximation to the study of the inequality

$$0 < q\alpha - p < 1/qf(q),$$

where f is some increasing function. When f is close to 1, or grows very slowly, this is the range where the difficulties occur.

3. *Proof of the Theorem.*—In reference 7, Schmidt uses a multiplicative recursion. We shall use here an additive one.

LEMMA. *Let B, q be positive integers, satisfying the inequalities*

$$0 < \left|q\alpha - p\right| < 1/q \qquad and \qquad 1 \leq q \leq B/\omega(B)^{1/2},$$

and assume q, p relatively prime. Then

$$\lambda(B) - \lambda(B - q) = \int_{B-q}^{} \psi(t)dt + \theta,$$

where $\left|\theta\right| \leq c_1$ (and c_1 is an absolute constant).

Proof: (Cf. Behnke[2] and Ostrowski.[5]) We note that $\lambda(B) - \lambda(B - q)$ is the number of integers n satisfying

$$0 < R(n\alpha) < \psi(n) \qquad and \qquad B - q < n \leq B. \tag{2}$$

A trivial computation shows that $0 < \psi(B - q) - \psi(B) \leq 1/q$, whence

$$\psi(B) \leq \psi(n) \leq \psi(B - q) \leq \psi(B) + \frac{1}{q}.$$

Thus, bounds for $\lambda(B) - \lambda(B - q)$ can be determined by replacing $\psi(n)$ in (2) by $\psi(B)$ and $\psi(B) + 1/q$. Since n ranges over q consecutive integers, and since $0 < \alpha - p/q < 1/q^2$, the number of solutions is equal to $q\psi(B) + O(1)$, and $q\psi(B)$ differs from the desired integral by a bounded term, thereby proving the lemma.

For B sufficiently large, select q such that

$$\frac{B}{\omega(B)^{1/2}g(B)} \leq q < \frac{B}{\omega(B)^{1/2}} \qquad and \qquad 0 < \left|q\alpha - p\right| < 1/q.$$

Since $\omega(t)^{1/2}g(t)/t$ is decreasing, we get, using the left inequality for q,

$$\int_{B-q}^{B} \frac{\omega(t)^{1/2}g(t)}{t}\, dt \geq \frac{q\omega(B)^{1/2}g(B)}{B} \geq 1.$$

By the lemma, it follows that

$$\lambda(B) - \lambda(B - q) = \int_{B-q}^{B} \psi(t)dt + \theta_{B,q} \int_{B-q}^{B} \frac{\omega(t)^{1/2}g(t)}{t}\, dt,$$

with $|\theta_{B,q}| \leq c_1$. Repeating our argument with $B - q$ instead of B, and taking the sum inductively, we find that

$$\lambda(B) = \Psi(B) + \theta \int_1^B \frac{\omega(t)^{1/2} g(t)}{t} \, dt + O(1),$$

with $|\theta| \leq c_1$. This proves our theorem.

The same method can be used to estimate various sums. Here we limit ourselves to one example, with α of type g such that $g(t)/t$ is decreasing, namely, the sum

$$S_N = \sum_{n=1}^N \left(R(n\alpha) - \frac{1}{2} \right).$$

Taking $N/g(N) \leq q < N$ and $|q\alpha - p| < 1/q$ with q, p relatively prime, we estimate $|S_N - S_{N-q}|$ and conclude inductively that

$$|S_N| = O\left(\int_1^N \frac{g(t)}{t} \, dt \right).$$

This generalizes results of references 2 and 5, and trivializes the proof.

* Partially supported by NSF grant GP 3426.

[1] Adams, W., "Asymptotic diophantine approximations to e," these Proceedings, **55**, 28 (1966).

[2] Behnke, H., "Zur Theorie der Diophantischen Approximationen," *Abh. Math. Sem. Univ. Hamburg*, **3**, 261–318 (1922).

[3] Lang, S., "Asymptotic approximations to quadratic irrationalities," *Am. J. Math.*, **87**, 481–495 (1965).

[4] Lang, S., "Report on diophantine approximations," *Bull. Soc. Math. France*, **93**, 177–192 (1965).

[5] Ostrowski, A., "Bemerkungen zur Theorie der Diophantischen Approximationen," *Abh. Math. Sem. Univ. Hamburg*, **2**, 77–98 (1922).

[6] Schmidt, W., "A metrical theorem in diophantine approximations," *Canad. J. Math.*, **11**, 619–631 (1959).

[7] Schmidt, W., "Simultaneous approximation to a basis of a real number field," to appear in *Am. J. Math.*

INTRODUCTION TO TRANSCENDENTAL NUMBERS

SERGE LANG

Columbia University, New York, New York

ADDISON-WESLEY PUBLISHING COMPANY

Reading, Massachusetts · Palo Alto · London · Don Mills, Ontario

Foreword

The theory of transcendental numbers is reaching a stage where it is ready to take its place as one of the most attractive branches of mathematics. It consists in determining the transcendence and algebraic independence of numbers obtained as values of classical functions, suitably normalized. (We shall make this more precise in its appropriate place in the book.)

There is one main method, with several variations, which consists of constructing a function with many zeros out of the functions whose values one is considering. By *many* zeros one may mean just one zero with high multiplicity, or many distinct zeros with no condition on the multiplicity, or many distinct zeros with high multiplicities. We shall see examples of all three cases.

The theory has applications ranging from a very elementary setting (concerning the function e^t), to rather sophisticated contexts, having to do with abelian functions and automorphic functions. The embedding of "elementary" results in broad coherent theories has been known to cause acute cases of paranoia to persons who prefer "simple" significant examples to the elaboration of the more extensive theories. I personally like both, but I have made an attempt to keep the more elementary portions separate from the others, and hence more accessible for the reader of limited mathematical background, by:

1. Treating separately the special case of the exponential function, which needs only standard facts from basic courses in complex variables, and elementary properties of algebraic numbers.

2. Summarizing at the beginning of the book, with proofs, the few facts about algebraic numbers which we shall need in most of what follows.

Historical notes at the end of each chapter serve as much to describe past work in the subject as to suggest further possibilities and conjectures.

The book can be used as a text for a one term course in the theory of transcendental numbers, at the graduate level, if only certain portions of the theory are covered, e.g. Chapter I, Chapter II (omitting §4), Chapter III (omitting §4), the beginning of Chapter V, and Chapter VII. The other parts of the book can be covered if more time is available, depending on the degree of erudition of the audience. Chapter VI involves somewhat more complicated techniques, which are absolutely essential to carry on further work in the direction of that chapter. The reader should keep in mind, however, that in order to obtain the most far-reaching and best possible results, it may be necessary to start with a substantially different point of view, and different structure for the proofs, more closely related to what has classically been called diophantine approximations.

In any case, it is remarkable that a mathematical theory as old as the theory of transcendental numbers (dating back to Hermite's first result of 1873, the transcendence of e) is still in what can only be called an under-developed state.

Berkeley, 1966   SERGE LANG

Contents

v

400

CHAPTER V

Finitely Generated Values

CHAPTER VI

Transcendence Measures

CHAPTER VII

Linear Differential Equations

CHAPTER I

Preliminaries

§1. *Algebraic integers*

A *finite* extension of the field of rational numbers is called a *number field*. According to this convention, the field of all algebraic numbers will not be called a number field.

Let K be a number field. An element $\alpha \in K$ is called an *algebraic integer* if it satisfies either one of the following two equivalent conditions:

INT 1. *There exists a polynomial*

$$f(X) = X^n + a_{n-1}X^{n-1} + \cdots + a_0$$

with $n \geqq 1$ and coefficients $a_i \in \mathbf{Z}$ such that $f(\alpha) = 0$.

INT 2. *There exists a finitely generated $\mathbf{Z}$-module $M \neq 0$ (contained in some algebraic extension of K) such that $\alpha M \subset M$.*

The equivalence of these conditions is easily proved. Assume INT 1. Let M be generated by $1, \alpha, \ldots, \alpha^{n-1}$. Then $\alpha M \subset M$. Conversely, assume INT 2, and say M is generated by $v_1, \ldots, v_n$. Then

$$\alpha v_i = \sum_{j=1}^{n} a_{ij} v_j$$

for some integers a_{ij}. Subtracting the left-hand side from the right, we conclude that the determinant

$$\begin{vmatrix} a_{11} - \alpha & & & & a_{ij} \\ & a_{22} - \alpha & & & \\ & & \ddots & & \\ a_{ij} & & & & \\ & & & & a_{nn} - \alpha \end{vmatrix}$$

annihilates the module M, whence this determinant must be equal to 0. In this way we obtain a polynomial with integer coefficients, leading coefficient 1, which has α as a root.

1

From condition INT 2, we see that *the set of algebraic integers in K is a ring.* Indeed, if α, α' are algebraic integers in K, and M, M' are finitely generated (non-zero) **Z**-modules in some algebraic extension of K such that $\alpha M \subset M$ and $\alpha' M' \subset M'$, then MM' is finitely generated, and is mapped into itself by multiplication with $\alpha + \alpha'$ and $\alpha\alpha'$. We denote the ring of algebraic integers in K by I_K.

Let α be any element of K, satisfying an irreducible equation (over **Z**)

$$a_m X^m + \cdots + a_0 = 0$$

with $a_i \in \mathbf{Z}$. We assume that the coefficients $a_0, \ldots, a_m$ are relatively prime, and $a_m > 0$. Then this equation is uniquely determined by α. A positive integer d will be called a *denominator* for α if $d\alpha$ is an algebraic integer. It is clear that a_m is a denominator for α, because if we multiply the above equation by a_m^{m-1} we find that $a_m\alpha$ satisfies the equation

$$(a_m\alpha)^m + a_{m-1}(a_m\alpha)^{m-1} + \cdots + a_m^{m-1}a_0 = 0,$$

with integer coefficients, and leading coefficient 1.

In particular, we see that K is the quotient field of I_K, and that every element of K can be written as a quotient of an algebraic integer and a positive (rational) integer.

Each embedding $\sigma: K \to \mathbf{C}$ of K into the complex numbers will be called a *conjugate* of K. If $\alpha \in K$, then we call $\sigma\alpha$ a *conjugate* of α. The number of conjugates of K is equal to the degree $[K : \mathbf{Q}]$ (dimension of K as vector space over $\mathbf{Q}$). This is a simple elementary fact of field theory.

Let $[K : \mathbf{Q}] = n$. We can map I_K into $\mathbf{C}^n$ by $\tau: \alpha \mapsto (\sigma_1\alpha, \ldots, \sigma_n\alpha)$. This is an additive embedding. In any bounded region of $\mathbf{C}^n$ $(= \mathbf{R}^{2n})$ there is only a finite number of elements of $\tau(I_K)$. Indeed, if we bound a certain region, then we bound the conjugates $\sigma_i\alpha$ of elements α in I_K. Each such α is a root of the polynomial

$$(X - \sigma_1\alpha) \cdots (X - \sigma_n\alpha) = X^n + a_{n-1}X^{n-1} + \cdots + a_0,$$

whose coefficients are integers (because they are algebraic integers, and rational numbers being symmetric in the conjugates). Consequently, it follows by an elementary result that *I_K is a free abelian group, whose rank must be precisely n* since K is the quotient field of I_K, and a basis for I_K over **Z** must at the same time be a basis of K over **Q**. (For a proof of the elementary result, which is standard, cf. for instance my book *Linear Algebra*.)

If B is real > 0, we shall say that $\text{size}(\alpha) \leq B$ if there exists a denominator d for α such that $\log d \leq B$, and if

$$\log \max_{\sigma} |\sigma\alpha| \leq B$$

for all embeddings σ of K into $\mathbf{C}$ (i.e. all conjugates of α have logs of absolute value bounded by B). Thus

$$\text{size}(\alpha) = \max(\log d, \log |\sigma\alpha|),$$

where d is the smallest denominator for α, and σ ranges over all embeddings of K into $\mathbf{C}$. If r is a positive integer, then $\text{size}(\alpha^r) \leq r \cdot \text{size}(\alpha)$.

Let $\alpha \in K$, $\alpha \neq 0$ satisfy the irreducible equation

$$a_m X^m + \cdots + a_0 = 0$$

with relatively prime integers $a_0, \ldots, a_m$. Then

$$\frac{|a_0|}{|a_m|} = |\mathbf{N}\alpha| = \prod_{i=1}^{m} |\sigma_i\alpha|.$$

But $a_m = d$ is a denominator for α. Hence

$$1 \leq |a_m| \prod_{i=1}^{m} |\sigma_i\alpha|.$$

From this we obtain the fundamental inequality, to be used many times in the book, namely

$$-(2[K : \mathbf{Q}] - 1)\text{size}(\alpha) \leq \log |\sigma\alpha|$$

for any embedding σ of K into $\mathbf{C}$.

§2. *Integral linear equations*

We shall prove lemmas due to Siegel, which are used constantly in the sequel. They show that under certain circumstances, one can solve homogeneous linear equations with integer coefficients by means of a solution whose size is approximately the same as the size of the coefficients.

LEMMA 1. *Let*

$$a_{11}x_1 + \cdots + a_{1n}x_n = 0$$
$$\cdots$$
$$a_{r1}x_1 + \cdots + a_{rn}x_n = 0$$

be a system of linear equations with integer coefficients a_{ij}, and $n > r$. Let A be a number ≥ 1 such that $|a_{ij}| \leq A$ for all i, j. Then there exists an integral, non-trivial solution with

$$|x_j| \leq 2(2nA)^{r/(n-r)}.$$

Proof. We view our system of linear equations as a linear equation $L(X) = 0$, where L is a linear map, $L : \mathbf{Z}^{(n)} \to \mathbf{Z}^{(r)}$, determined by the matrix of coefficients. If B is a number ≥ 1, we denote by $\mathbf{Z}^{(n)}(B)$ the set of vectors X in $\mathbf{Z}^{(n)}$ such that $|X| \leq B$ (where $|X|$ is the maximum of the absolute values of the coefficients of X). Then L maps $\mathbf{Z}^{(n)}(B)$ into $\mathbf{Z}^{(r)}(nBA)$. The number of elements in $\mathbf{Z}^{(n)}(B)$ is $\geq B^n$ and $\leq (2B + 1)^n$. We seek a value of B such that there will be two distinct elements X, Y in $\mathbf{Z}^{(n)}(B)$ having the same image, $L(X) = L(Y)$. For this, it will suffice that $B^n > (nBA)^r$, and thus it will suffice that $B = (2nA)^{r/(n-r)}$. We take $X - Y$ as the solution of our problem.

The trace Tr from K to $\mathbf{Q}$ establishes an isomorphism between K (as vector space over $\mathbf{Q}$) and its dual space, under the bilinear map $(x, y) \mapsto \mathrm{Tr}(xy)$. Indeed, since the trace is a non-zero linear functional, the kernel on the right and left of this bilinear map is 0.

Let $X = (x_1, \ldots, x_n)$ be a vector of elements of K. We write

$$\|X\| = \max_{i,\sigma} |\sigma x_i|,$$

that is $\|X\|$ is the maximum of the absolute values of all conjugates of the coordinates x_i.

Let $\omega_1, \ldots, \omega_M$ be a basis of I_K over $\mathbf{Z}$. Let $\alpha \in I_K$, and write

$$\alpha = a_1\omega_1 + \cdots + a_M\omega_M, \qquad\qquad a_j \in \mathbf{Z}.$$

Let $\omega_1', \ldots, \omega_M'$ be the dual basis of $\omega_1, \ldots, \omega_M$ with respect to the trace. Then we can express the (Fourier) coefficients a_j of α as a trace,

$$a_j = \mathrm{Tr}(\alpha\omega_j').$$

The trace is a sum over the conjugates. Hence the order of magnitude of these coefficients is bounded by that of α, times a fixed constant, depending on the elements ω_j'.

LEMMA 2. *Let K be a finite extension of* $\mathbf{Q}$. *Let*

$$\alpha_{11}x_1 + \cdots + \alpha_{1n}x_n = 0$$
$$\cdots$$
$$\alpha_{r1}x_1 + \cdots + \alpha_{rn}x_n = 0$$

be a system of linear equations with coefficients in I_K, and $n > r$. Let A be a number such that $\|\alpha_{ij}\| \leq A$, for all i, j. Then there exists a nontrivial solution X in I_K such that

$$\|X\| \leq C_1(C_2nA)^{r/(n-r)}$$

where C_1, C_2 are constants depending only on K.

Proof. Let $\omega_1, \ldots, \omega_M$ be a basis of I_K over $\mathbf{Z}$. Each x_j can be written

$$x_j = \xi_{j1}\omega_1 + \cdots + \xi_{jM}\omega_M$$

with unknowns $\xi_{j\lambda}$. Each α_{ij} can be written

$$\alpha_{ij} = a_{ij1}\omega_1 + \cdots + a_{ijM}\omega_M$$

with integers $a_{ij\lambda} \in \mathbf{Z}$. If we multiply out the $\alpha_{ij}x_j$, we find that our linear equations with coefficients in I_K are equivalent to a system of rM linear equations in the nM unknowns $\xi_{j\lambda}$, with coefficients in $\mathbf{Z}$, whose magnitude is bounded by CA, where C is a number depending only on M and the size of the elements ω_λ, together with the products $\omega_\lambda\omega_\mu$, in other words where C depends only on K. Applying Lemma 1, we obtain a solution in terms of the $\xi_{j\lambda}$, and hence a solution X in I_K, whose magnitude satisfies the desired bound.

In applying Lemma 2, we meet equations whose coefficients are not necessarily algebraic integers. However, this case is trivially reduced to Lemma 2 by clearing denominators. We formulate it separately.

LEMMA 3. *Let K be a number field. Let*

$$\alpha_{11}x_1 + \cdots + \alpha_{1n}x_n = 0$$
$$\cdots$$
$$\alpha_{r1}x_1 + \cdots + \alpha_{rn}x_n = 0$$

be a system of linear equations with coefficients in K, and $n > r$. Let A be a number such that $\|\alpha_{ij}\| \leqq A$ for all i, j. Let d_i $(i = 1, \ldots, r)$ be a common denominator for the coefficients of the i-th equation, and let

$$d = \max d_i.$$

Then there exists a non-trivial solution X in I_K such that

$$\|X\| \leq C_1(C_2 n\, dA)^{r/(n-r)}$$

where C_1, C_2 are constants depending only on K.

Proof. We multiply the i-th equation by d_i, and apply Lemma 2.

It is convenient to mention here the following notation. If

$$P(T) = a_n T^n + \cdots + a_0$$

is a polynomial with complex coefficients a_i, we let

$$|P| = \max |a_i|.$$

If the coefficients are algebraic numbers, we let

$$\|P\| = \max \|a_i\|.$$

In *every* application of Lemma 2, we shall deal with a situation when r is approximately equal to cn for some constant c with $0 < c < 1$. Then the expression

$$\frac{r}{n - r}$$

is just equal to $c/(1 - c)$, i.e. is essentially constant, independent of the magnitude of n or r which will be quite large. Thus in this circumstance, the solution is bounded by

$$CnA^{c'}$$

for some constant c'. It will also be the case that n is very small compared to A (essentially, n will be of the order of magnitude of $\log A$), and consequently we shall have solutions whose order of magnitude is $A^{c'}$.

§3. *Estimating*

In all our work, we shall estimate. To do this efficiently, we need various notation.

Let f, g be real valued functions defined for arbitrarily large real numbers, and assume $g \geq 0$. We write

$$f = O(g)$$

if there is a constant $C > 0$ such that $|f(x)| \leq Cg(x)$ for all sufficiently large x. We write

$$f = o(g)$$

if $\lim_{x \to \infty} f(x)/g(x) = 0$.

We write $f \ll g$ if there exists a constant $C > 0$ such that for all x sufficiently large we have $f(x) \leq Cg(x)$. It will always be made clear from the context what the constant C depends on. If f, g are both ≥ 0, we write $f \gg\ll g$ to mean $f \ll g$ and $g \ll f$.

In applications, f and g may be defined for all sufficiently large positive integers.

We shall also use the $\ll$ notation when f, g are defined on some set S (not a set of numbers). In that case, $f \ll g$ means that there exists a number $C > 0$ such that $f(x) \leq Cg(x)$ for all elements of the set.

We shall frequently estimate sums. This is always done in a relatively coarse manner, namely the absolute value of a sum is bounded by the number of terms in the sum, times the maximum value of the terms. It will also turn out in each case that the number of terms in the sum is very small compared to the value of the terms (something like the log), so that the number of terms in a sum can always be essentially disregarded in making our estimates.

In estimating sums, one deals most easily with the ordinary triangle inequality. However, all the estimates will be important only as they appear in exponents. Thus it is convenient to take the log to simplify the notation. The reader will no doubt need, as I do occasionally, to reformulate the estimates explicitly in the exponential form, but after having done this several times, he will appreciate the other notation as a means of achieving expository clarity.

Finally, we describe very briefly the kind of estimates which lead to transcendence proofs. They are all based on the following principle, at the present stage of the theory. One deals with entire, or meromorphic functions, say f, g, which are assumed algebraically independent, over the constant field, and are of finite order, in the analytic sense. It is known from complex variables, that such functions cannot have too many zeros.

We are then interested in those complex numbers z such that $f(z)$ and $g(z)$ are both algebraic. Our functions will usually have some additional property like satisfying an algebraic differential equation, or possessing an algebraic addition theorem, which, out of one such value, allows the generation of many such values, and the problem is to describe the restrictions on those numbers z in such a way as to end up with sharp criteria. The arithmetic problem involving algebraic values is reduced to an analytic one concerning *zeros* of an auxiliary function, a polynomial in f, g,

$$ F = \sum a_{ij} f^i g^j $$

with coefficients in a number field. From the assumption that f, g take on values in the number field at certain points, one can construct a function F having many zeros, using Siegel's lemma. By some form of a three circle theorem, one can then obtain a contradiction, since on the one hand, F has small absolute value at some point w where $f(w)$ and $g(w)$ are algebraic, and on the other hand, the construction of F could be achieved in such a way that the size of $F(w)$ was relatively small. The contradiction then comes from the fundamental inequality mentioned in §1, relating the size of an algebraic number with one of its absolute values. Each chapter will exhibit a variation of the general principle just described.

It should be pointed out that Gelfond was the first to realize explicitly the connection between transcendence problems and algebraic values of entire functions. Many years before he proved his α^β theorem, investigating special cases in 1929 (cf. [12]) he saw that such a problem was related with a question (I believe raised by Polya) concerning the possibility of an entire function taking integral values at integers, and the interpolation problem arising from it. A function like F above was first introduced by Siegel in connection with the transcendence of values of the Bessel function.

CHAPTER II

Meromorphic Functions

§1. Algebraic values of e^t

The function e^t is the simplest function to consider, and as promised in the foreword, we begin by treating it as a special case of more general functions to be handled later.

We note the trivial fact that

$$\max_{|t|=R} |e^t| \leq e^R.$$

This kind of growth condition is used all the time.

Let $\beta_1, \ldots, \beta_m$ be complex numbers, linearly independent over the rationals. Then the functions

$$e^{\beta_1 t}, \ldots, e^{\beta_m t}$$

are algebraically independent over the complex numbers. In fact, if we denote these functions by $x_1, \ldots, x_m$ then we note that they are multiplicatively independent: In any relation

$$x_1^{n_1} \cdots x_m^{n_m} = 1$$

with integer exponents $n_1, \ldots, n_m$ we must have $n_1 = \cdots = n_m = 0$. From this it follows easily that $x_1, \ldots, x_m$ are algebraically independent. One may view this as a special case of Artin's theorem on the independence of characters, or the reader may devise a simple proof using his own ingenuity.

THEOREM 1. *Let β_1, β_2 be complex numbers, linearly independent over* **Q**, *and let z_ν ($\nu = 1, 2, 3$) be complex numbers, also linearly independent over* **Q**. *Then at least one of the numbers*

$$e^{\beta_1 z_\nu}, e^{\beta_2 z_\nu} \qquad (\nu = 1, 2, 3)$$

is transcendental (over **Q**).

Before proving Theorem 1, we give some corollaries.

8

COROLLARY 1. *Let β be a complex number, and suppose that there exist three non-zero, multiplicatively independent algebraic numbers*

$$\alpha_\nu \quad (\nu = 1, 2, 3)$$

such that α_ν^β is algebraic. Then β is rational.

Proof. Suppose that β is irrational. Let $\beta_1 = 1, \beta_2 = \beta$, and $z_\nu = \log \alpha_\nu$ (with any determination of the logarithm). The corollary follows by a direct application of Theorem 1.

COROLLARY 2. *Let y be real, and x^y algebraic for all positive rational $x \neq 0$. Then y is rational.*

Proof. Special case of Corollary 1.

We view the function e^t as an entire function, i.e. a function which is analytic in the plane.

We recall that an entire function F is said to be of *order* $\leq \rho$ if there is a constant $C > 0$ such that

$$|F|_R = \max_{|t|=R} |F(t)| \leq C^{R^\rho}$$

for all R sufficiently large. (We omit the usual ϵ which the reader will find in texts on complex variables, because it is irrelevant for what follows.) With the notation of Chapter I, §3, we may write

$$\log |F|_R = O(R^\rho)$$

or also

$$\log |F|_R \ll R^\rho$$

for R sufficiently large. The implied constant then depends on F. In all books on complex variables, it is proved that the number of zeros of such a function F in a circle of radius R is $O(R^\rho)$, provided F is not identically zero.

We shall now prove Theorem 1. Suppose that the conclusion of Theorem 1 is false, and let K be a finite extension of $\mathbf{Q}$, containing

$$e^{\beta_1 z_\nu}, e^{\beta_2 z_\nu}$$

for $\nu = 1, 2, 3$. Let n be a large integer, assumed to be a square for convenience, which will tend to infinity later. Let $r = (4n)^{3/2}$. We can find algebraic integers a_{ij} not all zero in K such that the function

$$F(t) = \sum_{i,j=1}^{r} a_{ij} e^{i\beta_1 t} e^{j\beta_2 t}$$

has a zero at every point $k \cdot z = k_1 z_1 + k_2 z_2 + k_3 z_3$, where $k = (k_1, k_2, k_3)$ is a triple of integers such that $1 \leq k_\nu \leq n$. This amounts to solving linear equations in r^2 unknowns, with $r^2 = (4n)^3$, and

$$\text{number of equations} = n^3.$$

The coefficients of the equations are the values

$$e^{i\beta_1(k \cdot z)} e^{j\beta_2(k \cdot z)}$$

which are elements of K. For each equation (corresponding to some k), we have

$$\text{size of coefficients} \ll nr \ll n^{5/2},$$

the implied constant depending on the values $\exp(\beta_\mu z_\nu)$. For each k, we can find a common denominator d for the coefficients of the k-th equation, satisfying the bound

$$\log d \ll nr,$$

because the coefficients are powers of the fixed algebraic numbers $\exp(\beta_\mu z_\nu)$, and these powers are essentially bounded by nr. We can therefore apply Siegel's lemma, and we can in fact find the a_{ij} such that

$$\text{size } a_{ij} \ll n^{5/2}$$

for n sufficiently large.

Since β_1, β_2 are linearly independent over $\mathbf{Q}$, it follows that F is not identically zero, and takes on values in K for all linear combinations of z_1, z_2, z_3 with positive integer coefficients. On the other hand, F cannot vanish at all such linear combinations, because they are not discrete, or alternatively because F is entire of order 1, and in a circle of large radius R, there are more such linear combinations than the bound $O(R)$ for the number of possible zeros of F. Let s be the largest integer such that $F(k \cdot z) = 0$ for all k_ν with $1 \leq k_\nu \leq s$. Then $s \geq n$. Let

$$w = k_1 z_1 + k_2 z_2 + k_3 z_3$$

with some $k_\nu = s + 1$, and $1 \leq k_\nu \leq s + 1$ for all ν, and $F(w) \neq 0$. Then

$$\text{size } F(w) \ll s^{5/2}.$$

We now estimate $|F(w)|$, and we use the expression

$$F(w) = \frac{F(t)}{\prod(t - k \cdot z)} \prod (w - k \cdot z) \Big|_{t=w},$$

the products being taken over all k_ν with $1 \leq k_\nu \leq s$. There are s^3 terms in the product. The function on the right of this last equality is an entire function, and we apply the maximum modulus principle on a circle of radius $R = s^{3/2}$. Note that for $|t| = R$, we have $|t - k \cdot z| \geqq R/2$ (for s large), and also

$$\frac{|w - k \cdot z|}{|t - k \cdot z|} \leqq \frac{C_1 s}{R} \leqq \frac{C_1}{s^{1/2}}$$

for some constant C_1, and s large. Hence

$$\log |F(w)| \ll \log |F|_R + s^3 - \tfrac{1}{2}s^3 \log s.$$

A trivial estimate shows that

$$|F|_R \leqq r^2 C_2^{n^{5/2}} C_3^{rR} \leqq C_4^{s^3},$$

and hence

$$\log |F(w)| \ll s^3 - s^3 \log s.$$

This contradicts the lower bound

$$-\text{size } F(w) \ll \log |F(w)|,$$

if we let n, and therefore s, tend to infinity. Theorem 1 is proved.

Remark. One would like to shrink the number of z_ν from three to two, but the same pattern of proof in this case does not give the desired contradiction at the end (it just misses).

In investigating values of α^β when β is transcendental and α is algebraic, there may be one algebraic α such that α^β is algebraic. For instance,

$$2^{\frac{\log 3}{\log 2}} = 3.$$

It will be shown later (Gelfond-Schneider Theorem) that α^β is transcendental if α is algebraic $\neq 0$, 1 and β is irrational. Thus $(\log 3)/(\log 2)$ is transcendental, and constitutes an exceptional number β for which 2^β is algebraic. Theorem 1 shows that there are at most two multiplicatively independent possibilities, and our remark is that there should only be one. In any case, we see Theorem 1 as a complement of the Gelfond-Schneider theorem. It also shows that among the numbers 2^π, 3^π, 5^π, ... at most two are algebraic. Of course, in this case, one expects none of them to be algebraic.

§2. *Algebraic values of meromorphic functions*

We shall axiomatize the proof of Theorem 1 so that it applies to more general functions.

A meromorphic function f is said to be of *order* $\leqq \rho$ if it can be expressed as a quotient of entire functions of order $\leqq \rho$. If S is a set of complex numbers, we say that such a meromorphic function f is *defined* on S if we can write $f = g/h$ where g, h are entire of order $\leq \rho$ and no point of S is a zero of h. For simplicity, we shall *always* assume that ρ is an integer ≥ 1.

Let S be a set of complex numbers, expressed as a union, $S = \bigcup_{n=1}^{\infty} S_n$, such that $S_n \subset S_{n+1}$ for all n. We say that the subsets $\{S_n\}$ form a filtration of S. We shall assume throughout when dealing with such filtrations that there is a constant $C > 0$ such that for all n and all $z \in S_n$ we have $|z| \leq Cn$.

Let f be a meromorphic function defined on S, with values in a number field K. We shall say that f is of *arithmetic order* $\leq \rho$ on S (or more accurately, with respect to the filtration $\{S_n\}$) if there is a constant $C \geq 1$ such that the following conditions are satisfied:

AO 1. *For all n and $z \in S_n$ we have size $f(z) \leq Cn^\rho$.*

AO 2. *There is an entire function h of order $\leq \rho$, such that hf is entire, h has no zero in S, and for all n, $z \in S_n$,*

$$\log |1/h(z)| \leq Cn^\rho.$$

Using our notation $\ll$, we can write our conditions in the form

$$\text{size } f(z) \ll n^\rho \qquad \text{and} \qquad \log |1/h(z)| \ll n^\rho,$$

for $z \in S_n$, and $n \to \infty$.

THEOREM 2. *Let f, g be meromorphic functions of order $\leq \rho$. Let $S = \bigcup S_n$ be as above, and assume that f, g are defined on S, with values in the number field K, and of arithmetic order $\leq \rho$ on S. Let λ be a number > 2. If $\operatorname{card}(S_n) \gg\ll n^{\lambda\rho}$ for $n \to \infty$, then f, g are algebraically dependent over K.*

Proof. Let $C_1, C_2 > 0$ be such that

$$C_1 n^{\lambda\rho} \leq \operatorname{card}(S_n) \leq C_2 n^{\lambda\rho}$$

for all n. Without loss of generality, we may assume (for convenience) that C_2 is an integer, and is a square. We shall deal with a large integer n, taken to be a square, and which will tend to infinity later.

Let $r = 2C_2^{1/2} n^{\lambda\rho/2} \ll n^{\lambda\rho/2}$.

We can find algebraic integers a_{ij} not all zero in K, such that the function

$$F = \sum_{i,j=1}^{r} a_{ij} f^i g^j$$

has a zero at every point $z \in S_n$. This amounts to solving linear equations in r^2 unknowns, and we have

$$\text{number of equations} \leq C_2 n^{\lambda \rho} \leq r^2/4.$$

The coefficients of these equations are the values

$$f(z)^i g(z)^j$$

with $z \in S_n$. By hypothesis, we have

$$\text{size of coefficients} \ll n^\rho r.$$

For each z, we also note that there is a common denominator for the coefficients of the corresponding equation with the similar bound $\ll n^\rho r$. By Siegel's lemma, we can find the a_{ij} such that

$$\text{size } a_{ij} \ll n^\rho r$$

for all n sufficiently large.

If f, g are algebraically independent over K, then F is not identically zero. Let s be the smallest integer such that $f(z) = 0$ for all $z \in S_s$ but $F(w) \neq 0$ for some $w \in S_{s+1}$. (Such s exists because in a circle of radius s we know that F has $O(s^\rho)$ zeros for $s \to \infty$.) Then $s \geq n$ and trivial estimates yield

$$\text{size } F(w) \ll s^\rho r.$$

We note that $s^\rho r \ll s^{\lambda \rho}$.

We now estimate $|F(w)|$. Let h be the entire function satisfying condition AO 2 with respect to both f and g. (If h_1 satisfies this condition with respect to f, and h_2 satisfies it with respect to g, then $h = h_1 h_2$ will serve our purposes.) Then $h^{2r}F$ is an entire function having zeros at all elements of S_s. We use the expression

$$F(w) = \frac{h(t)^{2r}F(t)}{h(w)^{2r}\prod(t - z)} \prod (w - z)\bigg|_{t=w}$$

where the products are taken for z in S_s. We apply the maximum modulus principle, and bound the function on the right on a circle of radius $R = s^{\lambda/2}$. We estimate three things:

First, we estimate $h^{2r}F$ using the fact that h, hf, hg are entire of order $\leq \rho$, and using the expression

$$h^{2r}F = \sum_{i,j=1}^{r} a_{ij} h^{2r-i-j}(hf)^i(hg)^j.$$

We take the log of the absolute value, and find

$$\sup_{|t|=R} \log |h(t)^{2r}F(t)| \ll s^{\lambda\rho/2}r \ll s^{\lambda\rho}.$$

Second, we have the upper bound (by AO 2),

$$\log |1/h(w)^{2r}| \ll s^{\rho}r \ll s^{\lambda\rho}.$$

Third, and most important, we have

$$\log \sup_{|t|=R} \prod_{z\epsilon S_s} \frac{|w-z|}{|t-z|} \ll s^{\lambda\rho} - \frac{\lambda-2}{2} s^{\lambda\rho} \log s.$$

Combining these three estimates, we get the upper bound

$$\log |F(w)| \ll s^{\lambda\rho} - \frac{\lambda-2}{2} s^{\lambda\rho} \log s$$

(because $3s^{\lambda\rho} \ll s^{\lambda\rho}!$). For s sufficiently large, this contradicts the fact that

$$-\text{size } F(w) \ll \log |F(w)|,$$

thereby proving Theorem 2.

Remark. Theorem 2 remains valid if instead of assuming

$$\text{card}(S_n) \gg\ll n^{\lambda\rho}$$

we assume merely that $\text{card}(S_n) \gg n^{\lambda\rho}$. Indeed, in that case, we reduce the proof to the preceding one by constructing sets $S'_n \subset S_n$ such that $\text{card}(S'_n) \gg\ll n^{\lambda\rho}$ and $S'_n \subset S'_{n+1}$. This is trivially done inductively, and we can then apply Theorem 2 to the sets S'_n.

§3. *Application to linear groups*

We shall now assume that the reader is acquainted with group varieties, but will recall briefly the main properties of this type of object.

A group variety is a group in affine or projective space, which is also an algebraic variety (connected), i.e. its points are the set of solutions of algebraic equations, and such that the law of composition and inverse have graphs which are also algebraic varieties. An important example is the linear group $GL(m)$ of invertible $m \times m$ matrices.

If K is a subfield of the complex numbers, we say that the group variety is defined over K if all the above-mentioned algebraic equations can be chosen to have coefficients in K. If that is the case, then we denote by G_K the set of points of G having coordinates in K, and it follows that G_K

is a group. When K is a number field, we view G_K as a discrete group. When $K = \mathbf{C}$, we view $G_\mathbf{C}$ as a complex-analytic manifold, i.e. a complex analytic group.

Let G be a group variety. By a 1-*parameter subgroup* of G we mean a complex-analytic homomorphism $\varphi \colon \mathbf{C} \to G_\mathbf{C}$ of the complex line into $G_\mathbf{C}$ whose derivative at the origin is injective. Thus φ is an analytic curve in $G_\mathbf{C}$. However, it may well be that algebraically, φ has dimension > 1. Indeed, if, say, G is embedded in some projective space, and $(\varphi_0, \dots, \varphi_N)$ are the projective coordinates of φ, then by the *algebraic dimension* of φ we shall mean the maximum number of algebraically independent coordinate functions φ_i/φ_0 $(i = 1, \dots, N)$. This is the same as the dimension of the smallest group subvariety of G containing $\varphi(\mathbf{C})$, i.e. containing the 1-parameter subgroup.

THEOREM 3. *Let G be a linear group variety defined over the field of algebraic numbers. Let $\varphi \colon \mathbf{C} \to G_\mathbf{C}$ be a 1-parameter subgroup, of algebraic dimension ≥ 2. Let Γ be a subgroup of $\mathbf{C}$ having at least three linearly independent elements over $\mathbf{Q}$. Then $\varphi(\Gamma)$ cannot be contained in the group of algebraic points of G.*

Proof. We can represent G globally as a group of matrices, so that for some matrix M,

$$\varphi(t) = \exp(tM) = \sum t^\mu M^\mu / \mu!.$$

Let z_ν $(\nu = 1, 2, 3)$ be complex numbers linearly independent over $\mathbf{Q}$, such that $\varphi(z_\nu)$ is algebraic, say with components in a number field K. Suppose that M is an $m \times m$ matrix. Then we can write

$$\varphi(t) = (\varphi_{ij}(t)), \qquad\qquad 1 \leq i, j \leq m,$$

with entire functions φ_{ij} which are of order ≤ 1. If φ has algebraic dimension ≥ 2, then at least two among the functions φ_{ij} are algebraically independent over K, say f, g.

Let S_n be the set of linear combinations

$$k_1 z_1 + k_2 z_2 + k_3 z_3$$

with $1 \leq k_\nu \leq n$. Condition AO 1 is easily verified, because

$$\varphi(k_1 z_1 + k_2 z_2 + k_3 z_3) = \varphi(z_1)^{k_1} \varphi(z_2)^{k_2} \varphi(z_3)^{k_3},$$

and it is easy to make the necessary estimates on the size of a product of matrices to see that AO 1 is true. Condition AO 2 is always true for entire functions (take $h = 1$), and consequently, we can apply Theorem 2 to obtain the desired contradiction.

CorollarY. *Let G be a linear group variety defined over a number field K. Let Γ be a commutative subgroup of G_K, containing at least three elements independent over $\mathbf{Z}$. If Γ is contained in a 1-parameter subgroup of G, then this 1-parameter subgroup is a group subvariety (i.e. algebraic) of dimension 1.*

We observe that Theorem 1 of §1 is a special case of Theorem 3. Indeed, it involves the 1-parameter subgroup

$$t \longmapsto (e^{\beta_1 t}, e^{\beta_2 t})$$

of the product of the multiplicative group with itself, $\mathbf{C}^* \times \mathbf{C}^*$.

We note that the additive group of $m \times m$ matrices may be viewed as the tangent space at the origin of the general linear group $GL(m)$. Thus the map

$$\exp : \mathrm{Mat}_m\,(\mathbf{C}) \to GL(m, \mathbf{C})$$

given by the exponential power series is the usual exponential map of the differential geometer. When the group variety is the multiplicative group $\mathbf{C}^*$, then the exponential map is simply the ordinary function $t \mapsto e^t$. A 1-parameter subgroup is nothing but the image of a straight line through the origin under the exponential map.

In the next section, we consider another type of group variety.

§4. *Abelian varieties*

An abelian variety is a group variety in projective space. If A is an abelian variety, then $A_\mathbf{C}$ is a compact, complex Lie group, and consequently is a complex torus: We can represent $A_\mathbf{C}$ as a factor group of $\mathbf{C}^d$ $(d = \dim A)$ by a discrete subgroup of dimension $2d$ over $\mathbf{R}$. This representation can be achieved by theta functions in d variables, these being entire functions of order ≤ 2 (cf. [3] or [33]). We shall denote this representation by

$$\Theta : \mathbf{C}^d \to A_\mathbf{C}$$

If $\alpha \in \mathbf{C}^d$, and $\alpha \neq 0$, then the curve $\varphi : \mathbf{C} \to A_\mathbf{C}$ such that $\varphi(t) = \Theta(t\alpha)$ is a 1-parameter subgroup of A, and all 1-parameter subgroups of A can be so described. Furthermore, Θ is a representation of the exponential map on $A_\mathbf{C}$, and we may identify the tangent space at the origin of $A_\mathbf{C}$ with $\mathbf{C}^d$.

THEOREM 4. *Let A be an abelian variety defined over the field of algebraic numbers. Let $\varphi : \mathbf{C} \to A_\mathbf{C}$ be a 1-parameter subgroup of A. Let Γ be a subgroup of $\mathbf{C}$ having at least seven linearly independent elements over $\mathbf{Q}$. If φ has algebraic dimension ≥ 2, then $\varphi(\Gamma)$ cannot be contained in the group of algebraic points of A.*

Proof. Let $\Theta = (\theta_0, \ldots, \theta_N)$ be the representation of Θ by means of theta functions $\theta_0, \ldots, \theta_N$, giving an embedding of $A_\mathbf{C}$ into projective N-space. We induce the functions on the analytic curve φ, and obtain entire functions $(f_0, \ldots, f_N)$ in one variable, of order ≤ 2, which realize our 1-parameter map φ, that is $\varphi(t) = (f_0(t), \ldots, f_N(t))$. Without loss of generality, we may assume that the divisor of zeros of θ_0 does not pass through the origin, and that, in particular, f_0 is not identically zero. Let D be a small disc of radius $\delta > 0$ around the origin in $\mathbf{C}$, so that no zero of f_0 lies in D. Then $|f_0(t)|$ is bounded away from 0 on D, and hence $\log |1/f_0(t)|$ is bounded from above on D, thereby satisfying condition AO 2.

By hypothesis, at least two of the functions among $f_1/f_0, \ldots, f_N/f_0$ are algebraically independent over the algebraic numbers, say f, g.

We may assume that A is defined over a number field K, and that Γ is generated by seven elements z_ν ($\nu = 1, \ldots, 7$) linearly independent over $\mathbf{Q}$, such that $\varphi(z_\nu)$ is contained in A_K for $\nu = 1, \ldots, 7$. Let S_n be the set of linear combinations

$$k \cdot z = k_1 z_1 + \cdots + k_7 z_7$$

with $-n \leq k_\nu \leq n$, and such that $k \cdot z$ lies in D. By routine techniques (which will be recalled below), one verifies that $\mathrm{card}(S_n) \gg n^5$ or n^6, depending on whether at least two of the elements z are linearly independent over $\mathbf{R}$ or not. For definiteness, let us assume that we are in the case $\mathrm{card}(S_n) \gg n^5$. Let $S = \bigcup S_n$. We already know that our two functions f, g satisfy condition AO 2 on S, since in fact they have the bounded denominator $h = f_0$ on D. We shall prove in the next lemma that they satisfy condition AO 1, and thus are of arithmetic order ≤ 2 on S. Applying Theorem 2 then gives the contradiction which proves Theorem 4.

LEMMA 1. *Let A be an abelian variety defined over a number field K. Let $\varphi \colon \mathbf{C} \to A_\mathbf{C}$ be a 1-parameter subgroup, represented by projective coordinates $(f_0, \ldots, f_N)$. Say f_0 is not identically zero. Let $z_1, \ldots, z_m \in \mathbf{C}$ be linearly independent over $\mathbf{Q}$, such that $\varphi(z_\nu) \in A_K$ for $\nu = 1, \ldots, m$. Let S_n^* be the set of linear combinations*

$$k_1 z_1 + \cdots + k_m z_m, \qquad\qquad -n \leq k_\nu \leq n,$$

and let Z_n be the subset of S_n^* consisting of the zeros of f_0. Let S_n be the complement of Z_n in S_n^*. Let $S = \bigcup S_n$. Let $f = f_i/f_0$ for some index $i = 1, \ldots, N$. Then f satisfies condition AO 1 on S, with $\rho = 2$.

Proof. We shall use the quadratic form of Néron-Tate [26], [20]. We denote by h the logarithmic height, defined on the group of rational points A_K of A in K. Then $h = q + l + O(1)$, for some quadratic form q, and some linear form l. If we write $P_\nu = \varphi(z_\nu)$, and

$$P = k_1 P_1 + \cdots + k_m P_m = \varphi(k_1 z_1 + \cdots + k_m z_m)$$

with $-n \leqq k_\nu \leqq n$, then we see that

$$(1) \qquad\qquad h(P) \ll n^2 \quad \text{for} \quad n \to \infty.$$

What we need is to bound the size of a single coordinate function $f(z)$ for $z \in S_n$ in terms of the height. This is a trivial technical matter. Indeed, the height of a point $P = (\xi_0, \ldots, \xi_N)$ in projective space over K is defined by

$$h(P) = \sum_v \sup_j v(\xi_j)$$

where the sum is taken over the set of normalized absolute values on K, and $v(\xi_j) = \log \| \xi_j \|_v$ (notation as in [18], Chapter III, except that we take the logarithm). From this definition, it is clear that if, say $\xi_0 \neq 0$, and $\alpha = \xi_i / \xi_0$ is the i-th affine coordinate of our point, then

$$(2) \qquad\qquad h(\alpha) = h(\xi_0, \xi_i) \leq h(P).$$

Let $d(\alpha)$ be the leading coefficient of the irreducible polynomial satisfied by α over the *integers* $\mathbf{Z}$, with relatively prime coefficients. Then it is well known that $\log d(\alpha) \ll h(\alpha)$, for $\alpha \in K$. (See for instance [15], Proposition 4 of Chapter III, §2. The implied constant depends on the degree of α over $\mathbf{Q}$.) Furthermore, $d(\alpha)$ is a denominator for α, that is $d(\alpha)\alpha$ is integral over $\mathbf{Z}$. From this and the estimate for the height, we see that

$$(3) \qquad\qquad \text{size}(\alpha) \ll h(\alpha)$$

for $\alpha \in K$. Putting (1), (2), (3) together, we find that f satisfies condition AO 1 on S with $\rho = 2$, as was to be shown.

There remains but to justify that the number of linear combinations $k \cdot z$ lying in D has at least the order of magnitude of n^5. This is done by standard arguments. Suppose for instance that z_6, z_7 are linearly independent over $\mathbf{R}$, and let Λ be the lattice generated by them. We consider the combinations

$$k_1 z_1 + \cdots + k_5 z_5$$

on the torus $\mathbf{C}/\Lambda = \mathbf{R}^2/\Lambda$. We cut up a fundamental domain into approximately $1/\delta$ small squares of sides approximately equal to $\delta^{1/2}$, and use the Dirichlet box principle on the $(cn)^5$ elements consisting of linear combinations $k \cdot z$ with $|k| \leqq cn$ for some constant c depending on the z_ν. Subtracting one element from all the others crowded in a small box, we obtain essentially $\gg \delta n^5$ elements in D (up to a constant factor, depending only on the z_ν). This does what we wanted.

(By being slightly more careful, and using translations by z_6, z_7, using points lying outside small circles around the zeros of the entire function f_0, and using the minimum modulus principle for entire functions, one can probably avoid the drop of 7 to 5, and thus replace 7 by 5 in the theorem. However, this is a secondary matter here.)

COROLLARY 1. *Let A be an abelian variety defined over a number field K. Let Γ be a finitely generated subgroup of A_K, of rank $\geqq 7$. If Γ is contained in a 1-parameter subgroup of $A_{\mathbf{C}}$, then this 1-parameter subgroup is an abelian subvariety of dimension 1 (an elliptic curve) in A.*

COROLLARY 2. *Let G, G' be group varieties defined over a number field. Let G have dimension 1, and let G' be linear or abelian. Let $\varphi\colon G_{\mathbf{C}} \to G'_{\mathbf{C}}$ be a complex analytic homomorphism. If there exist 7 algebraic points α_ν ($\nu = 1, \ldots, 7$) on G, linearly independent over $\mathbf{Z}$, such that $\varphi(\alpha_\nu)$ is algebraic, then some power of φ is a rational homomorphism.*

Proof. Composing φ with the exponential map on G shows that the graph of φ is algebraic. That some power of φ is rational then follows from trivial facts about group varieties.

Remark. In the study of abelian varieties, certain theorems, like the Mordell-Weil theorem, and Siegel's theorem concerning integral points on curves, were originally proved using theta functions. However, these theorems dealt only with the algebraic-arithmetic aspects of the situation, and when an algebraic theory of abelian varieties was developed (by Weil), it became clear that one could give expositions for the proofs entirely within the algebraic context. (Néron did it for the Mordell-Weil theorem in his thesis, and I did it for the Siegel theorem, cf. [18].) In other words, the theta functions were used only as a convenient tool to write down the group law on abelian varieties, and this tool became superfluous for the preceding applications when one saw how to formulate the group law purely algebraically.

Besides purely algebraic theorems on abelian varieties, there are other types, for instance those concerned with the complex analytic structure (as in Weil's book on Kähler manifolds). In the present context, we deal still with a third aspect of abelian varieties, namely the direct study of the properties of their transcendental parametrizations, with an arithmetic point of view, in which this parametrization occupies a central position.

Historical note

The use of a function similar to our F occurs in Gelfond's and Schneider's proof of the transcendence of α^β when α is algebraic $\neq 0, 1$ and β is algebraic irrational. Theorem 1 was known to Siegel [cf. L. Alaoglu and P. Erdös, "On highly composite and similar numbers", *Trans. Am. Math.*

Soc. **56** (1944) p. 455]. Corollary 2 is the portion of Theorem 1 needed for applications. Its need had arisen in the paper just mentioned, and more recently was asked by Serre in connection with the determination of certain characters of idele classes of number fields. Unfortunately, Theorem 1 does not appear in any of the three books on transcendental numbers [13], [29], [32] and is not otherwise generally known. Theorem 2 is a variant of results of Schneider [28], who started to axiomatize the transcendence proofs, and showed more explicitly than Gelfond how the order of growth of an entire function restricts its algebraic values.

Further comments on the exponential function will be made at the end of the next chapter. The applications of §3 and §4 are taken from [17].

It is a problem to extend Theorem 2 and its applications to functions of several variables. Formally, the proof can be given a d-variable version, the only difficulty occurring at the estimate of an entire function with many zeros. One needs an analogue to the three circle theorem, i.e. an estimate of type

$$|F|_R \leqq \frac{|F|_{R_2} R^N}{R_2^N}$$

if F has N zeros more or less evenly distributed inside a polydisc of radius R_1, and $R_1 < R < R_2$. That the zeros must be more or less evenly distributed is clear, since a function of several variables always has a divisor of zeros (i.e. the zeros are not isolated). Consequently, in the multidimensional analogue of Theorems 3 and 4, taking linear combinations of vectors, it is necessary to assume that these combinations are evenly distributed. In applications, this leads to questions in diophantine approximations which at the moment appear to lie beyond the methods available. (For a special case when this problem does not arise, cf. Chapter IV.)

Finally, it would be worth while to extend Theorems 3 and 4 to arbitrary group varieties (which one may assume commutative, since the Zariski closure of the 1-parameter group is necessarily commutative). This is not a problem in transcendental numbers, since Theorem 2 reduces the question to general properties on group varieties. Thus one must show that Axioms AO 1 and AO 2 are satisfied. This involves first a purely algebraic question, i.e. an upper bound for the height of points on arbitrary commutative group varieties (to replace the upper bound derived from the presently available Néron-Tate form), and second, a mixed algebraic-transcendental problem, i.e. showing that the exponential map on arbitrary commutative group varieties can be parametrized by meromorphic functions of order $\leqq 2$, similar to theta functions. (This would involve systematizing Severi's quasi-abelian functions.)

CHAPTER III

Algebraic Differential Equations

§1. The main theorem

In the preceding chapter, we saw that there exist certain restrictions on the set of numbers at which algebraically independent meromorphic functions take on algebraic values. In the applications, we use very strongly the algebraic addition formula for our functions. In this chapter, we shall describe an analogous situation when the functions satisfy an algebraic differential equation. In that case, the set of points is much smaller.

THEOREM 1. *Let K be a number field. Let $f_1, \ldots, f_N$ be meromorphic functions of order $\leq \rho$. Assume that the field $K(f_1, \ldots, f_N)$ has transcendence degree ≥ 2 over K, and that the derivative $D = d/dt$ maps the ring $K[f_1, \ldots, f_N]$ into itself. Let $w_1, \ldots, w_m$ be distinct complex numbers not lying among the poles of the f_i, such that*

$$f_i(w_\nu) \in K$$

for all $i = 1, \ldots, N$, and $\nu = 1, \ldots, m$. Then $m \leq 20\rho\,[K : \mathbf{Q}]$.

As corollaries, we obtain the transcendence of several types of classical numbers.

COROLLARY 1 (Hermite-Lindemann). *Let α be an algebraic number $\neq 0$. Then e^α is transcendental.*

Proof. Suppose e^α is algebraic. Let K be the field generated by α and e^α over $\mathbf{Q}$. Let f, g be the two functions $f(t) = t$ and $g(t) = e^{\alpha t}$. Then the ring $K[f, g]$ is mapped into itself by $D = d/dt$, and f, g are obviously algebraically independent. We note that f, g also take on values in K at all numbers $\alpha, 2\alpha, \ldots, m\alpha$ for m arbitrary large. This contradicts Theorem 1.

COROLLARY 2 (Gelfond-Schneider). *Let α, β be algebraic, $\alpha \neq 0, 1$ and β irrational. Then α^β is transcendental.*

21

422

Proof. If α^β is algebraic, let K be the field generated over $\mathbf{Q}$ by α, β and α^β. We apply Theorem 1 to the two functions e^t and $e^{\beta t}$, and to the set of integral multiples of $\log \alpha$ (for any determination of the logarithm) to obtain a contradiction as in Corollary 1.

The next two corollaries may be omitted by those who don't know about elliptic modular functions.

COROLLARY 3 (Schneider).   *Let $\wp$ be a Weierstrass function, with algebraic g_2, g_3. If α is algebraic $\neq 0$, then $\wp(\alpha)$ is transcendental.*

Proof. The argument is similar to the proof of Corollary 1, using the functions t, $\wp(t)$, and the algebraic addition theorem for the $\wp$-function. If α happens to be a period, so a pole of $\wp$, then one may work with α/n for some positive integer n, and consider values of $\wp$ at integral multiples of α/n which are not periods, and at which $\wp$ is therefore defined.

COROLLARY 4 (Schneider).   *Let τ be algebraic with $\mathrm{Im}(\tau) > 0$, and let j be the elliptic modular function. If $j(\tau)$ is algebraic, then τ is a quadratic imaginary irrationality.*

Proof. One can find a Weierstrass $\wp$-function having algebraic g_2, g_3 and the corresponding value $j(\tau)$. Let K be a number field containing g_2, g_3, $j(\tau)$ and τ. Consider the two functions $f(t) = \wp(t)$ and $g(t) = \wp(\tau t)$. They cannot be algebraically independent over K. Indeed, we have $\tau = \omega_2/\omega_1$ (where ω_2, ω_1 are fundamental periods), and our functions take on algebraic values at all complex numbers $m\alpha$, where $\alpha = \omega_1/2$ and m is an odd integer. (We use this trick to avoid poles of $\wp$.) It is a routine matter from the theory of elliptic functions to conclude that multiplication by τ maps the vector space generated over $\mathbf{Q}$ by the period lattice (ω_1, ω_2) into itself, and consequently that τ is quadratic over $\mathbf{Q}$, and is not real.

The analysis needed to understand Corollaries 3 and 4 is covered in any text dealing with elliptic functions. We shall see later generalizations in higher dimensions.

We emphasize, however, that the proof of Theorem 1 in §2 and §3 is completely self-contained and elementary, using nothing more of the theory of complex variables than the maximum modulus principle. The main part of the proof is given in §3, and in §2 we shall prove some simple preliminary estimates involving derivatives.

The result of Theorem 1 is essentially best possible, without making further assumptions on the functions involved. For instance, $e^{Q(t)}$ takes on algebraic values at all zeros of a polynomial $Q(t)$ with algebraic coefficients, and there may well be a finite number of these.

§2. *Estimation of derivatives*

Let

$$P(T_1, \ldots, T_N) = \sum \alpha_{(\nu)} M_{(\nu)}(T)$$

be a polynomial with complex coefficients $\alpha_{(\nu)}$, in N variables $T_1, \ldots, T_N$, and let

$$Q(T_1, \ldots, T_N) = \sum \beta_{(\nu)} M_{(\nu)}(T)$$

be a polynomial with *real* coefficients ≥ 0. We say that Q *dominates* P if $|\alpha_{(\nu)}| \leq \beta_{(\nu)}$ for all (ν), and we write $P \prec Q$. The following properties are then immediately verified.

If Q_1 dominates P_1 and Q_2 dominates P_2 then

$$P_1 + P_2 \prec Q_1 + Q_2 \qquad \text{and} \qquad P_1 P_2 \prec Q_1 Q_2.$$

Furthermore, if $D_i \, (i = 1, \ldots, N)$ is the i-th derivative with respect to T_i, and if Q dominates P, then

$$D_i P \prec D_i Q.$$

Let P be a polynomial with complex coefficients as above. We let

$$|P| = \max |\alpha_{(\nu)}|$$

be the maximum of the absolute values of the coefficients. If the total degree of P is $\leq r$, then

$$P \prec |P|(1 + T_1 + \cdots + T_N)^r.$$

Since it is easy to take the derivative of the right-hand side with respect to any variable T_i, we have an easy way of estimating repeated formal derivatives of polynomials.

Let K be a number field.

Let S be a set of elements of K. We denote by $\|S\|$ the maximum of the absolute values of all conjugates of elements of S. By a *denominator* for S, we mean a positive integer which is a common denominator for all elements of S. If P is a polynomial with coefficients in K, we let $\|P\|$ and a denominator for P be defined in terms of the set of coefficients of the polynomial. We abbreviate denominator by den.

LEMMA 1. *Let K be a number field. Let $f_1, \ldots, f_N$ be functions, holomorphic on a neighborhood of a point $w \in \mathbf{C}$, and assume that $D = d/dt$ maps the ring $K[f_1, \ldots, f_N]$ into itself. Assume that $f_i(w) \in K$ for all i. Then there exists a number C_1 having the following property. Let $P(T_1, \ldots, T_N)$ be a polynomial with coefficients in K, of degree $\leq r$.*

If we set $f = P(f_1, \ldots, f_N)$, then we have, for all positive integers k,

$$\| D^k f(w) \| \leq \| P \| r^k k! C_1^{k+r}.$$

Furthermore, there is a denominator for $D^k f(w)$ bounded by $\mathrm{den}(P) C_1^{k+r}$.

Proof. There exist polynomials $P_i(T_1, \ldots, T_N)$ with coefficients in K such that

$$Df_i = P_i(f_1, \ldots, f_N).$$

Let δ be the maximum of their degrees. There exists a unique derivation $\overline{D}$ on $K[T_1, \ldots, T_N]$ such that $\overline{D} T_i = P_i(T_1, \ldots, T_N)$. For any polynomial P in $K[T]$, we have

$$\overline{D}(P(T_1, \ldots, T_N)) = \sum_{i=1}^{N} (D_i P)(T_1, \ldots, T_N) \cdot P_i(T_1, \ldots, T_N)$$

where $D_1, \ldots, D_N$ are the formal partial derivatives. The polynomial P is dominated by

$$\| P \| (1 + T_1 + \cdots + T_N)^r,$$

and each P_i is dominated by

$$\| P_i \| (1 + T_1 + \cdots + T_N)^\delta.$$

Thus $\overline{D} P$ is dominated by

$$\| P \| C_2 r (1 + T_1 + \cdots + T_N)^{r+\delta}.$$

Proceeding inductively, one sees that

$$\overline{D}^k P \prec \| P \| C_3^k r^k k! (1 + T_1 + \cdots + T_N)^{r+k\delta}.$$

Substituting values $f_i(w)$ for T_i, we have

$$D^k f(w) = \overline{D}^k P(f_1(w), \ldots, f_N(w)),$$

whence we obtain the desired bound on $D^k f(w)$. The second assertion concerning denominators is proved also by a trivial induction (even easier than the preceding one).

§3. *The main proof*

We now come to the proof of Theorem 1. Let f, g be two functions among $f_1, \ldots, f_N$ which are algebraically independent over K. Let r be a positive integer divisible by $2m$. We shall let r tend to infinity at the end of the proof.

Let

$$F = \sum_{i,j=1}^{r} a_{ij} f^i g^j$$

have coefficients a_{ij} in I_K. Let $n = r^2/2m$. We can select the a_{ij} not all equal to 0, and such that

$$D^k F(w_\nu) = 0$$

for $0 \leq k < n$ and $\nu = 1, \ldots, m$. Indeed, we have to solve a system of mn linear equations in $r^2 = 2mn$ unknowns. Note that

$$mn/(2mn - mn) = 1.$$

The size of the coefficients, and a denominator for them, are obtained as a result of Lemma 1. By Siegel's lemma, we see that we can take the a_{ij} such that

$$\begin{aligned}
\text{size}(a_{ij}) &\leq n \log r + n \log n + (n + r)C_1 \\
&\leq 2n \log n
\end{aligned}$$

for n sufficiently large.

Since f, g are algebraically independent over K, our function F is not identically zero. We let s be the smallest integer such that all derivatives of F up to order $s - 1$ vanish at all points $w_1, \ldots, w_m$ but such that $D^s F$ does not vanish at one of the w_ν, say w. Then $s \geq n$. Furthermore, using Lemma 1, we find at once

$$\text{size } D^s F(w) \leq 5s \log s$$

for n (and hence s) sufficiently large.

On the other hand, let θ be an entire function of order $\leq \rho$, such that θf and θg are entire, and $\theta(w) \neq 0$. Then $\theta^{2r} F$ is entire. We consider the entire function

$$E(t) = \frac{\theta(t)^{2r} F(t)}{\displaystyle\prod_{\nu=1}^{m} (t - w_\nu)^s}.$$

Then $E(w)$ differs from $D^s F(w)$ by obvious factors, bounded by $C_4^s s!$ for some constant C_4. We use the maximum modulus principle to estimate E on a circle of radius $R = s^{1/2\rho}$. On this circle, $|t - w_\nu|$ has approximately the same absolute value as R, and consequently,

$$|E(w)| \leq |E|_R \leq \frac{s^{3s} C_5^{2rR^\rho}}{R^{ms}}.$$

We therefore obtain the estimate

$$\log |D^s F(w)| \leq 4s \log s - \frac{1}{2\rho} ms \log s.$$

Comparing this with the estimate for the size of $D^s F(w)$, we obtain at once $m \leq 20\rho\,[K : \mathbf{Q}]$, as desired.

Note: We have made no particular effort to shrink the universal constant 20, and with a little care, one can obviously reduce it to a smaller constant. But this is essentially irrelevant, since the interesting thing is that in special cases, using additional structure on the functions $f_1, \ldots, f_N$, we can eliminate it altogether.

§4. *The exponential map on groups*

Let G be a group variety defined over a field K. Let $T_e(G)$ be its tangent space at the origin. We may view $T_e(G)$ as a vector space over any field containing K, but it has a basis defined over K, because one can define tangent vectors algebraically.

Over the complex numbers, we can identify the tangent space with $\mathbf{C}^d$ if $d = \dim G$. Thus the exponential map

$$\exp\colon \mathbf{C}^d \to G_{\mathbf{C}}$$

can be viewed as a complex analytic map from $\mathbf{C}^d$ into $G_{\mathbf{C}}$. Of course, we can perform any linear automorphism on $\mathbf{C}^d$ and still have an exponential map. However, if we normalize this exponential map so that $\mathbf{C}^d$ is extended from a vector space over K, then any further freedom we have in making linear automorphisms is restricted to such automorphisms over K. In terms of matrices, this means that the matrix of such a linear automorphism must have coefficients in K.

Finally, we observe that to normalize the exponential map as above is equivalent to saying that the partial derivatives $\partial/\partial t_1, \ldots, \partial/\partial t_d$ (of the d complex variables $t_1, \ldots, t_d$) are defined over K, in other words map the field of rational functions $K(G)$ into itself.

Let V be a variety (algebraic, irreducible) defined over a field K. If P is a point of V with affine coordinates $(x_1, \ldots, x_n)$ then we denote by $K(P)$ the field $K(x_1, \ldots, x_n)$. We say that P is *rational* over K if $K(P) = K$, *algebraic* over K if $K(P)$ is algebraic over K, and *transcendental* over K if $K(P)$ is not algebraic over K.

THEOREM 2. *Let G be a group variety defined over the field of algebraic numbers* **K**. *Let T_e be its tangent space at the origin, with its structure of* **K**-*vector space. Let α be a non-zero element of T_e which is rational over* **K** *(i.e. algebraic), and let* $\exp = \exp_G$ *be the exponential map on G, normalized to have algebraic derivative at the origin. Assume that* $\exp(t\alpha)$ *is not a rational function of t. Then* $\exp(\alpha)$ *is transcendental over* **K**.

Proof. Assume first that G is a linear group variety, thereby admitting a global matrix representation over K. A tangent vector at the origin is simply a matrix M, and $\exp(M)$ is given by the series

$$\exp(M) = \sum M^\mu/\mu! \,.$$

If B is an invertible matrix, then

$$\exp(B^{-1}MB) = B^{-1}\exp(M)B.$$

Assume that B is rational over $\mathbf{K}$. We observe that $\exp(M)$ is rational over $\mathbf{K}$ if and only if $\exp(B^{-1}MB)$ is rational over $\mathbf{K}$. Any matrix M rational over $\mathbf{K}$ can be conjugated into a matrix consisting of blocks

$$\begin{pmatrix} \alpha & & & \\ 0 & \alpha & & N \\ \vdots & & \ddots & \\ 0 & \cdots & & \alpha \end{pmatrix} = \alpha I + N,$$

where α is algebraic, and N is a rational nilpotent matrix. Therefore $\exp(\alpha I + N)$ is equal to $e^\alpha I$ times a rational matrix, and is transcendental if and only if $\alpha \neq 0$. Consequently, in the linear case, Theorem 2 is equivalent with the transcendence of e^α for algebraic $\alpha \neq 0$, which followed from Theorem 1.

In general, G contains a maximal linear subgroup L such that G/L is an abelian variety A. Let $\pi\colon G \to A$ be the canonical homomorphism. Then π is an algebraic mapping. Let π_* be the induced homomorphism on the tangent spaces at the origin. If $\pi_*\alpha = 0$, then α is contained in the Lie algebra of L, and the theorem reduces to the linear case. If $\pi_*\alpha \neq 0$, then it suffices to prove that

$$\pi \circ \exp_G(\alpha) = \exp_A \circ \pi_*(\alpha)$$

is transcendental on A, and our theorem is reduced to the case when G is an abelian variety. We note that in that case, the exponential map is always a transcendental function. For example, when $\dim A = 1$, we can represent this map by the $\wp$-function, so that viewing A in the projective plane, we have

$$\exp(t) = \big(1, \wp(t), \wp'(t)\big),$$

in terms of projective coordinates. Thus when $\dim A = 1$, our theorem was covered by Corollary 3 of Theorem 1. In this case, the normalization amounts to specifying that the invariants g_2, g_3 of $\wp$ must be algebraic.

For arbitrary abelian varieties, we have the representation

$$\Theta : \mathbf{C}^d \to A_{\mathbf{C}}$$

as a quotient of $\mathbf{C}^d$, by means of theta functions. A tangent vector at the origin is then represented by coordinates

$$\alpha = (\alpha_1, \ldots, \alpha_d),$$

and is algebraic if and only if all α_i are algebraic. Our normalization of the exponential map means that the partial derivatives $\partial/\partial t_i$ are defined over the field of algebraic numbers, and

$$\exp(t\alpha) = \Theta(t\alpha).$$

Thus in the case of abelian varieties, Theorem 2 can be formulated as follows.

THEOREM 3. *Let A be an abelian variety of dimension d, defined over the field of algebraic numbers* $\mathbf{K}$. *Let*

$$\Theta : \mathbf{C}^d \to A_{\mathbf{C}}$$

be the homomorphism given by the theta functions, inducing an isomorphism of the complex torus onto $A_{\mathbf{C}}$. Assume that the derivations

$$\partial/\partial t_i \ \ (i = 1, \ldots, d)$$

are defined over $\mathbf{K}$. If $\alpha \in \mathbf{C}^d$ is a complex vector $\neq 0$ such that all α_i are algebraic, then $\Theta(\alpha)$ is transcendental over $\mathbf{K}$.

Proof. All algebraic objects in Theorem 3 are in fact defined over a finitely generated extension of $\mathbf{Q}$, so that without loss of generality, we may replace $\mathbf{K}$ by a number field K (finite over $\mathbf{Q}$). In particular, we may assume $\exp(\alpha)$ is rational over K, and so is $\exp(t\alpha)$ for $t = 1, 2, \ldots, m$. We take $m > 40\,[K : \mathbf{Q}]$ (so that we can apply Theorem 1 later) and let these m points be $w_1, \ldots, w_m$. Let

$$\Theta = (\theta_0, \ldots, \theta_N)$$

be the projective coordinates of our map. We can find a suitable linear combination of $\theta_0, \ldots, \theta_N$ with coefficients in K which does not vanish at any of our m points, so that after a projective change of coordinates, we may assume that it is θ_0. We let f_j be the meromorphic function induced by θ_i/θ_0 on the straight line $t\alpha$ in $\mathbf{C}$, so that

$$f_j(t) = \theta_j/\theta_0(t\alpha)$$

for $j = 1, \ldots, N$. We note that each θ_j/θ_0 is an abelian function, holomorphic at our m points. Since the partial derivative of a function is holomorphic at every point where the function is holomorphic, it follows that the ring

$$K[\theta_1/\theta_0, \ldots, \theta_N/\theta_0]$$

is mapped into itself by the partial derivatives $\partial/\partial t_i$ $(i = 1, \ldots, d)$, because of our assumption that these partial derivatives are defined over K. But then $D = d/dt$ maps the ring,

$$K[f_1, \ldots, f_N]$$

into itself.

The exponential map Θ cannot be an algebraic function, and hence the transcendence degree of the field

$$K\big(t, f_1(t), \ldots, f_N(t)\big)$$

over K is at least 2. We are now in a position to apply Theorem 1, with $\rho = 2$ to conclude the proof of Theorem 3, and therefore the proof of Theorem

COROLLARY. *Notation being as in Theorem 3, the periods of Θ are transcendental.*

Remark. As is well known, if $\exp: \mathbf{C}^d \to G_{\mathbf{C}}$ is the exponential map, then its inverse mapping, which should also be called the log in general, is none other than the abelian integral of vector-differentials of the first kind on G. Thus if $P \in G_{\mathbf{C}}$, then $u = \log P \in \mathbf{C}^d$ means that $P = \exp(u)$. Different values of $\log P$ differ by periods of the exponential map. Varying G, e.g. taking for G the generalized Jacobian varieties of a curve, one obtains the various abelian integrals (of all three kinds), and Theorem 2 says that (under the hypotheses stated there), if P is an algebraic point on G, then $\log P$ is transcendental. However, $\log P$ is a vector, and similarly, if α is algebraic, $\alpha \neq 0$, then $\exp(\alpha)$ has d local coordinates. In either case, one knows that at least one coordinate is transcendental. It becomes a difficult problem to determine whether *all* the coordinates (belonging to local uniformizing parameters) are transcendental.

Theorem 2 expresses on group varieties a generalization of the transcendence of e^α for α algebraic $\neq 0$. We can also express a generalization for the transcendence of α^β.

THEOREM 4. *Let G be a group variety, defined over a number field, and assume that its exponential map can be represented by meromorphic functions of finite order. Let $\varphi: \mathbf{C} \to G_{\mathbf{C}}$ be a 1-parameter subgroup, normalized to have algebraic derivative at the origin. If there exists a point $u \in \mathbf{C}$,*

$u \neq 0$ *such that $\varphi(u)$ is an algebraic point of G, then φ has algebraic dimension* 1, *and $\varphi(\mathbf{C})$ is a* 1-*dimensional group subvariety of G.*

Proof. This follows from Theorem 1 like Theorems 2 and 3.

We note that Theorem 4 applies to linear groups, abelian varieties, and products of these. The representation of the exponential map by meromorphic functions of finite order is not explicitly in the literature for arbitrary group varieties, except in special cases (e.g. the Weierstrass zeta function which can be used to parametrize certain 2-dimensional group varieties, and Severi's quasi-abelian functions). We note that one needs only to consider commutative group varieties, since the Zariski closure of the 1-parameter subgroup is commutative.

Historical note

Hermite proved the transcendence of e, and several years later, Lindemann proved the transcendence of e^{α} for algebraic $\alpha \neq 0$ by an extension of the method of Hermite. This method involved constructing a function with high zeros, but was formally quite different from the one we have used, and we shall make comments on this again later.

The method we have used is essentially the method of Gelfond and Schneider, used to prove the transcendence of α^{β}, properly axiomatized and formulated. The differential equation and the addition theorem for the exponential function are used by both, and Schneider had made a partial attempt at axiomatization in [28]. However, Schneider's formulation there was too special to be applied, say to abelian functions, and the arrangement of his proof is still comparatively complicated.

Theorem 2 had been conjectured by Cartier, and was proved in [16]. The transcendence of the periods of abelian functions is due to Schneider [27], who had to devise an argument in several variables to get them. Because of our more general theorem on arbitrary algebraic differential equations, we can handle this result still as one involving only a single variable.

Lindemann proved also a theorem on algebraic independence, namely, if $\alpha_1, \ldots, \alpha_m$ are algebraic, and linearly independent over $\mathbf{Q}$, then

$$e^{\alpha_1}, \ldots, e^{\alpha_m}$$

are algebraically independent. This will be proved by Siegel's method in Chapter VII. For the ordinary exponential function, Schanuel has made a very general conjecture, to the effect that *if $\alpha_1, \ldots, \alpha_m$ are complex numbers, linearly independent over $\mathbf{Q}$, then the transcendence degree of*

$$\alpha_1, \ldots, \alpha_m, \; e^{\alpha_1}, \ldots, e^{\alpha_m}$$

is at least m. From this statement, one would get most statements about algebraic independence of values of e^t and $\log t$ which one feels to be true. For instance, it has been conjectured for a long time (by anybody who has looked at the subject) that if $\beta_1, \ldots, \beta_m$ are non-zero, multiplicatively independent algebraic numbers, then

$$\log \beta_1, \ldots, \log \beta_m$$

are algebraically independent. Similarly, one could deduce the algebraic independence of e and π, by considering

$$1, \ 2\pi i, \ e, \ e^{2\pi i}.$$

I would also conjecture that π cannot lie in the field obtained by starting with the algebraic numbers, adjoining values of the exponential function, taking algebraic closure, and iterating these two operations. It is an exercise to show that this follows from Schanuel's conjecture.

More generally, given a group variety G defined over a number field K, one can ask for the transcendence degree of the point $(\alpha, \exp(\alpha))$, where α is a tangent vector at the origin, and exp is normalized as in Theorem 2. We shall discuss this question again in the next chapter, since it involves problems in several variables.

Theorem 2 and Theorem 3 give a reasonably satisfactory beginning for the theory of transcendental points on group varieties—the main problem is to determine transcendence degrees. However, there is an interesting direction where the transcendence question is still open. Let V be a projective, non-singular curve (variety of dimension 1) defined over a number field, and of genus ≥ 2. Let U be a disc, centered at the origin in $\mathbf{C}$. Let $\varphi: U \to V_\mathbf{C}$ be the universal covering map, i.e. essentially the universal uniformizing map of the Riemann surface $V_\mathbf{C}$, normalized so that the origin goes to an algebraic point, and the tangent map $\varphi'(0)$ is algebraic. One conjectures that if P is an algebraic point of the disc, $\neq 0$, then $\varphi(P)$ is transcendental on V. As a side question, one can ask if the radius of the disc is transcendental. In this problem, as well as in Theorem 1, it is the normalization of the map which allows one to distinguish between algebraic and transcendental points.

When dealing with differential equations, the normalization can be taken to be algebraic initial conditions, and algebraic coefficients. We shall meet again such a situation in Chapter VII. However, not all classical maps satisfy a differential equation, and the study of automorphic functions from the point of view of transcendental numbers is barely begun.

We conclude by returning to Hermite's proof of the transcendence of e. This proof involves the continued rational fractions associated with the power series of e^t. It is by means of these fractions that Hermite constructs the function having a zero of high order. This is quite an interesting direction, which relates the theory of transcendental numbers with the theory of diophantine approximations. So far, continued fractions have been developed mostly as a one variable theory, and hence are still not very good for generalizing Hermite's proof to proofs of algebraic independence for other types of functions beside e^t. There is no doubt, however, that eventually one will have to have such a generalization, which will give quantitative results concerning classical transcendental numbers. These are known as measures of transcendence, and we shall return in greater detail to this point at the end of Chapter VI and Chapter VII, where such measures of transcendence will be obtained for some special numbers.

Independently of transcendence problems, one can raise an interesting question of algebraic-analytic nature, namely given a 1-parameter subgroup of an abelian variety (say Zariski dense), is its intersection with a hyperplane section necessarily non-empty, and infinite unless this subgroup is algebraic? This question arises in connection with Theorem 4 of Chapter II, and Theorem 3 of the present chapter.

CHAPTER IV

Functions of Several Variables

§1. *Partial differential equations*

We shall extend the main theorem of Chapter III to functions of several variables.

We recall that an entire function of d variables $g(U) = g(u_1, \ldots, u_d)$ is said to be of order $\leq \rho$ if there exists a constant $C > 0$ such that

$$|g(U)| \leq C^{R^\rho}$$

on the domain $|u_i| \leq R$ for all sufficiently large R. A meromorphic function will be said to be of order $\leq \rho$ if it can be written as a quotient of two entire functions of order $\leq \rho$. If P is a point in $\mathbf{C}^d$ and f a meromorphic function of order $\leq \rho$ we say that f is defined at P if we can write $f = g/h$ where g, h are entire of order $\leq \rho$ and $h(P) \neq 0$.

THEOREM 1. *Let K be a number field, $f_1, \ldots, f_N$ meromorphic functions of order $\leq \rho$ in d variables. Assume that the ring $K[f_1, \ldots, f_N]$ is mapped into itself by the partial derivatives $D_1, \ldots, D_d$, and that its transcendence degree over K is $\geq d + 1$. Let S be a finite set of points in $\mathbf{C}^d$ at which all the functions f_μ are defined, and such that $f_\mu(P) \in K$ for all μ and all $P \in S$. Suppose in addition that after some change of coordinate system, the set S becomes a product*

$$S_1 \times \cdots \times S_d,$$

where each S_i is a set of m distinct complex numbers. Then

$$m \leq b\rho \, [K : \mathbf{Q}]$$

with a constant b depending only on d, and easily estimated.

The proof of Theorem 1 will follow the pattern of the one dimensional proof. We begin by preliminary remarks on estimates of partial derivatives and integrals.

33

434

§2. *Estimates of derivatives and integrals*

We have the analogue in several variables for the estimate of Chapter III, §2.

LEMMA 1. *Let K be a number field. Let $f_1,\ldots,f_N$ be functions of d complex variables, holomorphic at a point A in $\mathbf{C}^d$, and such that the ring $K[f_1,\ldots,f_N]$ is mapped into itself by the partial derivatives $D_1,\ldots,D_d$. Assume that $f_j(A) \in K$ for $j = 1,\ldots,N$. There exists a number C_1 having the following property. If $Q(T_1,\ldots,T_N)$ is a polynomial in N variables with coefficients in I_K, of total degree $\leq r$, and if*

$$D = D_1^{k_1} \cdots D_d^{k_d}$$

is a differential operator of order $k = k_1 + \cdots + k_d$, then

$$\|D\big(Q(f_1,\ldots,f_N)\big)(A)\| \leq \|Q\|\, k!\, C_1^{k+r}.$$

Furthermore, there is a denominator for $D\big(Q(f_1,\ldots,f_N)\big)(A)$ bounded by $\operatorname{den}(Q)C_1^{k+r}$.

Proof. The proof uses exactly the same kind of estimates as the proof for one variable, and the same kind of induction. We may therefore leave it as a simple exercise to the reader.

Next, we discuss some applications of Cauchy's formula. When dealing with functions of one variable, it was convenient to make one estimate using the maximum modulus principle. This argument does not extend to functions of several variables, but in the application we have in mind, we can use Cauchy's integral theorem instead. For clarity, we recall some facts in one variable. Let S be a set of m distinct complex numbers, and let $E(z)$ be an entire function of one variable z. Let R be a positive number such that $|y| < R$ for all $y \in S$. For each $y \in S$, suppose given an integer $\lambda(y) \geq 0$. Let

$$Q(z) = \prod_{y \in S} (z - y)^{\lambda(y)},$$

and for each $y \in S$, let

$$Q_y^*(z) = \frac{Q(z)}{(z - y)^{\lambda(y)}}.$$

Then Q_y^* is defined and not zero at y. Let

$$\Delta_y = (d/dz)^{\lambda(y)-1}$$

if $\lambda(y) \geq 1$, and 0 otherwise. Then

$$\frac{1}{2\pi i} \int_{|z|=R} \frac{E(z)}{Q(z)}\, dz = \sum_{y \in S} \frac{1}{(\lambda(y) - 1)!} \Delta_y(E/Q_y^*)(y).$$

If we deal with a function of several variables, then we can form a repeated integral over a similar quotient, and obtain an analogous formula. We shall need a special case.

LEMMA 2. Let $E(z_1,\dots,z_d)$ be an entire function. Let each one of $S_1,\dots,S_d$ be a set of m complex numbers. Let

$$\Delta_i = \partial/\partial z_i.$$

Let σ be an integer > 0. Assume that for all differential operators Δ of order $\leq \sigma$ and all points $Y \in S_1 \times \cdots \times S_d$ we have

$$\Delta E(Y) = 0.$$

Let $(\lambda) = (\lambda_1, \dots, \lambda_d)$ be a vector of integers ≥ 0 such that

$$\lambda_1 + \cdots + \lambda_d = \sigma + 1,$$

and assume that

$$\Delta^{(\lambda)} E(A) \neq 0$$

for some A in $S_1 \times \cdots \times S_d$. Let $S_i' = S_i - \{a_i\}$. Let

$$Q(Z) = (z_1 - a_1)^{\lambda_1+1} \cdots (z_d - a_d)^{\lambda_d+1} \prod_i \prod_{y_i \in S_i'} (z_i - y_i)^{\lambda_i}.$$

Let $R > |Y|$ for all Y. Then

$$\int \cdots \int \frac{E(Z)}{Q(Z)}\, dZ = \frac{1}{\lambda!} \frac{1}{(2\pi i)^d} \frac{\Delta^{(\lambda)} E(A)}{Q_A^*(A)}$$

where the repeated integral is taken over $|z_i| = R$, and

$$Q_A^*(Z) = \frac{Q(Z)}{(z_1 - a_1)^{\lambda_1+1} \dots (z_d - a_d)^{\lambda_d+1}}.$$

Proof. Taking the repeated integral and using the remarks we made above for one variable, we obtain a sum over vectors Y involving differential operators. However, when evaluated at vectors Y, they will give only a zero contribution because of our assumption that $\Delta E(Y) = 0$, except for the term belonging to A, which is the only term for which a differential operator of order $\sigma + 1$ will occur. Up to powers of $2\pi i$ and factorials, this term is of type

$$\Delta^{(\lambda)}(E/Q_A^*)|_{Z=A}.$$

Our lemma now follows from the next remark.

LEMMA 3. *Let E_1, E_2 be two functions holomorphic at a point A. Assume that $\Delta E_1(A) = 0$ for all differential operators Δ of order $\leq \sigma$, and let $\Delta^{(\lambda)}$ be an operator of order $\sigma + 1$. Then*

$$\Delta^{(\lambda)}(E_1 E_2)(A) = \Delta^{(\lambda)} E_1(A) \cdot E_2(A).$$

Proof. Using repeatedly the rule for the derivative of a product, with respect to $\Delta_1, \ldots, \Delta_d$, we find that all terms will be 0 except the term which applies all the differential operations to E_1 and not to E_2. This yields what we wanted.

§3. *The main proof*

We carry out the proper part of the proof of Theorem 1, and follow the same pattern as the proof of Theorem 1, Chapter III, taking into account the additional parameter d. Constants depending only on d will be denoted by $b_1, b_2, \ldots$ Constants depending on $K, f_1, \ldots, f_N$ and S will be denoted by $c_1, c_2, \ldots$.

For any positive integer n, we have

$$b_1 n^d \leq \binom{n + d}{d} \leq b_2 n^d.$$

For convenience, we assume that b_2 is the d-th power of an integer.

Let $f_1, \ldots, f_N$ be our functions in d variables $U = (u_1, \ldots, u_d)$. We let $D_i = \partial/\partial u_i$. Say $f_1, \ldots, f_{d+1}$ are algebraically independent over K. Let r be a large integer, which will tend to infinity later, which is a d-th power of an integer, such that $r^{1/d}$ is divisible by $2m b_2$. Let

$$F = \sum a_{(i)} f_1^{i_1} \cdots f_{d+1}^{i_{d+1}}$$

have coefficients $a_{(i)}$ in I_K, the sum being taken for

$$0 \leq i_1, \ldots, i_{d+1} < r.$$

We can select the coefficients $a_{(i)}$ not all equal to 0, and such that

$$DF(P) = 0$$

for all differential operators D of order at most

$$n = \frac{r^{(d+1)/d}}{b_2^{1/d} 2m}$$

and all $P \in S$. We have to solve a system of linear equations in r^{d+1} unknowns, and the number of equations is at most $b_2 n^d m^d \leq r^{d+1}/2^d$. Using

Lemma 1, we see that the size of the coefficients, namely the size of the expressions

$$D(f_1^{i_1} \cdots f_{d+1}^{i_{d+1}})(P),$$

is bounded by $b_3 r \log r$ for r sufficiently large (with respect to the data of the theorem). By Siegel's lemma, we can solve our linear equations with

$$\text{size}(a_{(i)}) \leqq b_4 r \log r.$$

Since the functions $f_1, \ldots, f_{d+1}$ are algebraically independent over K, it follows that F is not identically zero.

Let σ be the smallest positive integer such that

$$DF(P) = 0$$

for all differential operators D of order $\leqq \sigma$ and all $P \in S$. Then $\sigma \geqq n$ and there exists a d-tuple $(s) = (s_1, \ldots, s_d)$ such that

$$s_1 + \cdots + s_d = \sigma + 1,$$

and

$$D^{(s)}F(\overline{A}) \neq 0$$

for some $\overline{A}$ in S. By Lemma 1,

$$\text{size } D^{(s)}F(\overline{A}) \leqq b_5 \sigma \log \sigma,$$

for n, and hence σ, sufficiently large.

We shall now estimate one conjugate of $D^{(s)}F(\overline{A})$ using Lemma 2.

Let B be the $d \times d$ matrix giving the change of coordinates mentioned in the theorem, so that $U = ZB$, and $Z = (z_1, \ldots, z_d)$, both U, Z being row vectors. Then

$$f_i(U) = g_i(Z)$$

and

$$F(U) = F(ZB) = G(Z).$$

We let $\Delta_i = \partial/\partial z_i$. Then there are linear combinations $L_1, \ldots, L_d$ of $\Delta_1, \ldots, \Delta_d$ with complex coefficients such that

$$D_i F(U) = L_i G(Z),$$

and hence

$$D^{(s)}F(U) = L^{(s)}G(Z).$$

We can write

$$L^{(s)} = L_1^{s_1} \cdots L_d^{s_d} = \sum \xi_{(\lambda)} \Delta_1^{\lambda_1} \cdots \Delta_d^{\lambda_d}$$

with complex coefficients $\xi_{(\lambda)}$ and $\lambda_1 + \cdots + \lambda_d = \sigma + 1$.

Let A be the point in $S_1 \times \cdots \times S_d$ corresponding to $\overline{A}$ under the linear transformation. We then have

$$D^{(s)}F(\overline{A}) = L^{(s)}G(A) = \sum \xi_{(\lambda)} \Delta_1^{\lambda_1} \cdots \Delta_d^{\lambda_d} G(A).$$

We let Y range over $S_1 \times \cdots \times S_d$ (the transform of S under our linear transformation). Since no point of Y lies in the poles of the functions g_i, we can find an entire function θ of order $\leqq \rho$ such that $\theta(A) \neq 0$ and θg_i is entire of order $\leqq \rho$. We let

$$H(Z) = \theta(Z)^{(d+1)r}.$$

Then

$$E(Z) = G(Z)H(Z)$$

is an entire function, and $H(A) \neq 0$. Furthermore,

$$\Delta E(Y) = 0$$

for all differential operators Δ of order $\leqq \sigma$, and by Lemma 3,

$$\Delta^{(\lambda)}E(Y) = \Delta^{(\lambda)}G(Y)H(Y).$$

By Lemma 2, we obtain

$$\Delta^{(\lambda)}G(A) \leq \lambda!(2\pi)^{-d}\frac{Q_A^*(A)|}{|H(A)|}\left| \int \cdots \int \frac{E(Z)}{Q(Z)}dZ \right|,$$

taking the repeated integral on the circles of radius

$$R = \sigma^{1/(d+1)\rho}.$$

We can easily estimate the integral, and see that $1/Q(Z)$ gives essentially a contribution bounded by $1/R^{m\sigma}$ as soon as σ is large enough with respect to the points Y. One can also estimate $E(Z)$ and one finds a bound of type

$$|E(Z)| \leqq c_1^{rR^\rho} \leqq c_2^\sigma$$

for σ sufficiently large.

We therefore find

$$\log |\Delta^{(\lambda)}G(A)| \leqq 2\sigma \log \sigma - m\sigma \log R,$$

whence, estimating the $\xi_{(\lambda)}$, the number of terms in the sum expressing $L^{(s)}G(A)$, and $H(A), Q_A^*(A)$ in a trivial manner, we find

$$\log |D^{(s)}F(\overline{A})| = \log |L^{(s)}G(A)| \leqq 2\sigma \log \sigma - \frac{m\sigma \log \sigma}{(d+1)\rho}.$$

Comparing this estimate with the size of the element $D^{(s)}F(\overline{A})$ in K, we conclude that

$$m \leq b\rho\,[K : \mathbf{Q}]$$

for some constant b depending only on d, thereby proving Theorem 1.

§4. Applications

We begin by a weak version in higher dimensions for an analogue of Theorem 4, Chapter II, §4.

THEOREM 2. *Let G be either an abelian variety, or a linear group variety, defined over a number field. Let*

$$\varphi\colon \mathbf{C}^d \to G_{\mathbf{C}}$$

be a d-parameter subgroup, normalized so that the derivative at the origin is algebraic. Let Γ be a subgroup of $\mathbf{C}^d$ containing at least d linearly independent points over $\mathbf{C}$. If $\varphi(\Gamma)$ is contained in the group of algebraic points of G, then φ has algebraic dimension d, and $\varphi(\mathbf{C}^d)$ is an algebraic subgroup of dimension d.

Proof. The assumption that the points of Γ are linearly independent over the complex numbers allows us to change coordinates in such a way that they become a product. In the present case, we obtain ∞^d points in $\mathbf{C}^d$ at which our functions take on algebraic values. In the case of abelian varieties, we dehomogenize the projective map at a product of m such points with m large, with respect to the new coordinate system. We can then apply Theorem 1 directly.

It would be very desirable to have a version of Theorem 2 in which the map φ is not necessarily normalized with respect to the derivative, but which allows a subgroup with sufficiently many independent points over the rationals. Cf. the historical note.

Next we shall obtain an analogue to the transcendence statement concerning the modular function given in Chapter III.

Let A be an abelian variety defined over the number field K. Let

$$\Theta\colon \mathbf{C}^d \to A_{\mathbf{C}}$$

be a representation as a quotient of complex d-space. We *assume* throughout that our representation is normalized so that the derivative of Θ at the origin is defined over K. Let $\Omega_1, \ldots, \Omega_{2d}$ be fundamental periods, and let Ω be the $d \times 2d$ matrix of periods, viewing the Ω_i as column vectors. Let $\mathbf{P}$ be a principal matrix such that Ω and $\mathbf{P}$ satisfy the Riemann relations. After making an *integral* change of coordinates, we can assume that

$\mathbf{P}$ has the usual canonical form

$$\mathbf{P} = \begin{pmatrix} 0 & \mathbf{D} \\ -\mathbf{D} & 0 \end{pmatrix}.$$

Then $\Omega = (W_1, W_2)$, where each W_1, W_2 is a $d \times d$ matrix, and W_1 is invertible. In the theory of moduli, one takes an integral matrix multiple of $W_1^{-1} W_2$ as the moduli point associated with the abelian variety in the Siegel upper half space H_d. Our next theorem is concerned however with $W_2 W_1^{-1}$. The relation between $W_1^{-1} W_2$ and $W_2 W_1^{-1}$ is not clear.

THEOREM 3. *Let A be as above, and also $\Theta \colon \mathbf{C}^d \to A_{\mathbf{C}}$, as well as the period matrix Ω, normalized so that the principal matrix has the usual canonical form. Let $T = W_2 W_1^{-1}$. If T is algebraic, then T (viewed as a linear transformation) maps the period lattice tensored with $\mathbf{Q}$ into itself.* (*When $d = 1$, then $T = \tau$, $W_1 = \omega_1$, $W_2 = \omega_2$ in the usual notation.*)

Proof. Let $\Phi(U) = \Theta(TU)$ and assume that T is algebraic. After extending K if necessary, we may assume that all components of T lie in K. We view U as a vertical vector for this proof. Then $\Omega_1, \ldots, \Omega_d$ are periods for Φ and Θ, and generate a lattice such that for any point P in this lattice, $\Phi(P)$ and $\Theta(P)$ are points of A rational over K. A change of basis in $\mathbf{C}^d$ transforms lattice points into a set of points as in Theorem 1, namely

$$(j_1, \ldots, j_d), \qquad\qquad 0 \leqq j_i \leqq m.$$

Taking m sufficiently large, we dehomogenize the projective maps Φ and Θ to obtain two rings of functions

$$K[f_1, \ldots, f_N] \qquad \text{and} \qquad K[g_1, \ldots, g_M]$$

giving corresponding maps into affine space and defined at the above points. Applying Theorem 1, we conclude that the product map

$$(\Phi, \Theta) \colon \mathbf{C}^d \to A_{\mathbf{C}} \times A_{\mathbf{C}}$$

has in fact dimension d, i.e. that Φ is algebraic over Θ. For some integer r, it follows that $\Phi(rU)$ is rational over $\Theta(U)$, and hence that *all* periods of $\Theta(U)$ are also periods of $\Phi(rU)$. Hence T maps the period lattice tensored with $\mathbf{Q}$ into itself, as was to be shown.

Historical note

Schneider was the first to consider the transcendence question for functions of several variables in a special case [27], and as we have said earlier, proved that the periods of algebraic abelian integrals of the first and second kind are transcendental. He uses Cauchy's formula, and the technique of repeated integrals is taken from his paper. The results of this chapter are taken from [17], and generalize Schneider's results.

He also makes a comment on the moduli point, and presumably had something like our Theorem 3 in mind ([27], page 113, top), but there is some confusion about the matter, in view of the fact that we find $W_2 W_1^{-1}$ and not $W_1^{-1} W_2$.

Functions of several variables have of course divisors of zeros (not isolated points), and thus in investigating other types of sets S on which such functions take on algebraic values, one meets genuine difficulties in putting natural conditions on S. For instance, the functions z_1, z_2, $e^{z_1 - z_2}$ take on algebraic values whenever z_1, z_2 are algebraic and $z_1 = z_2$. Furthermore, already in one variable, there may be certain collapsing in transcendence degree, since the functions

$$t, e^t, e^{t^2}, \ldots$$

have values in a field of transcendence degree 1 whenever t is an integer.

In extending Theorems 1, 2, 3, one has the following possibilities.

First, a suggestion of Nagata, that the set S in Theorem 1 may be described as contained in a hypersurface (algebraic). This would be a good way of eliminating the unnatural condition that S be a product. (One would also need to bound the degree of the hypersurface.)

Second, even though we know that the vector periods of abelian integrals are transcendental, this means only that at least *one* component is transcendental. It is therefore desirable to have stronger theorems, giving the transcendence of each component. In this direction, the Riemann relations give significant examples of degeneracy, as follows. Let $C_1, \ldots, C_m$ be elements of complex d-space, which are linearly independent over $\mathbf{Q}$. Let A be an abelian variety of dimension d defined over a number field K, and assume that its ring of endomorphisms is trivial. In particular, A is simple. Let

$$\Phi_m \colon \mathbf{C}^{md} \to A_{\mathbf{C}}^m$$

be the mapping given by $\Phi_m(t) = (\Theta(tC_1), \ldots, \Theta(tC_m))$. Thus Φ_m is the exponential map passing through the vector $(C_1, \ldots, C_m)$. *We contend that this map has algebraic dimension md, i.e. that it is Zariski-dense in A^m.* We prove this by induction on m. For $m = 1$, it follows from the simplicity of A. Assume it for m. If the dimension of the map with $m + 1$ factors is not $d(m + 1)$, then $\Phi_{m+1}(t)$ is contained in an abelian subvariety of A^{m+1} whose projection on the first m factors is A^m, and which is algebraic over A^m. There exists an integer $r \neq 0$ such that

$$\Theta(rtC_{m+1})$$

is rational over $\Phi_m(t)$, i.e.

$$\left(\Theta(tC_1),\ \ldots,\ \Theta(tC_m),\ \Theta(rtC_{m+1}) \right)$$

lies in an abelian subvariety B of A^{m+1} which is isomorphic to A^m under projection, and whose projection on the last factor is A. Hence there exists a homomorphism of A^m onto A, and since A has only trivial endomorphisms, there exist integers $r_1, \ldots, r_m$ such that

$$\Theta(trC_{m+1}) = r_1\Theta(tC_1) + \cdots + r_m\Theta(tC_m).$$

This implies that
$$rC_{m+1} = r_1C_1 + \cdots + r_mC_m,$$
a contradiction.

We apply the above to the $2d$ periods $\Omega_1, \ldots, \Omega_{2d}$. The dimension of the map Φ_{2d} is then $2d^2$, but the period $(\Omega_1, \ldots, \Omega_{2d})$ is mapped onto 0 by Φ_{2d}. According to the Riemann relations, the transcendence degree of the period matrix cannot be $2d^2$.

(On the other hand, observe that the Riemann relations do not give a counterexample for the transcendence of the components of a basic period vector.)

One may raise the question whether such degeneracy can occur in the case of a simple abelian variety, possibly under the additional assumption that the ring of endomorphisms is trivial. In other words, suppose that A is a simple abelian variety. Let α be a non-zero algebraic tangent vector (i.e. an element of $\mathbf{C}^d$ which is algebraic). Then one may ask whether the transcendence degree of the point $(\alpha, \Theta(\alpha))$ in $\mathbf{C}^d \times A_{\mathbf{C}}$ is $\geqq d$.

It would be desirable to have a general conjecture also for products. For instance, if A has dimension 1, and our mapping is represented by the Weierstrass function with algebraic g_2, g_3, let ω_1, ω_2 be two fundamental periods. If A has no complex multiplication, then it is reasonable to expect that ω_1, ω_2 are algebraically independent. One would see this by looking at the representation of $A \times A$ as a quotient of $\mathbf{C}^2$.

For the period matrix itself, Grothendieck has made a very interesting conjecture concerning its relations, and his conjecture applies to a general situation, as follows. Let V be a projective, non-singular variety defined over the rational numbers. One can define the cohomology of V with rational coefficients in two ways. First, by means of differential forms (de Rham), purely algebraically, thereby obtaining a vector space $H_{\mathrm{diff}}(V, \mathbf{Q})$ over $\mathbf{Q}$. Secondly, one can take the singular cohomology $H_{\mathrm{sing}}(V, \mathbf{Q})$ with rational coefficients, i.e. the singular cohomology of the complex manifold $V_{\mathbf{C}}$. Let us select a basis for each of these vector spaces over $\mathbf{Q}$, and let us tensor these spaces over $\mathbf{C}$. Then there is a unique (period) matrix Ω with complex coefficients which transforms one basis into the other. Any algebraic cycle on V or the products of V with itself will give rise to a polynomial relation with rational coefficients among the coefficients of this matrix. Grothendieck's conjecture is that the ideal generated by these relations is an ideal of definition for the period matrix.

We observe that one can also consider group varieties which are not complete, and that the transcendental parametrizations of these are the inverse maps of abelian integrals of various kinds. The transcendence and algebraic independence of values of abelian integrals thus appears as a question on the inverse mapping of the general exponential map on group varieties, i.e. as the algebraic independence of logarithms in an extended sense. This would be the case for instance for the integral

$$\int_0^1 \frac{1}{1+t^3}\, dt = \frac{1}{3}\left(\log 2 + \frac{\pi}{\sqrt{3}}\right)$$

which appears at the end of Siegel's book. Of course, this integral involves only the algebraic independence of ordinary logarithms ($2\pi i = \log 1$ and $\log 2$). More generally, one can consider an intermediate case related to the Weierstrass zeta function ($\zeta' = -\wp$), which, together with the $\wp$-function, parametrizes the group variety associated with integrals of the second kind by the map $\mathbf{C} \times \mathbf{C} \to G_{\mathbf{C}}$ given by

$$(t,\, u) \mapsto \left(1,\, \wp(t),\, \wp'(t),\, u - \zeta(t)\right).$$

This map has periods $(\omega_1,\, \eta_1)$ and $(\omega_2,\, \eta_2)$ (classical notation), and the Legendre relation

$$\eta_1\omega_2 - \eta_2\omega_1 = 2\pi i$$

is nothing but the Riemann relation for that case.

Schanuel's conjecture mentioned in Chapter III asserts that such exceptions do not occur for the ordinary exponential function. It would be interesting to determine if there are other examples of degeneracy for the transcendence degree of the point $(\alpha, \exp(\alpha))$ whenever the analytic curve $\exp(t\alpha)$ has the correct dimension, other than those given by period relations.

For abelian varieties, the transcendence degree is affected not only by the Riemann relations, but also by the existence of non-trivial endomorphisms, or of algebraic cycles on the variety and its products. In any case, one is also led in this way to investigate the transcendence degree of points $(\alpha, \varphi(\alpha))$, where

$$\varphi : H_d \times \mathbf{C}^d \to V_{\mathbf{C}}$$

is the uniformization of the variety of moduli V by means of the Siegel upper half space H_d, and complex d-space.

In any case, the Riemann relations show that one cannot prove the algebraic independence of logarithms of multiplicatively independent algebraic numbers by "general" statements about entire functions. One must make use of the special properties of e^t.

Third, as we have already mentioned, one would like a higher dimensional version of Theorems 3 and 4, Chapter I, §3 and §4, corresponding to Theorem 2 of this chapter, but without assuming any normalization on the parametrization of the d-parameter subgroup. One then meets two difficulties in trying to extend the one-dimensional proof. First, a purely analytic difficulty which consists in formulating a suitable interpolation estimate. The use of Cauchy's formula does not seem to generalize without some new idea. In any case, it is obviously necessary to assume that the linear combinations of the log vectors of algebraic points with integer coefficients are essentially equidistributed. This leads to the second difficulty, much deeper than the first, namely to prove that such combinations satisfy this equidistribution property. This is a problem which by definition belongs to the theory of diophantine approximations, and at present is very much beyond any result which has ever been obtained in this theory.

CHAPTER V

Finitely Generated Values

§1. The inductive technique

Up to now, we have considered the problem of proving the transcendence of values of certain functions over the field of rational numbers. However, in cases of numbers like

$$e + \pi \qquad \text{or} \qquad e + \log 2$$

we find that the number is a sum of values of *two* different types of functions (namely exp and log), so that the natural approach to the transcendence of such values is to prove that say e, π are algebraically independent. So far, such a result has eluded available techniques, but we shall point out a good possibility in this direction, and obtain partial results. We work out a special case first. The general case will involve only additional technical complications.

Let x be a complex number, transcendental over $\mathbf{Q}$. For any polynomial $P(x) \in \mathbf{Z}[x]$ with integer coefficients, we define

$$\text{size } P(x) \;=\; \max(\deg P, \log |P|).$$

Let τ be a number ≥ 2. We shall say that x is of *transcendence type* $\leq \tau$ if for all polynomials with integral coefficients, we have

$$-(\text{size } P(x))^{\tau} \ll \log |P(x)|.$$

This is a natural extension of the formula stated in Chapter I for algebraic numbers (in that case, $\tau = 1$). From Dirichlet's box principle, one sees that one can never take $\tau < 2$. (Our definition is adjusted to the method for which it will be useful. Since our results are only partial, a refinement of the method will require a corresponding refinement of the definition, to prove more refined results.)

To estimate products of polynomials, we note that

$$\text{size}(P_1 \cdots P_n) \;\leq\; 3 \sum_{i=1}^{n} \text{size } P_i.$$

45

446

This is trivially proved, because $P_1 \cdots P_n$ is dominated by

$$|P_1| \cdots |P_n|(1 + T)^{d_1 + \cdots + d_n},$$

where $d_i = \deg P_i$, and hence

$$|P_1 \cdots P_n| \leqq |P_1| \cdots |P_n| 2^{d_1 + \cdots + d_n},$$

whence our estimate follows. (The absolute value around P_i means of course the maximum of the coefficients, and is to be distinguished from $|P_i(x)|$, taken in the complex numbers.)

We illustrate the generalization of the technique of Chapter I by a simple statement.

THEOREM 1. *Let $\beta_1, \ldots, \beta_d$ be complex numbers, linearly independent over $\mathbf{Q}$. Let $z_1, \ldots, z_m$ be complex numbers linearly independent over $\mathbf{Q}$. Let x be a complex number of type $\leqq \tau$. If $dm \geqq \tau(m + d)$ then not all values*

$$e^{\beta_i z_\nu} \quad (i = 1, \ldots, d \text{ and } \nu = 1, \ldots, m)$$

can lie in the field $\mathbf{Q}(x)$.

Proof. For simplicity, assume that all values above are polynomials, in $\mathbf{Z}[x]$. (In the general case, one has merely to clear denominators throughout the proof.) Let n be a large integer, and let r be approximately equal to $n^{m/d}$, up to a constant factor. We form the function

$$F(t) = \sum a_{(i)} x^{i_{d+1}} e^{i_1 \beta_1 t} \cdots e^{i_d \beta_d t}$$

with

$$1 \leqq i_1, \ldots, i_d \leqq r \quad \text{and} \quad 1 \leqq i_{d+1} \leqq nr,$$

with integer coefficients $a_{(i)}$, not all zero, such that F has a zero at all numbers

$$k \cdot z = k_1 z_1 + \cdots + k_m z_m$$

with

$$1 \leqq k_\nu \leqq n.$$

We have to solve linear equations, and for each (k), substituting $k \cdot z$ for t in $F(t)$, we obtain a polynomial in x with integer coefficients. Trivial estimates show that the degrees of these polynomials are $\ll rn$. Thus the number of linear equations, bounded by the number of (k) times a bound for the degrees, satisfies

$$\text{number of equations} \ll rn^{m+1}.$$

On the other hand,

$$\text{number of variables} = nr^{d+1}.$$

Thus our choice of r guarantees that the exponent in Siegel's lemma is constant. Simple estimates show that the coefficients of our linear equations satisfy

$$\text{size of coefficients} \ll nr.$$

Hence we can solve our equations with integers $a_{(i)}$ not all 0, of size $\ll nr$. We then let s be the largest integer such that $F(k \cdot z) = 0$ for $1 \leq k_\nu \leq s$, and let $w = k \cdot z$ with some $k_\nu = s + 1$ be such that $F(w) \neq 0$. Then we estimate

$$F(w) = \left. \frac{F(t)}{\prod(t - k \cdot z)} \prod (w - k \cdot z) \right|_{t=w},$$

using the maximum principle on the circle of radius $R = s^{1+1/d}$. We find that the quotient of the products pulls towards zero with a strength at least

$$\frac{1}{d} s^m \log s,$$

whereas the numerator pulls to infinity with a strength at most

$$s^{m/d} s R = s^{m/d + 2 + 1/d}.$$

This is safely smaller than the strength of the denominator, and we obtain

$$\log |F(w)| \ll -s^m \log s.$$

On the other hand, simple estimates show that

$$\text{size } F(w) \ll sr \ll s^{m/d+1},$$

the size being of course the size of the polynomial in x. By assumption, we must have

$$-(s^{m/d+1})^\tau \ll -s^m \log s,$$

which contradicts our assumption on m and d.

We observe that in Theorem 1, we must always have $d > \tau$, and that if we do, then we can choose m sufficiently large so that the condition $dm \geq \tau(m + d)$ is satisfied. For instance, if $\tau = 2$, we can take $d = 3$ and $m = 6$.

We shall see in the next sections that one may deal with any finite extension of a field $\mathbf{Q}(x)$ and obtain a similar result. This is a matter only of technique, as is the generalization to finitely generated extensions $\mathbf{Q}(x_1, \ldots, x_q, y)$ with algebraically independent $x_1, \ldots, x_q$, and y algebraic over $\mathbf{Q}(x)$. One then has to assume a transcendence type for the elements (x), and the pattern of the proof is the same.

If instead of $\mathbf{Q}(x)$ we deal with a finite extension of $\mathbf{Q}(x)$, then we see that Theorem 1 yields a result on algebraic independence, namely one of the values

$$e^{\beta_i z_\nu}$$

is algebraically independent from x. Thus a theorem like Theorem 1 yields an inductive procedure for algebraic independence results. However, it is very imprecise, even more so than the analogous Theorem 1 of Chapter II, where we already mentioned the desirability of shrinking 3 to 2. It seems that to go deeper, one must refine the notion of transcendence type, and load the induction hypotheses much more than we have done, in addition to changing something in the structure of the proof. A discussion of this problem is best postponed until the end of Chapter VI, in the light of Feldman's results.

§2. *Transcendence types*

Let K be a subfield of the complex numbers, finitely generated over $\mathbf{Q}$. Suppose that we write

$$K = \mathbf{Q}(x_1, \ldots, x_q, y),$$

where $x_1, \ldots, x_q$ are algebraically independent, y is algebraic over $\mathbf{Q}(x)$, and is integral over $\mathbf{Z}[x_1, \ldots, x_q] = \mathbf{Z}[x]$. One can always find such generators, and a set (x, y) satisfying these conditions will be called a *proper set of generators*.

With respect to such a proper set (x, y), we define the *size* of an element of K as follows. We do it first for polynomials, i.e. elements of $\mathbf{Z}[x]$. If

$$\alpha = \sum c_{(i)} x_1^{i_1} \cdots x_q^{i_q} = P(x)$$

with coefficients $c_{(i)} \in \mathbf{Z}$, we let

$$\text{size}(\alpha) = \max(\deg P, \log |c_{(i)}|).$$

Thus we let the size be the maximum of the degree of P and the logs of the absolute values of the coefficients. We shall also call it the size of P, and write it $\sigma(P)$.

Next, let $\alpha \in \mathbf{Z}[x, y]$. Then we can write

$$\alpha = P_0(x) + P_1(x)y + \cdots + P_{N-1}(x)y^{N-1},$$

where $P_0(x), \ldots, P_{N-1}(x) \in \mathbf{Z}[x]$, if N is the degree of y over $\mathbf{Q}(x)$. We define

$$\text{size}(\alpha) = \max \text{ sizes } P_0(x), \ldots, P_{N-1}(x).$$

Finally, if γ is an element of K, we say

$$\text{size}(\gamma) \leqq B$$

if γ can be written as a quotient $\gamma = \alpha/\beta$ with $\alpha \in \mathbf{Z}[x, y]$, $\beta \in \mathbf{Z}[x]$, such that $\text{size}(\alpha)$, $\text{size}(\beta) \leq B$. Thus the size of an element of K is defined in a natural way, similar to our definition of the size of an algebraic number.

LEMMA 1. *Let the notation be as above. There exists a number $c > 0$ (depending on x, y) such that, if $\alpha_1, \ldots, \alpha_m$ are elements of K, then*

$$\text{size}(\alpha_1 \cdots \alpha_m) \leqq c(\text{size}(\alpha_1) + \cdots + \text{size}(\alpha_m)),$$

$$\text{size}(\alpha_1 + \cdots + \alpha_m) \leqq c(\text{size}(\alpha_1) + \cdots + \text{size}(\alpha_m)).$$

Proof. We first prove the first statement for polynomials in $\mathbf{Z}[x]$. If $P_1(x), \ldots, P_m(x)$ are such polynomials, of degrees $d_1, \ldots, d_m$ respectively, then their product is dominated by

$$P_1(T) \cdots P_m(T) \prec |P_1| \cdots |P_m|(1 + T_1 + \cdots + T_q)^{d_1 + \cdots + d_m},$$

and if we expand out the power of $(1 + T_1 + \cdots + T_q)$ on the right, we see that the coefficients are bounded by

$$c^{d_1 + \cdots + d_m}$$

for some constant c. Taking the log yields what we want.

Next, suppose that our elements $\alpha_1, \ldots, \alpha_m$ are written

$$\alpha_k = P_{k1}(x) + \cdots + P_{k,N-1}(x)y^{N-1}$$

for $k = 1, \ldots, m$. Introduce a new variable T_{q+1}, and let

$$A_k = P_{k1}(T) + \cdots + P_{k,N-1}(T)T_{q+1}^{N-1}.$$

Then the product $A_1(T) \cdots A_m(T)$ is dominated in the same way that we argued previously for polynomials, by a sum

$$|A_1| \cdots |A_m|c^{d_1 + \cdots + d_m} \sum T_1^{i_1} \cdots T_q^{i_q} T_{q+1}^{i_{q+1}},$$

and $i_1, \ldots, i_{q+1} \leqq d_1 + \cdots + d_m + N$. Let $g(Y)$ be the irreducible polynomial of y over $\mathbf{Z}[x]$, with leading coefficient 1 since we assumed y integral over $\mathbf{Z}[x]$. For each i we can write

$$Y^i = g(Y)h_i(Y) + r_i(Y)$$

with polynomial h_i, r_i in $\mathbf{Z}[Y]$. A simple induction, using long division, shows that the size of the coefficients of the remainder term $r_i(Y)$ is bounded by cd for some constant c. We substitute each such remainder term $r_i(T_{q+1})$ for the corresponding power of T_{q+1}^i in our sum above, and thus see again that if we substitute $(x_1, \dots, x_q)$ for $(T_1, \dots, T_q)$ and y for T_{q+1}, we obtain a bound for the size of the desired type.

The second statement concerning the size of a sum is trivially proved from the first, putting all terms in the sum over a common denominator.

It will be convenient to use the resultant in the next lemma. If

$$f(Y) = a_n Y^n + \cdots + a_0$$

$$g(Y) = b_m Y^m + \cdots + b_0$$

with indeterminate coefficients a_i, b_j, then

$$R(f, g) = \left. \begin{vmatrix} a_n \cdots\cdots a_0 & & & \\ & a_n \cdots\cdots a_0 & & \\ & & \cdots\cdots\cdots & \\ & & a_n \cdots\cdots a_0 \\ b_m \cdots\cdots b_0 & & & \\ & b_m \cdots\cdots b_0 & & \\ & & \cdots\cdots\cdots & \\ & & b_m \cdots\cdots b_0 \end{vmatrix} \right\} \begin{matrix} m \\ \\ \\ \\ n \end{matrix}$$

and this resultant is also equal to

$$R(f, g) = \begin{vmatrix} a_n \cdots\cdots a_0 & & & Y^{m-1}f \\ & a_n \cdots\cdots a_0 & & Y^{m-2}f \\ & & \cdots\cdots\cdots & \vdots \\ & a_n \cdots\cdots a_0 & & Yf \\ & & a_n \cdots\cdots a_1 & f \\ b_m \cdots\cdots b_0 & & & Y^{n-1}g \\ & b_m \cdots\cdots b_0 & & Y^{n-2}g \\ & & \cdots\cdots\cdots & \vdots \\ & b_m \cdots\cdots b_0 & & Yg \\ & & b_m \cdots\cdots b_1 & g \end{vmatrix}$$

If w is a complex number, and the coefficients a_i, b_j of f, g are taken to be complex numbers, with $a_n b_m \neq 0$, then we obtain, substituting w for Y in the second determinant,

$$|R(f, g)| \leqq (1 + |w|)^{m+n} (|f(w)| + |g(w)|) |f|^m |g|^n (m + n)^{m+n}.$$

LEMMA 2. *There exists a constant $c > 0$ such that for all $\alpha \neq 0$ in K we have* $\mathrm{size}(\alpha) \leqq c \cdot \mathrm{size}(1/\alpha)$.

Proof. Without loss of generality, we may assume that

$$\alpha = P_1(x) + \cdots + P_{N-1}(x)y^{N-1} = f(y),$$

where $P_1, \ldots, P_{N-1}$ are polynomials with integer coefficients. Let $g(Y)$ be the irreducible polynomial of y over $\mathbf{Z}[x]$. Since $f(y) \neq 0$, the resultant of g and f, which is an element of $\mathbf{Z}[x]$, is not zero. Let $P(x)$ be this resultant. Then

$$P(x) = R(f, g) = Af + Bg,$$

where A, B are polynomials, elements of $\mathbf{Z}[x][Y]$. From the expression of the resultant as a determinant, and from Lemma 1, it is trivial to see that the size of the coefficients of A and B (in $\mathbf{Z}[x]$) is bounded by $c \cdot \mathrm{size}(\alpha)$ for some constant c. Furthermore, the degree of A in y is bounded by a constant depending on N. Substituting y for Y in our expression for the resultant, we find that

$$P(x) = A(y)f(y),$$

and thus $\alpha^{-1} = A(y)/P(x)$. From the preceding remarks, we find that $\mathrm{size}(\alpha^{-1}) \ll \mathrm{size}(\alpha)$.

We shall say that K has *transcendence type* $\leqq \tau$ (for some number $\tau \geqq 2$) if there exists a proper set of generators (x, y) such that, with respect to this set, for every non-zero element $\alpha \in K$, we have

$$-(\mathrm{size}\ \alpha)^\tau \ll \log |\alpha|.$$

LEMMA 3. *Assume that* $\mathbf{Q}(x)$ *has transcendence type* $\leqq \tau$, *with respect to* (x). *If y is a generator of K over $\mathbf{Q}(x)$, and y is integral over $\mathbf{Z}[x]$, then K has transcendence type* $\leqq \tau$ *with respect to* (x, y).

Proof. Write each $\alpha \in K$, $\alpha \neq 0$ in the form

$$\alpha(x, y) = \frac{1}{P(x)}\left(P_0(x) + \cdots + P_{N-1}(x)y^{N-1}\right) = \frac{1}{P(x)}f(y),$$

where $P, P_0, \ldots, P_{N-1}$ have size $\leqq \sigma_\alpha$, and

$$\log |\alpha| \ll -c_\alpha \sigma_\alpha^\tau$$

for some positive number c_α. If the assertion of our lemma is false, then we can find a sequence of numbers α for which c_α tends to infinity. Multiplying α by $P(x)$, and observing that $|P(x)| \leq c^{\sigma_\alpha}$ for some constant c, we see that, without loss of generality, we may assume that $P(x) = 1$. Let g be the irreducible polynomial of y over $\mathbf{Z}[x]$, and let $Q_\alpha(x) = R(f, g)$

be the resultant of f and g. Using the estimate for the resultant, resulting from the determinant expression, we find

$$|Q_\alpha(x)| \leq |\alpha| c^{\sigma_\alpha}$$

for some constant c. Again, the determinant expression of the resultant shows that

$$\text{size } Q_\alpha(x) \ll \text{size } \alpha.$$

Hence we obtain

$$\log |Q_\alpha(x)| \ll -c_\alpha \sigma_\alpha^\tau \ll -c_\alpha (\text{size } Q_\alpha(x))^\tau.$$

This contradicts the fact that the field $\mathbf{Q}(x)$ has transcendence type $\leq \tau$ with respect to (x), and proves our lemma.

The preceding lemmas show that working with a finite extension of the pure transcendental field $\mathbf{Q}(x)$, from the present point of view, is essentially no harder than working with $\mathbf{Q}(x)$ itself, involving only routine polynomial technique. In particular, the result of the preceding section would hold just as well for such a field K. We shall formulate it more generally in the next section.

§3. *Algebraic independence*

We have all the tools to extend the definitions and results of Chapter II, §2 to the case of functions taking values in a finitely generated field K, of transcendence degree q and transcendence type $\leq \tau$. We always assume that $\tau \geq q + 1 \geq 2$.

Let

$$S = \bigcup S_n$$

be again a set expressed as a union of subsets S_n, with $S_n \subset S_{n+1}$ for all $n = 1, 2, \ldots$ We assume again that $|z| \leq Cn$ for all $z \in S_n$. A meromorphic function f defined on S, of order $\leq \rho$, with values in K, is said to be of arithmetic order $\leq \rho$ on S if there is a constant $C \geq 1$ such that the following conditions are satisfied:

AO 1. *For all n and $z \in S_n$ we have* $\text{size } f(z) \leq Cn^\rho$.

AO 2. *There is an entire function h of order $\leq \rho$, such that hf is entire, h has no zero in S, and for all n, $z \in S_n$,*

$$\log |1/h(z)| \leq Cn^\rho.$$

Observe that the statements of AO 1 and AO 2 are identical with those of Chapter II.

THEOREM 2.   *Let $f_1,\ldots,f_d$ be meromorphic functions, of order $\leq \rho$, defined on the set S as above, with values in K, and of arithmetic order $\leq \rho$ on S. Let m be a number such that $(m + d)\tau \leq md$. Asume that $\mathrm{card}(S_n) \gg\ll n^{m\rho}$, for $n \to \infty$. Then $f_1,\ldots,f_d$ are algebraically dependent over K.*

Proof. Suppose that $f_1, \ldots, f_d$ are independent over K. Let n be as usual a large integer which will tend to infinity. Let r be approximately equal to

$$n^{(m-q+1)\rho/(d+q-1)}.$$

We form the function

$$F = \sum a_{(i)} x_1^{i_{d+1}} \cdots x_q^{i_{d+q}} f_1^{i_1} \cdots f_d^{i_d},$$

with

$$1 \leq i_1, \ldots, i_d \leq r \qquad \text{and} \qquad 1 \leq i_{d+1}, \ldots, i_{d+q} \leq rn^\rho.$$

We require that $F(z) = 0$ for all $z \in S_n$. This amounts to solving the usual linear equations, setting the coefficient of each monomial in (x) equal to 0. We have

$$\text{number of unknowns} = r^d(rn^\rho)^q.$$

The degrees in (x) of the elements $F(z)$ with $z \in S_n$ is $\ll rn^\rho$. Hence

$$\text{number of equations} \ll n^{m\rho}(rn^\rho)^q.$$

Our assumption on r means that the number of unknowns is approximately equal to the number of equations, and thus that the exponent in Siegel's lemma is bounded by a constant. The size of each $F(z)$, with $z \in S_n$, is also bounded by rn^ρ, up to a constant factor. Hence we can solve for the coefficients $a_{(i)}$ so that they have size $\ll rn^\rho$.

Let s be the largest integer such that $F(z) = 0$ for all $z \in S_s$, and let $w \in S_{s+1}$ be such that $F(w) \neq 0$. Then $s \geq n$ and

$$\text{size } F(w) \ll rs^\rho.$$

We estimate $|F(w)|$ as usual, from the expression

$$F(w) = \frac{h(t)^{dr}F(t)}{h(w)^{dr}\prod(t-z)} \prod (w-z) \Bigg|_{t=w},$$

where h is a function satisfying AO 2 for each one of $f_1, \ldots, f_d$. We take the circle of radius $R = s^{1+1/dq}$ (any $s^{1+\epsilon}$ would do). Then the entire function $h(t)^{dr}F(t)$ on the circle of radius R is bounded by

$$\log |h^{dr}F|_R \ll rR^\rho \ll rs^{(1+1/dq)\rho} \ll s^{m\rho}.$$

On the other hand, the quotient of the two products in the estimate for $F(w)$ pulls towards zero, and in fact

$$\log \max_{|t|=R} \left| \frac{\prod(w-z)}{\prod(t-z)} \right| \ll -s^{m\rho} \log s.$$

The power $h(w)^{d\tau}$ does not affect anything by hypothesis. Hence essentially only the product counts in the estimate for $|F(w)|$, and we get

$$\log |F(w)| \ll -s^{m\rho} \log s.$$

Now we get a contradiction of the inequality

$$-(\text{size } F(w))^\tau \ll \log |F(w)|,$$

using our hypothesis relating m, d and τ (the hypothesis was in fact made up so that at this point in the proof, we get the desired contradiction). This proves Theorem 2.

COROLLARY 1. *Assume that we are given $d > \tau$, and let m be such that $m(d - \tau) \geq d\tau$. Then with the other hypotheses of Theorem 2, the functions $f_1, \ldots, f_d$ are algebraically dependent over K.*

COROLLARY 2. *Let $\mathbf{Q}(x)$ be a purely transcendental extension of $\mathbf{Q}$, of transcendence type $\leq \tau$ for some integer $\tau \geq 2$. Let G be the general linear group, of some dimension, and $\mathbf{K}$ the algebraic closure of $\mathbf{Q}(x)$. Let $\varphi \colon \mathbf{C} \to G_\mathbf{C}$ be a 1-parameter subgroup of G, of algebraic dimension d. Let Γ be a subgroup of $\mathbf{C}$, containing at least m elements linearly independent over $\mathbf{Q}$, such that $\varphi(\Gamma) \subset G_\mathbf{K}$. If $m \geq d\tau$, then $d \leq \tau$.*

Proof. As in Chapter I, the estimates for the exponential series are easily carried out, to show that Corollary 1 applies.

Remark. One also wants a statement similar to Corollary 2 for abelian varieties. Everything goes through as in Chapter II, except that we used the Néron-Tate form to verify that condition AO 1 is satisfied, in the analogous situation. Here, the necessary verification has not yet been made. However we observe that the Néron-Tate form was a much more powerful tool than actually needed. All we need are upper estimates on the size of a sum of points on an abelian variety. This type of consideration belongs in a separate treatment of the problem on abelian varieties, which is purely algebraic. Note that it would admit the possibility that the abelian variety is defined over a finitely generated field, of finite transcendence type!

Aside from the algebraic independence of e, π or e, $\log 2$, or $\log 2$, $\log 3$, the next simplest case to treat is $\wp(\alpha_1)$, $\wp(\alpha_2)$, where α_1, α_2 are linearly independent over $\mathbf{Q}$, and $\wp$ is a Weierstrass function with algebraic g_2, g_3

such that the corresponding elliptic curve has only trivial endomorphisms. It may be that this would come from pushing further the ideas of Feldman [4], some of which bear a vague analogy with Siegel's, except that Feldman deals with the algebraic differential equation of the $\wp$-function. Feldman's success with a certain inversion associated with the $\wp$-function, and this analogy, indicate that one may reach a proof of the similar result on abelian varieties, by a deepening of the method, still modelled on the usual pattern of constructing the function F with a lot of zeros.

Historical note

Gelfond proved a statement analogous to our Theorem 1 [13], but with the following substantial differences. To begin with, he does not make the assumption on the transcendence type of the field of values, and he uses the differential equation of the exponential function. Furthermore, he uses in an incidental way a measure of irrationality for exponents β_i. Finally, he uses a very special feature about the exponential function, which makes it very unclear how to extend his proof to more general func-tions. On the other hand, in place of the transcendence type, he has the following theorem, valid for *all* numbers. (For a proof, cf. [21].)

THEOREM. *Let x be a complex number. Let σ be a strictly monotone increasing real function tending to infinity, and assume that there is a number $a_0 > 1$ such that $\sigma(N + 1) < a_0\sigma(N)$ for all integers $N > N_0$. Assume that for each integer $N > N_0$ there exists a non-zero polynomial F_N with integer coefficients, such that*

$$|F_N(x)| < e^{-C\sigma^2(N)}$$

where C is a sufficiently large constant, and

$$\max(\deg F_N, \log |F_N|) \leqq \sigma(N).$$

Then x is algebraic.

The difficulty about applying this theorem is that one needs very many polynomials F_N having small values at x. This means that, in an analogous situation to our Theorem 1, he must have a way of showing that the integer s is not too large, and in fact is about the same order of magnitude as the integer n. It is here that he uses the above-mentioned special properties of the function e^t. In addition, other technical complications arise in the course of the proof. Thus in spite of much greater complica-tions, he still ends up only with partial results, as we do here.

The method I use in this chapter is much simpler than Gelfond's method, and has the additional advantage of showing clearly the inductive relationship between transcendence types and the possibility of proving results of algebraic independence. For instance, we could state Theorem 2 for fields of transcendence degree > 1, whereas Gelfond could not state a similar theorem. Of course, the problem remains of proving that certain numbers have definite types, and to solve this problem, one expects a higher order of complication, of the nature encountered by Feldman in his papers (cf. Chapter VI). It then becomes clear that one must refine the notion of transcendence type, in a manner to be discussed at the end of Chapter VI.

A transcendence measure for a number x is any function g of two variables such that for all polynomials P with integer coefficients, $P \neq 0$, of degree $\leq d$ and height $\leq h$ one has

$$\log |P(x)| \geq g(d, h).$$

The problem of determining the best possible transcendence measures for classical numbers is by definition a problem in diophantine approximations, which is thereby shown to be inseparable from the theory of transcendental numbers.

As we said above, the result of this chapter is quite weak, and is only intended to show in a simple case the first example of an inductive procedure which could eventually be used to obtain best possible results. It will then be necessary to refine the method of proof. This can come from the following directions:

(1) Use linear inequalities, rather than the linear equations, which are extremely wasteful in the number of variables. Cf. Chapter VI, §3.

(2) Use an improved transcendence measure. It seems that the whole inductive procedure is set up in such a way that only an extremely refined inductive assumption can lead to the proper result. We shall discuss in Chapter VI what such assumptions could be like.

(3) Even using such assumptions, there is still something missing in the present structure of the proof, even using an algebraic differential equation, or for concreteness the functions t, e^t. As far as I can tell, making any and all of these assumptions, it is still not possible with the present structure of the proof to derive a contradiction leading to the best possible conjecturable results.

CHAPTER VI

Transcendence Measures

§1. *The Liouville estimate*

We shall reformulate a somewhat more general result than the fundamental inequality of Chapter I, §1, to deal with algebraic numbers whose degree is not necessarily fixed.

If P is a polynomial with integer coefficients, we define its *height* $h(P)$ to be

$$h(P) = \log |P|,$$

i.e. it is the log of the maximum of the absolute values of its coefficients. Similarly, if ξ is an algebraic number, and P is its irreducible polynomial over $\mathbf{Z}$, then we define its height,

$$h(\xi) = h(P).$$

We define its *absolute size* $\sigma(\xi)$ to be

$$\sigma(\xi) = \max(\deg \xi, h(\xi)).$$

We shall now reformulate the fundamental inequality of Chapter I, §1 to deal with the absolute size of an algebraic number.

LEMMA 1. *Let*

$$P(X) = a_d(X - \alpha_1) \cdots (X - \alpha_d), \qquad a_d \neq 0$$

be a polynomial with complex coefficients. Then

$$|a_d| \prod_{i=1}^{d} \max(1, |\alpha_i|) \leq 2^d |P|.$$

Proof. Dividing both sides by $|a_d|$, we may assume without loss of generality that $a_d = 1$. We now use induction on the number of indices i such that $|\alpha_i| > 2$. If $|\alpha_i| \leq 2$ for all i, our assertion is obvious. Suppose now that

$$P(X) = g(X)(X - \alpha)$$

57

458

with $|\alpha| > 2$, and suppose that our assertion is true for

$$g(X) = X^d + b_{d-1}X^{d-1} + \cdots + b_0.$$

We have

$$P(X) = X^{d+1} + (b_{d-1} - \alpha)X^d$$
$$+ (b_{d-2} - \alpha b_{d-1})X^{d-1} + \cdots + (-\alpha)b_0.$$

We can assume $|g| = |b_i| \geq |b_{i-1}|$ for some i, $0 \leq i \leq d$ (with the convention $b_d = 1$, $b_{-1} = 0$). Then

$$|P| \geq |\alpha b_i - b_{i-1}| \geq |\alpha|\,|b_i| - |b_{i-1}|$$
$$\geq |\alpha|\,|b_i| - |b_i| = (|\alpha| - 1)|b_i|$$
$$\geq \tfrac{1}{2}|\alpha|\,|b_i| = \tfrac{1}{2}|\alpha|\,|g|,$$

and our lemma is now obvious, since $|\alpha| > 2$.

Remark. If in Lemma 1 we take a polynomial whose coefficients lie in a field with a non-archimedean valuation, then we have a similar inequality *without the factor* 2^d. This is nothing but the Gauss lemma for valuations, and is trivially proved. In fact, we have the *equality*

$$|a_d|_p \prod_{i=1}^{d} \max(1, |\alpha_i|_p) = |P|_p$$

if $| \ |_p$ denotes a p-adic valuation.

THEOREM 1. *Let* $\xi_1, \ldots, \xi_m$ *be algebraic numbers, of degrees* $d_1, \ldots, d_m$ *and heights* $h_1, \ldots, h_m$ *respectively. Let*

$$d = [\mathbf{Q}(\xi_1, \ldots, \xi_m) : \mathbf{Q}].$$

Let P be a polynomial in m variables $X_1, \ldots, X_m$, *with integer coefficients, of degree* N_i *in* X_i. *If* $P(\xi_1, \ldots, \xi_m) \neq 0$, *then*

$$-d\left[h(P) + \sum_{i=1}^{m} \frac{N_i h_i}{d_i} + 2\sum_{i=1}^{m} N_i\right] \leq \log|P(\xi_1, \ldots, \xi_m)|.$$

Proof. Let P_i be the irreducible polynomial of ξ_i over $\mathbf{Z}$, and let a_i be its leading coefficient. Assume $P(\xi) \neq 0$. Let $j = 1, \ldots, d$ be indices for the embeddings of $\mathbf{Q}(\xi)$ into the complex numbers. Since $P(X_1, \ldots, X_m)$ is dominated by

$$|P|(1 + X_1)^{N_1} \cdots (1 + X_m)^{N_m},$$

it follows that for each j,

$$|P(\xi_1^{(j)}, \ldots, \xi_m^{(j)})| \leq |P|(1 + |\xi_1^{(j)}|)^{N_1} \cdots (1 + |\xi_m^{(j)}|)^{N_m}.$$

Taking the product over $j = 1, \ldots, d$ and using Lemma 1, applied to the irreducible polynomials of $\xi_1, \ldots, \xi_m$ respectively, we find

$$\prod_{j=1}^{d} |P(\xi)^{(j)}| \leq |P(\xi)| \, |P|^d \prod_{i=1}^{m} [4^{d_i} a_i^{-1} |P_i|]^{dN_i/d_i}.$$

On the other hand, we let $|\ |_p$ denote the absolute value on the algebraic closure of the p-adic field $\mathbf{Q}_p$, and apply Lemma 1 in the p-adic case, using (j) to denote p-adic conjugates. We then obtain the similar estimate,

$$\prod_{j=1}^{d} |P(\xi)^{(j)}|_p \leq \prod_{i=1}^{m} \prod_{j=1}^{d} \sup(1, |\xi_i^{(j)}|_p)^{N_i}$$

$$\leq \prod_{i=1}^{m} |a_i^{-1} P_i|_p^{dN_i/d_i}.$$

Since P_i has integral coefficients, $|P_i|_p \leq 1$, we can delete it from our p-adic estimate. Taking the product over all p, and over the ordinary absolute value, and using the product formula, we obtain

$$1 \leq |P(\xi)| \, |P|^d \prod_{i=1}^{m} [4^{d_i} |P_i|]^{dN_i/d_i}.$$

Taking the log gives the estimate of the theorem.

We shall call the estimate of Theorem 1 a *Liouville estimate*.

CoROLLARY. *Let P be a polynomial with integer coefficients, and ξ an algebraic number. If $P(\xi) \neq 0$, then*

$$-[\deg(\xi)h(P) + (\deg P)h(\xi) + 2 \deg P] \leq \log |P(\xi)|.$$

§2. *Polynomial and algebraic approximations*

Our next task is to investigate the relationship between polynomial and algebraic approximations.

LEMMA 2. *Let P be a polynomial of degree d with integer coefficients. Then every factor Q of P over $\mathbf{Z}$ satisfies the inequality*

$$|Q| \leq 4^d |P|.$$

Proof. Any factor Q of P over $\mathbf{Z}$ can be written in the form

$$Q(X) = b_{d_1} \prod_{\nu=1}^{d_1} (X - \alpha_{i_\nu}),$$

where α_{i_ν} $(\nu = 1, \ldots, d_1)$ are roots of P, and b_{d_1} is an integer dividing a_d, the leading coefficient of P. Then

$$|Q| \leqq |b_{d_1}| 4^d \left| \frac{1}{a_d} P \right| \leqq 4^d |P|,$$

as was to be shown.

COROLLARY. *Notation as in the lemma, we have* $\sigma(Q) \leqq 3\sigma(P)$, *where* σ *is the size,* $\sigma(P) = \max(\deg P, h(P))$.

Proof. Obvious.

The next lemma shows that given a polynomial with integer coefficients taking a small value at a number w, we can find an irreducible factor which also has a small value at w.

LEMMA 3. *Let* $P(X)$ *be a polynomial with integer coefficients, of degree* $d \geqq 1$, *and let* λ *be a positive number. If*

$$\log |P(w)| \leqq -\lambda d \sigma(P),$$

then there exists an irreducible factor P_1 *of* P *over* $\mathbf{Z}$ *of degree* d_1 *such that*

$$\log |P_1(w)| \leqq -\tfrac{1}{3}\lambda d_1 \sigma(P_1).$$

Proof. Factor P into a product of irreducible polynomials,

$$P = P_1 \cdots P_s.$$

Suppose that our assertion is false. Say

$$\log |P_i(w)| > -\tfrac{1}{3}\lambda d_i \sigma(P_i)$$

for $i = 1, \ldots, s$. Taking the sum, and using the Corollary of Lemma 2, we find

$$\log |P(w)| > -\tfrac{1}{3}\lambda (d_1 \sigma(P_1) + \cdots + d_s \sigma(P_s))$$

$$> -\lambda \sigma(P)(d_1 + \cdots + d_s)$$

$$> -\lambda d \sigma(P),$$

a contradiction which proves the lemma.

LEMMA 4. *Let*

$$P(X) = a_d X^d + \cdots + a_0$$

be a polynomial with integer coefficients, $a_d \neq 0$, and without multiple roots. Let w be a complex number, and let $\xi_1, \ldots, \xi_d$ be the roots of P. Then

$$\min_i |w - \xi_i| \leqq |P(w)| e^{5d(\log d + h(P) + 4)}.$$

Proof. Let $\xi_1, \ldots, \xi_d$ be so ordered that

$$|w - \xi_1| \leqq \cdots \leqq |w - \xi_d|.$$

Then for all $i \neq 1$, we have

$$|w - \xi_i| \geqq \tfrac{1}{2}|\xi_1 - \xi_i|.$$

Otherwise,

$$|w - \xi_i| < \tfrac{1}{2}|\xi_1 - w| + \tfrac{1}{2}|w - \xi_i|,$$

which is impossible. Now we have

$$|P(w)| = |a_d|\,|w - \xi_1| \cdots |w - \xi_d| \geqq |w - \xi_1| 2^{-d+1} |P'(\xi_1)|.$$

We must therefore estimate $|P'(\xi_1)|$ from below. Note that $|\xi_1| \leqq d|P|$ trivially. Using the fundamental estimate for the resultant, we get

$$1 \leqq |R(P, P')| \leqq (1 + d|P|)^{2d} |P'(\xi_1)|\,|P|^{d-1}|P'|^d (2d)^{2d},$$

whence the desired estimate follows at once.

Remark. The estimate of the lemma is quite sharp in its dependence on d. Except for the constant 5 it is also sharp in its dependence on h. Variations of the proof can be given to eliminate the constant 5, as a coefficient of $h(P)$, at the cost of making the dependence on d somewhat worse.

THEOREM 2. *Let w be a transcendental number, and let $\lambda \geqq 1$ be a positive function of two real variables. Then the following conditions are equivalent:*

TM 1. *For all algebraic numbers ξ of degree $\leqq d$ and absolute size $\leqq \sigma$, we have*

$$\log |w - \xi| \gg -\lambda(d, \sigma)\, d\sigma.$$

TM 2. *For all polynomials P of degree $\leqq d$, with integer coefficients, and size $\leqq \sigma$, we have*

$$\log |P(w)| \gg -\lambda(d, \sigma)\, d\sigma.$$

Proof. Assume TM 1. To prove TM 2, we may assume by Lemma 2 that P is irreducible over $\mathbf{Z}$. By Lemma 4, we find

$$\log \min |w - \xi_i| \ll \log |P(w)| + d\sigma.$$

Using TM 1 to get an inequality on the left, we see that $\log |P(w)|$ satisfies the desired inequality. Conversely, assume TM 2. To prove TM 1, we need only a more trivial inequality than that of Lemma 4, namely if P is the irreducible polynomial of ξ over $\mathbf{Z}$, then

$$|P(w)| \leqq |a_d|\,|w - \xi_1| \cdots |w - \xi_d|,$$

and since $|\xi_i| \leqq d|P|$ for all i, we obtain trivially

$$\log |P(w)| \ll \log |a_d| + \log |w - \xi| + d\sigma.$$

Applying TM 2 yields TM 1.

§3. *Linear inequalities*

In order to make a function have small values at certain points, we shall need a substitute for the Siegel lemma, allowing us to solve linear inequalities instead of linear equations. This yields finer estimates than Siegel's lemma.

LEMMA 5. *Let*

$$
\begin{aligned}
L_1(X) &= \alpha_{11}x_1 + \cdots + \alpha_{1n}x_n \\
&\ \ \vdots \qquad\qquad\qquad \vdots \\
L_r(X) &= \alpha_{r1}x_1 + \cdots + \alpha_{rn}x_n
\end{aligned}
$$

be a system of linear forms with complex coefficients α_{ij}, and $n > 2r$. Let $A \geqq 1$ and $|\alpha_{ij}| \leqq A$ for all i, j. Let B be a number $\geqq 1$. Then there exists a non-trivial integral solution X of the inequalities

$$|L_i(X)| \leqq \frac{1}{B} \qquad and \qquad |X| \leqq 2(8nAB)^{2r/(n-2r)}$$

for all i.

Proof. Assume first that the coefficients are real numbers, and only that $n > r$. We shall prove that we can satisfy the inequalities

$$|L_i(X)| \leqq \frac{1}{B} \qquad and \qquad |X| \leqq 2(4nAB)^{r/(n-r)}.$$

For any number $C \geqq 1$, the linear map $L: \mathbf{Z}^n \to \mathbf{R}^r$ given by our linear forms maps $\mathbf{Z}^n(C)$ into $\mathbf{R}^r(nAC)$. Cut the interval

$$-nAC \leqq t \leqq nAC$$

into $[4nACB]$ segments of equal length. Note that

$$[4nACB] \geq 2nACB.$$

Then each small segment has length

$$\frac{2nAC}{[4nACB]} \leq \frac{2nAC}{2nACB} \leq \frac{1}{B}.$$

Then $\mathbf{R}^r(nAC)$ is decomposed into $[4nACB]^r$ small cubes of sides $\leq 1/B$. In $\mathbf{Z}^n(C)$ we have at least C^n integral vectors. If

$$C^n > [4nACB]^r,$$

then there exist two distinct vectors $Y, Y' \in \mathbf{Z}^n(C)$ such that $L(Y)$ and $L(Y')$ lie in the same small cube. For this, it suffices that

$$C > (4nAB)^{r/(n-r)}.$$

Then $X = Y - Y'$ satisfies our requirements.

Now in the complex case, for each linear form with complex coefficients, we write down two linear forms, the real and imaginary parts, separately. This yields twice as many equations, whence the needed assumption $n > 2r$. Furthermore, if L' is the real part of L, and L'' its imaginary part, we solve

$$|L'(X)| \leq 1/2B \qquad \text{and} \qquad |L''(X)| \leq 1/2B,$$

so that we must replace B by $2B$ in our real result to obtain the desired bound in the complex case.

§4. Interpolation estimates

In the proof of this chapter, we shall not do as before, construct a function with many zeros, but rather, we construct a function having very small values at many points. This means that the maximum principle cannot be used any more, and that we shall use an interpolation method to estimate the function, obtained from an application of Cauchy's theorem. We state it as a separate lemma.

LEMMA 6. *Let E be an entire function, and let*

$$Q(t) = [(t - z_1) \cdots (t - z_m)]^l$$

be a polynomial with distinct roots z_ν ($\nu = 1, \ldots, m$) of multiplicity l. For each $\nu = 1, \ldots, m$ we let

$$Q_\nu^*(t) = [(t - z_1) \cdots \overbrace{(t - z_\nu)} \cdots (t - z_m)]^l,$$

i.e. we omit the factor $(t - z_\nu)^l$. *Let R be a positive number such that* $|z_\nu| < R$ *for all ν, and let z be a number inside this circle, unequal to any* z_ν. *Let Γ be the circle of radius R. Then*

$$\frac{E(z)}{Q(z)} = \frac{1}{2\pi i} \int_\Gamma \frac{E(\zeta)}{Q(\zeta)} \frac{d\zeta}{\zeta - z}$$

$$- \frac{1}{2\pi i} \sum_{\nu=1}^{m} \sum_{k=0}^{l-1} \frac{1}{k!} D^k E(z_\nu) \int_{\Gamma_\nu} \frac{1}{Q_\nu^*(\zeta)(\zeta - z)} \frac{d\zeta}{(\zeta - z_\nu)^{l-k}},$$

where Γ_ν is a circle around z_ν, not containing z, and not containing any other z_μ.

Proof. We start with the formula given for one variable in Chapter IV, §2, and then expand the multiple derivative $D^{l-1}(EG)$ as a sum

$$\sum_{k=0}^{l-1} \binom{l-1}{k} D^k E \cdot D^{l-1-k} G,$$

where G is an obvious function. We then express each $D^{l-1-k}G(z_\nu)$ by the usual Cauchy integral, and the desired formula comes out trivially.

From our expression, we see that if $D^k E(z_\nu)$ is small, then $E(z)$ will also be small, i.e. we have obtained the technical equivalent of the estimates which we would have if E had zeros of multiplicity l at all points $z_1, \ldots, z_m$. Thus $E(z)$ is bounded essentially in terms of $1/Q(\zeta)$, and $D^k E(z_\nu)$. The other expressions entering in our formula play a secondary role in practice. We shall put this formula in a form which is best adapted for the applications we have in mind, also giving an estimate for derivatives of E (easily achieved using Cauchy's formula).

LEMMA 7. *Let the hypotheses be as in Lemma 6. Let $2 < R_1 < R$. Let z' be a complex number, distinct from $z_1, \ldots, z_m$, and assume that $z', z_1, \ldots, z_m$ lie inside a circle of radius $R_1/2$. Let δ be less than the minimum of the distances of any pair of distinct numbers among*

$$z', z_1, \ldots, z_m,$$

and also $0 < \delta < 1$. Then for any integer $r \geq 0$ we have

$$|D^r E(z')| \leq \frac{2^r r!}{R_1^r} [(C_1 R_1/R)^{ml} |E|_R + (C_1 R_1/\delta)^{ml} \max_{k,\nu} |D^k E(z_\nu)|]$$

where C_1 is an absolute constant.

Proof. We have

$$|D^r E(z')| \leq |D^r E|_{R_1}.$$

We use the preceding lemma to estimate $|D^r E|_{R_1}$, and begin by estimating $|E|_{R_1}$, taking z on the circle of radius R_1 in Lemma 6. We estimate $Q(z)$ on the circle of radius R_1, and $Q(\zeta)$ on the circle of radius R. Then for some universal constant C_1,

$$|E|_{R_1} \leqq C_1^{ml}(R_1/R)^{ml}|E|_R + mlC_1^{ml}(1/\delta)^{ml}\max_{k,\nu}|D^k E(z_\nu)|.$$

To get the r-th derivative, we use Cauchy's formula,

$$D^r E(z') = \frac{r!}{2\pi i}\int_{|t|=R_1}\frac{E(t)\,dt}{(t-z')^{r+1}}$$

and apply our estimate for $|E|_{R_1}$ to prove our lemma.

In the applications, we must make two estimates to apply Lemma 7, corresponding to the two terms in the sum.

The first will be small because of $(R_1/R)^{ml}$, and will be determined by this expression. It is then necessary to verify that $r!$ and C_1^{ml}, as well as $|E|_R$, do not tend to infinity faster than $(R_1/R)^{ml}$ tends to 0.

The second will be small because of $\max_{k,\nu}|D^k E(z_\nu)|$, and will be determined by these derivatives. It is then necessary to show that $r!$, C_1^{ml}, and $(1/\delta)^{ml}$ again do not tend to infinity faster than the absolute value of the derivatives involved.

§5. *A determinant*

Let f, g be two meromorphic functions of a complex variable. Let φ_{ij} be complex numbers, and let us form the function

$$F = \sum_{i=0}^{m-1}\sum_{j=0}^{n-1}\varphi_{ij}f^i g^j.$$

Then

$$D^k F = \sum_{i=0}^{m-1}\sum_{j=0}^{n-1}\varphi_{ij}D^k(f^i g^j),$$

and

$$D^k(f^i g^j) = \sum_{\kappa=0}^{k}\binom{k}{\kappa}D^\kappa f^i \cdot D^{k-\kappa}g^j.$$

If x is a complex number, μ an integer, such that f, g and their derivatives are defined at μx, we let

$$A_{i,j}^{k,\mu}(t) = \sum_{\kappa=0}^{k}\binom{k}{\kappa}D^\kappa f^i(\mu t) \cdot D^{k-\kappa}g^j(\mu x).$$

We shall assume from now on that $f(t) = t$, and that x is a period of g and of its derivatives. We let $g_{k,j} = D^k g^j(0)$. Then

$$A_{i,j}^{k,\mu}(t) = \sum_{\kappa=0}^{k} \binom{k}{\kappa} \mu^{i-\kappa} i(i-1) \cdots (i - \kappa + 1) t^{i-\kappa} g_{k-\kappa,j}$$

is a polynomial in t, and

$$D^k F(\mu x) = \sum_{i=0}^{m-1} \sum_{j=0}^{n-1} \varphi_{ij} A_{i,j}^{k,\mu}(x).$$

THEOREM 3. *Let k, μ be such that*

$$0 \leq k \leq n - 1 \qquad and \qquad 0 \leq \mu \leq m - 1.$$

Let $\Delta_g = \mathrm{Det}(g_{kj})$, and let

$$\Delta_1 = \begin{vmatrix} 1 & 0 & \cdots & 0 \\ 1 & 1 & \cdots & 1 \\ 1 & 2 & \cdots & 2^{m-1} \\ \vdots & \vdots & & \vdots \\ 1 & m-1 & \cdots & (m-1)^{m-1} \end{vmatrix}$$

be the Vandermonde determinant. Then

$$\mathrm{Det}\, |A_{i,j}^{k,\mu}(t)| = t^{\frac{1}{2}mn(m-1)} \Delta_1^n \Delta_g^m.$$

(Here, (i, j) indexes rows, and (k, μ) indexes columns.)

Proof. Since each $A_{i,j}^{k,\mu}$ is a polynomial of degree $\leq i$, it follows that

$$\Delta(t) = \mathrm{Det}\, A_{i,j}^{k,\mu}(t)$$

is a polynomial of degree $\leq \frac{1}{2}mn(m-1)$. We shall prove that

$$\Delta(t) = ct^{\frac{1}{2}mn(m-1)}$$

for some constant c, and determine this constant.

Let $H = (H^1, \ldots, H^M)$ be an $M \times M$ matrix of functions, with column vectors $H^1, \ldots, H^M$. Then for any integer $s \geq 0$, we have

$$D^s \mathrm{Det}(H) = \sum_{(\sigma)} \mathrm{Det}(D^{\sigma_1} H^1, \ldots, D^{\sigma_M} H^M),$$

where the sum is taken over all (σ) such that $\sigma_1 + \cdots + \sigma_M = s$. This is trivially proved by induction.

We contend that $D^s\Delta(0) = 0$ if $s < \frac{1}{2}mn(m-1)$. *Proof:* Let $M = mn$. For any s with $0 \le s \le \frac{1}{2}mn(m-1)$, we have

$$D^s\Delta(0) = \sum_{(\sigma)} \text{Det}(\ldots, D^{\sigma_{k\mu}}A^{k\mu}(0), \ldots) = \sum_{(\sigma)} \text{Det}_{(\sigma)},$$

where the sum is taken over (σ) such that $\sum\sigma_{k\mu} = s$. We have, for any integer $\sigma \ge 0$,

$$D^\sigma A_{ij}^{k\mu}(0) = \begin{cases} \binom{k}{i-\sigma}\mu^{\sigma}i!\,g_{k-i+\sigma,j} & \text{if } \sigma \le i \\ \\ 0 & \text{if } \sigma > i. \end{cases}$$

If a term $\text{Det}_{(\sigma)}$ is such that, for some k, there exist $\mu_1 \ne \mu_2$ such that $\sigma_{k\mu_1} = \sigma_{k\mu_2} = \tau$, then we can factor out μ_1^{τ} and μ_2 from the (k, μ_1) and (k, μ_2) columns respectively, and obtain a determinant in which two distinct columns are equal. Thus $\text{Det}_{(\sigma)} = 0$. Hence the only non-zero determinants $\text{Det}_{(\sigma)}$ are such that $\sigma_{k\mu_1} \ne \sigma_{k\mu_2}$ if $\mu_1 \ne \mu_2$. For such (σ),

$$\sum_{\mu=0}^{m-1} \sigma_{k\mu} \ge \frac{m(m-1)}{2}.$$

Taking the sum over k, we find that if $D^s\Delta(0) \ne 0$, then $s \ge \frac{1}{2}mn(m-1)$. This proves the first part of our theorem, that $\Delta(t) = ct^{\frac{1}{2}mn(m-1)}$.

We must now determine the constant c. For this, we need a formal lemma on determinants.

LEMMA. *Let*

$$X = (x_{kj}), \qquad\qquad k, j = 0, \ldots, n-1$$

and

$$Y = (y_{i\mu}), \qquad\qquad i, \mu = 0, \ldots, m-1$$

be two matrices of independent variables, and let

$$X * Y = (x_{kj}y_{i\mu})$$

be the corresponding $mn \times mn$ matrix. Then

$$\text{Det}(X * Y) = \text{Det}(X)^m \, \text{Det}(Y)^n.$$

Proof. We may write

$$\text{Det}(X * Y) = \text{Det} \begin{vmatrix} x_{00}Y & x_{01}Y & \cdots & x_{0,n-1}Y \\ x_{10}Y & x_{11}Y & \cdots & x_{1,n-1}Y \\ \vdots & \vdots & & \vdots \\ x_{n-1,0}Y & x_{n-1,1}Y & \cdots & x_{n-1,n-1}Y \end{vmatrix}$$

Subtracting a multiple of the first column from each other column, we find that the determinant of $X * Y$ is equal to the determinant

$$\text{Det} \begin{vmatrix} x_{00}Y & 0 & \cdots & 0 \\ x_{10}Y & x'_{11}Y & \cdots & x'_{1,n-1}Y \\ \vdots & \vdots & & \vdots \\ x_{n-1,0}Y & x'_{n-1,1}Y & \cdots & x'_{n-1,n-1}Y \end{vmatrix}$$

and by induction, we find

$$\text{Det}(X * Y) = G(X)\,\text{Det}(Y)^n,$$

where $G(X)$ does not depend on Y. By symmetry, it follows that

$$\text{Det}(X * Y) = c_0\,\text{Det}(X)^m\,\text{Det}(Y)^n$$

for some constant c_0. Letting X, Y be the unit matrices, we see that $c_0 = 1$. This proves our lemma.

Theorem 3 now follows at once from the lemma, if we simply take the determinant of the matrix obtained from $A_{ij}^{k\mu}$ by replacing each polynomial by its term of highest degree. It is then clear that the determinant c is of the type considered in the lemma.

COROLLARY. *Let $f(t) = t$ and $g(t) = e^t$. Let $x = 2\pi\sqrt{-1}$. Then*

$$\text{Det}\,|D^k(f^i g^j)(\mu x)| \neq 0.$$

§6. *A transcendence measure for logarithms*

We shall illustrate a general method of Feldman by a special case, to see how accurate a result one can obtain under the most special hypotheses.

THEOREM 4. *For all d, $h \geq 3$ and all algebraic numbers ξ of degree $\leq d$ and absolute height $\leq h$, we have*

$$\log|2\pi i - \xi| \gg -d(d\log d + h)\log(d\log d + h).$$

(The constant implicit in the symbol $\gg$ is an absolute constant.)

Proof. Let c be a constant, taken sufficiently large with respect to the absolute constant C_1 of Lemma 7, and with respect to 2π. Let λ be a parameter, of which we need only that it is sufficiently large with respect to c. Finally, let

$$N = \max\left(d, \frac{h}{\log h}\right).$$

We may assume N sufficiently large with respect to λ.

Now let
$$B = e^{c\lambda^4 dN(\log N)^2}.$$

For convenience, let $x = 2\pi\sqrt{-1}$, and suppose that

$$(1) \qquad \log|x - \xi| \leqq -\lambda^6 dN(\log N)^2.$$

We shall reach a contradiction. Let d_0 be the precise degree of ξ. We consider a function in the usual manner,

$$F(t) = \sum_{i_0=0}^{d_0-1} \sum_{i=0}^{r_1-1} \sum_{j=0}^{r_2-1} a_{(i)} \xi^{i_0} t^i e^{jt}$$

with integer coefficients $a_{(i)} = a_{i_0 ij}$. We take

$$r_1 = [\lambda^3 d \log N] \qquad \text{and} \qquad r_2 = [\lambda^2 N \log N].$$

We want

$$(2) \qquad |D^k F(\nu x)| < \frac{1}{B}$$

for

$$0 \leqq k \leqq [\lambda^3 N \log N] \qquad \text{and} \qquad 0 \leqq \nu \leqq [\lambda d \log N].$$

These conditions amount to solving linear inequalities, with

$$\text{number of variables} \gg dr_1 r_2 \gg \lambda^5 d^2 N(\log N)^2,$$
$$\text{number of inequalities} \ll \lambda^4 dN(\log N)^2.$$

Furthermore, easy estimates of the usual type show that

$$\log|\text{coefficients}| \ll \lambda^3 N(\log N)^2.$$

Hence by Lemma 5 on linear inequalities, there exists a solution in integers $a_{(i)}$ not all zero such that

$$\log|a_{(i)}| \ll \frac{1}{\lambda d} \log B \ll \lambda^3 N(\log N)^2.$$

We may consider $D^k F(\nu x)$ as a polynomial in two variables, say

$$D^k F(\nu x) = P_{k,\nu}(\xi, x).$$

Estimates of the usual type yield:

$$\log|P_{k,\nu}| \ll \lambda^3 N(\log N)^2,$$
$$\deg P_{k,\nu} \text{ in } \xi \ll d,$$
$$\deg P_{k,\nu} \text{ in } x \ll r_1 \ll \lambda^3 d \log N.$$

By Theorem 1, if $P_{k,\nu}(\xi,\ \xi) \neq 0$, then

$$(3) \qquad\qquad -\lambda^3 \, dN (\log N)^2 \ll \log |P_{k,\nu}(\xi,\ \xi)|.$$

On the other hand,

$$(4) \qquad\qquad P_{k,\nu}(\xi,\ \xi) = P_{k,\nu}(\xi,\ x) + \int_x^\xi P'_{k,\nu}(\xi,\ z)\, dz.$$

Hence

$$|P_{k,\nu}(\xi,\ \xi)| \leq |P_{k,\nu}(\xi,\ x)| + |x - \xi| M_{k,\nu}$$

where $M_{k,\nu}$ is a bound for the expression inside the integral sign, estimated
by the same type of bound that we obtained above for the coefficients of
our linear inequalities, namely

$$\log M_{k,\nu} \ll \lambda^3 N (\log N)^2.$$

Using (1) and (2), we find

$$\log |P_{k,\nu}(\xi,\ \xi)| \ll -\lambda^4 \, dN (\log N)^2,$$

which contradicts (3) as soon as λ is sufficiently large (with respect to an
absolute constant). Hence

$$P_{k,\nu}(\xi,\ \xi) = 0$$

for all $k,\ \nu$.

We now wish to prove that

$$P_{r,\mu}(\xi,\ \xi) = 0$$

for

$$0 \leq r \leq r_2 - 1 \qquad \text{and} \qquad 0 \leq \mu \leq r_1 - 1.$$

If we can do this, then with the notation of the preceding section, we have

$$0 = P_{r,\mu}(\xi,\ \xi) = \sum_{i=0}^{r_1-1} \sum_{j=0}^{r_2-1} \varphi_{ij}(\xi) A_{ij}^{r\mu}(\xi).$$

By the Corollary of Theorem 3, the determinant of this system of linear
equations is not 0, whence $\varphi_{ij}(\xi) = 0$ for all i, j, and we obtain a contradiction. We carry out this program in two steps.

First step. Let τ be a positive number such that $3 + \tau < 4$ but
$3 + \tau + 1 + \tau > 5$. For instance $\tau = 3/4$. We shall prove that

$$P_{k',\nu'}(\xi,\ \xi) = 0$$

for

$$0 \leq k' \leq [\lambda^{3+\tau} N \log N] \qquad \text{and} \qquad 0 \leq \nu' \leq [\lambda^{1+\tau} d \log N].$$

We apply the interpolation Lemma 7, with $z' = v'x$, $r = k'$, $z_1, \ldots, z_m$ equal to the numbers vx with $0 \leq v \leq [\lambda d \log N]$, omitting z', and

$$R_1 = 4\lambda^{1+\tau} d(\log N)|x|,$$

$$R = 3C_1 R_1.$$

Since $r \log r \ll \lambda^{3+\tau} N(\log N)^2$, and

$$\log |F|_R \ll \lambda^{3+\tau} dN(\log N)^2,$$

the first term in the estimate of Lemma 7 is dominated by $(C_1 R_1/R)^{ml}$, whose logarithm is $\ll -\lambda^4 dN(\log N)^2$. The second term satisfies a similar estimate, because of our choice of B, with a large constant c, so that in the product

$$\frac{2^r r!}{R_1^r} C_1^{ml} ml(1/\delta)^{ml} \max_{k,v} |D^k F(vx)|$$

the term involving derivatives dominates the estimate. Thus finally,

$$\log |D^{k'} F(v'x)| \ll -\lambda^4 dN(\log N)^2.$$

Now we use again the estimate

$$|P_{k',v'}(\xi, \xi)| \leq |P_{k',v'}(\xi, x)| + |x - \xi| M_{k',v'}$$

and argue as before. If $P_{k',v'}(\xi, \xi) \neq 0$, we get the inequalities

$$-\lambda^{3+\tau} dN(\log N)^2 \ll \log |P_{k',v'}(\xi, \xi)| \ll -\lambda^4 dN(\log N)^2,$$

a contradiction which shows that $P_{k',v'}(\xi, \xi) = 0$.

We now obtain from (4) (with k', v' instead of k, v):

$$|P_{k',v'}(\xi, x)| \leq |x - \xi| M_{k',v'},$$

whence

(5) $$\log |D^{k'} F(v'x)| \ll -\lambda^6 dN(\log N)^2.$$

Second step. We now prove that

$$P_{r,\mu}(\xi, \xi) = 0$$

for

$$0 \leq r \leq r_2 - 1 \quad \text{and} \quad 0 \leq \mu \leq r_1 - 1.$$

We use Lemma 7 again to estimate $D^r F(\mu x)$. We take $z' = \mu x$. We let

$z_1, \ldots, z_m$ be the numbers $\nu'x$, omitting z', and

$$R_1 = 4r_1|x| = 4[\lambda^3 d \log N]|x|,$$

$$R = 3C_1 R_1.$$

Then $\log |F|_R \ll \lambda^5 \, dN(\log N)^2$ and hence by our choice of τ, the first term is dominated by $(C_1 R_1/R)^{ml}$, whose logarithm is

$$\ll -\lambda^{4+2\tau} \, dN(\log N)^2.$$

For the second term, we use (5), and find that the logarithm of the second term is dominated by (5). Hence

$$\log |D^r F(\mu x)| \ll -\lambda^{4+2\tau} \, dN(\log N)^2$$

whence

$$\log |P_{r,\mu}(\xi, \xi)| \ll -\lambda^{4+2\tau} \, dN(\log N)^2.$$

On the other hand, estimating the degree and height of $P_{r,\mu}$, if

$$P_{r,\mu}(\xi, \xi) \neq 0,$$

then

$$-\lambda^3 \, dN(\log N)^2 \ll \log |P_{r,\mu}(\xi, \xi)|,$$

which is a contradiction. This proves the theorem.

Using the same method as that for Theorem 1, Feldman obtains:

THEOREM 5. *Let α be algebraic, and $x = \log \alpha \neq 0$. For all $d, h \geq 3$, and all algebraic numbers ξ of degree $\leq d$ and absolute height $\leq h$, we have*

$$\log |x - \xi| \gg - d^2(d \log d + h)(\log d) \log(d \log d + h).$$

The main difference is the appearance of d^2, due to the fact that one parameter is lost since $e^x = \alpha$, and this introduces an extra term, so that instead of having a polynomial $P_{k,\nu}$ in two variables, we have a polynomial $P_{k,\nu}(\xi, x, \alpha)$ in three variables. It is then necessary to adjust the values for i, j, k, ν accordingly. Also, the symbol $\gg$ now depends on the given α.

Feldman also obtains analogous results for the Weierstrass $\wp$-function. In that case, the fact that such a function is of order 2 simply introduces another parameter, and makes the final result (using the same method) correspondingly worse. It should be noted, however, that Feldman proves the corresponding non-vanishing of the determinant of §5. It is an interesting problem, independent of the theory of transcendental numbers, to investigate such determinants for abelian functions and other generalized exponential functions. Finally, it should be mentioned that when

h is large compared to d, then Feldman improves the dependence of the inequality on h, and obtains the following typical result.

THEOREM 6. *Let α be algebraic $\neq 0$, and $x = \log \alpha \neq 0$. For all $d \geq 3$, and for all algebraic numbers ξ of degree $\leq d$ and absolute height $\leq h$, with $h > d^4$, we have*

$$\log |x - \xi| \gg -h\, d^2 (\log d)^2,$$

where $\gg$ depends only on α.

Historical note

This entire chapter is due to Feldman who obtained the results, and related ones, in the series of papers listed in the bibliography. These call for a number of comments.

We shall proceed systematically, and first make some very general comments on transcendental numbers and diophantine approximations.

The theory of transcendental numbers determines which classical numbers are linearly independent or algebraically independent (over the rationals). This requires a definition of the notion of classical number, and essentially, a classical number is one which appears as a value of a classical function suitably normalized. Let us give examples. In a classical situation, one meets an open subset of some complex space, say U, and a map $f: U \to V$ of U into an algebraic variety. Given such a map, we can generate a field of numbers, by taking the smallest field Ω generated from the rational numbers by performing the following operations inductively, and iterating them:

Taking algebraic closure.

Adjoining values of f and its inverse function with the argument in the field obtained inductively after a finite number of steps.

Examples of maps f are given by exponential maps, uniformizing maps, solutions of algebraic differential equations, zeta functions (including L-series and gamma functions), etc.

Given say real numbers $x_1, \ldots, x_m$ in a field Ω, one wishes to study the inequality

$$(*) \qquad |q_0 + q_1 x_1 + \cdots + q_m x_m| < \frac{1}{q^m g(q)} \qquad (q = \max |q_i|)$$

with integers q_i, and some function g, positive and increasing. Assuming that $1, x_1, \ldots, x_m$ are linearly independent over the rationals, one defines $x_1, \ldots, x_m$ to be of *type* $\leq g$ if the above inequality has only a finite number of solutions. Similarly, if x is a given transcendental number, one should say that it has transcendence type $\leqq g$ if the inequality (*) has only a finite number of solutions uniformly for $x_i = x^i$, and all m. Similarly, we can make a definition with respect to approximation by an algebraic number of degree d, using the inequality

$$(**) \qquad |x - \xi| < \frac{1}{H(\xi)^{d+1} g\left(H(\xi), d(\xi)\right)} \qquad \text{with } H(\xi) = e^{h(\xi)}.$$

The results of [20] show that the theory of diophantine approximations of a number, say by rational numbers, achieves coherence and simplicity only if one measures the order of approximation not in the exponent but as a factor of H^{d+1}. (It is not entirely clear conjecturally to what extent g should be independent of d.)

The general problem is now to determine inductively the type of a classical number obtained as value of classical function. One must of course first determine a type for algebraic numbers, and in this respect, the Thue-Siegel-Roth theorem appears as rather weak, in spite of the difficulties which one has encountered historically in reaching a proof for it.

Even in the case of algebraic numbers, no result is known to take into account varying degrees, say a result of type

$$|\alpha - \xi| \gg \frac{1}{H(\xi)^{d+1+\epsilon}}$$

or some such exponent as $d + 1 + \epsilon$, let alone more refined results. Such a result is not even known if the degree d of ξ is kept *fixed*.

The Feldman result may be seen as a first step in the inductive procedure, and the Liouville estimate (Theorem 1) is the induction hypothesis. However, given the present structure of the proof, this Feldman estimate is weaker than what should be expected. Thus it is good in that the exponent for h and d (in Theorem 4) is precisely equal to 1, but bad in that an extra log appears, so that one does not even get an estimate like

$$\log |x - \xi| \gg dh.$$

This, however, would only be a first step towards determining the refined types as in (*) and (**) above.

Even assuming a very good type for a number x, and using the more complicated techniques of Feldman, applied to the inductive procedure of Chapter V, I still do not see how to achieve a best possible result which would for instance yield the algebraic independence of e and π, assuming that e or π has a very good transcendence type. It is very hard to tell whether this is because of a superficial defect in the proof, or whether one needs an entirely different structure for the algebraic independence proof.

It should be noted that Feldman's results (and a subsequent one by Gelfond [13], following Feldman's method) are the only ones which exhibit a good dependence on the degree d. For instance, the dependence on d in Mahler's papers [24], using similar techniques to Siegel's, give a much worse dependence on d. We shall mention this again in the next chapter. The dependence on h is much better in all known cases.

It may be that to achieve the best possible dependence on d, one must first determine a best possible type for the approximation by rational numbers, and *then* use a best possible type for approximation of algebraic numbers by rational numbers. Thus the induction must start with loaded hypotheses, not only going from numbers to values of a function, but also from the rational numbers to algebraic numbers. The few examples which one has now do suggest an absolutely fantastic rigidity in the entire theory.

CHAPTER VII

Linear Differential Equations

This chapter will deal with a method of Siegel, which is particularly efficient when the functions under consideration, aside from satisfying a (linear) differential equation, have a power series expansion of a special type. We shall begin by describing this type.

§1. E-functions

An *E-function* is a function which admits a power series expansion

$$f(z) = \sum_{n=0}^{\infty} \alpha_n \frac{z^n}{n!}$$

with complex coefficients α_n, belonging to some number field K, satisfying the following conditions:

E 1. *We have $\|\alpha_n\| \leqq c^n$ for some constant c.*

E 2. *There exists a sequence of integers $d_n \in \mathbf{Z}$, $d_n > 0$ such that d_n is a denominator for α_k ($k = 0, \ldots, n$) and*

$$d_n \leqq c^n.$$

(*Note:* We take the bound c^n for convenience. Actually, everything goes through if we define E-functions using the bound $O(n^{\epsilon n})$ for every $\epsilon > 0$. However, all the examples satisfy the stronger conditions stated above, and I see no point here in introducing an extra parameter.)

The ordinary exponential function e^z is an E-function. So is the Bessel function

$$J_0(z) = \sum \frac{z^{2n}}{(n!)^2}$$

or the Bessel function J_λ with algebraic parameter λ. Similarly, hypergeometric functions (with algebraic parameter) are also examples of E-functions. We refer the reader to Siegel's book for such examples. Of course, polynomials with algebraic coefficients are E-functions.

76

477

In what follows, we assume that all E-functions mentioned have coefficients in the number field K. If f is an E-function as above, we define

$$\text{size}_n(f) = \text{size}(\alpha_0, \ldots, \alpha_n)$$

to be the size of its first $n + 1$ coefficients (i.e. coefficients of $z^n/n!$). If f' is the derivative of f, then

$$\text{size}_n(f') \leqq \text{size}_{n+1}(f),$$

whence in particular, f' is also an E-function.

Let

$$g(z) = \sum \beta_n \frac{z^n}{n!}$$

be an E-function. Then $f + g$ is an E-function, and

$$\text{size}_n(f + g) \leqq \text{size}_n(f) + \text{size}_n(g) + 2.$$

Furthermore,

$$f(z)g(z) = \sum \gamma_n \frac{z^n}{n!}$$

where

$$\gamma_n = \sum_{k=0}^{n} \binom{n}{k} \alpha_k \beta_{n-k},$$

whence fg is an E-function, with

$$\text{size}_n(fg) \leqq \text{size}_n(f) + \text{size}_n(g) + 2n,$$

because if d_n is a denominator for $\alpha_0, \ldots, \alpha_n$ and d'_n is a denominator for $\beta_0, \ldots, \beta_n$ then $d_n d'_n$ is a denominator for $\gamma_0, \ldots, \gamma_n$.

Finally, if α is in K and f is as above, an E-function, then $f(\alpha z)$ is also an E-function, with

$$\text{size}_n f(\alpha z) \leqq \text{size}_n(f) + n \cdot \text{size}(\alpha).$$

§2. *The Lindemann theorem*

We shall carry out a special case of the Siegel method to prove the Lindemann theorem.

THEOREM 1. *Let $\alpha_1, \ldots, \alpha_s$ be algebraic numbers, linearly independent over the rationals. Then $e^{\alpha_1}, \ldots, e^{\alpha_s}$ are algebraically independent.*

The proof will use several lemmas.

Let $K = \mathbf{Q}(\alpha_1, \ldots, \alpha_s)$. Let $\beta_1, \ldots, \beta_m$ be distinct non-zero elements of K, and let E_j $(j = 1, \ldots, m)$ be the m functions

$$E_j(z) = e^{\beta_j z}.$$

We shall form a new function

$$F_1(z) = P_1(z)e^{\beta_1 z} + \cdots + P_m(z)e^{\beta_m z}$$

with polynomials $P_1, \ldots, P_m$ having coefficients in I_K, not all zero, such that F_1 has a high zero at 0. We regard m as given, and we shall deal with a parameter n. Constants $c, c_1, c_2, \ldots$ thus depend on the β_j and m.

LEMMA 1. *Given an integer* $n > 0$ *we can find polynomials* $P_j \in I_K[z]$ *not all zero, such that:*
   (i) $\deg P_j < 2n$ *and* $\|P_j\| \leq c^n n^{2n}$.
   (ii) *The function* F_1 *has a zero of order* $\geq (2m - 1)n$ *at 0.*
   (iii) *If*

$$F_1(z) = \sum_{\nu=0}^{\infty} a_\nu \frac{z^\nu}{\nu!}$$

then $|a_\nu| \leq c^\nu c^n n^{2n}$.

Proof. We write P_j with unknown coefficients, namely

$$P_j(z) = (2n - 1)! \sum_{\mu=0}^{2n-1} x_{j\mu} \frac{z^\mu}{\mu!}$$

and have

$$E_j(z) = \sum_{\mu=0}^{\infty} \beta_j^\nu \frac{z^\nu}{\nu!} = \sum_{\nu=0}^{\infty} \beta_{j\nu} \frac{z^\nu}{\nu!}.$$

Then

$$P_j E_j(z) = (2n - 1)! \sum_{\nu=0}^{\infty} b_{j\nu} \frac{z^\nu}{\nu!}$$

where

$$b_{j\nu} = \sum_{k=0}^{2n-1} \binom{\nu}{k} x_{jk} \beta_{j,\nu-k}.$$

We obtain

$$P_1 E_1 + \cdots + P_m E_m = \sum_{\nu=0}^{\infty} a_\nu(x) \frac{z^\nu}{\nu!}$$

where

$$a_\nu(x) = (2n - 1)!(b_{1\nu} + \cdots + b_{m\nu}).$$

We must solve for (x) the linear equations $a_\nu(x) = 0$, with $\nu < (2m - 1)n$.

We have $2mn$ unknowns

$$x_{j\mu} \quad \text{with } j = 1, \ldots, m \text{ and } \mu = 0, \ldots, 2n - 1$$

and $(2m - 1)n$ equations. The coefficients of these equations are in K, bounded by

$$\|\text{coefficients}\| \leq m \cdot \max \binom{\nu}{k} \max \|\beta_{j,\nu-k}\|$$
$$\leq c_1^n,$$

since the binomial coefficient is bounded by $2^{(2m-1)n}$. A denominator for these coefficients is also bounded by c_1^n. By Siegel's lemma on linear equations, we can find a non-trivial solution with $x_{j\mu} \in I_K$ satisfying $\|x_{j\mu}\| \leq c^n$. It is now an easy matter to estimate the coefficients a_ν for all ν to get the estimate (iii). Of course, only those a_ν may be $\neq 0$ for $\nu \geq (2m - 1)n$. This proves our lemma.

LEMMA 2. *Let* E_j, P_j, F_1 *be as in Lemma 1. Let*

$$F_{k+1} = D^k F_1$$

where D is the derivative, $k = 1, 2, \ldots$ Write

$$F_k = P_{k1}E_1 + \cdots + P_{km}E_m$$

with polynomials P_{kj}. Then the rank of the matrix (P_{kj}) $(k, j = 1, \ldots, m)$ is equal to m.

Proof. Let Y be the vertical vector of functions $(E_1, \ldots, E_m)$. Then Y satisfies the linear differential equation

$$Y' = QY$$

where Q is the matrix

$$Q = \begin{pmatrix} \beta_1 & \cdots & & 0 \\ \vdots & \beta_2 & & \vdots \\ & & \ddots & \\ 0 & \cdots & & \beta_m \end{pmatrix}$$

In fact, we have

$$D^k F_1 = (D + \beta_1)^k P_1 E_1 + \cdots + (D + \beta_m)^k P_m E_m.$$

Thus the matrix (P_{kj}) is none other than

$$\begin{pmatrix} P_1 & \cdots & P_m \\ (D + \beta_1)P_1 & \cdots & (D + \beta_m)P_m \\ \vdots & \cdots & \vdots \\ (D + \beta_1)^{m-1}P_1 & \cdots & (D + \beta_m)^{m-1}P_m \end{pmatrix}$$

Let $\Delta = \Delta(z)$ be its determinant. Then Δ is a polynomial in z, and we shall prove that it is not zero by looking at the highest power of z occurring in it. In fact, let $u_1, \ldots, u_m$ be the leading coefficients of the polynomials $P_1, \ldots, P_m$ respectively, and let $d_1, \ldots, d_m$ be their degrees. Then our determinant has one term of type

$$\begin{vmatrix} u_1 & \cdots & u_m \\ \beta_1 u_1 & \cdots & \beta_m u_m \\ \vdots & \cdots & \vdots \\ \beta_1^{m-1} u_1 & \cdots & \beta_m^{m-1} u_m \end{vmatrix} \, z^{d_1 + \cdots + d_m}$$

plus other terms in the expansion which have lower degree. We factor out $u_1 \cdots u_m$ from the determinant, and see that the remaining constant is a Vandermonde determinant which is not 0. This proves Lemma 2.

As in the proof of Lemma 2, let Δ be the determinant

$$\Delta = \det(P_{kj}) \qquad (k, j = 1, \ldots, m).$$

Then

$$\deg \Delta \leq (2n - 1)m.$$

Let P be the matrix (P_{kj}) $(k, j = 1, \ldots, m)$, and let $\widetilde{P}$ be the matrix such that

$$\widetilde{P}P = \Delta I.$$

Thus $\widetilde{P}$ is the transpose of the matrix of minors of Δ. Let F be the column vector of $(F_1, \ldots, F_m)$ and let Y be as before, the column vector of $(E_1, \ldots, E_m)$. Then

$$F = PY \quad \text{and} \quad \Delta Y = \widetilde{P}F.$$

Each F_j $(j = 1, \ldots, m)$ has a zero at 0 of order $\geq (2m - 1)n - m$, and since none of the components of Y vanishes at 0, we conclude that

$$\operatorname{ord} \Delta \geq (2m - 1)n - m.$$

(Here, ord means order at 0.) Comparing this order with the degree of Δ, it follows that if ξ is any complex number $\neq 0$, then

$$\operatorname{ord}_\xi \Delta \leq \deg \Delta - \operatorname{ord} \Delta$$

$$\leq n.$$

LEMMA 3. *For any complex number $\xi \neq 0$, the matrix*

$$(P_{kj}(\xi)) \qquad (k = 1, \ldots, m + n \text{ and } j = 1, \ldots, m)$$

has rank m.

Proof. We have $\widetilde{P}P = \Delta I$. For any $k = 1, 2, \ldots$ we have

$$(D + Q)^k P_{(1)} = P_{(k)},$$

where $P_{(k)}$ is the k-th row of P viewed as column vector. Let $r = \operatorname{ord}_\xi \Delta$. We apply $(D + Q)^r$ to ΔI, and find

$$(D + Q)^r (\Delta I) = \sum_{\mu=0}^{r} C_\mu (D + Q)^\mu {}^t P,$$

where C_μ are matrices of polynomials. Evaluating these expressions at ξ, we see that in the expansion on the left, all terms will vanish except $D^r \Delta(\xi) I$, whence

$$D^r \Delta(\xi) I = \sum_{\mu=0}^{r} C_\mu(\xi)(D + Q)^\mu {}^t P(\xi).$$

On the left we have a non-zero scalar matrix. On the right, the columns of the matrix $(D + Q)^\mu {}^t P(\xi)$ are simply the columns

$$P_{(k)}(\xi) = {}^t(P_{k1}(\xi), \ldots, P_{km}(\xi))$$

with $k \leqq m + n$. It follows that these columns have rank at least m, thereby proving our lemma.

Let $\alpha \in K$, $\alpha \neq 0$. We shall estimate $|F_k(\alpha)|$ and $\|P_{kj}(\alpha)\|$. Further constants depend on the size of α.

LEMMA 4. *Let $k \leqq m + n$. Assume $n \geqq c_2(\alpha)$. Then*

$$|F_k(\alpha)| \leqq c_3^n n^{3n} n^{-(2m-2)n}$$

$$\|P_{kj}(\alpha)\| \leqq c_3^n n^{3n}, \quad \text{and} \quad \operatorname{den}(P_{kj}(\alpha)) \leqq c_3^n.$$

Proof. The first inequality will come from the estimates of the coefficients of F_1, applying k derivatives, and using the fact that $D^k F_1$ begins with a high power of z, and hence a high factorial in the denominator which contributes the term $n^{-(2m-2)n}$ tending to 0 with n. We do this in detail.

The power series for F_1 is dominated term by term by

$$F_1(z) < c^n n^{2n} \sum_{\nu=(2m-1)n}^{\infty} c^\nu \frac{z^\nu}{\nu!}$$

and that of F_k is therefore dominated by

$$F_k(z) < c^n n^{2n} c^k \sum_{\nu=(2m-1)n}^{\infty} c^{\nu-k} \frac{z^{\nu-k}}{(\nu - k)!}.$$

Therefore,

$$|F_k(\alpha)| \leqq c_4^n n^{2n} \sum_{\nu=(2m-1)n}^{\infty} \frac{(c|\alpha|)^{\nu-k}}{(\nu-k)!}.$$

We observe that for any integer $r > 0$,

$$\sum_{\nu=r}^{\infty} \frac{w^{\nu}}{\nu!} = \frac{w^r}{r!}\left(1 + \frac{w}{r+1} + \frac{w^2}{(r+1)(r+2)} + \cdots\right).$$

Here we take $r = (2m-1)n - k$. In making our estimate, we take the maximum value of r when estimating numerators, and its minimum value when estimating denominators, for $0 \leqq k \leqq m + n$. We have

$$(2m-2)(n-1) \leqq (2m-2)n - m \leqq r \leqq (2m-1)n.$$

Since we took n large compared to α, the sum in parentheses is $\leqq 2$. Also, $r!$ is approximately equal to $r^r e^{-r}$. Putting all this together, we obtain the desired estimate for $|F_k(\alpha)|$. In fact, we get an exponent n^{2n} instead of n^{3n}, but we have nevertheless put n^{3n} to fit some later generalization.

To estimate $\|P_{kj}(\alpha)\|$, we estimate the size of the coefficients of P_{kj}. We know that

$$P_{(k)} = (D + Q)^k P_{(1)}.$$

It is easy to estimate this by induction, using the same technique as in Chapter III, §2, except that the situation here is easier. A given polynomial P_j $(j = 1, \ldots, m)$ is dominated by

$$P_j(z) \prec c^n n^{2n}(1+z)^{2n-1},$$

and applying $(D + \beta_j)^k$ to this polynomial is easily seen by induction to be dominated by

$$(D + \beta_j)^k P_j \prec c_5^n n^{3n}(1+z)^{2n-1}.$$

(The extra power of n comes from a term bounded by $(2n-1)^{m+n}$ arising from successive derivatives.) Substituting α for z then gives the desired result. The estimate for the denominator of $P_{kj}(\alpha)$ is even more trivial.

The next, and final, lemma is the decisive step in the proof. From the linear independence of the *functions* $E_1, \ldots, E_m$ over the polynomials, it gives a lower bound for the rank of the *numbers* $E_1(\alpha), \ldots, E_m(\alpha)$ over K.

LEMMA 5. *The rank of* $E_1(\alpha), \ldots, E_m(\alpha)$ *over* K *is* $\geqq m/2[K : \mathbf{Q}]$.

Proof. Let r be this rank. Let

$$
\begin{aligned}
0 &= \lambda_{11}E_1(\alpha) + \cdots + \lambda_{1m}E_m(\alpha) \\
&\ \vdots \qquad\qquad \vdots \qquad\qquad\qquad \vdots \\
0 &= \lambda_{m-r,1}E_1(\alpha) + \cdots + \lambda_{m-r,m}E_m(\alpha)
\end{aligned}
$$

be $m - r$ linearly independent relations with coefficients $\lambda_{kj} \in I_K$. By Lemma 3, we can find r functions among the F_k, $k \leq m + n$, such that, if we put

$$
\begin{aligned}
F_{k_1}(\alpha) &= P_{k_11}(\alpha)E_1(\alpha) + \cdots + P_{k_1m}(\alpha)E_m(\alpha) \\
&\ \vdots \qquad\qquad \vdots \qquad\qquad\qquad \vdots \\
F_{k_r}(\alpha) &= P_{k_r1}(\alpha)E_1(\alpha) + \cdots + P_{k_rm}(\alpha)E_m(\alpha)
\end{aligned}
$$

then the matrix consisting of the (λ) and the $(P_{k_ij}(\alpha))$ has rank m. Let δ be the determinant of this matrix. Then δ is an element of K, $\delta \neq 0$. We obtain

$$
\delta E_1(\alpha) = B_1 F_{k_1}(\alpha) + \cdots + B_r F_{k_r}(\alpha)
$$

where $B_1, \ldots, B_r$ are minors of the determinant δ. From Lemma 4, we then have

$$
\operatorname{size}(\delta) \leq 3nr \log n + O(n) \qquad \text{and} \qquad \operatorname{den}(\delta) \leq O(n).
$$

Again by Lemma 4, we obtain the upper bounds

$$
|B_1|, \ldots, |B_r| \leq c_6^n n^{3n(r-1)}.
$$

Lemma 4 also gives us an upper bound for $|F_k(\alpha)|$. We therefore obtain the upper bound

$$
\log |\delta| \leq 3n(r - 1) \log n + 3n \log n - (2m - 2)n \log n + O(n).
$$

$$
\leq 3rn \log n - (2m - 2)n \log n + O(n).
$$

We compare this with the size, divide by $n \log n$ throughout, get rid of $O(n)$, and find

$$
2m - 2 \leq 3r\,[K : \mathbf{Q}].
$$

The assertion of the lemma follows trivially.

To prove Theorem 1, we start with the s functions

$$
f_1(t) = e^{\alpha_1 t}, \ldots, f_s(t) = e^{\alpha_s t}.
$$

Let g be a polynomial with coefficients in K, not all zero, of degree d. We must show that for $\alpha = 1$,

$$
g(e^{\alpha_1}, \ldots, e^{\alpha_s}) = g(f_1(\alpha), \ldots, f_s(\alpha)) \neq 0.
$$

Let ν be a large integer, and let

$$m_\nu = \binom{\nu + s}{s}$$

be the binomial coefficient.

There are precisely m_ν monomials

$$f_1^{\nu_1} \cdots f_s^{\nu_s}, \qquad \nu_1 + \cdots + \nu_s \leqq \nu.$$

We let $m = m_\nu$ and let $E_1, \ldots, E_m$ be these monomials. (Thus, in our special case of the exponential function, the numbers $\beta_1, \ldots, \beta_m$ are the linear combinations

$$\nu_1 \alpha_1 + \cdots + \nu_s \alpha_s, \qquad \nu_1 + \cdots + \nu_s \leqq \nu.)$$

If $g\big(f_1(\alpha), \ldots, f_s(\alpha)\big) = 0$, then for $\nu_1 + \cdots + \nu_s \leqq \nu - d$, we have

$$f_1(\alpha)^{\nu_1} \cdots f_s(\alpha)^{\nu_s} g\big(f_1(\alpha), \ldots, f_s(\alpha)\big) = 0.$$

In this way, we obtain relations with coefficients in K among the m_ν monomials

$$f_1(\alpha)^{\nu_1} \cdots f_s(\alpha)^{\nu_s}, \qquad \nu_1 + \cdots + \nu_s \leqq \nu.$$

It is immediately seen that these relations are linearly independent over K, and we have

$$m_{\nu-d} = \binom{\nu + s - d}{s}$$

such relations. By Lemma 5, we must have

$$m_\nu - m_{\nu-d} \geqq m_\nu / 2 \, [K : \mathbf{Q}].$$

This is impossible, because m_ν and $m_{\nu-d}$ are both polynomials in ν, starting with the same term $\nu^m/m!$. This contradicts the assumption that $g\big(f_1(\alpha), \ldots, f_s(\alpha)\big) = 0$, thereby proving Theorem 1.

§3. *Shidlovsky's lemma*

In order to extend the proof of Theorem 1 to arbitrary E-functions satisfying a linear differential equation with rational functions as coefficients, we must state and prove a lemma which allows us to generalize Lemma 2 above. This is the only difficult point in extending the proof of Theorem 1, but involves only linear algebra, no arithmetic. Thus in this section, we may assume that K is an arbitrary field of characteristic 0, and we deal with power series in $K[[z]]$.

We begin by some remarks on linear differential equations. Let

$$Y = {}^t(y_1, \ldots, y_m)$$

be a column vector of power series, and assume that Y satisfies the linear differential equation

$$Y' = QY$$

where Q is a matrix (Q_{ij}) of rational functions in $K(z)$. Let $T = T(z)$ be the polynomial which is the greatest common denominator of the Q_{ij}. We call T a polynomial denominator for Q (or for the Q_{ij}). If $P_1, \ldots, P_m$ are polynomials in $K[z]$, we let $P_{(1)}$ be their column vector. Let

$$F_1 = P_1 y_1 + \cdots + P_m y_m = \langle P_{(1)}, y \rangle$$

be the scalar product of $P_{(1)}$ and Y. We construct F_k inductively by taking

$$F_{k+1} = TDF_k = (TD)^k F_1.$$

Here, $D = d/dz$ is the formal derivative of power series with respect to z. We see trivially that

$$\begin{aligned}
D\langle P_{(1)}, Y \rangle &= \langle DP_{(1)}, Y \rangle + \langle P_{(1)}, DY \rangle \\
&= \langle DP_{(1)}, Y \rangle + \langle P_{(1)}, QY \rangle \\
&= \langle DP_{(1)}, Y \rangle + \langle {}^tQP_{(1)}, Y \rangle \\
&= \langle (D + {}^tQ)P_{(1)}, Y \rangle.
\end{aligned}$$

Hence

$$TDF_1 = \langle T(D + {}^tQ)P_{(1)}, Y \rangle$$

and, inductively,

$$F_k = \langle (T(D + {}^tQ))^k P_{(1)}, Y \rangle.$$

Thus we can write

$$F_k = P_{k1}y_1 + \cdots + P_{km}y_m$$

with polynomials P_{kj}.

SHIDLOVSKY'S LEMMA. *Let* $y_1, \ldots, y_m$ *be as above, formal power series linearly independent over* $K(z)$, *and satisfying the linear differential equation* $Y' = QY$, *where* Q *is a matrix of rational functions. Let* $P_1, \ldots, P_m$ *be polynomials in* $K[z]$, *and let*

$$F_1 = P_1 y_1 + \cdots + P_m y_m.$$

Let T *be a polynomial denominator for* Q, *and define inductively*

$$F_k = TDF_{k-1} = P_{k1}y_1 + \cdots + P_{km}y_m.$$

Let r be the rank of the matrix (P_{kj}) $(k, j = 1, \ldots, m)$ and suppose $r < m$. Then

$$\operatorname{ord} F_1 \leqq r(\max \deg P_j) + c_0,$$

where c_0 is a positive number depending only on $y_1, \ldots, y_m$, Q and not on the P_j.

Although the proof of the lemma is rather long, we shall not use any part of it later. We use only the statement of Shidlovsky's lemma, and then only in the proof of Lemma 2, §4. Hence the reader may omit the rest of this section without impairing his understanding of the rest of the chapter.

The proof will involve lemmas, numbered as 2.1, 2.2, $\ldots$

LEMMA 2.1. *Let $\varphi_1, \ldots, \varphi_n$ be power series in $K[[z]]$, and let d be a positive integer. There exists an integer N (depending on φ and d) such that, if $P_1, \ldots, P_n$ are polynomials in $K[z]$ of degrees $\leqq d$, then either*

$$F = P_1\varphi_1 + \cdots + P_n\varphi_n$$

is equal to 0, or

$$\operatorname{ord} F \leqq N.$$

Proof. Consider first the case when we take the $P_1, \ldots, P_n$ to be constants in K. Write the column vector of functions Φ:

$$
\begin{aligned}
\varphi_1 &= a_{10} + a_{11}z + a_{12}z_2 + \cdots \\
&\ \vdots \qquad \vdots \qquad \vdots \qquad \vdots \\
\varphi_n &= a_{n0} + a_{n1}z + a_{n2}z_2 + \cdots
\end{aligned}
$$

Let $C = (c_1, \ldots, c_n)$ be a constant vector, and let $A^0, A^1, \ldots$ be the column vectors of the matrix of coefficients of $\varphi_1, \ldots, \varphi_n$. We note that

$$C \cdot \Phi = c_1\varphi_1 + \cdots + c_n\varphi_n = 0$$

if and only if $C \cdot A^k = 0$ for $k = 0, 1, 2, \ldots$

Let $A^0, A^1, \ldots, A^M$ generate the space of column vectors. If $C \cdot \Phi \neq 0$, then at least one of the dot products $C \cdot A^k$ is not 0 for $0 \leqq k \leqq M$. This means that

$$\operatorname{ord} C \cdot \Phi \leqq M,$$

and proves our lemma in case the polynomials $P_1, \ldots, P_m$ are constant. The general case is reduced to this one by replacing $\varphi_1, \ldots, \varphi_m$ by $z^i\varphi_j$ $(i = 0, \ldots, d$ and $j = 1, \ldots, m)$.

LEMMA 2.2. *Let V, W be two vector spaces consisting of power series, finite dimensional over the constant field K. There exists an integer N*

such that if φ, ψ are non-zero elements of V, W respectively such that φ/ψ is a rational function, then $\deg \varphi/\psi \leqq N$.

(By the *degree* of a rational function, we mean the maximum of the degree of its numerator and denominator.)

Proof. Let $\{\varphi_1, \ldots, \varphi_n\}$ be a basis of V over K, and let $\{\psi_1, \ldots, \psi_r\}$ be a basis of W over K. Consider the set of constant vectors

$$C = (c_1, \ldots, c_n)$$

for which there exists a constant vector $C' = (c'_1, \ldots, c'_r)$ and a rational function R such that

$$C \cdot \Phi = RC' \cdot \Psi.$$

Let $C_1, \ldots, C_s$ be a maximal set of linearly independent vectors C over K. We can write

$$C_k \cdot \Phi = R_k C'_k \cdot \Psi \qquad\qquad (k = 1, \ldots, s)$$

with rational functions R_k. Given any C in our set, written as

$$C = x_1 C_1 + \cdots + x_s C_s$$

with coefficients $x_i \in K$, we see that

$$\begin{aligned}
C \cdot \Phi &= (x_1 R_1 C'_1 + \cdots + x_s R_s C'_s) \cdot \Psi \\
&= RC' \cdot \Psi.
\end{aligned}$$

From this it is clear that the degree of R is bounded in terms of the degrees of $R_1, \ldots, R_s$.

Lemma 2.3. *Let* $\varphi_1, \ldots, \varphi_m$ *be power series, linearly independent over the constants. Then the Wronskian determinant*

$$W(\varphi_1, \ldots, \varphi_m) = \begin{vmatrix} \varphi_1 & \varphi_2 & \cdots & \varphi_m \\ \varphi'_1 & \varphi'_2 & \cdots & \varphi'_m \\ \vdots & \vdots & & \vdots \\ \varphi_1^{(m-1)} & \varphi_2^{(m-1)} & \cdots & \varphi_m^{(m-1)} \end{vmatrix}$$

is not 0.

Proof. This is a standard easy lemma on derivatives, which is proved by induction. We leave it to the reader.

Corollary. *Let* $m > r$, *and let* $\varphi_1, \ldots, \varphi_m$ *be solutions in* $K[[z]]$ *of the differential equation*

$$\psi_r D^r y + \psi_{r-1} D^{r-1} y + \cdots + \psi_1 y = 0,$$

where $\psi_1, \ldots, \psi_r \in K[[z]]$ and $\psi_r \neq 0$. Then $\varphi_1, \ldots, \varphi_m$ are linearly dependent over the constants.

We return to the differential equation, determined by the matrix Q. Let $P_1, \ldots, P_m$ be polynomials whose column vector is denoted by $P_{(1)}$. We form inductively

$$P_{(k+1)} = T(D + {}^tQ)P_{(k)}$$

and obtain the matrix $(P_{(1)}, \ldots, P_{(m)})$ whose transpose is written

$$\begin{pmatrix} P_{11} & \cdots & P_{1m} \\ \vdots & & \vdots \\ P_{m1} & \cdots & P_{mm} \end{pmatrix}.$$

Suppose that for some integer $r < m$ the vectors $P_{(1)}, \ldots, P_{(r)}$ are linearly independent, but $P_{(r+1)}$ can be written as a linear combination

$$P_{(r+1)} = g_1 P_{(1)} + \cdots + g_r P_{(r)}$$

with rational functions $g_1, \ldots, g_r$. Applying $T(D + {}^tQ)$ to this expression, we find that $P_{(r+2)}$ can also be written as a linear combination of $P_{(1)}, \ldots, P_{(r)}$ with rational functions as coefficients. Inductively, it follows that the rank of the matrix (P_{kj}) $(k, j = 1, \ldots, m)$ is equal to r, and that its first r rows are linearly independent.

LEMMA 2.4. *Given the matrix Q, there exists an integer N having the following property. Let $P_1, \ldots, P_m$ be polynomials such that the rank of the matrix (P_{kj}) is $r < m$. After renumbering the indices, if necessary, suppose that*

$$\begin{pmatrix} P_{11} & \cdots & P_{1m} \\ \vdots & & \vdots \\ P_{r1} & \cdots & P_{rm} \end{pmatrix} = \begin{pmatrix} P_{11} & \cdots & P_{1r} & P_{1,r+1} & \cdots & P_{1m} \\ \vdots & & \vdots & \vdots & & \vdots \\ P_{r1} & \cdots & P_{rr} & P_{r,r+1} & \cdots & P_{rm} \end{pmatrix}$$

is written in terms of two blocks, say

$$(P_{\mathrm{I}}, P_{\mathrm{II}}),$$

such that P_{I} is $r \times r$, P_{II} is $r \times (m - r)$, and P_{I} is non-singular. Then there is a matrix A of rational functions such that

$$P_{\mathrm{II}} = P_{\mathrm{I}}A$$

and such that the degrees of the rational functions are $\leq N$.

Proof. Observe that this lemma is independent of any solutions of the differential equation. Consequently, after making a change in local parameter from z to $z - z_0$, where z_0 is not a pole of the coefficients of Q, we may assume that we deal with solutions of the differential equation in the power series ring $K[[z]]$ and that 0 is not a pole of the coefficients of Q. Let S_Q be the K-space of solutions of the differential equation $Y' = QY$ in vectors Y with components in $K[[z]]$. Then S_Q has dimension m over the constant field K. In fact, each solution is determined uniquely by its initial condition (i.e. its value at 0), as is easily shown recursively, and the map $Y \mapsto Y(0)$ establishes the K-isomorphism between S_Q and the space of m-tuples over K. Let $U = (U^{(1)}, \ldots, U^{(m)})$ be a fixed system of linearly independent column vectors of solutions, say those corresponding to the initial conditions

$$\begin{pmatrix} 1 \\ 0 \\ \vdots \\ 0 \end{pmatrix} \cdots \begin{pmatrix} 0 \\ 0 \\ \vdots \\ 1 \end{pmatrix}.$$

We have, for any solution Y of $Y' = QY$,

$$\begin{aligned}
(TD)^r \langle P_{(1)}, Y \rangle &= \langle P_{(r+1)}, Y \rangle \\
&= g_1 \langle P_{(1)}, Y \rangle + \cdots + g_r \langle P_{(r)}, Y \rangle \\
&= g_1 \langle P_{(1)}, Y \rangle + \cdots + g_r (TD)^{r-1} \langle P_{(1)}, Y \rangle.
\end{aligned}$$

This shows that $\langle P_{(1)}, Y \rangle$ satisfies a linear equation of order r, to which we can apply the corollary of Lemma 2.3. We conclude that the map

$$Y \mapsto \langle P_{(1)}, Y \rangle$$

is a K-linear map from the space S_Q into a K-space of dimension $\leqq r$. Its kernel consists of those Y which are orthogonal to $P_{(1)}$ (and hence to $P_{(k)}$ for all k). We can find a basis

$$\{\overline{Y}^{(1)}, \ldots, \overline{Y}^{(m)}\}$$

of S_Q such that $\overline{Y}^{(1)}, \ldots, \overline{Y}^{(m-r)}$ consist of vectors orthogonal to

$$P_{(1)}, \ldots, P_{(r)}.$$

Hence we have

$$(P_{\mathrm{I}}, P_{\mathrm{II}})(\overline{Y}^{(1)}, \ldots, \overline{Y}^{(m-r)}) = 0.$$

There certainly is some matrix A of rational functions such that

$$P_{\mathrm{II}} = P_{\mathrm{I}} A,$$

and we note that A is $r \times (m - r)$, because the columns of P_{II} depend on the columns of P_{I}. We shall prove that A is uniquely determined and can be expressed in a special way in terms of $\overline{Y}$, where $\overline{Y}$ is the matrix $(\overline{Y}^{(1)}, \ldots, \overline{Y}^{(m)})$.

Decompose $(\overline{Y}^{(1)}, \ldots, \overline{Y}^{(m-r)})$ into two vertical blocks,

$$(\overline{Y}^{(1)}, \ldots, \overline{Y}^{(m-r)}) = \begin{pmatrix} \overline{Y}_{\mathrm{I}}^{(1)} & \cdots & \overline{Y}_{\mathrm{I}}^{(m-r)} \\ \overline{Y}_{\mathrm{II}}^{(1)} & \cdots & \overline{Y}_{\mathrm{II}}^{(m-r)} \end{pmatrix},$$

such that the top block has r rows and the bottom block has $m - r$ rows, corresponding to the decomposition of $(P_{\mathrm{I}}, P_{\mathrm{II}})$. Then we obtain

$$P_{\mathrm{I}} \overline{Y}_{\mathrm{I}}^{(k)} + P_{\mathrm{I}} A \, \overline{Y}_{\mathrm{II}}^{(k)} = 0 \qquad \text{for } k = 1, \ldots, m - r,$$

whence

$$(*) \qquad \overline{Y}_{\mathrm{I}}^{(k)} + A \, \overline{Y}_{\mathrm{II}}^{(k)} = 0 \qquad \text{for } k = 1, \ldots, m - r.$$

On the other hand, the columns

$$\overline{Y}_{\mathrm{II}}^{(1)}, \ldots, \overline{Y}_{\mathrm{II}}^{(m-r)}$$

are linearly independent, because the $m \times m$ matrix

$$\begin{pmatrix} \overline{Y}_{\mathrm{I}}^{(1)} & \cdots & \overline{Y}_{\mathrm{I}}^{(m-r)} & \cdots & \overline{Y}_{\mathrm{I}}^{(m)} \\ \overline{Y}_{\mathrm{II}}^{(1)} & \cdots & \overline{Y}_{\mathrm{II}}^{(m-r)} & \cdots & \overline{Y}_{\mathrm{II}}^{(m)} \end{pmatrix}$$

is non-singular, and a linear combination of the above-mentioned columns, together with relations $(*)$, would yield a contradiction. Hence the matrix A is uniquely determined. Relations $(*)$ can then be viewed as a system of $r(m - r)$ linear equations for the components of A, having a unique solution.

There exists a constant $m \times m$ matrix C such that

$$\overline{Y} = UC,$$

where U is our fixed system of basic solutions of the original differential equation. If we substitute UC for $\overline{Y}$ in the linear equations $(*)$, we see that A depends only on the polynomials $P_1, \ldots, P_m$ by means of the constant matrix C. In solving the system of linear equations for the components of A, we meet certain determinants involving the components of $\overline{Y}$. Let

$$\varphi_1, \ldots, \varphi_n$$

consist of all monomials of degree $\leq m^2$ in the components of U. Then each component of A can be expressed as a quotient of linear combinations with constant coefficients of $\varphi_1, \ldots, \varphi_n$. We can therefore apply Lemma 2.2 to conclude the proof of Lemma 2.4.

We shall now conclude the proof of Shidlovsky's lemma.

Let $(y_1, \ldots, y_m)$ be our solution of $Y' = QY$, linearly independent over $K(z)$, in the power series ring $K[[z]]$. Let

$$F_1 = P_1 y_1 + \cdots + P_m y_m$$

and

$$F_k = P_{k1} y_1 + \cdots + P_{km} y_m = (TD)^{k-1} F_1$$

as usual. Let r be the rank of the matrix (P_{kj}) and assume $r < m$. Let $A = (A_{ij})$ be the matrix of Lemma 2.4, and let T_1 be a denominator for A, of bounded degree. By Lemma 2.4, we can write, for $k = 1, \ldots, r$ and $j = r + 1, \ldots, m$,

$$P_{kj} = \sum_{\nu=1}^{r} P_{k\nu} A_{\nu j}$$

and

$$F_k = P_{k1} y_1 + \cdots + P_{kr} y_r + \sum_{j=r+1}^{m} P_{kj} y_j,$$

whence

$$(**) \qquad F_k = \sum_{\nu=1}^{r} P_{k\nu} \left(y_\nu + \sum_{j=r+1}^{m} A_{\nu j} y_j \right).$$

Let $\Delta_0 = \det(P_{k\nu})$ $(k, \nu = 1, \ldots, r)$. Note trivially that by induction,

$$\deg P_{k\nu} \leqq \max \deg P_j + rq,$$

where q is a constant depending only on the degrees of Q_{ij}. Consequently,

$$\operatorname{ord} \Delta_0 \leqq \deg \Delta_0 \leqq r \cdot \max \deg P_j + N_0$$

for some constant N_0. On the other hand, solving the linear equations $(**)$ yields

$$\Delta_0 \left(y_\nu + \sum_{j=r+1}^{m} A_{\nu j} y_j \right) = \sum_{k=1}^{r} \Delta_{\nu k} F_k$$

where $\Delta_{\nu k}$ are subdeterminants of Δ_0, and in any case are polynomials.

We multiply throughout by T_1 to clear denominators. We note that (from successive derivatives),

$$\operatorname{ord} F_1 - r \leqq \operatorname{ord} F_k,$$

and consequently

$$\operatorname{ord} F_1 - r \leqq \operatorname{ord} \Delta_0 + \operatorname{ord} \left(T_1 y + \sum_{j=r+1}^{m} T_1 A_{\nu j} y_j \right).$$

We can use Lemma 1 on the expression in parentheses on the right, to see that its order is bounded from above by a constant. Combining this with the upper bound for ord Δ_0 in terms of its degree, we have finally

$$\text{ord } F_1 \leq r \cdot \max \deg P_j + N_1$$

for some constant N_1, thereby proving Shidlovsky's lemma.

§4. *The general theorem*

We wish to extend Theorem 1 to arbitrary E-functions, satisfying a linear differential equation with rational functions as coefficients. Let $f_1, \ldots, f_s$ be E-functions, satisfying the differential equation

$$X' = Q^*X,$$

where X is a column vector of $(X_1, \ldots, X_s)$, and

$$Q^* = (Q^*_{ij}) \qquad\qquad (i, j = 1, \ldots, s)$$

is a square matrix of rational functions in $K(z)$ over the number field K. We shall prove the theorem of Siegel-Shidlovsky:

THEOREM 2. *Assume that $f_1, \ldots, f_s$ are algebraically independent over $K(z)$, and satisfy a linear differential equation as above. Let $\alpha \in K$ be distinct from 0, and from the poles of the rational functions Q^*_{ij}. Then the values*

$$f_1(\alpha), \ldots, f_s(\alpha)$$

are algebraically independent.

Examples of functions as in Theorem 2, besides the exponential function, are given by the Bessel functions, solutions of the differential equation

$$y'' + \frac{1}{z}\, y' + \left(1 - \frac{\lambda^2}{z^2}\right) y = 0,$$

the constant λ being taken in a number field. If J_λ, J'_λ are two linearly independent solutions, then it is known that they are algebraically independent (over $\mathbf{C}(z)$) if 2λ is not an odd integer. The proof of this fact can be found for instance in Siegel's book, and involves only function theory, no arithmetic. The proof is by no means dull, but the main point of the theory of transcendental numbers is to reduce the algebraic independence of *values* to the algebraic independence of *functions*, so that it is not unreasonable to omit such a function-theoretic proof in the present book, in view of its easy availability elsewhere. It is easy to verify that

J_λ, J'_λ are E-functions, and thus we obtain a special case of Theorem 2, originally proved by Siegel, namely the algebraic independence of $J_\lambda(\alpha)$, $J'_\lambda(\alpha)$ whenever α is an algebraic number $\neq 0$, and 2λ is distinct from an odd integer.

We also refer to Siegel [32] for examples of hypergeometric functions satisfying the hypotheses of Theorem 2.

The proof of Theorem 2 follows very closely that of Theorem 1. There will be essentially no change in Lemma 1. For Lemma 2, we quote Shidlovsky's lemma, and then there is no further difficulty. Lemma 3 is essentially the same as before, and the estimates of Lemma 4 are only very slightly more difficult, since we must take polynomial denominators into account. We devote a small amount of space to these estimates. The final Lemma 5 does not change, and the last clinching arguments are exactly as before.

We shall repeat the lemmas in the general context, and preserve their numbering.

LEMMA 1. *Let $E_1, \ldots, E_m$ be E-functions, with coefficients in K, and let n be a positive integer. There exist polynomials $P_1, \ldots, P_m \in I_K[z]$ not all zero, such that*

(i) *$\deg P_j < 2n$ and $\|P_j\| \leq c^n n^{2n}$.*
(ii) *The function*

$$F_1 = P_1 E_1 + \cdots + P_m E_m = \sum_{\nu=0}^{\infty} a_\nu \frac{z^\nu}{\nu!}$$

has a zero of order $\geq (2m - 1)n$ at 0.
(iii) *We have $|a_\nu| \leq c^\nu c^n n^{2n}$.*

Proof. There is no change from the previous case.

Let $E_1, \ldots, E_m$ be E-functions with coefficients in K, and assume that they are linearly independent over $K(z)$. Assume also that they satisfy the linear differential equation

$$Y' = QY,$$

where Q is a matrix (Q_{ij}) of rational functions in $K(z)$. Let T be a polynomial denominator for Q, say $T \in I_K[z]$, and let

$$F_1 = P_1 E_1 + \cdots + P_m E_m$$

be as in Lemma 1. We may then form inductively

$$F_k = TDF_{k-1} = P_{k1} E_1 + \cdots + P_{km} E_m$$

with polynomials P_{kj}.

LEMMA 2. *Let E_j, P_j, F_1 be as in Lemma 1, and assume that*

$$(E_1, \ldots, E_m)$$

satisfy the linear differential equation

$$Y' = QY$$

as above. If $n > c_0'$ (where c_0' depends only on $E_1, \ldots, E_m, Q$), then the rank of the matrix (P_{kj}) $(k, j = 1, \ldots, m)$ is equal to m.

Proof. If the rank r of (P_{kj}) is $< m$, we have, by Shidlovsky's lemma,

$$(2m - 1)n \leqq r(2n - 1) + c_0,$$

from which the assertion of Lemma 2 is obvious.

We let Δ be the determinant

$$\Delta = \det(P_{kj}) \qquad\qquad (k, j = 1, \ldots, m).$$

For some constant c_1 depending on $\deg T$ and $\deg TQ_{ij}$, we have

$$\deg \Delta \leqq (2n - 1)m + c_1.$$

Let P be the matrix (P_{kj}) $(k, j = 1, \ldots, m)$ and let $\tilde{P}$ be the matrix such that

$$\tilde{P}P = \Delta I.$$

Let F be the column vector of $(F_1, \ldots, F_m)$. Then

$$F = PY \qquad \text{and} \qquad \Delta Y = \tilde{P}F.$$

Each F_j $(j = 1, \ldots, m)$ has a zero at 0 of order $\geqq (2m - 1)n - m$, and $\operatorname{ord} E_j \leqq c_2$ for some constant c_2. Consequently

$$\operatorname{ord} \Delta \geqq (2m - 1)n - c_3.$$

Comparing this order with the degree of Δ, it follows that if ξ is any complex number $\neq 0$ then

$$\operatorname{ord}_\xi \Delta \leqq \deg \Delta - \operatorname{ord} \Delta$$

$$\leqq n + c_4.$$

LEMMA 3. *For any complex number $\xi \neq 0$, and ξ not equal to any zero of T, the matrix*

$$(P_{kj}(\xi)) \qquad (k = 1, \ldots, n + c_5, j = 1, \ldots, m)$$

has rank m.

Proof. For any k we have

$$(T(D + {}'Q))^k P_{(1)} = P_{(k)}.$$

Let $r = \mathrm{ord}_\xi \Delta$. We apply $(T(D + {}'Q))^r$ to $\Delta I = {}^t(\widetilde{P}P)$ and find as before an expression

$$(T(D + {}'Q))^r(\Delta I) = \sum C_\mu (T(D + {}'Q))^\mu {}^t P$$

where C_μ are matrices of polynomials. Evaluating these expressions at ξ, we have again

$$T^r(\xi) D^r \Delta(\xi) I = \sum_{\mu=0}^{r} C_\mu(\xi)(T(D + {}'Q))^\mu {}^t P(\xi).$$

On the left we have a non-zero scalar matrix. On the right, the columns of the matrix $(T(D + {}'Q))^\mu {}^t P(\xi)$ are simply the columns

$$P_{(k)}(\xi) = {}^t(P_{k1}(\xi), \ldots, P_{km}(\xi)),$$

with $k \leqq m + r$. It follows that these columns have rank at least m, thereby proving the lemma.

Let α be our element of K, distinct from 0 and the zeros of T. We shall estimate $|F_k(\alpha)|$ and $\|P_{kj}(\alpha)\|$.

LEMMA 4. *Let* $k \leqq n + c_5$. *Assume* $n \geqq c_6$. *Then*

$$|F_k(\alpha)| \leqq c_7^n n^{3n} n^{-(2m-2)n},$$
$$\|P_{kj}(\alpha)\| \leqq c_8^n n^{3n} \qquad and \qquad \mathrm{den}\, P_{kj}(\alpha) \leqq c_8^n.$$

Proof. First, we note that by induction, one proves easily that

$$(TD)^k = \sum_{\mu=0}^{k} T_\mu D^\mu$$

for all k, where T_μ is a polynomial dominated by

$$T_\mu(z) \;<\; c^k k! (1 + z)^{k + \deg T}.$$

The constant c depends on T, of course. We apply $(TD)^k$ to F_1 for $k \leqq n + c_5$. We can easily estimate the derivative $D^\mu F_1(\alpha)$ as we did previously, to find an estimate of type

$$|D^\mu F_1(\alpha)| \leqq c^n n^{2n} n^{-(2m-2)n}.$$

The $k!$ contributes one more power n^n giving a total of n^{3n}. The $|T_\mu(\alpha)|$ are trivially estimated, and we obtain the estimate for $|F_k(\alpha)|$ as stated in the lemma. The estimates for the $P_{kj}(\alpha)$, which don't involve a power series, only polynomials, is even easier and is left to the reader.

LEMMA 5. *The rank of $E_1(\alpha), \ldots, E_m(\alpha)$ over K is $\geqq m/2\,[K : \mathbf{Q}]$.*

Proof. The proof is identical with the proof in the special case. There is no need to repeat it.

To prove Theorem 2, we repeat essentially verbatim the final arguments of §2, with our given functions $f_1, \ldots, f_s$. We let again $E_1, \ldots, E_m$ be the monomials

$$f_1^{\nu_1} \cdots f_s^{\nu_s}, \qquad\qquad \nu_1 + \cdots + \nu_s \leqq \nu,$$

and let $\nu \to \infty$. If g is a polynomial with coefficients in K, not all zero, of degree d, and if

$$g\big(f_1(\alpha), \ldots, f_s(\alpha)\big) = 0,$$

then we have

$$f_1^{\nu_1}(\alpha) \cdots f_s^{\nu_s}(\alpha) g\big(f_1(\alpha), \ldots, f_s(\alpha)\big) = 0$$

for $\nu_1 + \cdots + \nu_s \leqq \nu - d$. This is a system of linear relations, with coefficients in K, among the $m = m_\nu$ monomials

$$f_1^{\nu_1}(\alpha) \cdots f_s^{\nu_s}(\alpha), \qquad\qquad \nu_1 + \cdots + \nu_s \leqq \nu.$$

These relations are linearly independent over K, and we have $m_{\nu-d}$ such relations. Since $m_\nu - m_{\nu-d} \geqq m_\nu/2\,[K : \mathbf{Q}]$, we get the contradiction again by Lemma 5, thereby proving Theorem 2.

§5. *A transcendence measure*

We shall see that by modifying the arguments at the end of the proof, one can obtain a result concerning a transcendence measure for the numbers $f_1(\alpha), \ldots, f_s(\alpha)$.

THEOREM 3. *Let $f_1, \ldots, f_s$ be E-functions, algebraically independent over $K(z)$, and satisfying a linear differential equation as at the beginning of §4. Let α be an algebraic number, distinct from 0 and from the poles of the rational functions Q_{ij}^*. Let $g(X_1, \ldots, X_s)$ be a polynomial with coefficients in $\mathbf{Z}$, of degree d. Then*

$$\big|g\big(f_1(\alpha), \ldots, f_s(\alpha)\big)\big| \geqq c|g|^{-bd^s}$$

where c is a number > 0, depending on the f_j, s, Q^*, α, and (unfortunately) d, while b depends on the degree $N = [K(\alpha) : \mathbf{Q}]$ and on s.

Proof. We start after Lemma 4. With ulterior motives, we select an integer $l > 0$ such that

$$1 > 4N\left(1 - \left[\frac{lN-1}{lN}\right]^\sigma\right) \qquad \text{for all } \sigma, \ 1 \le \sigma \le s.$$

We let

$$m = \binom{s + lNd}{s} \qquad \text{and} \qquad v = \binom{s + lNd - d}{s}.$$

Let $w = m - v$. Then there are exactly m monomials

$$f_1^{\nu_1} \cdots f_s^{\nu_s}$$

of degree $\le lNd$, and those monomials are E-functions, which we denote by $E_1, \ldots, E_m$. The differential equation $X' = Q{*}X$ gives rise to a differential equation $Y' = QY$ with a matrix of rational functions Q.

There are exactly v polynomials of type

$$f_1^{\nu_1} \cdots f_s^{\nu_s} g(f_1, \ldots, f_s),$$

with $\nu_1 + \cdots + \nu_s \le \lambda Nd - d$. We denote these functions by

$$\psi_1, \ldots, \psi_v.$$

We have inductively, using the construction of Lemma 1, with $n > c_0'$,

$$F_k = P_{k1}E_1 + \cdots + P_{km}E_m,$$

and Lemma 3 states that the matrix $(P_{kj}(\alpha))$ with $j = 1, \ldots, m$ and $k \le n + c_5$ has rank m for some constant c_5. We can write

$$\psi_1 = \lambda_{11}E_1 + \cdots + \lambda_{1m}E_m$$
$$\vdots \qquad \vdots \qquad\qquad \vdots$$
$$\psi_v = \lambda_{v1}E_1 + \cdots + \lambda_{vm}E_m,$$

with coefficients λ_{ij} which can be taken to be coefficients of the original polynomial g. Their absolute values are therefore bounded by $|g|$. We can find then $m - v = w$ functions among the F_k ($k \le n + c_5$), say

$$\varphi_1 = P_{k_11}(\alpha)E_1 + \cdots + P_{k_1m}(\alpha)E_m$$
$$\vdots \qquad \vdots \qquad\qquad \vdots$$
$$\varphi_w = P_{k_w1}(\alpha)E_1 + \cdots + P_{k_wm}(\alpha)E_m$$

such that the determinant δ of the matrix consisting of the (λ) and the $P_{k_ij}(\alpha)$ is not zero. We may assume without loss of generality that $E_1 = 1$, namely the monomial of degree 0. Then

$$\delta = A_1\psi_1(\alpha) + \cdots + A_v\psi_v(\alpha) + B_1\varphi_1(\alpha) + \cdots + B_w\varphi_w(\alpha),$$

where the coefficients (A) and (B) are obvious subdeterminants of δ. Each P_{kj} has degree $\leq 2n - 1 + (k - 1)q$ where q depends only on Q. It is easy to estimate the coefficients (A) and (B), and we find

$$|A_1\psi_1(\alpha) + \cdots + A_v\psi_v(\alpha)| \leq c_9 g(f(\alpha))|g|^{v-1}n^{4nw},$$
$$|B_1\varphi_1(\alpha) + \cdots + B_w\varphi_w(\alpha)| \leq c_{10}|g|^v n^{4nw}n^{-(2m-2)n}.$$

One sees easily that

$$\|\delta\| \leq |g|^v n^{4nw},$$

and that a denominator for δ is bounded by c^n. By the usual lemma, we find

$$(*) \qquad 1 \leq c_{11}|g|^{Nv}n^{4nwN}\left[\frac{|g(f(\alpha))|}{|g|} + \frac{1}{n^{n(2m-2)}}\right].$$

We take n to be the smallest integer $> c_0'$, and such that

$$n^n > 2c_{11}|g|^N.$$

Recall that $w = m - v$. We contend that

$$2m - 2 - 4N(m - v) > v.$$

It will certainly suffice to prove that $m > 4N(m - v)$, in view of the inequality

$$2m - 2 - 4N(m - v) - v = m - 2 - (4N - 1)(m - v).$$

We have

$$s!m = (lNd + 1) \cdots (lNd + s) = (lN)^s d^s + \xi_{s-1}(lN)^{s-1} d^{s-1} + \cdots,$$
$$s!v = (lN - 1)^s d^s + \xi_{s-1}(lN - 1)^{s-1} d^{s-1} + \cdots,$$

where $\xi_{s-1}, \ldots, \xi_0$ are integers depending only on s. We estimate each

$$(lN)^\sigma - (lN - 1)^\sigma$$

using the definition of l, and our contention follows at once.

If we now substitute $2c_{11}|g|^N$ for n^n we get an upper bound of $\frac{1}{2}$ for the second term on the right of $(*)$. Transposing this $\frac{1}{2}$ to the left of $(*)$ yields

$$|g(f(\alpha))| \geq \left(|g|^{Nv}n^{4nwN}\right)^{-1}$$

under our assumption that $n^n > 2c_{11}|g|^N$, or rewriting this in the log notation,

$$\log |g(f(\alpha))| \geqq -Nv \cdot \log |g| - 4nwN \cdot \log n.$$

Since n is chosen smallest satisfying the inequality $n^n > 2c_{11}|g|^N$, we see that n^n is of the order of magnitude of $|g|^N \log g$. Both v and w are obviously of type $b\,d^s$ with a suitable constant b, and hence

$$\log |g(f(\alpha))| \geqq -bN\,d^s \log |g| - bN^2\,d^s \log |g| - c_{12}.$$

This proves our theorem.

It is in fact easy to see that w has degree $s - 1$ in both N and d, and consequently that our constant b is of type $b_0 N^{s+1}$ where b_0 depends only on s.

Historical note

This entire chapter is due to Siegel (cf. [31] and [32]), except for Shidlovsky's lemma [30]. In his original paper, Siegel proved the non-vanishing of the crucial determinant only in special cases (including the case of the Bessel function), but left the general case open. He axiomatizes the situation in his book [32]. Shidlovsky saw how to prove the non-vanishing in general, and consequently closed the last gap in obtaining the general theorem stated here as Theorem 2.

In his original paper, Siegel also obtains a transcendence measure for the values of his functions, of the same type as Theorem 3. Once Shidlovsky proved his lemma, it was clear that Siegel's argument could be extended to the general case also (cf. [19]).

This estimate does not give a transcendence type as we defined it in Chapter V, because we view d as fixed. I have checked that the proof of Theorem 1 in fact carries with it an explicit determination of the constant as a function of the degree, i.e. a function of m. In the estimates, these involve expressions like m^{2m} or m^{3m} (or a similar low exponent of m^m). This is of course very unfortunate. As far as I know, there is no result known on the approximation of e^α (α algebraic) by algebraic numbers ξ which depend on d in a manner similar to the Feldman results of Chapter VI.

It seems probable that to obtain good dependence on d, one will have first to improve (or rather change completely) the known methods used to prove approximation theorems concerning algebraic numbers, and apply similar methods to values of transcendental functions.

Finally, we note that Lindemann actually proves something slightly stronger than the algebraic independence of $e^{\alpha_1}, \ldots, e^{\alpha_s}$ if $\alpha_1, \ldots, \alpha_s$

are linearly independent over $\mathbf{Q}$. He proves that if $\beta_1, \ldots, \beta_m$ are distinct non-zero algebraic numbers, then $e^{\beta_1}, \ldots, e^{\beta_m}$ are linearly independent over the field of algebraic numbers. This does not come out of the Siegel method as it stands: The discrepancy is apparent in Lemma 5. Just to round out the theory, it would be worth while to see if one could not adjust the Siegel method so that Lemma 5 yields this stronger result, in the general case of E-functions satisfying a linear differential equation, in other words, if the functions are linearly independent, then their values are also linearly independent.

One also wishes to investigate transcendental numbers from the point of view of diophantine approximations. A general discussion is given in [21], and also in my book *Introduction to Diophantine Approximations*.

APPENDIX

The *p*-adic Case

The theory of transcendental numbers can also be developed over p-adic fields. We let $\mathbf{C}_p$ be the completion of the algebraic closure of the p-adic field $\mathbf{Q}_p$. Then $\mathbf{C}_p$ plays the role of the complex numbers.

In considering values of functions, the theorems are local, concerned with values of power series converging in some neighborhood of the origin. Extensions of theorems from the complex case to the p-adic case have proved themselves to be either fairly easily accessible in the past, with a proof which closely parallels the complex case, or completely out of reach. We shall now give examples of both cases.

The counting of zeros has an analogue in a theorem of Mahler [23]:

THEOREM. *Let* $f(t) = \sum a_\nu t^\nu$ *be a power series such that* $a_\nu \in \mathbf{C}_p$, $|a_\nu|_p \leq 1$, *and* $\lim a_\nu = 0$. *Let* m *be a positive integer, and*

$$x_1, \ldots, x_n \in \mathbf{C}_p$$

such that $|x_i|_p \leq 1/p^m$. *Assume that* $f(x_i) = 0$ *for* $i = 1, \ldots, n$. *Then given* $x \in \mathbf{C}_p$ *with* $|x|_p \leq 1/p^m$, *we have*

$$|f(x)|_p \leq 1/p^{mn}.$$

Furthermore, for the p-adic absolute value on the number field K, *we have the estimate*

$$-\text{size}(\alpha) \ll \log |\alpha|_p$$

for all $\alpha \neq 0$ *in* K.

The exponential function is defined by the usual series, and converges in the open disc of radius $p^{-1/(p-1)}$ in $\mathbf{C}_p$, where it satisfies the functional equation.

The results of Chapter I then go over without difficulty. We give Theorem 1 as an example.

THEOREM 1. *Let* β_1, $\beta_2 \in \mathbf{C}_p$ *be linearly independent over* $\mathbf{Q}$, *and* z_ν ($\nu = 1, 2, 3$) *in* $\mathbf{C}_p$ *be also linearly independent over* $\mathbf{Q}$. *Assume that*

101

502

β_1, β_2 *have p-adic value* ≤ 1 *and that* $|z_\nu|_p < 1/p$. *Then at least one of the numbers*

$$\exp(\beta_1 z_\nu), \qquad \exp(\beta_2 z_\nu) \qquad\qquad (\nu = 1, 2, 3)$$

is transcendental (over $\mathbf{Q}$).

Proof. We take a large integer c and let

$$f(t) = \exp(p^c \beta_1 t), \qquad g(t) = \exp(p^c \beta_2 t).$$

Then the power series for f, g have p-integral coefficients, tending to 0, and it suffices to prove that not all values $f(z_\nu)$, $g(z_\nu)$ lie in K (using the functional equation). Suppose the contrary. We form

$$F = \sum_{i,j=1}^{r} a_{ij} f^i g^j$$

with the same values of r and n as before, requiring F to have zeros as before. Nothing is changed in this part of the proof which occurs entirely within the number field, and we obtain the same upper estimate for the coefficients a_{ij}, and the values of F. We let s be as before, and now estimate $|F(w)|_p$. Applying the theorem of Mahler, we find

$$-s^{5/2} \ll \log |F(w)|_p \ll -s^3 \log p,$$

which gives the contradiction when s is sufficiently large.

The theorems concerning algebraic groups go over in a similar manner. The exponential map can be represented locally by power series, with integral coefficients, tending to 0. We do not have any global conditions when dealing with abelian varieties. We state the theorem to give an example to the reader.

THEOREM 2. *Let G be a linear group or an abelian variety defined over the field of algebraic numbers. Let $\varphi\colon D \to G_{\mathbf{C}_p}$ be a 1-parameter subgroup defined on a disc around the origin in $\mathbf{C}_p$. Let Γ be a subgroup of D having at least 3 linearly independent elements over $\mathbf{Q}$ in the linear case, and 5 such elements in the abelian case. If $\varphi(\Gamma)$ is contained in the group of algebraic points of G, then φ parametrizes locally an algebraic subgroup of G of dimension 1. In other words, there exists an open subdisc D_0 of D containing 0, and an algebraic subgroup H of G of dimension 1 such that $\varphi(D_0)$ is an open subgroup of $H_{\mathbf{C}_p}$ containing the origin.*

In the situation of Theorem 2, one may say that the one parameter subgroup is locally algebraic. In the complex case, essentially by analytic continuation, the corresponding subgroup is algebraic.

In Chapter II, the situation is not quite as good. The results concerning e^α and α^β are still valid p-adically, and locally, wherever they make sense. Thus we must read $\exp(\alpha)$ instead of e^α, and $\exp(\beta \log \alpha)$ instead of α^β. Under the obvious conditions that β and α lie in the proper domains for convergence of the exp and log functions, we have the same statements as in the complex case. This is due to Mahler [23]. The general theorem on differential equations is also true, as shown by Adams [1]. Here, one begins to feel some difficulties, since instead of a finite number of points $z_1, \ldots, z_m$, one must take an infinite sequence.

However, the analogue for the Weierstrass $\wp$-function is not known, and the difficulty here lies in the fact that it is of arithmetic order 2. In the complex case, we could take a large radius R to estimate our function, but in the local p-adic case, this is impossible, and thus the question remains open.

An analogous difficulty arises for the transcendence of the Bessel function $J(\alpha)$ for algebraic $\alpha \neq 0$. The power series for the Bessel function converges just like the exponential series, but the technique of the factorials used by Siegel breaks down, and no other technique is known at present to replace it.

The p-adic theory of transcendental numbers has applications to various problems in algebraic number theory. We mention two of these. First, Leopoldt has defined a regulator in the p-adic case. Contrary to the complex case, no proof is known showing that his regulator is non-zero. This would follow from the algebraic independence of (p-adic) logarithms of multiplicatively independent algebraic numbers.

Second, Corollary 2 of Theorem 1, Chapter I, §1 answered a question put to me by Serre, which he encountered in his study of characters of idele classes, taking algebraic values. Of course, the same problem had occurred earlier (cf. the historical note of Chapter II, i.e. the paper of Alaoglu-Erdös).

Bibliography

[1] W. Adams, "Transcendental numbers in the p-adic domain," *Am. J. Math.* (to appear).

[2] ——, "Asymptotic approximations to e," *Proc. Nat. Acad. Sciences USA* (1966), pp. 28–31.

[3] W. Baily, "On the theory of theta functions, the moduli of abelian varieties, and the moduli of curves," *Ann. of Math.* **75** (1962), pp. 342–381.

[4] N. I. Feldman, "The approximation of certain transcendental numbers, I," *Izvestia Akad. Nauk SSSR Ser. Math.* **15** (1951), pp. 53–74.

[5] ——, *idem*, II, *loc. cit.*, pp. 153–176.

[6] ——, "Arithmetic properties of values of elliptic functions," *Izvestia Akad. Nauk SSSR Ser. Math.* **22** (1958), pp. 563–576.

[7] ——, "Arithmetic properties of logarithms of algebraic numbers," *Izvestia Akad. Nauk SSSR Ser. Math.* **24** (1960), pp. 475–492.

[8] ——, "On the transcendental number π," *Izvestia Akad. Nauk SSSR* **24** (1960), pp. 357–368.

[9] ——, "Arithmetic properties of solutions of certain transcendental equations," *Vestnik*, Moscow, No. 1 (1964), pp. 13–20.

[10] —— and A. O. Gelfond, "On lower bounds of forms in three logarithms of algebraic numbers," *Vestnik*, Moscow, No. 5, (1949).

[11] —— and A. O. Gelfond, "On the measure of the relative transcendence of certain numbers," *Izvestia Akad. Nauk SSSR Ser. Math.* **14** (1950), pp. 493–500.

[12] A. O. Gelfond, "Sur les propriétés arithmétiques des fonctions entières," *Tohoku Math. J.* **30** (1929), pp. 280–285.

[13] ——, *Transcendental and algebraic numbers*, Dover, New York, 1960.

[14] C. Hermite, "Sur la fonction exponentielle," *Oeuvres III*, pp. 150–181.

[15] S. Lang, *Diophantine Geometry*, Interscience, New York, 1962.

[16] ——, "Transcendental points on group varieties," *Topology* **1** (1962), pp. 313–318.

[17] ——, "Algebraic values of meromorphic functions I," *Topology* **3** (1965), pp. 313–318, and II, to appear.

[18] ——, "Diophantine approximations on toruses," *Amer. J. of Math.* **86** (1964), pp. 521–533.

[19] ——, "A transcendence measure for E-functions," *Mathematika* **9** (1962), pp. 157–161.

104

[20] ——, "Asymptotic diophantine approximations," *Proc. Nat. Acad. Sciences US.1* (1966), pp. 31–34.

[21] ——, "Report on diophantine approximations," *Bull. Soc. Math. France* **93** (1965), pp. 177–192.

[22] F. LINDEMANN, "Über die Zahl π," *Math. Annalen* **20** (1882), pp. 213–225.

[23] K. MAHLER, "Über transzendente p-adische Zahlen," *Compositio Mathematica* **2** (1935), pp. 259–275.

[24] ——, "On the approximation of logarithms of algebraic numbers," *Philos. Trans. Roy. Soc. Londond Ser. A* **245** (1953), pp. 371–398.

[25] ——, "On the approximation of π," *Proceedings Koninklijke Nederl. Adak. Wetensch. Amsterdam. Ser. A* **56** (1953), pp. 30–42.

[26] A. NÉRON, "Quasi-fonctions et hauteur sur les variétés abeliennes," *Annals of Math.* (1965), pp. 249–331.

[27] T. SCHNEIDER, "Zur Theorie der Abelschen Funktionen und Integrale," *J. reine angew. Math.* (1941), pp. 110–128.

[28] ——, "Ein Satz über ganzwertige Funktionen als Prinzip fur Transzendenzbeweise," *Math. Ann.* **121** (1949–1950), pp. 131–140.

[29] ——, *Einführung in die transzendenten Zahlen*, Springer Verlag, Berlin, 1957.

[30] A. V. SHIDLOVSKY, "On a criterion of algebraic independence," *Izvestia Akad. Nauk SSSR* **23** (1959), pp. 35–66.

[31] C. L. SIEGEL, "Über einige Anwendungen diophantischer Approximationen," *Abh. Preuss. Akad. Wiss.* (1929), pp. 1–41.

[32] ——, *Transcendental Numbers*, Annals of Math. Studies **16**, Princeton, 1949.

[33] A. WEIL, *Variétés Kählériennes*, Hermann, Paris, 1958.

Note: The reader will find much more complete bibliographies in Gelfond's and Schneider's books. We have included here only the most significant papers.

Inventiones math. 11, 1–14 (1970)

Analytic Subgroups of Group Varieties

Enrico Bombieri (Pisa) and Serge Lang (New York)

Certain results in the theory of transcendental numbers can be formulated as criteria when an analytic subgroup of a group variety is a closed subvariety. Essentially this happens when the subgroup contains "sufficiently many" algebraic points. The theorems stated in [4], concerning 1-parameter subgroups, are extended here to arbitrary dimensions. The difficulty in making this extension lay in finding an appropriate substitute for the maximum modulus estimate of an analytic function with many zeros. When dealing with functions of several variables, one must have some means of proving the existence of a low maximum modulus in the presence of many points at which the function is zero. If the points are sufficiently far apart, then they have the effect of increasing considerably the area of the divisor of zeros of the function, and result in lowering the modulus. We call this the Schwarz Lemma.

Our theorems are local. Using the above estimate, we first give a criterion for the algebraic dependence of functions taking algebraic values at "sufficiently many" points, suitably distributed in a ball around the origin. We then apply this to the functions which represent an analytic subgroup

$$\varphi \colon \mathbf{C}^d \to G_{\mathbf{C}}$$

of a group variety G, defined over a number field. We assume that $\mathbf{C}^d$ contains a finitely generated subgroup Γ, whose points are sufficiently well distributed in the given ball, and such that $\varphi(\Gamma)$ is contained in the group of algebraic points of G. In that case, $\varphi(\mathbf{C}^d)$ is closed, and is a group subvariety of dimension d.

One expects the condition of "good distribution" to be satisfied in applications when the elements of Γ are, say, logarithms of algebraic points on groups. To prove this appears to be a very hard problem in diophantine approximations. In the simplest case of logarithms of algebraic numbers, recent work of Baker and Feldman constitutes significant progress in this direction.

§1. The Mass and Density of a Hypersurface

This section recalls for completeness certain results concerning complex analytic divisors, and we refer to the literature for the proofs. Analogous facts also hold for lower dimensional analytic subsets [10].

Let W be a divisor (analytic) in $\mathbf{C}^n$. This means that locally at each point, W is defined by one equation $f=0$, where f is a power series. We denote by A the $(2n-2)$-dimensional area, which we could take to be the $(2n-2)$-Hausdorff measure in Euclidean space (cf. for instance Federer [3]). Let a be a point in $\mathbf{C}^n$ and let $r>0$. Let $B_r(a)$ be the closed ball of radius r centered at a, and let $W_r(a)=W\cap B_r(a)$. We define the *average mass* of W in the ball $B_r(A)$ to be

$$\Theta_W(r, a) = \frac{A\big(W_r(a)\big)}{v_{2n-2}\, r^{2n-2}},$$

where $v_{2n-2}=\pi^{n-1}/(n-1)!$ is the Lebesgue measure of the unit ball in real $(2n-2)$-space. This average mass is defined for $r<\operatorname{dist}(a, \partial W)$.

Proposition 1. *The function $r\mapsto \Theta_W(r, a)$ is increasing.*

There are a number of proofs why this is so. One of them runs along the same lines as the arguments of the next section, so we sketch it here. It depends on the fact that the average mass is the integral of a positive density (which makes the monotonicity obvious). In the case when W is defined locally by the function f, and say $a=0$, this density is given by the formula

$$v = \frac{1}{\pi^{n-1}}\, \frac{i}{2}\, \partial\bar{\partial} \log |f| \wedge (i\, \partial\bar{\partial} \log |z|)^{n-1}.$$

Then it can be verified that for $r<R$, we have

$$\int_{B_R \setminus B_r} v = \Theta_W(R, 0) - \Theta_W(r, 0).$$

(Cf. Lelong [7], Federer [3], Theorems 5.4.3 and 4.3.19.) We also refer the reader to Bombieri [1] for the case when $\partial\bar{\partial} \log |f|$ is replaced by a $\bar{\partial}$-closed positive current.

In view of Proposition 1, the limit

$$\Theta_W(a) = \lim_{r\to 0} \Theta_W(r, a)$$

exists, and will be called the *density* of W at a. If W is a cone C with vertex at a, then we note that its average mass $\Theta_C(r, a)$ is constant, i.e.

$$\Theta_C(r, a) = \Theta_C(a).$$

This special case is of importance to handle the general situation.

The next two propositions tell us how to compute the average mass and the density.

Proposition 2. *Let W be an algebraic hypersurface in $\mathbf{C}^n$, of degree k. Then*

$$\lim_{r \to \infty} \Theta_W(r, a) = k.$$

For the proof, we refer to Stoll [10], and Federer loc. cit.

Suppose that W is defined by the equation $f = 0$ near the point a, where f is a power series. Let P_k be the first non-vanishing homogeneous term in this power series. By definition, the *tangent cone* C of W at a is the algebraic (affine) hypersurface defined by $P_k = 0$, and k is called the *multiplicity* of a in W.

Proposition 3. *Let W be a divisor in $\mathbf{C}^n$, and let a be a point of multiplicity k in W. Let C be the tangent cone of W at a. Then the density is equal to the multiplicity, that is,*

$$\Theta_W(a) = \Theta_C(a) = k.$$

We refer to Federer and Stoll for the proof of the equality $\Theta_W(a) = \Theta_C(a)$. The second equality then follows from Proposition 2.

Corollary. *Let $z_1, \ldots, z_N$ be points with $|z_j| \leq R/2$ for all j. Let W be a divisor in $\mathbf{C}^n$, and assume that each z_j lies in W, with multiplicity m_j. For each j let*

$$0 < r_j \leqq R/2,$$

and assume that the balls $B_{r_j}(z_j)$ are disjoint. Then the area of W satisfies the inequality

$$A\big(W_R(0)\big) \geqq v_{2n-2} \sum_{j=1}^{N} m_j\, r_j^{2n-2}.$$

Proof. Immediate from the proposition.

§ 2. Growth Estimates for a Holomorphic Function

Throughout this section, we let f be holomorphic on a neighborhood of the closed ball B_R centered at the origin in $\mathbf{C}^n$. We let W be its divisor of zeros. Let $0 < r \leq R$, and let $B_r = B_r(0)$. Let $W_r = W \cap B_r$, and let

$$A(r) = A(W_r)$$

be the $(2n-2)$-dimensional area of W_r. Let

$$|f|_r = \max_{|z|=r} |f(z)|,$$

where $|\ |$ is the Euclidean norm, $|z| = (\sum z_\alpha \bar{z}_\alpha)^{\frac{1}{2}}$.

1*

Let μ be Lebesgue measure in $\mathbf{C}^n = \mathbf{R}^{2n}$. Let Δ be the Laplacian. Define the functional

$$\sigma = \left(\frac{1}{2\pi} \Delta \log |f|\right) d\mu,$$

so that for any C^∞ function φ with compact support in a ball of radius slightly bigger than R, where f is defined, we have

$$\sigma(\varphi) = \int_{\mathbf{C}^n} \varphi \, \sigma = \frac{1}{2\pi} \int_{\mathbf{C}^n} (\Delta\varphi) \log |f| \, d\mu.$$

Then σ is a positive functional, and we can see this as follows. We have trivially

$$\sigma(\varphi) = \lim_{\varepsilon \to 0} \frac{1}{4\pi} \int_{\mathbf{C}^n} \varphi \cdot \Delta \log(|f|^2 + \varepsilon) \, d\mu.$$

We must see that Δ applied to the log term is a positive function. For any function g we have

$$\Delta g \, d\mu = \frac{1}{(n-1)!} \frac{1}{2} \partial\bar\partial g \wedge \left(\frac{i}{2} \sum dz_\alpha \wedge d\bar z_\alpha\right)^{n-1},$$

the power on the right being the exterior power. Also

$$d\mu = \left(\frac{i}{2}\right)^n dz_1 \wedge d\bar z_1 \wedge \cdots \wedge dz_n \wedge d\bar z_n.$$

If f is holomorphic, then $\bar\partial f = 0$ and $\partial \bar f = 0$. We compute easily

$$\partial\bar\partial \log(f\bar f + \varepsilon) = \varepsilon \, \frac{\partial f \wedge \bar\partial \bar f}{(f\bar f + \varepsilon)^2},$$

which we see is positive, as desired.

We note that σ is precisely the measure giving the area of the divisor W, namely

$$A(r) = \int_{B_r} \sigma.$$

We refer to Lelong [5] for this fact, which is essentially classical, going back to Poincaré.

The next theorem and its proof are also contained in Bombieri [1].

Observe that with our present notation, we have

$$\Theta_W(r, 0) = \frac{A(r)}{v_{2n-2}\, r^{2n-2}}.$$

Schwarz Lemma. *Let f be holomorphic on a open neighborhood of the closed ball B_{4R} of radius $4R$. Let $0 < \tau < 1$. There exists a number $c = c(n)$, depending only on n, such that for any $w \in B_{\tau R}$ and $w \neq 0$, we have*

$$\log |f(w)| \leq \log |f|_{4R} - (1 - \tau) \, \Theta_w(|w|, 0) \log \frac{\tau R/c}{|w|}.$$

Proof. The case $n = 1$ is a standard result of functions of one variable, so we assume $n \geq 2$. Let

$$\sigma_R = \begin{cases} \sigma & \text{in } B_R \\ 0 & \text{outside } B_R. \end{cases}$$

The equation

$$\Delta g \, d\mu = \sigma_R$$

has a solution (for $n \geq 2$) given by

$$g(z) = \frac{(n-2)!}{4 \pi^n} \int_{\mathbf{C}^n} - \frac{1}{|\zeta - z|^{2n-2}} \, \sigma_R(\zeta).$$

(Cf. Lelong [5].) Hence

$$\Delta \left(\frac{1}{2\pi} \log |f| - g \right) d\mu = \sigma - \sigma_R \geq 0 \quad \text{in } \mathbf{C}^n.$$

Hence $\Delta \left(\dfrac{1}{2\pi} \log |f| - g \right) \geq 0$, and

$$\frac{1}{2\pi} \log |f| - g$$

is subharmonic and satisfies the maximum modulus principle. Let $R' = 4R$. We find:

$$\frac{1}{2\pi} \log |f(w)| - g(w) \leq \max_{|z| = R'} \left[\frac{1}{2\pi} \log f(z) - g(z) \right],$$

so that

$$\log |f(w)| \leq \log |f|_{R'} - 2\pi \cdot \min_{|z| = R'} \left[g(z) - g(w) \right].$$

We must estimate $g(z) - g(w)$ from below, for $|z| = R'$. We have:

$$\int_{B_R} \left[\frac{1}{|\zeta - w|^{2n-2}} - \frac{1}{|\zeta - z|^{2n-2}} \right] \sigma(\zeta)$$

$$\geq \int_{B_R} \left[\frac{1}{(|\zeta| + |w|)^{2n-2}} - \frac{1}{(R' - |\zeta|)^{2n-2}} \right] \sigma(\zeta)$$

$$= \int_0^R \left[\frac{1}{(t + |w|)^{2n-2}} - \frac{1}{(R' - t)^{2n-2}} \right] dA(t),$$

which we integrate by parts, using $A(0)=0$, to get

$$= A(R) \left[\frac{1}{(R+|w|)^{2n-2}} - \frac{1}{(R'-R)^{2n-2}} \right]$$

$$+ (2n-2) \int_0^R A(t) \left[\frac{1}{(t+|w|)^{2n-2}} + \frac{1}{(R'-t)^{2n-2}} \right] dt$$

$$\geq (2n-2) \int_0^R \frac{A(t)}{(t+|w|)^{2n-1}} \, dt.$$

Hence we find

$$2\pi [g(z)-g(w)] \geq \frac{1}{v_{2n-2}} \int_0^R \frac{A(t)}{(t+|w|)^{2n-1}} \, dt$$

$$= \int_0^R \left(\frac{t}{t+|w|} \right)^{2n-2} \frac{\Theta_w(t,0)}{t+|w|} \, dt.$$

Replacing $\Theta_w(t,0)$ by $\Theta_w(|w|,0)$ only makes the integral smaller, by the monotonicity of Θ_w. Furthermore, it is easy to see that

$$\int_1^x \left(\frac{t}{t+1} \right)^{2n-2} \frac{dt}{t} \geq (1-\tau) \log \frac{\tau x}{2n}.$$

Changing the variable in our last integral and using this estimate, we conclude the proof of the Schwarz Lemma.

In our applications, we use the estimate without any need of precise information concerning the constants. Hence we reformulate a weaker version which suffices for our purposes.

Corollary 1. *There exist numbers c_1 and τ $(0<\tau<1)$ depending only on n such that for $w \in B_{\tau R}$ and $w \neq 0$ we get:*

$$\log |f(w)| \leq \log |f|_{4R} - c_1 \frac{A(|w|)}{|w|^{2n-2}} \log \frac{\tau R}{|w|}.$$

We also have an application to functions of finite order. An entire function will be called of strict order $\leq \rho$ if we have

$$\log |f|_R \ll R^\rho.$$

The sign $X \ll Y$ means that there exists a number $c>0$ such that $X \leq c Y$. We usually specify what c depends on. For instance, in the inequality defining the order of f, the constant depends on f, but of course not on R.

Corollary 2. *Let f be entire of* strict order $\leq \rho$. *Then*

$$\Theta_w(R,0) = \frac{A(R)}{v_{2n-2} R^{2n-2}} \ll R^\rho.$$

Proof. Obvious from the Schwarz Lemma.

Remark. We use the notion of "strict order" instead of the usual order which involves an additional epsilon in the exponent, because in the considerations of this paper, this epsilon plays no role. Furthermore, the exponential functions occurring in practice in fact have strict order ≤ 1 or 2. Finally, in Corollary 2, observe that the bound R^ρ for the density (in one variable, the bound for the number of zeros) actually does not have the epsilon in it.

§ 3. A Criterion for Algebraic Dependence of Functions

Let K be a number field (finite extension of the rational numbers $\mathbf{Q}$), and let I_K be the ring of algebraic integers. If $[K:\mathbf{Q}]=n$, then there are n distinct embeddings of K in $\mathbf{C}$, called the conjugates of K. If $\alpha \in K$, we say that a positive integer d is a denominator for α if $d\alpha \in I_K$. If B is a real number >0, we shall say that $\text{size}(\alpha) \leq B$ if there exists a denominator d for α such that $\log d \leq B$ and

$$\log \max_{\sigma} |\sigma \alpha| \leq B$$

for all embeddings σ of K in $\mathbf{C}$. Thus

$$\text{size}(\alpha) = \max(\log d, \log |\sigma \alpha|),$$

where d is the smallest denominator for α, and σ ranges over all embeddings of K in $\mathbf{C}$. If r is a positive integer, then

$$\text{size } \alpha^r \leq r \cdot \text{size } \alpha.$$

Let $\alpha \in K$, and assume $\alpha \neq 0$. Let d be a denominator for α. Then $d\alpha$ is an algebraic integer. If $\sigma_1, \ldots, \sigma_n$ are the distinct embeddings of $\mathbf{Q}(\alpha)$ in $\mathbf{C}$, then the norm of $d\alpha$ satisfies the inequality

$$1 \leq |\mathbf{N}(d\alpha)| = \prod_{i=1}^{n} |\sigma_i(d\alpha)| = d^n \prod_{i=1}^{n} |\sigma_i \alpha|,$$

because the norm of $d\alpha$ is an algebraic integer and a rational number, whence an ordinary integer $\neq 0$. From this we get the fundamental inequality

$$-2[K:\mathbf{Q}] \text{ size } \alpha \leq \log |\sigma \alpha|,$$

for any embedding σ of K into $\mathbf{C}$.

Let B_r be as before, the closed ball of radius r centered at the origin in $\mathbf{C}^d$. Let S be a set of points in B_r, and suppose that S is expressed as a union of subsets,

$$S = \bigcup_{n=1}^{\infty} S_n.$$

Let $\lambda \geq 2$. We shall say that the family $\{S_n\}$ is λ-*distributed* in B_r if given $w \in B_r$ and a sufficiently large integer n, there exists an element $u \in S_n$ such that

$$|w - u| < \frac{1}{2n^\lambda}.$$

Observe that if S satisfies this property, then S is dense in B_r for the ordinary topology.

Let f be a holomorphic function on B_R, with $R = 4r/\tau$, and τ is a number satisfying $0 < \tau < 1$ such that the Schwarz Lemma can be applied. Assume that the values of f on S lie in a number field K. We shall say that f is of *arithmetic order* $\leq \rho$ with respect to the family $\{S_n\}$ if there exists a constant $C \geq 1$ such that the following condition is satisfied.

AO. *For all n and $z \in S_n$ we have size $f(z) \leq C n^\rho$.*

Using our notation $\ll$, we can write our condition in the form

$$\text{size } f(z) \ll n^\rho,$$

for $z \in S_n$, and $n \to \infty$.

The next theorem gives an extension of results of Schneider [8] and Lang [4].

Theorem 1. *Let $f_1, \ldots, f_{d+1}$ be holomorphic functions on the ball B_R. Let $r = \tau R/4$, with $0 < \tau < 1$. Let $\{S_n\}$ be a family of subsets of B_r, and assume that $f_1, \ldots, f_{d+1}$ map the set $S = \bigcup S_n$ into a number field K, and are of arithmetic order $\leq \rho$ with respect to the family $\{S_n\}$. Assume that the family $\{S_n\}$ is λ-distributed in B_r. If $\lambda > \rho(d+1)/2$, then $f_1, \ldots, f_{d+1}$ are algebraically dependent over K.*

Proof. Let

$$F = \sum a_{(j)} f_1^{j_1} \ldots f_{d+1}^{j_{d+1}}$$

have coefficients $a_{(j)}$ not all 0 in I_K. The sum is taken for

$$0 \leq j_1, \ldots, j_{d+1} < J.$$

We shall select J and find the coefficients $a_{(j)}$ so as to satisfy appropriate conditions which we now describe.

For each point l of the lattice $\mathbf{Z}^{2d}/n^\lambda$ in $\mathbf{R}^{2d} = \mathbf{C}^d$, such that $l \in B_r$, we select a point $v \in S_n$ such that

$$|v_l - l| < \frac{1}{4n^\lambda}.$$

We let S_n' be the set of all such points $\{v_l\}$. Then card $S_n' \gg n^{2d\lambda}$. If $u, v \in S_n'$ then

$$|u - v| > \frac{1}{2n^\lambda}.$$

We require that

$$F(v)=0, \quad \text{all } v\in S'_n.$$

This condition amounts to a system of linear equations in the $a_{(j)}$, treated as unknowns. We have:

$$\text{number of unknowns} = J^{d+1},$$

$$\text{number of equations} = C_2\, n^{2d\lambda} = \text{card } S'_n,$$

where C_2 is a number bounded by a constant independent of n. We let J be approximately equal to $n^{2d\lambda/(d+1)}$. For definiteness, say

$$J = 2\, C_2\, n^{2d\lambda/(d+1)}.$$

Then the number of unknowns is approximately equal to the number of equations. The coefficients of these equations are the values

$$f_1(z)^{j_1}\dots f_{d+1}(z)^{j_{d+1}}$$

with $z\in S'_n$. By hypothesis, we have

$$\text{size of coefficients} \ll J\, n^{\rho}.$$

By a standard lemma of Siegel [4], we can find $a_{(j)}$ not all 0, satisfying

$$\text{size } a_{(j)} \ll J\, n^{\rho}.$$

The above construction is performed for a large integer n. If $s\geq n$, then simple estimates show that

$$\text{size } F(u) \ll J\, s^{\rho}, \quad \text{all } u\in S_s.$$

We shall prove by induction that $F(u)=0$ for all $u\in S'_s$ and $s\geq n$. Since the union of the sets $S'_s\,(s\geq n)$ is dense in B_r, it follows that $F=0$, and therefore that the functions $f_1,\dots,f_{d+1}$ are algebraically dependent over K.

Assume the inductive step for s. We wish to prove it for $s+1$. Let $u\in S_{s+1}$. Then

$$|F(u)|\leq |F|_{2r},$$

by the ordinary maximum modulus principle. Thus all we need to estimate is $|F(w)|$, for $|w|=2r$. We use the estimate of the Schwarz Lemma, and obtain:

$$\log|F(w)| \leq \log|F|_R - c_3\, s^{2d\lambda}\left(\frac{1}{s^{\lambda}}\right)^{2d-2}$$

$$\leq c_4\, s^{\frac{2d\lambda}{d+1}}\, s^{\rho} - c_3\, s^{2\lambda}$$

$$\ll -s^{2\lambda},$$

provided $\dfrac{2d\lambda}{d+1}+\rho<2\lambda$. But λ was chosen precisely in such a way that this inequality would be satisfied.

Comparing this estimate with the size of $F(u)$, if $F(u)\neq 0$, we obtain

$$-J(s+1)^\rho \ll -\operatorname{size} F(u)\ll \log|F(u)|\ll -s^{2\lambda},$$

whence

$$s^{2\lambda}\ll J\,s^\rho \ll s^{\frac{2d}{d+1}+\rho},$$

which is a contradiction, proving our theorem.

§ 4. Application to d-Parameter Subgroups

Let G be a group variety over the complex numbers. If G is defined over a field K, we denote by G_K its group of points rational over K.

By a *d-parameter subgroup* of G we mean a complex analytic homomorphism

$$\varphi\colon\ \mathbf{C}^d\to G_{\mathbf{C}}$$

whose differential at the origin is injective (the differential is a linear map of $\mathbf{C}^d$ into the tangent space of $G_{\mathbf{C}}$ at the origin).

Usually, the image $\varphi(\mathbf{C}^d)$ is not closed in $G_{\mathbf{C}}$ but winds around. We shall find arithmetic conditions under which this image is closed, extending to d-parameter subgroups the results of [4], Chapter II. If G is embedded in some affine or projective space, and if φ is represented by affine functions $\varphi_1,\ldots,\varphi_M$, then by the *algebraic dimension* of φ we shall mean the maximum number of algebraically independent coordinate functions φ_i $(i=1,\ldots,M)$. This is the same as the dimension of the smallest group subvariety of G containing $\varphi(\mathbf{C}^d)$, i.e. the algebraic dimension of the Zariski-closure of $\varphi(\mathbf{C}^d)$.

Let $\varGamma$ be a finitely generated subgroup of $\mathbf{C}^d$, of rank m, say

$$\varGamma=\{u_1,\ldots,u_m\}$$

where $u_1,\ldots,u_m$ is a basis of $\varGamma$ over $\mathbf{Z}$. We can filter $\varGamma$ by subsets S_N consisting of all linear combinations

$$k_1u_1+\cdots+k_mu_m,\qquad |k_i|\leq N.$$

Let $\lambda\geq 1$. We shall say that $\varGamma$ is *λ-distributed* in the ball B_r, if there exists an integer N_0 satisfying the following condition: Given $w\in B_r$ and $N\geq N_0$, there exists an element $u\in S_N$ such that

$$|w-u|\leq\frac{1}{2N^\lambda}.$$

Observe that if $\varGamma$ satisfies this property, then $\varGamma\cap B_r$ is dense in B_r for the ordinary topology.

Theorem 2. *Let G be a linear group variety defined over a number field K. Let*
$$\varphi: \mathbf{C}^d \to G_{\mathbf{C}}$$
be a d-parameter subgroup. Let Γ be a finitely generated subgroup of $\mathbf{C}^d$, which is λ-distributed in a ball B_r. Assume that $\lambda > (d+1)/2$, and that $\varphi(\Gamma)$ is contained in G_K. Then $\varphi(\mathbf{C}^d)$ is a closed algebraic subgroup of $G_{\mathbf{C}}$, of algebraic dimension d.

Proof. The map φ may be viewed as a representation of the exponential map from the tangent space of $\varphi(\mathbf{C}^d)$, and as such, it can be factored through the tangent space of the general linear group, which is none other than the additive space of matrices. We can therefore identify the space on which φ is defined as a d-dimensional subspace of this space of matrices. Then φ is represented by the usual exponential series,

$$\varphi(u) = \sum u^n/n!,$$

where u is a matrix, and $\varphi(u)$ is an invertible matrix with coordinate functions $\varphi_{ij}(u)$. Each such function is clearly entire of strict order ≤ 1. The group Γ has rank d over $\mathbf{C}$, and we can select d elements among them, say $u_1, \ldots, u_d$, linearly independent over $\mathbf{C}$. Thus the functions φ_{ij} may be viewed as functions of d complex variables.

Let S_N be defined as above. Simple estimates involving the multiplication of matrices and the definition of the size show that, if $M_1, \ldots, M_r$ are matrices with coefficients in K, then

$$\text{size}(M_1 \ldots M_r) \ll \sum_{j=1}^{r} \text{size } M_j.$$

(Cf. for instance [4], Chapter V, § 2, Lemma 1 for estimates of a similar nature, which we may apply to generators of Γ.) In particular, we conclude that
$$\text{size } \varphi_{ij}(u) \ll N, \quad \text{for } u \in S_N.$$

In other words, the functions φ_{ij} are of arithmetic order ≤ 1, and we can apply Theorem 1. We conclude that they have transcendence degree d over K, and from this it follows that the Zariski closure of $\varphi(\mathbf{C}^d)$ has dimension d. It is then clear that $\varphi(\mathbf{C}^d)$ itself is equal to its own Zariski closure (they are analytic subgroups, connected, and having the same tangent space at the origin). This concludes the proof.

Theorem 3. *Let A be an abelian variety defined over a number field K. Let $\varphi: \mathbf{C}^d \to A_{\mathbf{C}}$ be a d-parameter subgroup of A. Let Γ be a finitely generated subgroup of $\mathbf{C}^d$, which is λ-distributed in a ball B_r. Assume that $\lambda > d+1$, and that $\varphi(\Gamma)$ is contained in A_K. Then $\varphi(\mathbf{C}^d)$ is an abelian subvariety of $A_{\mathbf{C}}$ (and so closed), of dimension d.*

Proof. The proof follows exactly the same pattern as that of Theorem 2. We represent the mapping φ by theta functions $(\Theta_0, \ldots, \Theta_M)$ giving the embedding of A in projective space. Without loss of generality, we may assume that the divisor of zeros of Θ_0 does not pass through the origin. We note that we can find r_1 sufficiently small so that the ball B_{r_1} does not intersect this divisor of zeros. Then Γ is λ-distributed in B_{r_1}. The functions $f_i = \Theta_i/\Theta_0$ are of arithmetic order ≤ 2, as one proves easily using the quadratic form of Neron-Tate. As the proof is given in detail in [4], Theorem 4, we omit it here. We can therefore apply Theorem 1 to conclude the proof.

§ 5. Various Comments

The condition of λ-distribution in Theorems 1, 2, 3 is not unnatural, but nevertheless is still undesirable. For one thing, as stated, these theorems do not include the 1-dimensional case. This is not serious, however, since the proof does include that case, and a minor rewording could take care of this objection.

More seriously, it becomes a substantial problem to show that the condition of λ-distribution is satisfied in various applications. Let us view elements $u_1, \ldots, u_m$ in $\mathbf{C}^d$ as vectors in $\mathbf{R}^{2d}$. Let

$$\delta(m, 2d) = \frac{m - 2d}{2d}$$

be the "Dirichlet" exponent. The usual Dirichlet box principle shows that given a positive integer N we can always find integers $k_1, \ldots, k_m$ not all 0, satisfying the inequalities

$$|k_1 u_1 + \cdots + k_m u_m| \ll \frac{1}{N^{\delta(m, 2d)}} \quad \text{and} \quad |k_i| \leq N. \qquad (*)$$

On the other hand, for almost all m-tuples of vectors $u_1, \ldots, u_m$, given ε there is only a finite number of solutions $k_1, \ldots, k_m$ of the inequalities

$$|k_1 u_1 + \cdots + k_m u_m| \leq \frac{1}{N^{\delta(m, 2d) + \varepsilon}} \quad \text{and} \quad |k_i| \leq N. \qquad (**)$$

(For this and much stronger quantitive results in the same direction, see for instance Schmidt [9].)

A special (but most significant) case of a standard transference principle (e.g. as in Cassels [2], Chapter V, Theorem VI) allows us to conclude that whenever $u_1, \ldots, u_m$ satisfy $(**)$, which is a negative homogeneous condition of diophantine approximation at the origin, then they also satisfy a positive inhomogeneous condition everywhere, namely:

Given $v \in \mathbf{C}^d$ and ε, there exist integers $k_1, \ldots, k_m$ not all 0, such that $|k_i| \leqq N$ and

$$|k_1 u_1 + \cdots + k_m u_m - v| \ll \frac{1}{N^{\delta(m,\, 2d) - \varepsilon}}. \qquad (\ast\ast\ast)$$

This is essentially what we would call $\big(\delta(m, 2d) - \varepsilon\big)$-distributivity. In particular, for such m-tuples to satisfy λ-distributivity with $\lambda > d + 1$, say, we obtain a condition involving only m, namely

$$m > 2d(d+1) + 2d + \varepsilon. \qquad (\ast\ast\ast\ast)$$

This sort of condition is of course the most desirable one.

At this point, one can try to reduce m as a function of d, but this leads into very deep considerations at the heart of the theory of transcendental numbers.

This fourth condition $(\ast\ast\ast\ast)$ also has the advantage that it is analytically invariant, and would allow the possibility of phrasing results analogous to Theorems 2 and 3 for analytic homomorphisms of one group variety (commutative) into another, as could be done in the one dimensional case ([4], Corollary 2 of Theorem 4, Chapter II, § 4). In this direction, we suggest that the condition of λ-distribution could be replaced by the condition that Γ be a direct sum

$$\Gamma = \Gamma_1 \oplus \cdots \oplus \Gamma_m,$$

where each Γ_i has rank d over $\mathbf{C}$, and m is large as a function of d only. Again, this yields an analytically invariant condition, which also has the advantage of covering a situation where $\mathbf{C}^d$ is expressed as a product,

$$\mathbf{C}^d = \mathbf{C}^r \times \mathbf{C}^s,$$

the subgroups Γ_i are contained in each factor, and some of them may be contained in the real part of one of the factors. For instance, in the one dimensional case, the points may be all real, but they still have rank 1 over $\mathbf{C}$. The difficulty again lies in the proof of the Schwarz lemma. If, say, the points are real, and we impose a condition of λ-distribution, then the points have to be close together, thus making the radii r_j small in Corollary 2 of Proposition 3, § 1, which weakens correspondingly the application of the Schwarz Lemma and at present prevents the extension of Theorem 1 to the more general case.

Finally, we observe that the theorems can be extended to finitely generated extensions of the rationals having a suitable transcendence type, in the same manner as the one dimensional case (cf. [4], Chapter V, § 3).

References

1. Bombieri, E.: Algebraic values of meromorphic maps. Inventiones Math. **10**, 267 – 287 (1970).
2. Cassels, J.W.S.: An introduction to diophantine approximation. Cambridge Tracts 45 (1957).
3. Federer, H.: Geometric measure theory. Berlin-Heidelberg-New York: Springer 1969.
4. Lang, S.: Introduction to transcendental numbers. Reading, Mass.: Addison Wesley 1966.
5. Lelong, P.: Fonctions entières (n variables) et fonctions plurisousharmoniques d'ordre fini dans $\mathbf{C}^n$. J. d'Analyse Math. **12**, 365 – 407 (1964).
6. — Propriétés métriques des variétés analytiques complexes définies par une equation. Ann. E.N.S. **67**, 393 – 419 (1950).
7. — Intégration sur un ensemble analytique complexe. Bull. Soc. Math. France **85**, 239 – 262 (1957).
8. Schneider, T.: Ein Satz über ganzwertige Funktionen als Prinzip für Transzendenzbeweise. Math. Annalen **121**, 131 – 140 (1949 – 1950).
9. Schmidt, W.: A metrical theorem in diophantine approximation. Canadian J. Math. **12**, 619 – 631 (1960).
10. Stoll, W.: Mehrfache Integrale auf komplexen Mannigfaltigkeiten. Math. Z. **57**, 116 – 154 (1952).
11. Thie, P.R.: The Lelong number of a point of a complex analytic set. Math. Ann. **172**, 269 – 312 (1967).

Enrico Bombieri
Università di Pisa
Istituto Matematico "Leonida Tonelli"
Pisa/Italia

Serge Lang
Department of Mathematics
Columbia University
New York, N.Y. 10027, USA

(Received June 29, 1970)

Review of L. J. Mordell's *Diophantine Equations*, by S. Lang, Bull. Amer. Math. Soc. **76** (1970), 1230–1234.

Diophantine equations, by L. J. Mordell, Academic Press, New York and London, 1969.

The theory of diophantine equations is one of the oldest in mathematics, one of its most attractive, and also at the moment one which is still fairly undeveloped as being exceptionally hard. One reason for this is perhaps that in the full generality of the Hilbert problem, it cannot be effectively dealt with. Nevertheless, I personally would expect a wide class of diophantine problems to be effectively solvable (e.g. those on curves or abelian varieties), and in any case, many special cases are solvable.

Because of difficulties which have been encountered historically, a portion of the subject has developed as an accumulation of special diophantine equations, mostly in two variables, i.e. curves. It was well understood in the nineteenth century that nonsingular cubic curves have a group law on them, parametrized by the elliptic functions from a complex torus, but Poincaré was the first to draw attention to the special group of rational points when this curve is defined by an equation with rational coefficients, and he guessed that this group might be finitely generated. Mordell proved this fact in 1922, and thereby provided the first opportunity to behold the beginnings of a much broader approach to this type of equation. He also conjectured that a curve of genus ≥ 2 has only a finite number of rational points, and this magnificent conjecture remains unproved today. These matters, which are perhaps Mordell's greatest contributions to the subject, are treated in Chapters 16 and 17 of the present book.

The other parts of the book are roughly distributed as follows. A number of concrete special equations of degrees 2, 3 and 4 are discussed at the beginning, mostly with the method of congruences. Chapter 7 gives a discussion of the fundamental theorem concerning quadratic forms over the rationals (solvability globally is equivalent to solvability locally everywhere).

Chapter 8 deals with Pell's equation, which essentially solves effectively for the units of a real quadratic field. The treatment is classical. Next comes a sequence of chapters on surfaces, mostly cubic and quartic, dealing with special cases when rational or integral points can be found. A brief chapter mentions the role of units as affecting certain equations in number fields, and examples are worked out. After the general discussion already mentioned on curves of genus 1 or >1, we return to special cases which can be handled without the general theory, somewhat more effectively using Minkowski's theorem on convex bodies and congruence methods, applied to the representation of numbers by quadratic and cubic forms. Next we have Thue's theorem on diophantine approximations (the weak version, not Roth's version), and its application to equations of type $f(x, y) = m$ where f is a homogeneous polynomial of higher degree. The next chapter mentions Skolem's method by p-adic analysis, but does not go into details of proofs.

We then return to cubic and quartic forms, or rather special cases, involving the explicit determination of integral solutions. The discriminant and other covariants of these forms are discussed (indispensable means to get at the solutions effectively). The scene shifts back once more to cubic and quartic curves of the elliptic type. like $y^2 = x^3 + k$ and $y^2 = f(x)$, where f is a cubic polynomial with no multiple roots, looking for integral points rather than rational points. Mordell's original proof that the number of these is finite is given, using Thue's theorem. The next chapter indicates the extension of this result to the case when f has arbitrary degree (proved by Siegel in 1926), but refers to an earlier Siegel paper for the stronger version of the analogue of Thue's theorem needed to make the proof go through. Mordell also states Siegel's general theorem of 1929, that a curve of genus at least 1 has only a finite number of integral points, giving explicitly the exceptional cases of genus 0 when infinitely many such points may occur. The book concludes with other special equations of higher degree, for instance special results on $ax^n - by^n = c$ and the Fermat curve, proving the nonexistence of integral solutions in a few simple cases, while assuming assorted facts of algebraic number theory, both of the standard variety, but also more specialized, like those involving regular primes.

As can be seen from this sketch, the contents of the book are jumpy, and some comments are now in order concerning the broader implications of Mordell's style, his point of view, and the context in which he writes.

Special concrete cases like cubic curves have provided much of the testing ground for experimentation, methods, theorems, and conjectures in diophantine analysis, and hence it is very welcome to have some of these cases brought together, as Mordell has done. I emphasize: He collects together special cases, without particular unifying order, or any design that I could make out, that might tie them together or make their succession in 30 chapters more than what appears to be an arbitrary succession. That is Mordell's taste, and I cannot quarrel with it. The book, as it is, will be very useful to those interested in diophantine equations, and wishing to work out special cases with essentially elementary techniques. I personally had bought a copy of the book before being sent the review copy, now given to the library.

But the reader must be aware of the limitations of Mordell's exposition. For one thing, Mordell clings systematically to the chronological development of the subject throughout the book. Even when an important development has taken place, e.g. Roth's theorem on diophantine approximations, subsuming previous results in the subject (by Thue and Siegel, say), Mordell gives the earliest theorem, namely Thue's, and only briefly refers to Roth's paper for the extension, when only a few additional pages at an equally elementary level would have been needed to get the full result. When a stronger version is needed to handle the equation $y^2 = f(x)$ with f of higher degree, Mordell refers to an earlier paper of Siegel (Math. Zeitschrift, 1921. a misprint in the reference gives the date erroneously as 1961), which also treats the number field case needed for this particular application. However, the inexperienced reader will have to figure out for himself that a single formulation of Roth's theorem in number fields can be used effectively for all these applications.

Even though I find the succession of equations treated somewhat arbitrary, there seems to be one thread which runs through them, suggested by the "List of Equations and Congruences" appearing at the end of the book in lieu of an index. This list is ordered according to degree (degree 1, degree 2, degree 3, degree 4, degree > 4) and then according as to whether the equation is homogeneous or not. Of course, one's first attempt in dealing with diophantine equations is to experiment with equations of low degree and small coefficients. But it soon becomes apparent that the degree is not a good invariant for the behavior of these equations, whether searching for rational points or integral points, and the classification by degree is to a large extent misleading. However, Mordell's taste when faced with a theorem like Siegel's on curves of higher genus is just to say: "The proof is of a very advanced character." And leave it at that.

Nor does Mordell tell us of Weil's generalization in 1929 concerning the finite generation of the group of rational points in the higher dimensional case; which is a pity, because this is one of the approaches which gives a method of attack for the Mordell conjecture on curves of genus ≥ 2: We embed them in their Jacobians, and look at their intersection with the finitely generated group of rational points on this Jacobian. By this method, and the positive definite quadratic form of Néron–Tate on this group, Mumford was for instance able to show that the gaps between rational points of ascending height become exponentially large.

It is also possible to connect both results and methods of diophantine analysis with algebraic geometry, and I found it interesting in my book *Diophantine Geometry* to present the known results which allowed us to make this connection coherently (e.g. as it applies to Severi's theorem of the base, following work of Néron, and with Néron). The intense dislike which Mordell has for this kind of exposition is clearly evidenced by his famous review of the book (Bull. Amer. Math. Soc. **70** (1964), 491–498). (If the review is not famous, it should be.) In this connection, I can do no better than to reproduce an exchange of letters with him in November 1966, shortly after I had sent him a copy of some of my other books. He kindly wrote me:

> Dear Professor Lang, Thank you very much for the textbooks which I shall be glad to read. I hope I shall not have to struggle with them as I did with D. G. You may be interested to know that I found your *Algebra* quite readable and very useful. It was obviously meant to be understood. (You may quote this if you wish to.) . . ."

And after a few other kind remarks, he expressed the hope to meet me at a talk of his on diophantine equations to be given shortly in New York. (We

met, and I enjoyed it.) Still, I answered Mordell on the substantial points
raised both by his review and his letter:

> Dear Professor Mordell, Thanks for your letter. What you write there prompts
> me to clarify some points about book writing.
>
> I see no reason why it should be prohibited to write very advanced mono-
> graphs, presupposing substantial knowledge in some fields, and thus allowing
> certain expositions at a level which may be appreciated only by a few, but achieves
> a certain coherence which would not otherwise be possible.
>
> This of course does not preclude the writing of elementary monographs. For
> instance, I could rewrite Diophantine Geometry by working entirely on elliptic
> curves, and thus make the book understandable to any first year graduate student
> (not mentioning you · · ·). Both books would then coexist amicably, and neither
> would be better than the other. Each would achieve different ends.
>
> When you write of any book that it is "obviously meant to be understood",
> whether as a compliment for one book or blame for another, you are still missing
> the point: I never meant Diophantine Geometry to be understood specifically
> by you, or anyone who did not have the rather vast background required for
> its reading. All my books are meant to be understood by readers having the
> prerequisites for the level at which the books are written. These prerequisites
> vary from book to book, depending on the subject matter, my mood, and other
> aesthetic feelings which I have at the moment of writing. When I write a standard
> text in Algebra, I attempt something very different from writing a book which
> for the first time gives a systematic point of view on the relations of diophantine
> equations and the advanced contexts of algebraic geometry. The purpose of the
> latter *is* to jazz things up as much as possible. The purpose of the former is to
> educate someone in the first steps which might eventually culminate in his know-
> ing the jazz too, if his tastes allow him that path. And if his tastes don't, then
> my blessings to him also. This is known as aesthetic tolerance. But just as a
> composer of music (be it Bach or the Beatles), I have to take my responsibility
> as to what *I* consider to be beautiful, and write my books accordingly, not just
> with the intent of pleasing one segment of the population. Let pleasure then fall
> where it may. With best regards, Serge Lang.
>
> Serge Lang

Note: Mordell in his review quotes from the preface of the first edition of my
calculus book, but with an elision. The full text of my sentence runs as
follows:

> One writes an advanced monograph for oneself, because one wants to give
> permanent form to one's vision of some beautiful part of mathematics, not
> otherwise accessible, somewhat in the manner of a composer setting down his
> symphony in musical notation.

I stand by the text as written, not as quoted. The musical analogy is an
essential part of what I meant.

MIX
Papier aus verantwortungsvollen Quellen
Paper from responsible sources
FSC® C105338

If you have any concerns about our products,
you can contact us on
ProductSafety@springernature.com

In case Publisher is established outside the EU,
the EU authorized representative is:
Springer Nature Customer Service Center GmbH
Europaplatz 3, 69115 Heidelberg, Germany

Printed by Libri Plureos GmbH
in Hamburg, Germany